IEEE Technology and Engineering Management Society Body of Knowledge (TEMSBOK)

IEEE Technology and Engineering Management Society Body of Knowledge (TEMSBOK)

Edited by

Gustavo Giannattasio (Editor in Chief)

Elif Kongar

Marina Dabić

Celia Desmond

Michael Condry

Sudeendra Koushik

Roberto Saracco

IEEE Press Series on Technology Management, Innovation, and Leadership

Library of Congress Cataloging-in-Publication Data Applied for:
Hardback ISBN: 9781119987604

Cover Design: Wiley
Cover Image: © GIGISTOCK/Shutterstock

Set in 9.5/12.5pt STIXTwoText by Straive, Pondicherry, India

Contents

A Note from the Series Editor

Welcome to the Wiley–IEEE Press Series on Technology Management, Innovation, and Leadership!

The IEEE Press imprint of John Wiley & Sons is well known for its books on technical and engineering topics. This new series extends the reach of the imprint, from engineering and scientific developments to innovation and business models, policy and regulation, and ultimately to societal impact. For those who are seeking to make a positive difference for themselves, their organization, and the world, technology management, innovation, and leadership are essential skills to home.

The world today is increasingly technological in many ways. Yet, while scientific and technical breakthroughs remain important, it is connecting the dots from invention to innovation to the betterment of humanity and our ecosphere that has become increasingly critical. Whether it is climate change or water management or space exploration or global healthcare, a technological breakthrough is just the first step. Further requirements can include prototyping and validation, system or ecosystem integration, intellectual property protection, supply/value chain set-up, manufacturing capacity, regulatory and certification compliance, market studies, distribution channels, cost estimation and revenue projection, environmental sustainability assessment, and more. The time, effort, and funding required for realizing real-world impact dwarf what was expended on the invention. There are no generic answers to the big-picture questions either, the considerations vary by industry sector, technology area, geography, and other factors.

Volumes in the series will address related topics both in general – e.g. frameworks that can be applied across many industry sectors – and in the context of one or more application domains. Examples of the latter include transportation and energy, smart cities and infrastructure, and biomedicine and healthcare. The series scope also covers the role of government and policy, particularly in an international technological context.

With 30 years of corporate experience behind me and about five years now in the role of leading a Management of Technology program at a university, I see a broad-based need for this series that extends across industry, academia, government, and nongovernmental organization. We expect to produce titles that are relevant for researchers, practitioners, educators, and others.

I am honored to be leading this important and timely publication venture.

Tariq Samad
Senior Fellow and Honeywell/W.R. Sweatt Chair in Technology Management
Director of Graduate Studies, M.S. Management of Technology
Technological Leadership Institute | University of Minnesota
samad@iccc.org

About the Editors

Gustavo Giannattasio, Eng. MBA, PMP (Editor in Chief). He is a senior member of the IEEE Technology and Engineering Management Board of Directors, IEEE Smart Cities Education Committee Vice-Chair, member of the IEEE Future Directions, Industry and Digital Adoption Consultant, invited expert to the ITU United for Sustainable Smart Cities program U4SSC, editor in chief of the IEEE COMSOC Wireless Body of Knowledge WEBOK first edition for the Wireless Engineering Certification, former IEEE Region 9 director, speaker at Smart Cities conferences on IoT security, IOTA DLT at Blockchain Summit, ITU Kaleidoscope on Industry 4.0, ITU Smart Cities webinars, and Data Center virtualization at Data Center Congress. He is the technical program chair and tutorial chair of IEEE Smart Cities Conferences ISC2 Trento, Wuxi, Casablanca, Cyprus, Project Management Institute Director of Programs in Chapter Montevideo. He is professor of Data Communications, Routing and Switching at UCUDAL and ORT Universities and online instructor for International Telecommunications Union on New Generation Networks. He is also the project manager in Communications and Data Center deployments.

Elif Kongar, PhD (Co-editor and Steering Committee member). Her research interest is reverse supply chain management, reverse logistics systems, economically and environmentally sustainable waste recovery systems and operations, sustainable operations, disassembly sequencing and planning, electronic waste, performance evaluation, big data and data analytics, Society 5.0, social sustainability, engineering education programs, female participation and efficiency in engineering disciplines, and K-12 STEM education. Her teaching interest lies in supply chain management, logistics, operations research, decision theory, forecasting techniques, data analytics, mathematical statistics and reliability, service management and engineering, data visualization and analysis, statistical quality control, machine learning and artificial intelligence for business, simulation, and modeling.

Marina Dabić, PhD. She is a fully tenured professor at the University of Zagreb, Faculty of Economics and Business, Croatia; University of Dubrovnik, Croatia; and the University of Ljubljana School of Economics and Business. From 2013 to 2021, she had a dual affiliation with Nottingham Trent University in the United Kingdom. Professor Dabić is Strategic Chair for all accreditations at the University of Zagreb Faculty of Economics and Business, and she is an AACSB member. Dabić has been a member of the peer review team for EFMD accreditations since 2011. She has published over 200 papers in indexed journals, including a wide variety of international journals like the *Journal of International Business Studies, Journal of World Business, Journal of Business Ethics, Technological Forecasting and Social Change, Small Business Economics, International Journal of Human Resource Management, IEEE – Transactions on Engineering Management, Technovation, Journal of Business Research*, and *Journal of Small Business Management.* She has additionally served as editor of books published by Routledge, Springer, and De Gruyter.

Celia Desmond, MSc (Expert consultants, Co-editor, and Steering Committee member). She has held numerous IEEE positions including secretary, IEEE VP – Technical Activities, IEEE Division III director, IEEE COMSOC president, IEEE Canada president, VP Technology and Engineering Management Society, and project director for COMSOC Wireless Communication Engineering Technologies certification (WCET) end Wireless Body of Knowledge (WEBOK). She is president of World Class – Telecommunications, providing telecommunications management training. Celia established and ran the PMO for Echologics. Her team managed all proposals, contracts, and project management processes, and she has lectured internationally on programs for success in today's changing environment. As director at Stentor Resource Center Inc., she was instrumental in establishing governance, culture, and service/product development processes. At Bell Canada, Celia provided strategic direction to corporate planners, ran technology/service trials, standardized equipment, and supported large business clients.

Michael Condry (Life Fellow, IEEE). He received the BS degree in mathematics from West Virginia University, Morgantown, WV, USA, in 1969, and the MS and PhD degrees in computer science from Yale University, New Haven, CT, USA, in 1975 and 1980, respectively. He is currently leading a consulting firm on technology future directions in business with a recent focus on digital health technologies and their data. Also, he chairs the Advisory Board for ClinicAI, a digital health startup. His career spans both academic and industry positions, mostly in industry. This includes senior leadership roles in Intel, Sun, and AT&T Bell Laboratories. At Intel, he retired after being the chief technical officer in the Client Division. He has held industry leadership positions with Intel, Sun Microsystems, and AT&T Bell Laboratories. He held teaching and research positions at Princeton University, Princeton, NJ, USA, and the University of Illinois, Urbana-Champaign, Champaign,

IL, USA. He has authored or coauthored multiple patents and publications. His background includes projects in the development of the Internet, computer architecture, software, operating systems, IoT standards, computer security, and recently digital health technologies. Dr. Condry is an IEEE Life Fellow and has many years engaging in the IEEE. He is an IEEE Industrial Electronics Society (IES) Board Member, past president of the IEEE Technology and Engineering Management Society (TEMS), and a member of the Computer, Consumer Technologies, and Engineering Medicine and Biology Societies. He started multiple efforts to bridge between industry and research with the IEEE programs.

Sudeendra Koushik, PhD (Co-editor and Steering Committee member). He is an innovator, speaker, and author, with expertise in intellectual property and patent generation and an immense insight into start-up to end-up and HR strategy for innovation. He holds bachelor's in Electrical Engineering from PES College, Mandya. An MBA and Postgraduate Strategy Management from IIM-K and a PhD in Innovation. A senior member of IEEE, he is also a senior member of Institution of Engineers Certified Independent Director, Vice-Chair IEEE Bangalore and Co-Chair IEEE Ad Hoc Committee on Innovation, and Co-Chair IEEE SYWL Congress Asia Pacific 2016. He has authored a book on innovation – *Conversation with the Innovator in You*. Sudeendra Koushik has also ventured into entrepreneurship by being the Founder, Director PraSu, and also Director at Founder Institute Bangalore. He has 24 years of international experience in product innovation and development and 20+ national and international patents in various stages.

Roberto Saracco (Expert consultants, Co-editor, and Steering Committee member). He fell in love with technology and its implications long time ago. His background is in math and computer science. Until April 2017 he led the EIT Digital Italian Node and then was head of the Industrial Doctoral School of EIT Digital up to September 2018. Previously, up to December 2011 he was the Director of the Telecom Italia Future Centre in Venice, looking at the interplay of technology evolution, economics, and society. At the turn of the century, he led a World Bank-Infodev project to stimulate entrepreneurship in Latin America. He is a senior member of IEEE where he leads the 2022 New Initiative Committee and co-chairs the Digital Reality Initiative. He is a member of the IEEE in 2050 Ad Hoc Committee. He teaches master's course both on Technology Forecasting and Market impact at the University of Trento and on Digital Transformation at the University of Cassino. He has published over 200 papers in journals and magazines and 20+ books.

TEMSBOK Co-editors and Steering Committee

Mark Wehde, MSc, MBA (Co-editor and Steering Committee member). He is chair of the Mayo Clinic Division of Engineering, assistant professor of Biomedical Engineering in the Mayo Clinic College of Medicine and Science, fellow in the Mayo Clinic Academy of Educational Excellence, and associate lecturer for the University of Wisconsin MBA Consortium program.

Mark is on the board of governors for the IEEE Technology and Engineering Management Society and the IEEE Systems Council. He is a technical committee member for the IEEE International Symposium on Medical Measurements and Applications and an affiliate for the University of Minnesota Medical Industry Leadership Institute.

Mark is a juror for the Medical Design Excellence Awards, the R&D 100 Awards, and the Edison Awards. He is also a member of the South Dakota State University Electrical Engineering Industry Advisory Board.

Shiyan Hu, PhD (Co-editor and Steering Committee). He received his PhD in Computer Engineering from Texas A&M University in 2008. Currently, he is a professor (chair in Cyber-Physical System Security) and director of Cyber Security Academy at University of Southampton. His research interests include Cyber-Physical Systems and Cyber-Physical System Security, where he has published more than 150 refereed papers, including 60+ in IEEE Transactions. He is an ACM Distinguished Speaker, an IEEE Systems Council Distinguished Lecturer, a recipient of the 2017 IEEE Computer Society TCSC Middle Career Researcher Award, and a recipient of the 2014 US National Science Foundation (NSF) CAREER Award. His publications have received distinctions such as the 2018 IEEE Systems Journal Best Paper Award, the 2018 IEEE TCSC Most Influential Paper Award, the 2017 Keynote Paper in IEEE Transactions on Computer-Aided Design, the Front Cover Paper in IEEE Transactions on Nanobioscience in March 2014, Multiple Thomson Reuters.

Dario Petri, PhD (Co-editor and Fellow, IEEE). He is currently a full professor in Measurement Science and Electronic Instrumentation and the chair of the Quality Assurance Committee of the University of Trento, Trento, Italy. He is the recipient of the 2020 IEEE Joseph F. Keithley Award "for contributions to measurement fundamentals and signal processing techniques in instrumentation and measurement." Dario Petri is an associate editor in chief of the IEEE Transactions on Instrumentation and Measurement. He is the chair of the IEEE Smart Cities Initiative in Trento since 2015. During his research career, he has been an author of about 350 papers published in international journals or in proceedings of peer-reviewed international conferences.

Tariq Samad, PhD (Co-editor and IEEE Fellow). Dr. Samad holds the W.R. Sweatt Chair and is a senior fellow in the Technological Leadership Institute (TIL), University of Minnesota. He joined TLI in 2016 after a 30-year career with Honeywell, retiring as Corporate Fellow. At Honeywell he led technology developments in automation and control for the process industries, homes and buildings, advanced manufacturing, aerospace, clean energy, and automotive sectors. He is a past president of the American Automatic Control Council and the IEEE Control Systems Society. He is a fellow of IEEE and IFAC and the founding chair of the IFAC Industry Committee. He is an elected member of the Board of Governors of the IEEE Technology and Engineering Management Society and services as the vice president of Publications for TEMS. His publications include the Encyclopedia of Systems and Control (co-editor-in-chief, Springer). He is the editor for a book series on "Technology Management, Innovation, and Leadership" (John Wiley & Sons/ IEEE Press).

Gus Gaynor Gerard (Gus), PhD (Steering Committee). He is an IEEE life fellow, who brings a record of technical accomplishments in instrumentation, control systems, automation, and product and process development. Gaynor's successful engineering career was followed by an entrepreneurial venture and more than 30 years in managing major operations in engineering, technology, innovation, and business revitalization. He was the vice president of IEEE TEMS Publications and past president of IEEE TEMS Society. He has served double three-year terms on the Publications Services and Product Board (PSPB), six years on the Technical Activities Board's (TAB) Society Review Committee, two terms on the IEEE New Initiatives Committee, chair of the IEEE Career Services Committee, two-year member of the Educational Activities Board Lifelong Learning Committee, and active participant in the development of Expert Now. Gaynor's other principal IEEE activities included founding editor and editor-in-chief of *Today's Engineer* magazine, EMS executive vice president, EMS president, EMS VP of Publications, EMS Newsletter editor, Chair EMS FinCom, Chair Strategic Planning Committee, Chair Editor Search Committee, and many other IEEE and EMS committee activities.

Eduardo Ahumada-Tello, PhD (Steering Committee). He is research professor at the Autonomous University of Baja California (UABC) and visiting professor at the San Pablo Catholic University in Bolivia and at the Global Humanistic University in Curaçao. He is currently a member of the Board of Governors in the IEEE Technology and Engineering Management Society (TEMS).

German Moya, MBA (Steering Committee). He is senior member of IEEE, past president of IEEE Section Costa Rica, entrepreneur, industry consultant, member of the TEMS Region 9 Conference organizers, and member of the Organizing committee of the IEEE TEMS Latin American Forum.

About the Contributors

Chapter 1

Mikael Collan, Doc., D. Sc. (Econ. & Bus. Adm.), is the Director General of the Finnish State Institute for Economic Research in Helsinki and a professor of Strategic Finance at the LUT-University in Lappeenranta, Finland. His research concentrates on enhancing and creating new profitability analysis methods for industrial investments and especially on real options analysis. He is the author of over 200 scientific publications. Dr. Collan is a life member of the Finnish Society of Sciences and Letters, the oldest academy of science in Finland, and a past president of the Finnish OR society. He is an investor and a board member in many Finnish and international SMEs.

Jyrki Savolainen, D.Sc. (Econ. & Bus. Adm.), M.Sc. (Eng), is a post-doctoral researcher in the School of Business and Management at LUT-University, Lappeenranta, Finland. His research focuses on profitability and real options analysis of project investments with system dynamic simulation models. Dr. Savolainen has close to 10 years of industrial experience in the mining industry and is a founding partner in an industrial analytics consultancy company.

Chapter 2

Reyna Virginia Barragán Quintero (Member, IEEE) received a bachelor's degree in Business Administration and a master's degree in Corporate Finance from CETYS Universidad, Tijuana, México, and the Ph.D. degree in Management Science from Universidad Autónoma de Baja California, Tijuana, in 2017, with a dissertation on the innovation capabilities of the wine industry in Guadalupe Valley, México. She worked in the private industry for more than 20 years and has held management positions in global companies, such as HSBC Bank and Scotiabank Mexico. During her practice, she was certified on Ethics for Financial Markets, as an Investment Markets Promoter, and as a Stockbroker by the Mexican Association of Stock Market Intermediaries. She is also certified in Human Resource Development, Foreign Trade for IMMEX Companies, and recently has been certified as a Professional Junior Trainer for Atlas.ti in Berlin for Qualitative Studies. She is currently the MBA coordinator with the School of Engineering, Management and Social Sciences, Universidad Autónoma de Baja California, and a professor in the fields of business strategy, finance, and management. Her research interests include finance, innovation for regional development, innovation management, strategic management, and people and organizations and is a published author and a lecturer in international forums.

Fernando Barragán Quintero graduated as a Lawyer from Universidad Iberoamericana, Mexico City, Mexico, and as a Classical Philologist from Universidad Nacional Autónoma de México (UNAM), Mexico City. His graduate studies in social sciences include a master's degree in Asian and African Studies, specializing in India and Southeast Asia, and a master's degree in Middle Eastern Studies, both from El Colegio de México (Colmex), Mexico City, Mexico. He has done research work on Roman law, legal history, international law, specifically international trade law, Fintech and Fintech law, and globalization.

Chapter 3

Eduardo Ahumada-Tello is a Research Professor at the Universidad Autónoma de Baja California (UABC), Mexico, and Visiting Professor at San Pablo Catholic University, Bolivia, and the Global Humanistic University, Curaçao. He is a Member-at-Large of the IEEE Technology and Engineering Management Society (TEMS) and has over 20 years of experience helping to create technology-based companies. Prof. Ahumada-Tello holds a Ph.D. in Management Sciences from the UABC, a Ph.D. in Education from the Universidad Iberoamericana, a master's degree in Administration from the UABC, a master's degree in Psychology from CIDH University, and a BSc in Computer Engineering from the UABC. He is also a member of the National System of Researchers in Mexico at Level 1. His research interests span Organizational Intelligence and Knowledge Management, Engineering Management, Subjective

Well-being and Happiness Management, and Complex Systems. He has authored numerous scientific papers in peer-reviewed journals and conference proceedings, including *Quality and Quantity, Engineering Management Review, Sustainability,* and *Corporate Governance.*

Oscar Omar Ovalle-Osuna joined the Universidad Autónoma de Baja California in 2010 and is currently a Research Professor and the Dean of the Faculty of Engineering, Administrative and Social Sciences. He holds a Ph.D. in Administrative Science Studies from the UABC, a master's degree in Organizational Engineering and Administration from the Universidad de Castilla la Mancha Spain, a master's degree in Administration (with a focus on Management) from CETYS Universidad, and a bachelor's degree in Mechatronic Engineering from the UABC. Prof. Ovalle has been a member of the National System of Researchers in Mexico since 2021 and has earned the recognition of Professor with a Desirable Profile (PRODEP) since 2012. He has also been awarded an Academic Certification in Intellectual Property and Technology Transfer. His research interests span innovation, technological change, digital strategy, and product knowledge. Prof. Ovalle has authored numerous peer-reviewed articles and is the author of the book *"Analysis of Economic Complexity and Innovation for Productive Capacities Management."*

Richard D. Evans is an Associate Professor in Digital Innovation within the Faculty of Computer Science at Dalhousie University, Canada, where he also acts as Program Director for the Bachelor of Applied Computer Science. He holds a Ph.D. in Knowledge Management for Collaborative Product Development from the University of Greenwich, working in partnership with BAE Systems plc., a master's degree in e-commerce, and a bachelor's degree in computing. Prof. Evans is an established and award-winning academic with research and teaching interests spanning digital innovation and transformation, design and engineering management, and entrepreneurship. He has authored/coauthored over 150 scientific papers in peer-reviewed journals and conference proceedings, including *Technological Forecasting and Social Change, IEEE Transactions on Engineering Management, Computers in Human Behavior,* and the *International Journal of Production Research.*

Chapter 4

Scott Ambler is a Consulting Methodologist with Ambysoft Inc. Scott leads the evolution of the Disciplined Agile (DA) tool kit and is an international keynote speaker. Scott is the (co)-creator of the DA tool kit as well as the Agile Modeling (AM) and Agile Data (AD) methodologies. He is the (co-)author of 28 books, including *Choose Your WoW! (Way of Working)* [9], *An Executive's Guide to the Disciplined Agile Framework, Refactoring Databases, Agile Modeling,* and *The Object Primer 3rd Edition.*

The agile movement began in 2001 with the publication of The Manifesto for Agile Software Development [26]. It started with the desire to improve software development practice, but over the years people, teams, and organizations around the world have applied agile and lean concepts in a wide range of contexts and domains. Although many have benefitted from adopting agile strategies, it is clear there is more work required to fulfill the promise of the agile movement.

This paper addresses the following ideas:

1) Agile methods/frameworks are only a good start
2) In organizational transformation, context counts
3) Continuous improvement (CI)
4) Guided continuous improvement (GCI)

Applying GCI in practice

Chapter 5

Marcelo Da Costa Porto, Bachelor of Administration – Accountant graduated from the Economic Science University – Universidad de la Republica. Graduated from the first generation in Uruguay from the Program for Management Development of the ORT University – ESADE Spain. PMP® certified by the Project Management Institute (USA). Accredited in Belbin® methodology (United Kingdom). Master in Strategic Management from the Miguel de Cervantes University (Spain) – Uneatlántico.

Consultant, teacher, and speaker with extensive experience in strategic planning, project management, and organizational development processes for public and private sectors at a national and international level. Founding partner of Avanza Consulting.

The contributions made to the Uruguay Central Bank, the Inter-American Development Bank (USA), the Community of Portuguese Language Countries (Angola, Mozambique, Portugal), the Directorate of National Taxes and Customs (Colombia), the International Monetary Fund (USA) stand out. Municipality of Montevideo (Uruguay), Mercado Libre (Argentina), Project Management Institute (USA), World Customs Organization (Belgium), Universidad de la Republica – Uruguay, CLAEH University.

He has led projects and made presentations on strategic planning, project management, and organizational assessment systems in Angola, Argentina, Belgium, Colombia, Costa Rica, Guatemala, Honduras, Mozambique, Panama, Paraguay, Portugal, and Uruguay.

At a managerial level, he was Director of the Strategic Planning Advisory of the National Customs Directorate of Uruguay between 2014 and 2017, Executive Secretary of the National Institute of Professional Training (INEFOP) during 2008, and Advisor to the Board of the National Postal Administration (ANC) between 2005 and 2009.

At teaching level, he is responsible for the Strategic Planning subject at the Postgraduate Center of the Economic Sciences and Administration University. He served as head of the General Administration, Strategy and Competence, and Organizational Behavior Chairs (ORT University), Academic Coordinator at the National School of Public Administration (ENAP), and mentor for the CEDDET (Spain), ENDEAVOR (USA), and Empretec Foundations (USA).

Chapter 6

Joseph Tidd is a physicist with subsequent degrees in technology policy and business administration. He is a Professor of technology and innovation management at the Science Policy Research Unit (SPRU), visiting Professor at University College London, and previously at Imperial College, Cass Business School, Copenhagen Business School, and Rotterdam School of Management. Dr Tidd was previously Deputy Director of SPRU, Head of the *Innovation Group,* and Director of the *Executive MBA* Programme at Imperial College. He has worked as a policy adviser to the CBI (Confederation of British Industry), presented expert evidence to three Select Committee Enquiries held by the House of Commons and House of Lords, and was the only academic member of the *UK Government Innovation Review*. He is a founding partner of Management Masters LLP.

He was a researcher for the five-year *International Motor Vehicle Program* of the Massachusetts Institute of Technology (MIT), which identified Lean Production, and has worked on technology and innovation management projects for consultants Arthur D. Little, CAP Gemini and McKinsey, and numerous technology-based firms, including American Express Technology, Applied Materials, ASML, BOC Edwards, BT, Marconi, National Power, NKT, Nortel Networks and Petrobras, and international agencies such as UNESCO in Africa. He is the winner of the Price Waterhouse Urwick Medal for contribution to management teaching and research, and the Epton Prize from the R&D Society.

He has written 9 books and over 90 papers on the management of technology and innovation, including *Managing Innovation* (seventh edition, 2020, with John Bessant), has 33,000 research citations (Google Scholar), hosts a popular YouTube channel, and the Innovation Portal which has in excess of a million-page visits (Google Analytics). He is part of the *Intrapreneurship Hub,* a collaborative venture between Sussex, Bocconi, and Renmin business schools. He is the founder and Managing Editor of the *International Journal of Innovation Management,* which is the official journal of the International Society of Professional Innovation Management (*ISPIM*), and Managing Editor of the research series on Technology Management for *Imperial College Press*, currently with more than 40 titles.

Chapter 7

Sudeendra Koushik, Innovation consultant at Innovation by Design, Bangalore, India, holds a degree in Electronics & Communication, MBA in Marketing, and PhD in Innovation. Dr Koushik has worked in several countries in Innovation & technology Management. His research area has been in developing Innovators, Intrapreneurs, and bringing Innovation culture in Technology centres of large organisations. Dr Koushik was head of Innovation at several companies including Volvo Group India, TTK Prestige, etc., and has consulted for companies like Bosch, Mercedes Benz, Continental, General Motors, ZF among others. He is also active in IEEE and is the chairman of committee on Ethics awareness at Region 10 and is VP Conferences at Technology and Engineering Management Society (TEMS) at IEEE Globally.

Chapter 8

Elif Kongar, PhD, is a Professor of Economics & Business Analytics at the Pompea College of Business, University of New Haven. Before joining UNH, she served as Associate Dean for Graduate Studies and Research in the School of Engineering at Fairfield University between September 2021 and December 2022. Prior to this appointment, Dr. Kongar was Professor of Technology Management and Mechanical Engineering and Chair of the Technology Management Department at the School of Engineering at the University of Bridgeport (UB), and held a Lecturer position at Yale School of Management. During her tenure at UB, she established several research programs, graduate concentrations and coursework in interdisciplinary areas including predictive analytics, data visualization, statistical analysis, supply chain management and logistics, simulation and modeling, economic and environmental sustainability, and service management and engineering. Dr. Kongar is Founding Director of the Ph.D. Program in Technology Management at UB – the only Ph.D. program of its kind in the Northeast; and one out of a handful of such programs in the U.S.; and has served in this role since the inauguration of the program in September 2014. Dr. Kongar utilizes quantitative and qualitative methods in her interdisciplinary and technology-driven research. She has published over one hundred well-cited papers in scholarly journals and conference proceedings. Her research areas include economically and environmentally sustainable waste recovery systems and operations, disassembly sequencing and planning, electronic waste, performance evaluation, big data and data analytics, quality improvement, engineering education programs, female participation and efficiency in engineering disciplines and in K-12 STEM education. Dr. Kongar is the Department Editor of the IEEE Transactions on Engineering Management and an active member of various national and international technical committees, advisory boards, program committees, and editorial boards. She is an elected member of the IEEE Technology and Engineering Management Society (TEMS) Board of Governors and has served as Chair of the IEEE TEMS Professional Training and Education Adhoc Committee. Dr. Kongar earned her Ph.D. in industrial engineering from Northeastern University and her BS and MS in industrial engineering from Yildiz Technical University, Turkey.

Tarek Sobh, PhD, received the B.Sc. in Engineering degree with honors in Computer Science and Automatic Control from the Faculty of Engineering, Alexandria University, Egypt in 1988, and M.S. and Ph.D. degrees in Computer and Information Science from the School of Engineering, University of Pennsylvania in 1989 and 1991, respectively. He is currently the President and a Professor of Electrical and Computer Engineering at Lawrence Technological University (LTU), Michigan. He is Distinguished Professor and Dean of Engineering Emeritus at the University of Bridgeport (UB), Connecticut.

Previously, he served as LTU's Provost (2020–2021). He also served as UB's Executive Vice President (2018–2020), Interim Provost (2020), the Founding Director of the Interdisciplinary Robotics, Intelligent Sensing, and Control (RISC) laboratory (1995–2020), Distinguished Professor of Engineering and Computer Science (2010–2020), Senior Vice President, Founding Dean of the College of Engineering, Business, and Education, and Dean of the School of Engineering. At UB, he was Professor of Computer, Electrical and Mechanical Engineering and Computer Science (2000–2010), and an Associate Professor (1995–1999). He also served as a Research Assistant Professor of Computer Science at the Department of Computer Science, School of Engineering, University of Utah (1992–1995) and was a Research Fellow at the General Robotics and Active Sensory Perception (GRASP) Laboratory of the University of Pennsylvania (1989–1991).

His background is in the fields of robotics, automation, computer science and engineering, STEM Education, manufacturing, AI, and computer vision. He has published over 250 refereed journal and conference papers, and book chapters in these areas, in addition to 27 books. Dr. Sobh served on the editorial boards of 18 journals, and has served as Chair, Technical Program Chair and on the program committees of over 300 international conferences and workshops in the Robotics, Automation, Sensing, Computing, and Engineering Education areas.

Dr. Sobh is a Fellow of the African Academy of Sciences, a Member of the Connecticut Academy of Science and Engineering, and a Fellow of the Engineering Society of Detroit. Dr. Sobh is a recipient of the ASEE Northeastern U.S. Distinguished Engineering Professor of the Year award, the IEEE Northeast Technological Innovation Research Award, an ACE Higher Education Award and several other merits in recognition of his educational, research, scholarly and service activities in engineering, education, computing, and diversity initiatives.

Chapter 9

Arun Tanksali is the co-founder and CTO of Nearex Pte Ltd, a leading payment technology startup. He was previously the CTO of Mahindra Comviva and, earlier, of Jataayu Software. He is the Chair of IEEE TEMS India. He holds a Master of Engineering in Electronics.

Chapter 10

Eldon Glen Caldwell Marin is full professor/cathedraticus and researcher of the Industrial Engineering Department at the University of Costa Rica. Also, he serves as Vice-Chair of IEEE IAS Society Costa Rica, Distinguished Professor of IEEE Computer Society US and President-Elect, Fellow, and Director of the Global Council of the Industrial Engineering and Operations Management Society, IEOM, US.

Dr. Caldwell earned his B.Sc. and Master's degree in Industrial Engineering at the University of Costa Rica and he earned a Master's degree in Service Marketing, as well in Financial Analysis at the Interamerican University of Costa Rica; M.Sc. Health Management Systems at UNED, Costa Rica, and a specialization in Operations Management at Veracruzana University, Mexico.

Professor Caldwell earned a Ph.D. in Computer Science/Industrial Engineering with a major in Lean Operations Engineering and Artificial Intelligence (AI) at the University of Nevada/ Autonomous University of Central America. Moreover, he earned another Ph.D. (Sc.D.) in Informatics/Robotics and Automation at the University of Alicante, Spain and, recently he earned a third Ph.D. in Education (Dr.Ed.) at the University of Costa Rica.

Dr. Caldwell has been recognized by IEEE, IEOM, and UCR for his career of over 30 years as an educator, researcher, and promoter of the development of engineering. He is the author of many scientific articles and two books: "Marketing of Social Products & Services" and "Lean Manufacturing: Fundamentals and techniques for cycle time reduction".

Finally, Dr. Caldwell served as Chair of the Industrial Engineering Department at the University of Costa Rica, Operations Manager at MASECA, CA; Lean Manufacturing Project Manager at Eaton Corp. Costa Rica, General Manager at Quirós & Cía-Bandag Inc. and General Manager at Lean Systems Intl., US. His experience as advisor and consultant covers many projects at the Interamerican Bank for Development, WHO, UN, World Bank, Ministry of Health, Costa Rica, Honduras and Panama, Costa Rican Institute for Electricity, RTC-Peru, Plan International-Honduras, Costa Rica Popular and Community Development Bank, Ministry of Education, Costa Rica, Bournes Group Inc., DOLE Latin America, and many others.

Chapter 11

Celia Desmond is President of World Class – Telecommunications, providing telecommunications management training. Celia established and ran the PMO for Echologics. Her team managed all proposals, contracts, and project management processes. She has lectured internationally on programs for success in today's changing environment. As Director at Stentor Resource Center Inc., she was instrumental in establishing governance, culture, and service/ product development processes. At Bell Canada, Celia provided strategic direction to corporate planners, ran technology/service trials, standardized equipment, and supported large business clients.

Celia is IEEE Canada Parliamentarian and Chair of Awards and Recognition Committee, along with Working Group Chair in Industry Engagement Committee and chair of multiple committees in IEEE OU's. She has held numerous IEEE positions including, Secretary, IEEE VP – Technical Activities, IEEE Division III Director. ComSoc President, IEEE Canada President, VP of Technology and Engineering Management Society, and Project Director for ComSoc's Wireless Engineering certification.

Celia holds MSc. Engineering, B.Sc. Mathematics and Psychology, Ontario Teaching Certificate and PMP certification. Celia has taught kindergarten, high school, and at five universities. She is the author of two Project Management books.

Chapter 12

Andrea Molinari. Since 1988 Andrea Molinari has been involved in R&D in ICT and re-engineering of processes, Project Management, and e-learning, with about 250 publications in books, journals, and scientific conferences. Since 1990 he has been a lecturer in ICT and Project Management at the University of Trento (Italy), since 2003 at the University of Bolzano (Italy), and since 2007 at Turku University and Abo Akademi (Finland). Since their debut, Andrea has been an early adopter and beta tester of MS Project and Project Server. He works on Enterprise Project & Portfolio Management tools and techniques in public, private, and EU funding projects. Since 1996, he has acted as Project Manager in various European projects in the various frameworks, from FP4 to H2020. NET software architect for large-scale software projects; he has been working since 2002 on the evolution and extension of Enterprise Project Management implementations in large organizations and public administrations, with SME Agile and predictive methodologies, using PMI, PMI Agile standards, Hybrid PM, and Scrum Body of Knowledge. From 2012 to 2014, he was in charge of the ICT and Project Management part of the Organizing Committee of the XXVI Winter Universiade. Since 2010 he has been co-founder and CTO of Okkam, a spin-off of the University of Trento and FBK that deals with semantic technologies and data integration through AI approaches.

Chapter 13

Cristina Bueti is the ITU Focal Point on Smart Sustainable Cities. She is also the Counsellor of ITU-T Study Group 20 "Internet of things (IoT) and smart cities and communities (SC&C)" and the Counsellor of the Focus Group on metaverse at the International Telecommunication Union (ITU). She also serves as TSB/ITU focal point for Latin America. Cristina Bueti graduated from the Faculty of Political Science, Law and International Cooperation and Development of the University of Florence, where she completed postgraduate studies in International Cooperation and Telecommunications Law in Europe. She also holds a specialization in Environmental Law with a special focus on Telecommunications. In 2003, Ms. Bueti built on her academic credentials by completing a specialized course in peace keeping and international cooperation with special focus on telecommunications at the Faculty of Laws, University of Malta, before joining the International Telecommunication Union in Geneva in January 2004. As part of the International Women's Day 2016, she was named as one of the twenty Geneva-based inspirational women working to protect the environment. She has authored over 40 reports on telecommunication issues. A native Italian speaker, Cristina is also fluent in English, French, and Spanish.

Mythili Menon works as Study Group Project Officer at the International Telecommunication Union (ITU) in the domain of standardization for new and emerging technologies. She serves as the Advisor of the ITU/WMO/UNEP Focus Group on Artificial Intelligence for Natural Disaster Management (FG-AI4NDM) and the ITU/FAO Focus Group on Artificial Intelligence (AI) and Internet of Things (IoT) for Digital Agriculture (FG-AI4A). Mythili possesses a strong background in environmental science, standardization, and the policy frameworks underlying sustainable development issues. Between 2015 and 2017, she worked for the ITU-T Study Group 20 on "Internet of Things and Smart Cities and Communities" and ITU-T Study Group 5 on "Environment, Climate Change, and Circular Economy." Previously, she worked as the Standardization Lead at Mandat International for various EU Horizon 2020 projects. Mythili holds a Masters in "Standardization, Social Regulation and Sustainable Development" from the University of Geneva. She completed her Bachelors in 'Environmental Science and Chemistry' from University College Utrecht in the Netherlands. The areas relating to the sustainable development goals, artificial intelligence, blockchain, intelligent transport systems, disaster management, data management, smart cities, and IoT are of enduring interest to her and she has co-authored several publications on these subjects.

Chapter 14

Cristina Emilia Costa received the M.S. degree in Electronic Engineering from the University of Genoa, Italy, in 1996, and the Ph.D. in ICT from the University of Trento in 2005. After graduation, she joined the Telecom Italia Lab (Turin, Italy) in the Internet Services group, providing technology transfer and scientific support to operational Telecom Italia divisions. She worked as a researcher at the CREATE-NET research centre, Trento, Italy, from 2004 to 2019. She was involved in several research projects both at the national and international level, gaining experience both as a researcher and as a project coordinator in the fields of Wireless and Mobile Networks, Multimedia Communications, Interfaces and Interaction, 5G, and Edge Computing. Presently, she is Head of the Smart Networks and Services (SENSE) Research Unit at Fondazione Bruno Kessler (FBK – Trento, Italy). She served as a member in the organizing committees of various conferences (Intetain, UCMedia, and SecureComm). She is an IEEE Senior Member and secretary of the IEEE WIE (Women In Engineering) A.G. Italy section.

Fabrizio Granelli is Full Professor at the University of Trento (Italy), IEEE ComSoc Distinguished Lecturer and Chair of the Joint IEEE VTS/ComSoc Italian Chapter. He received the "Laurea" (M.Sc.) and a Ph.D. degree from the University of Genoa, Italy, in 1997 and 2001, respectively. He was visiting professor at the State University of Campinas (Brasil) and at the University of Tokyo (Japan). He was IEEE ComSoc Distinguished Lecturer for the period 2012–2015 (two terms), ComSoc Director for Online Content in 2016–2017, Delegate for Education at DISI in 2015–2017, and IEEE ComSoc Director for Educational Services (2018–2019). He is TPC Co-Chair of the IEEE

NFV SDN conference in 2018 and 2019 and TPC Chair for IEEE Globecom 2022. Since 2019, he is the Founding Chair of the Aerial Communication Emerging Technology Initiative. He is the author or co-author of more than 250 papers published in international journals, books, and conferences. He is Associate Editor in Chief of IEEE Communications Surveys and Tutorials, and Area Editor of the IEEE Transactions on Green Communications and Networking.

Chapter 15

Francesco Pilati is Assistant Professor at the Department of Industrial Engineering of the University of Trento for the scientific field Industrial plants, Production systems, and Logistics. Coordinator of master degree program in Management and Industrial Systems Engineering. Invited Lecturer since 2019 at the University of Göttingen (Germany) for the Faculty of Economics and Management Sciences. Ph.D. in Mechatronics and Product Mechanical Innovation – Curriculum in Industrial Plants and Logistic Systems at the University of Padua, obtained in 2016. Visiting Scholar during 2015 at the Department of Industrial Engineering, Oklahoma State University (USA).

Main research activities concern the design and management of production and logistics systems, through mathematical models and optimization techniques with particular attention to the digitalization of production processes. Principles and technologies of Industry 4.0: human-machine interaction through augmented reality tools, Internet of Things for the digitalization of assembly processes, Digital twin for the simulation of material flows in industrial plants and layout of production systems. Collaboration with national and international companies for industrial research projects.

Daniele Fontanelli (M'10, SM'19) received an M.S. degree in Information Engineering in 2001, and a Ph.D. degree in Automation, Robotics, and Bioengineering in 2006, both from the University of Pisa, Pisa, Italy. He was a Visiting Scientist with the Vision Lab of the University of California at Los Angeles, Los Angeles, US, from 2006 to 2007. From 2007 to 2008, he has been an Associate Researcher with the Interdepartmental Research Center "E. Piaggio," University of Pisa. From 2008 to 2013, he joined as an Associate Researcher the Department of Information Engineering and Computer Science and from 2014 the Department of Industrial Engineering, both at the University of Trento, Trento, Italy, where he is now an Associate Professor, and he is leading the EIT-Digital International Master on Autonomous Systems. He has authored and co-authored more than 150 scientific papers in peer-reviewed top journals and conference proceedings. He is currently an Associate Editor in Chief for the IEEE Transactions on Instrumentation and Measurement and an Associate Editor for the IET Science, Measurement & Technology Journal. He has also served in the technical program committee of numerous conferences in the area of measurements and robotics. His research interests include distributed and real-time estimation and control, localization algorithms, synchrophasor estimation, clock synchronization algorithms, resource-aware control, wheeled mobile robots control, and service robotics.

Davide Brunelli is Associate Professor at University of Trento. He holds a Degree in Electrical Engineering (2002) and a Ph.D. in Electrical Engineering (2007) from the University of Bologna. From 2005 to 2007, he has been visiting researcher at ETH Zurich studying to develop methodologies for Energy Harvesting aware embedded design. He was scientific supervisor of several EU projects: H2020-GreenDatanet, EEB-ENV-3ENCULT, ICT-GENESI and FP7 Network of Excellence – ARTISTDESIGN – Design for embedded systems; ERASMUS+ "IoT Rapid Proto Lab." He was leading industrial cooperation activities with Telecom Italia and STMicroelectronics from 2007. Member of several Technical Program Committees of conferences in the field of sensor networks and energy management, including The Design, Automation, and Test in Europe (DATE) conference and the IEEE International Conference on Emerging Technologies and Factory Automation, the International Workshop on Energy in Wireless Sensor Networks. He is senior member of the IEEE. He is author of more than 220 papers in international journals and conferences.

Chapter 16

Gastón Lefranc was born in Chile. He obtained the title of Electrical Engineer (Bachelor) at Universidad Técnica Federico Santa María in Valparaíso, Chile (1969). Then, he obtained the title of Civil Electrical Engineer at Universidad de Chile (1976). He obtained Master's Degree and PhD (R) at Northwestern University in Evanston, USA (1979). He received the Title of Professor Honoris Causa at Agora University, Rumania (2022). He worked at the Electrical Engineering Department of the University of Chile for seven years and then in the Pontificia Universidad Católica de Valparaíso, Chile. He is full professor and have participated in several International and Chilean Research Projects. He has about 80 publications in International Journals and about 200 in International Conferences. He has received numerous distinctions as IEEE Eminent Engineer of Latin America in 1998, Millennium Medal IEEE in 2000, Outstanding Chapter Award of IEEE Control Systems Society (1992 and 2019), Medal of Academic Merit, Pontificia Universidad Católica de Valparaíso, Chile (2012) and many others. His fields of interest and work are in Automatic control, Robotics, Artificial Intelligence, and Biomedical Engineering.

He has been the first Chair of the IEEE Chilean Chapter in Control Systems in 1985. In 1992 and 2019, IEEE Outstanding Chapter Award was received from the Control Systems Society (it was granted for the first time in 1992). In 2015 he returns to the Chair of the Chapter. He has been IEEE Chile Section Chair for 4 times and IEEE Cono Sur Council Chair (Argentina, Uruguay, Paraguay Chile Sections) in 2017 and 2019. He is Life Senior Member of IEEE. He has been President and member of Board of Director of Chilean Association of Automatic Control (ACCA), NMO of IFAC. He is member in TC5.4 of IFAC. With IEEE Chilean CS Chapter and ACCA, has organized many activities every year, such as the co-organization of IEEE Conferences in Chile: IEEE Chilecon Conference and IEEE ICA ACCA Conference. (ACCA Congress), alternately every year. More than 300 papers are received, from more than 28 countries. The audience is around 300–400 people. He has been several times Visiting Professor at Ecole Centrale de Lille, France, at UNAM Mexico, Santa Catarina University in Brazil, Panamá, Perú Argentina, and Rumania, among others.

Chapter 17

Parag Chatterjee received his BSc and MSc degrees in Computer Science from the University of Calcutta, India, and a PhD from the National Technological University in Buenos Aires, Argentina, and is currently working as an Assistant Professor at the Department of Biological Engineering, University of the Republic *(Universidad de la República)*, Uruguay. He is the editor of several books including *The Internet of Things: Foundation for Smart Cities, eHealth, and Ubiquitous Computing* by CRC Press (Taylor & Francis), *Applied Approach to Privacy and Security for the Internet of Things by IGI Global, 5G, and Beyond – The Futuristic IoT* by CRC Press, *AI in Healthcare and COVID-19* by Elsevier. Chatterjee also authored several scientific papers published in international journals and conferences. He also delivered invited talks at institutions like the Indian Institute of Science (IISc Bangalore, India) and the University of Rome Tor Vergata (Italy), and also at international conferences like IoT Week 2017 Geneva (Switzerland), SICC-2017 Rome (Italy), and ExpoInternet LatinoAmerica (Argentina). He is a member of the editorial and reviewer boards of 10+ international journals including *Internet of Things Journal* (Elsevier), *Array* (Elsevier), and member of program committee in several IEEE international conferences. Currently, his work is focused to the domain of Artificial Intelligence applied to predictive and preventive healthcare, especially in cardiometabolic diseases.

Ricardo Armentano received his Engineering degree in 1984. He obtained his PhD in Physiological Sciences (1994) from the University of Buenos Aires and from Université de Paris VII Diderot in Mechanics of Biological Systems (1999). In 2020 was appointed as Fellow International Academy of Medical and Biological Engineering. Currently, he is Distinguished Professor of Biomedical Engineering and member of the IEEE EMBS Technical Committee on Cardiopulmonary Systems and Physiology-based Engineering. In 2019, Ricardo Armentano was conferred the IEEE R9 Eminent Engineer Award and he served as EMBS Distinguished Lecturer. Currently, he is Distinguished Professor of Biomedical Engineering, Director of the Biological Engineering Department, and Principal Investigator of UNPD/84/002 at *Universidad de la República* (Uruguay). He has acquired international recognition in the field of cardiovascular hemodynamics and arterial hypertension. He has taught in the fields of cardiovascular dynamics and in the broad area of engineering in medicine and biology and has extensive experience in PhD supervision and examination. He is on the editorial board of journals of cardiovascular research and is a reviewer for 15+ international scientific journals. He has 350+ publications including a book, book chapters, and peer-reviewed articles.

Chapter 18

Mariel Feder Szafir is a Software Engineer and Master in Computing and Information Systems from ORT University, with a postgraduate Diploma in Computing from the University of Oxford, and a Big Data postgraduate specialization title.

She has worked as a consultant in regional and international projects since 1995 in Unión Fenosa and Soluziona, as IT Manager in Codere, and currently works as a Leader of Digital Transformation and Software Solutions at Satellogic, an aerospace company.

Mariel is a university professor of Project Management and Software Engineering in undergraduate and postgraduate courses at ORT University, and she is the coordinator of the Project Management Area of the Software Engineering Department of that university.

She was the Program Director of PMI Montevideo Chapter and member of the board.

She is an honorary member of the committee of experts for the drafting of project management standards at UNIT/ISO.

She has also worked as an advisor in Interamerican Development Bank (IDB) projects, is an expert advisor to the Judiciary in Uruguay, and advisor in Agesic, the Agency for Electronic Government and for the Information and Knowledge Society in Uruguay.

She has published several articles and has participated in different national and international events as lecturer, including congresses and events at PMI and IEEE, and has served as an evaluator of projects for funding through different transformation and innovation software programs by the Uruguayan Government.

She owns and functions as Director at her own Software Development Company.

Chapter 19

Nicholas Napp is a Senior Member of IEEE, a Steering Committee Member for IEEE's Digital Reality Initiative and an active participant in multiple IEEE Standards activities. He is also the CEO of Xmark Labs, a consulting and product Development Company focused on the application of Digital Reality tools to digital transformation, and Board Chair at FabNewport, a non-profit provider of maker-centered learning programs and services for middle school and older students. He is particularly interested in the use of Blockchain, IoT, Augmented Reality, Mixed and Virtual Reality, and AI to redefine and improve business processes and products. Nicholas has been working with VR and AR systems since the 1990s, when his team at Rainbow Studios pioneered the use of VR in 3D animation production. He has launched over 40 products across a wide range of consumer and enterprise markets. His prior roles include leading multiple startups; Lead External Technology Scout, North America for Sony Ericsson; and Vice-President, Animation, Rainbow Studios (sold to THQ, Inc.). Nicholas has contributed to projects for companies such as Apple, AT&T, Hasbro, Hilti, HTC, IEEE, Microsoft, NREL, Tencent, and multiple divisions of Sony. He is also a frequent speaker at conferences such as AWE, COMPSAC, and Digital Innovation in Mental Health.

Louis Nisiotis (m) is a lecturer in Computing and the Course Leader for BSc Computing at the University of Central Lancashire, Cyprus campus. He is responsible for the delivery of the Games Development and Computer Graphics modules, and he is the Lead of the computing research on "Immersive Technologies" Research Pillar. He is engaging in multidisciplinary research specializing in Immersive Technologies, and their application in Cyber-Physical-Social Systems and in Education among other industrial and societal domains. He has been awarded his PhD in 2015 (Sheffield Hallam University, UK) for investigating the use of 3D Virtual Worlds to Support Students Experiencing Barriers Accessing Education. His research interests include VR/AR/MR/XR, Intelligent Reality Environments, Immersive Learning, and Human-Computer Interaction.

Currently leading research in innovation and application of immersive technologies and robotics to support the digital transformation of a plethora of domains. His current research work focuses on emerging a new type of Cyber-Physical-Social Eco-Society of systems that seamlessly blends the real with virtual worlds influenced by humans and their behavior, using XR, Robots, AI, and Social Networking technologies.

Dr. Nisiotis is also leading research focusing on the use of immersive learning technologies and intelligent XR environments to support and enhance education, specifically investigating student collaboration, engagement, motivation, and their learning experience. Over the years, he has developed a series of guidelines on how to design virtual environments and collaborative educational activities and learning tasks to effectively support online learning through virtual worlds.

Dr. Nisiotis is the General Chair of the second IEEE International Conference on Intelligent Reality (ICIR 2022). He is a member of the IEEE Digital Reality steering committee and a member of the Immersive Learning Research Network. He is also a Fellow of the Higher Education Academy, UK, member of IEEE and ACM SIGCSE.

Chapter 20

Roberto Saracco fell in love with technology and its implications long time ago. His background is in math and computer science. Until April 2017, he led the EIT Digital Italian Node and then was head of the Industrial Doctoral School of EIT Digital up to September 2018. Previously, up to December 2011 he was the Director of the Telecom Italia Future Centre in Venice, looking at the interplay of technology evolution, economics, and society. At the turn of the century, he led a World Bank-Infodev project to stimulate entrepreneurship in Latin America. He is a senior member of IEEE where he leads the 2022 New Initiative Committee and co-chairs the Digital Reality Initiative. He is a member of the IEEE in 2050 Ad Hoc Committee. He teaches master's course both on Technology Forecasting and Market impact at the University of Trento and on Digital Transformation at the University of Cassino. He has published over 200 papers in journals and magazines and 20+ books.

Chapter 21

Ignacio Varese is a Computer Engineer, MBA, certified as Project Management Professional (PMP), and Scrum Master. More than 20 years of experience in the software industry. Solid experience working with national and international companies from both the private and government sectors.

He has developed digital transformation solutions in multiple sectors such as finance, education, food, fashion, retail, logistics, and government. I have led the implementation of blockchain solutions, ERPs and Business Intelligence, Consulting, and hosting projects.

Founder of the companies Blockbear Software Solutions, Blockfashion, and Blockchain Summit Global, the first international blockchain conference in Uruguay and the most important in LATAM. International speaker, on blockchain and project management topics.

Chapter 22

Jon Clay has worked in the cybersecurity space for over 25 years.

Jon uses his industry experience to educate and share insights on all Trend Micro externally published threat research and intelligence. He delivers webinars, writes blogs, and engages customers and the general public on the state of cybersecurity around the world.

An accomplished public speaker, Jon has delivered hundreds of speaking sessions globally. He focuses on the threat landscape, cybercriminal undergrounds, the attack lifecycle, and the use of advanced detection technologies in protecting against today's sophisticated threats.

Within Trend Micro, Jon has held roles as sales engineer, sales engineering manager, training manager, product marketing manager, and director of global threat communications before becoming Vice President, Threat Intelligence. Jon has a Bachelor of Science in Electrical Engineering with an emphasis in Computer Engineering from Michigan State University.

Jon is a volunteer speaker for our Internet Safety for Kids and Families program. This experience has given him a broad technical background, an understanding of business security requirements, and an excellent understanding of the threat landscape. In his spare time, Jon enjoys fly fishing, fly tying, golf, tennis, and spending time with his family.

Chapter 23

Santiago Paz, Msc, is a Senior Cybersecurity Specialist at the Interamerican Development Bank advising on projects and collaborating with the different countries of the region for the development of their capacities. Previously, he was Director of the Information Security area at the Digital Government Agency of Uruguay (AGESIC), having founded the National CERT there and set up the electronic signature and identification systems. He is a Telecommunications Engineer, has a master degree in Computer Security and is certified as an Information Systems Auditor by ISACA (CISA). He has taken executive courses on the subject at Florida International University (FIU), The Hebrew University (HU), and Harvard University. He has publications related to cybersecurity, in the Institute of Electrical and Electronic Engineers (IEEE), in the Organization of American States (OAS), and the International Development Bank (IDB).

Chapter 24

Dario Petri (Fellow, IEEE) is currently a Full Professor in Measurement Science and Electronic Instrumentation and the chair of the Quality Assurance Committee of the University of Trento, Trento, Italy. He is the recipient of the 2020 IEEE Joseph F. Keithley Award "for contributions to measurement fundamentals and signal processing techniques in instrumentation and measurement." Dario Petri is an Associate Editor in Chief of the IEEE Transactions on Instrumentation and Measurement. He is the chair of the IEEE Smart Cities Initiative in Trento since 2015. He also chaired the Italian scientific Association of Researchers working in the area of Electrical and Electronic Measurement from 2013 to 2016 and the IEEE Italy Section from 2012 to 2014. He received the M.Sc. degree (summa cum laude) and the Ph.D. degree in electronics engineering from the University of Padua, Padua, Italy, in 1986 and 1990, respectively. During his research career, he has been the author of about 350 papers published in international journals or in proceedings of peer-reviewed international conferences. His research activities are focused on fundamentals of measurement theory, digital signal processing applied to measurement problems, data acquisition systems, embedded systems, instrumentation for smart energy grids, and measurement for quality management.

Luca Mari is currently a Full Professor in Measurement Science with Università Cattaneo - LIUC, Castellanza, Italy, where he teaches courses about measurement science, statistical data analysis, and systems theory. He has been the Chair of the TC7 (Measurement Science) of the International Measurement Confederation (IMEKO). He was also the Chair of the TC1 (Terminology) and the Secretary of the TC25 (Quantities and units) of the International Electrotechnical Commission (IEC). He is an IEC Expert in the WG2 (VIM) of the Joint Committee for Guides in Metrology (JCGM).

Matteo Brunelli received the Ph.D. degree from Åbo Akademi University, Finland. He spent five years as a Postdoctoral Researcher with the Systems Analysis Laboratory, Aalto University, Finland. He has been visiting researcher at the University of California, Berkeley (USA), Auckland University of Technology (New Zealand), and Japan Advanced Institute of Science and Technology (Japan). He is currently an Associate Professor of mathematical methods for economics with the Department of Industrial Engineering, University of Trento, Italy. His research interests include decision analysis, applied optimization, and mathematical representations of uncertainty. He is in the editorial board of Mathematical and Computational Applications and International Journal of General Systems.

Paolo Carbone (Fellow, IEEE) received the Laurea and Ph.D. degrees from the University of Padova, Padua, Italy, in 1990 and 1994, respectively. From 1994 to 1997, he was a Researcher with the Third University of Rome, Rome, Italy. From 1997 to 2002, he was a Researcher with the University of Perugia, Perugia, Italy. Since 2002, he has been a Full Professor with the University of Perugia, where he teaches courses in instrumentation and measurement and statistical signal processing. He has authored or co-authored more than 200 papers, appeared in international journals and conference proceedings. His research objective is to develop knowledge, models, and systems for the advancement of instrumentation and measurement technology. He was the President of the IEEE Systems Council and co-founded the IEEE International Symposium on Systems Engineering. He is the Editor-in-Chief of the journal *Measurement* and the President-Elect of the International Measurement Confederation (IMEKO).

Chapter 25

Cecilia Poittevin, President of DAMA Uruguay, a local chapter of DAMA International.

She is currently responsible for Data Management in a financial institution. Former CTO of Plan Ceibal, where she implemented the Data Governance and Data Privacy programs.

Computer Engineer graduate from the ORT University of Uruguay, Data Management and Data Governance Certifications from DAMA International. She also has a Master's Degree in Business Administration from IEEM (Montevideo University Business School).

Javier Barreiro, Vice President of DAMA Uruguay, a local chapter of DAMA International. ICT consultant and manager with more than 15 years of experience both in the public and private sectors.

Former CTO of Agesic, responsible for the implementation of different initiatives at the country level, such as the Artificial Intelligence Strategy for Digital Government and the Public Data Policy for Digital Transformation, among others. Member of several OECD working groups: Data-Driven Public Sector, ICT Procurement, Emerging Technologies, as well as the group of Blockchain experts and advisers (Blockchain Expert Policy Advisory Board [BEPAB]).

Member of the Group of Experts of fAIrLAC, an initiative of the Inter-American Development Bank that seeks to ensure an ethical and responsible adoption of Artificial Intelligence in Latin America.

Computer Engineer graduate from the University of the Republic of Uruguay and a Project Management Professional (PMP) graduate from the Project Management Institute (PMI). He also has postgraduate studies in Technology Management and Innovation. At the academic level, he is a professor of Software Engineering at the University of the Republic and at the University of Montevideo, as well as tutor and evaluator of final degree projects.

Gustavo Mesa, VP of Professional Development at DAMA Uruguay, a local chapter of DAMA Uruguay.

Data Management consultant with more than 10 years of experience in DWH, Data Quality, and Data Governance. He works in several projects of consultancy services for financial institutions in Latam and Caribe, both in the private and public sectors.

Computer Engineer graduate from the University of the Republic of Uruguay and Certified Data Management Professional (CDMP) with specializations on Data Governance and Data Quality certified by DAMA International.

Laura Rodriguez Mendaro, Board Member of DAMA Uruguay, a local chapter of DAMA International.

Advisory board at "Patterns – The science of data," an open-access journal of Cell Press, which publishes groundbreaking original research in data science.

Chief Technology Officer and IT Chief Architect at AGESIC, Uruguay eGovernment Agency at the Presidential Office. Leads the development of the national data strategy and data policy in the Uruguayan government as part of a digital government strategy. Leads the Data 360° working group, which aims to develop a holistic approach of data management in public administration, as part of the Digital Nations group along with governments of Canada, Denmark, Estonia, Israel, Mexico, New Zealand, Portugal, South Korea, and the UK.

ICT consultant with vast experience in the design and development of information systems specialized in interoperability and integration, enterprise architecture, and data management. Participation as speaker and keynote in several international seminars and published articles in reference journals on data and technology.

Chapter 26

Germán Stalker has a law degree from UNL, "Universidad Nacional del Litoral" Argentina, and specialized in ethics, innovation and technology. He is research coordinator of Center for Intellectual Property and Innovation Law Department, Universidad de San Andrés. He works for the National Scientific and Technical Research Council of Argentina ("https://www.conicet.gov.ar/programas/ciencia-y-justicia/"CONICET) as Head of the Science and Justice National Program in Buenos Aires, Argentina.

Stalker's research focuses on the fields of anti-corruption, open government, ethics, compliance and technology transfer. He is a consultant of startups and tech companies, San Andres University.

Prologues of the *TEMSBOK* Sections and Chapters

The following prologues are included to serve as condensed guidance for the section and chapter contents. Each chapter includes a specific overview. Brief biosketches of authors are included elsewhere.

TEMSBOK, Prologue of Section "Business Analysis"

The engineering profession creates not just technologies but also sustainable businesses. Basic analyses of businesses and business opportunities should be among the first priorities for managers and innovators. The scope of business analysis covers different aspects of organizations. In a nutshell, business analysis determines business needs and explores solutions to business problems. Financial analysis is of particular importance and one that often is not top of mind for the engineering profession. Yet innovation processes and the development of new services require financial understanding and discipline. Business analysis can also help in building strong business cases to justify key changes in organizations.

The section includes two chapters, summarized as follows.

Chapter: Profitability Analysis and Financial Evaluation of Projects
Authors: Mikael Collan and Jyrki Savolainen

This chapter introduces practical tools for the financial analysis and evaluation of different kinds of projects. In particular, it covers profitability analysis for engineering projects. The presentation is intended for engineering professionals; prior accounting and financial background is not assumed. Real examples are included in detail to validate the tools presented and reinforce the understanding of the key concepts. Key takeaways are included in the summary and conclusions. Extensive references to further reading are included to guide the reader on the subject.

Chapter: Fintech and Consumer Expectations: A Global Perspective
Authors: Reyna Virginia Barragán Quintero and Fernando Barragán Quintero

These past decades, especially the years following the financial crisis of 2008, have witnessed a change in the general modus operandi of the financial sector, stemming from the general public's increasing mistrust of traditional institutions, particularly financial ones. The consumers' need for reassurance, the need for inflows of cash for small businesses, and the public's attitude toward banking institutions paved the way for new players to participate in the financial sector. Fintech startups, through ingenuity and innovation, were able to respond to the consumers' concerns; this phenomenon gave rise to a series of new dynamics in the financial industry, mostly related to consumer demand, that are still being developed today.

A second crisis, the Covid-19 pandemic, has had the effect of boosting these developments. Consumers are expecting constant updates and ever-increasing adaptation of financial services to go hand in hand with the most recent technological advances. Thus, the financial industry sector has changed, but, more importantly, so have the expectations of consumers; their preferences have and still are changing in what is now a flowing process of financial innovation and consumer expectation. The chapter develops such concepts, includes key realistic case examples and ends with takeaway concepts and suggested readings.

Gustavo Giannattasio
IEEE TEMS Board of Governors

TEMSBOK, **Prologue of Section "Strategy"**

The concept of strategy is key to the decision process that engineering leaders must adopt to identify and select options for realizing the company's objectives. As an organization identifies its needs and vision, it must find a path from its current state to its desired future state. Resources are not unlimited, so prioritization is essential. Strategy is further complicated by the dynamics of the business, market, and societal environment – the world around the company will change as the company tries to change. Uncertainty is pervasive, to the point that little can be known with confidence over the planning horizons of interest, and yet decisions have to be made today. Strategy must face all of this complexity and identify the best path forward.

The section includes the following chapters.

Chapter: Strategic Roadmapping for Technological Change
Authors: Eduardo Ahumada-Tello, Oscar Omar Ovalle-Osuna, and Richard D. Evans

In this chapter, the authors describe the fundamental relationship between strategy and competitiveness and consider the roadmap to be one of the tools organizations can use to position them for success. They also discuss how society and the social context, especially during times of dramatic change, can influence the competitive environment.

They go on to note that as we are now moving well into the Fourth Industrial Revolution, change in technology and change in business processes are occurring at a dramatically increasing rate. Organizations no longer have time to "wait and see" but must aggressively design and pursue their future.

Mark Wehde
IEEE TEMS Board of Governors

TEMSBOK, **Prologue of Section "Leadership"**

Engineers have been trained in the technical aspects of designing solutions to problems leveraging various types of technologies. Across all engineering disciplines, it is problem-solving and creativity that make the best engineers stand out. These engineers are often chosen for leadership roles based on their exceptional technical expertise. But the leadership and management of people take a very different skill set, with an emphasis on team building, conflict resolution, and facilitation. As such, the very best engineering leaders apply themselves to the study of leadership with the same passion they might have applied to circuit or mechanical design. Leadership is an emergent skill. Some come by it naturally; in others it develops with time and attention. In all cases, it is incumbent on all who are leaders or who aspire to lead within their organization to nurture those skills most relevant for these positions. Best practices in leadership have evolved dramatically over the last century, and a leader within an organization will see these skills continue to evolve throughout their career. As such, a fundamental characteristic of any good leader is a commitment to lifetime learning

The section includes the following chapters.

Chapter: Guided Continuous Improvement (GCI): The Key to Becoming a Disciplined Agile Enterprise
Author: Scott Ambler

The adoption of agile methods/frameworks is often seen as the path to becoming an agile enterprise, but in practice, they prove to be only a good start at best. A viable organizational transformation will follow a fit-for-purpose strategy that reflects the organizational context, with the end goal of becoming a learning organization and not just adopting an agile framework. Traditionally, learning organizations adopt a continuous improvement (CI) strategy, but this chapter recommends a guided continuous improvement (GCI) strategy where the learnings of others are applied to improve faster. GCI can be applied to improve at the team, cross-team, value stream, and organizational levels [Author's abstract].

Chapter: Leading and Managing Optimal Working Groups, the Belbin Methodology
Author: Marcelo Da Costa Porto

What are the keys to high-performance teams? This chapter describes the Belbin methodology, which presents an answer to this question. Central to Belbin is the identification of team roles: Nine roles are identified and their characteristics are defined. Based on these definitions, teams can be appropriately composed, and team members can be given their preferred team roles. A historical overview of the methodology is included along with recent examples of its applications. The chapter includes comparisons with other approaches to identifying traits relevant to team performance. In particular, the compatibility of Belbin with agile project management is discussed.

Mark Wehde
IEEE TEMS Board of Governors

TEMSBOK, **Prologue of Section "Innovation"**

Innovation is not a choice but an imperative. Innovation helps us develop new technologies and processes which help improve how we live and work. The entire progress of humanity in the modern era has been due to application of various new technologies. A carbon composite fiber used in an aircraft made the airline industry more sustainable and efficient. A non-invasive technique to detect insulin levels in a human body made diabetes more manageable.

While R&D develops new technologies, the application of technologies is where the true benefit for humanity lies. There are cases where an improper application or inadequate management of technology has led to disaster and accidents, which arguably could have been prevented. Hence while we develop innovation, the management of technology and how it is applied and used are key.

The section includes the following chapters.

Chapter: Managing Innovation
Author: Joseph Tidd

Innovative organizations outperform those that fail to innovate, in terms of productivity, growth, sales and service, and the broader social benefits are even more important. However, the management of innovation is not simply the application of standard management methods to adopting or commercializing technology. Managing innovation is a distinct body of knowledge, developed over decades of systematic research and codified practice. In this process, skills and knowledge, which largely differ from the standard management experience, are needed.

This chapter proposes an integrated process approach to managing innovation, which incorporates the interactions between changes in organizations, technology, and markets. The process model consists of two contextual influences – strategy and organization, and four core process components – search, select, implement, and create value. For each of these six factors, the author provides empirical evidence, real-life case examples, and practical tools to apply the knowledge [Author's abstract].

Chapter: Innovation and R&D
Author: Sudeendra Koushik

The chapter presents the role of innovation in companies and why it should be addressed and managed. Innovation happens not only internally in an organization but also through ecosystems and open innovation. The speed of change in technology is forcing companies to innovate or disappear in the market.

Innovation is not just for big companies but for all kinds of businesses that need to adapt to changes that globalization and digital adoption has brought to every market. The relationship among organizations, individuals, and the environment plays a key role in innovation outcomes. The connection of business and innovation is presented as well as the role of research in the acquisition of knowledge that drives innovation. Concepts of top-line and bottom-line explain the link between increasing revenues and decreasing costs; innovation can impact companies and businesses through both means.

Chapter: Women in Technology and Innovation
Authors: Elif Kongar and Tarek Sobh

Despite many efforts to address the issue, women remain underrepresented in science, engineering, technology, and mathematics (STEM), with the disparity increasing as levels of responsibility and authority increase. Personal, societal, and financial considerations steer women away from STEM professions, and legislation, regulation, and other incentives introduced have mostly been inadequate. Gender inequity in technology and innovation is an economic drag as well. One comprehensive study (McKinsey) estimates that global GDP would increase by $12 trillion if all countries matched their best-in-region country. The chapter discusses the challenges faced by women in entering into and advancing within technological professions, the value for society of greater engagement of women in designing and engineering new products and services, and the cost to society as a whole of women's lack of participation. Recommendations on how to close the gap are offered, with an emphasis on collaborations among government, industry, and academic stakeholders.

Sudeendra Koushik
IEEE TEMS Board of Directors

TEMSBOK, **Prologue of Section "Entrepreneurship"**

Recent advancements in engineering and technology have brought with them several opportunities for new business creation. Highlighting the growing significance of small and medium-sized businesses in the economy, the Entrepreneurship section of the *TEMSBOK* provides in-depth knowledge regarding the latest developments and trends in the field.

The section includes the following chapters.

Chapter: Roadmap for Entrepreneurial Success
Author: Arun Tanksali

Tanksali proposes five stages of an entrepreneurial journey: preparation, initiation, traction, growth, and exit. Together, these stages create a roadmap for prospective and early-stage entrepreneurs to obtain better insight. For each stage, several concepts and tools are introduced and the cross-cutting relevance of these are noted. The author concludes that persistence and adaptability are two primary differentiators for a successful journey.

Chapter: Toward Smart Manufacturing and Supply Chain Logistics
Author: Eldon Glen Caldwell Marin

Caldwell discusses entrepreneurship from the lens of smart manufacturing and logistics (SMLS) and provides a clear distinction between cyber-physical and other systems based on flow development. The section also includes methodological approaches and various technological tools to manage the implementation of SMLS. Related case studies are presented. The risk factors that need to be considered based on the business ecosystem are also detailed.

Elif Kongar
Bridgeport University

TEMSBOK, **Prologue of Section "Project Management"**

Projects are the units of work in all organizations and must be managed for the organization to succeed. All projects have objectives, goals, constraints, and limitations. They also have multiple aspects, including time, scope, cost, risks, communications, staffing and people management, stakeholders, quality requirements, and possibly procurement.

Fortunately, there are organizations around the world developing methodologies to enhance the quality of project planning and execution. This section discusses many basic project management methodologies and practices, with examples of technical projects. As project managers and teams adopt these methodologies, their overall level of success will rise.

This section has two chapters. Both cover recommendations for project management from the Project Management Institute (PMI).

Chapter: Project Management
Author: Celia Desmond

The chapter by Celia Desmond describes the PMI structure for project management. The chapter covers the basic recommendations from PMI, with a focus on Critical Path (CP) methodology. This methodology is recommended for projects that have clear objectives and for which the path to project completion can be foreseen (CP can be contrasted with Agile in these respects). The chapter describes the process areas of CP, noting a number of tools that can be applied. Examples from different industry sectors are presented to illustrate the concepts and methods.

Chapter: Project Plan Structure: The Gantt Chart and Beyond
Author: Andrea Molinari

The chapter by Andrea Molinari also addresses PMI recommendations and mentions some of the other organizations that also recommend project management methodologies. The chapter discusses Agile management in detail, which has gained popularity lately for research and software projects in particular. The chapter highlights the project plan as a crucial artifact in project management. Tools for project planning are noted and examples illustrate the application of the methods described.

Celia Desmond
IEEE Industry Engagement Committee R&D G&O Working Group Chair

TEMSBOK, **Prologue of Section "Digital Disruption"**

This section deals with the disruptions to industry and business as consequences of digital adoption. It covers different visions for the future as well as new and emerging technologies. Examples of disruptive technologies are internet of things, big data, eHealth, 5G telecommunications, robotics, and business automation. One theme of the section, however, is that technology by itself does not necessarily lead to digital disruption; other factors also contribute, like market demands, customer behavior trends, and ecosystems. These can be critical for creating new business models and enhancing customer value.

The section includes the following chapters.

Chapter: The Evolution of Smart Sustainable: Exploring the Standardization Nexus
Authors: Cristina Bueti and Mythili Menon

Smart sustainable cities (SSCs) leverage digital infrastructures to improve the quality of life of citizens and the efficiency of urban operations and services, without compromising the ability to meet the needs of future generations. After introducing a definition of a smart sustainable city, the chapter illustrates the crucial role of international standards for the implementation of digital technologies, with a focus on the main activities to be undertaken by cities after the Covid-19 pandemic. In particular, the main concepts contained in the standards related to the lifecycle, the main subsystems, the stakeholder identification, and the entrepreneurial ecosystem of SSCs are illustrated. Selected references are provided.

Chapter: Wireless 5G (The 5G Mobile Network Standard)
Authors: Cristina Emilia Costa and Fabrizio Granelli

The 5G mobile network standard will bridge the gap between the mobile and internet worlds. Thus, it is expected to represent a disruptive technology in mobile communications. The chapter provides a brief introduction to clarify the general architecture of a mobile network and the fundamentals of the 5G system. Then the most relevant technology components and the different phases of deployment are described. Skills required to successfully manage the involved technology and related certifications are illustrated to support practitioners and companies to understand their needs and potential knowledge gaps. The chapter also includes some examples of deployment of 5G technology and associated solutions for modern information and communication technology (ICT) services. Finally, best practices and recommendations on the 5G technology as well as emerging markets and business cases are briefly discussed.

Chapter: The Internet of Things and Its Potential for Industrial Processes
Authors: Francesco Pilati, Daniele Fontanelli, and Davide Brunelli

Internet of things (IoT) is expected to be a disruptive technology, especially in manufacturing environments. This chapter provides a structured analysis of different IoT hardware and software components that will be useful for practitioners and researchers in the assessment, comparison, and selection of the most suitable IoT solution for a given industrial problem. The fundamentals of IoT technology are briefly presented, and the potential advantages provided by the availability of vast quantities of data are highlighted. A specific focus is dedicated to the digital-twin technology, which enables the achievement of optimal plant performances. Then, a set of enabling technologies of the fourth industrial revolution is described, and some case studies are presented. A further section describes the relevant approaches and techniques to leverage value from the vast amount of unstructured data provided by IoT and a framework supporting indoor locating system implementation is provided. The last section proposes some realistic examples of signal processing for indoor localization and different industrial case studies.

Chapter: Trends in Robotics Management and Business Automation
Author: Gastón Lefranc

Companies can gain in efficiency and cost reduction, and consumers are served better by the use of robots. The main characteristics of robots are described in the chapter. Different applications of robots are presented, including for agriculture, in food chains, in manufacturing, and for natural disaster assistance. Trends in mobile robots and cobots are described in detail. Examples of selection criteria for selecting optimal robot manipulators are presented. An extensive list of references for further readings is included in the chapter.

Chapter: Healthcare Through Data Science – A Transdisciplinary Perspective from Latin America
Authors: Parag Chatterjee and Ricardo Armentano

This chapter details the immense potential that data science technologies such as artificial intelligence (AI) and data analytics, together with other technological developments such as the internet of things, hold for applications in healthcare. The emerging field of eHealth brings together a variety of technological tools and human resources in a transdisciplinary approach spanning, medical, engineering, biology, and computer science. Reliable healthcare data in vast amounts are needed for data intelligence via automation and AI. A case study of cardiometabolic diseases in liver transplantation is presented, in the context of the National Liver Transplantation program in Uruguay.

Dario Petri
Chair, IEEE Smart Cities Initiative, Trento, Italy

TEMSBOK, **Prologue of Section "Digital Transformation"**

Today, managers and engineers are facing the challenge of adopting and adapting to digital transformation in their organizations. The pervasiveness of the digital world is affecting not only product/service design and development but all aspects of operations and strategy. From a technological perspective, it is developments such as big data and analytics, digital twins, artificial intelligence and machine learning, the internet of things, and any number of "smart" systems that have been disrupting companies individually, industry sectors as a whole, and society in the large. New visions for smart cities, for example, are leveraging digital transformation concepts.

The section includes the following chapters, covering the topics of big data, digital reality, and digital twins.

Chapter: Digital Transformation Enabled by Big Data
Author: Mariel Feder Szafir

The management of big data has become a requirement in many organizations and a key tool for digital transformation. It is necessary to implement solutions that allow organizations to acquire, store, manipulate, and obtain benefits from the available data, turning it into useful information for decision-making and for the execution of activities related to digital transformation. This new threshold in the use of data implies the need for new technologies, new methodologies, and even new management strategies that allow the implementation of successful transformation solutions in an organization and ensure that they generate the expected results [Author's abstract].

Chapter: Digital Reality Technology, Challenges, and Human Factors
Author: Nicholas Napp and Louis Nisiotis

This chapter explores the concept of digital reality by reviewing its fundamentals including its history, technological components, and current trends. It discusses the broader components of a digital reality project and goes beyond the technological components by including business factors and related considerations. The chapter also identifies and discusses common challenges, examples of practical solutions, and presents a generalized framework for how to approach digital reality projects. This includes how to manage the project, a summary of good practices, and key recommendations. Multiple examples are presented with use cases drawn from different disciplines. The chapter concludes with a summary of IEEE's Digital Reality Exchange and the role it can play in digital reality projects [Author's abstract].

Chapter: Digital Reality – Digital Twins
Author: Roberto Saracco

The technology of digital twins represents an overlapping of physical space and cyberspace. An analogy is drawn to distinguish between the economy of scarcity (physical world) and the economy of abundance (cyberspace); digital reality is the convergence of these two domains. A five-stage evolution for digital twins is outlined, resulting in the concept of an autonomous community of digital twins. Two varieties of digital twins are discussed in particular: cognitive and personal. Industrial examples of digital twins that are presented include a simulator for crane operators and the manufacturing and operation of heavy machinery for mining environments. A new platform from a startup company to support digital twin applications is also discussed.

The concluding section reinforces key concepts.

Gustavo Giannattasio
IEEE TEMS Board of Governors

TEMSBOK, Prologue of Section "Security"

Today most services depend almost totally on software and supporting hardware servers – in the cloud, in the premises of the organization, and in third parties. The extensive networked digital environment creates challenges for securing applications, data, and infrastructure. Cyber risks are pervasive across all business, government, and consumer sectors.

To help understand and address the challenges of cybersecurity and digital attacks, this section develops concise information to make professionals aware of the risks and possible preventive and remedial solutions. Different complementary visions are developed in the chapters of this section, enabling the reader to see a broad spectrum of the complex issue of security.

Chapter: Bitcoin, Blockchain, Smart Contracts, and Real Use Cases
Author: Ignacio Varese

Blockchain is a digital, distributed, decentralized ledger that operates across a computer network. Blockchain is the foundation for Bitcoin and other digital currencies, but its applications extend beyond electronic money to many other types of business transactions, such as smart contracts. This chapter describes the concepts behind and the features of blockchain. Different types of blockchains are discussed: open public, private, and permissioned public. The chapter outlines a flowchart, comprising six questions, to help readers determine whether a blockchain solution is warranted. The steps involved in developing a blockchain application are also reviewed. Use cases for blockchains are presented, including nonfungible tokens (NFTs). The dramatic disruption to current financial and business transactions that blockchain represents implies several risks and challenges in using the technology; these are also noted.

Chapter: Cybersecurity
Author: Jon Clay

This chapter reviews the history of cybersecurity and the threat landscape, highlighting trends toward financially motivated cybercrime, advanced zero-day exploits, and changing attack surfaces. Technological fundamentals are discussed that can form the basis of best practices. Examples include advanced firewalls, antimalware, cloud security, and advanced detection technologies. Cyberattacks rely on technology as well, but organizations need to also recognize the importance of people and processes; a cohesive cybersecurity strategy must encompass all three of these dimensions. A number of standardized frameworks are now available to help organizations be cybersecure. The chapter notes some of these and provides some details of the framework developed by the US National Institute of Standards and Technology (NIST). New developments in cybersecurity are outlined.

Chapter: Cybersecurity Standards and Frameworks
Author: Santiago Paz

This chapter presents most of the standards that are recommended to manage security. The list includes ISO 27002, NIST SP 800-53, SANS CIS, Open Web Application Security, and Project OWASP. The role and importance of the chief information security officer (CISO) are discussed. The concept of a cybersecurity operations center (SOC) to monitor network systems and respond to possible attacks is presented. The chapter also highlights the importance of establishing a computer security incident response team (CSIRT). The description of CSIRT included guides for the reader on key tools necessary for the control of security. Specific references are included at the end of the chapter to help the reader go deeper in the never-ending and increasingly challenging world of security.

Gustavo Giannattasio
IEEE TEMS Board of Governors

TEMSBOK, Prologue of Section "Data Science"
The new era of big data is characterized by significant advances in all aspects of data and information. The acquisition, storage, processing, and transmission of large-scale data is an essential enabler of digital transformation. Data scientists have developed new data management frameworks, algorithmic approaches, and application scenarios to leverage the massive amounts of data that organizations can now access.

The section includes the following chapters that discuss several data science concepts and technologies relevant for big data application.

Chapter: Information-Enabled Decision-Making in Big Data Scenarios
Authors: Dario Petri, Luca Mari, Matteo Brunelli, and Paolo Carbone

Information-enabled decision-making (IEDM) leverages the availability of big data and data science tools to help organizations reduce the risk of wrong decisions being made, in comparison to decision-making based on intuition or untested beliefs. IEDM requires that information of sufficient quantity and quality is available and that analytics and problem-solving processes have been carefully and rigorously developed. Several methods for implementing IEDM strategies are discussed. The difference between correlational and causal knowledge is highlighted, and an application of IEDM for causal modeling of an industrial process is presented. The chapter also introduces a Quality of Measurement Information (QoMI) framework relying on three layers of information based on the semiotic categorization of syntax, semantics, and pragmatics. The chapter includes an extensive reference list.

Chapter: Data Management: An Enabler of Data-Driven Decisions
Authors: Cecilia Poittevin, Javier Barreiro, Gustavo Mesa, and Laura Rodriguez Mendaro

According to the *Data Management Book of Knowledge (DMBoK)*, "Data management consists of the development, execution and supervision of plans, policies, programs and practices that allow to deliver, control, protect and increase the value of data during its life cycle." This chapter analyzes this definition and highlights the importance of frameworks for effective data management. The reference framework of *DMBoK* and its main artifacts are discussed: knowledge areas of data management; the relationships among the three pillars of people, process, and technology; and context diagrams that interrelate the first two elements. Real-world examples from the food and banking industries and from e-government are presented. Initiatives in different countries related to data ethics are noted. Adaptations of the *DMBoK* framework for different environments are also outlined.

Shiyan Hu
IEEE TEMS Vice President of Technical Activities

TEMSBOK Prologue of Section "Legal and Ethics"

A business driven by innovation also needs to be supported by company values. Values are bound up with culture. In established companies, innovation initiatives require cultural change. In entrepreneurial ventures, a culture is established along with the business. In either case, ensuring that the culture of an enterprise or organization reflects and supports core values requires the explicit attention of leadership. Ethical and legal considerations are at the forefront in this context. The rapidly evolving market of digital products and services continues to bring new challenges related to ethics and legal matters, requiring sustained vigilance by technology managers and leaders.

The section consists of the following chapter.

Chapter: Innovating with Values. Ethics and Integrity for Tech Startups
Author: Germán Stalker

Economic development can be achieved if, besides innovation, companies have values when doing business. This article focuses on startups as key actors for a cultural change. It describes how innovative ecosystems work, its features, and main stakeholders. The article also explains how startups emerge and why they should design and launch integrity programs. Tech enterprises that are guided by values build trust and improve their reputation. They create an environment of innovation and integrity that surpasses in the world of business [From the author].

Gustavo Giannattasio
IEEE TEMS Board of Governors

Introduction to *TEMSBOK*

Gustavo Giannattasio[1] and Tariq Samad[2]

[1] *Member of TEMS Board of Governors*
[2] *Technological Leadership Institute, University of Minnesota, Minneapolis, MN, USA*

The *IEEE Technology and Engineering Management Society Body of Knowledge (TEMSBOK)* is a guide to the broad field of technology management and innovation. With the continuing emergence of new technologies, the shortening of the research-to-deployment cycle, the development of new business models, and the dramatic transformation in innovation ecosystems, the need for a comprehensive resource is greater than ever before. Researchers, innovators, educators, and students alike stand to benefit.

TEMSBOK is an initiative of the IEEE Technology and Engineering Management Society (TEMS), a unit of IEEE – "the world's largest technical professional organization dedicated to advancing technology for the benefit of humanity" (www.ieee.org). The intent of the publication is to capture key topics and developments in the field, in a form that will engage a variety of stakeholders – the TEMS membership, the broad IEEE constituency, and the academic, business, and entrepreneurial communities.

These audiences have interests that are both common and differentiated, and *TEMSBOK* has been designed to serve all. It is thus intended to provide basic educational information for students, both beginning and advanced; best practices for technology managers in established organizations, whether large or small, commercial or nonprofit; guidance for entrepreneurs in all stages of their journey from idea to impact; advice and tools for intrapreneurs seeking to bring innovation inhouse to their employers; and assessments of the state of the art and gaps therein for researchers and scholars.

TEMSBOK is thus a handbook for technology and engineering management for all those with an interest in the field. As befits such an aspiration, a diverse yet cohesive set of topics is covered:

- Market research and business analysis
- Leadership in technology and innovation
- Strategy and change management
- Managing innovation and entrepreneurship
- Project and data management
- Digital disruption and digital transformation
- Ethics and legal issues
- Key information and communication technologies such as data science, blockchain, and cybersecurity

Structure of the Publication

The *TEMSBOK* structure follows a common model for technology management and innovation, with primary sections aligned with key functions and sequenced accordingly. This structure is depicted in Figure I.1, where typical iterations and feedback pathways are also shown.

The structure is based on the notion that a business vision will drive an innovation or product/ service development effort. A solid business case, including market analysis, should be consolidated before proceeding to next steps. Strategy, leadership, and innovation follow, in which the overall direction of the organization, the identification of leaders and teams, and the detailing of potential offerings are elaborated. The next step is labeled Entrepreneurship in the figure, but it is equally applicable to innovation in established companies – in this context, the label encompasses "Intrapreneurship" as well. Project management cannot be overemphasized in taking ideas to market, and a section is devoted to it. The final section, titled "Legal and Ethics," contains a chapter focusing on ethical considerations.

A caveat is in order with the order outlined above, however: A discipline as broad as the one under purview cannot be accurately captured in a unique process. Depending on various factors, both technological and related to business needs and opportunities, variations in the flow of activities may be seen, and more extensive iteration, as well as parallel pathways, will often be needed.

As shown in the figure, the book also includes chapters that provide basic information about several key digital technologies, an understanding of which is *de rigueur* for contemporary technology managers.

As noted earlier, technology and engineering management is a dynamic and extensive discipline. This edition of *TEMSBOK* does not cover all topics within the field. In addition, new developments are ongoing, in tools and methodologies as well as in the emergence of novel. IEEE TEMS plans on producing expanded editions of this volume in the future.

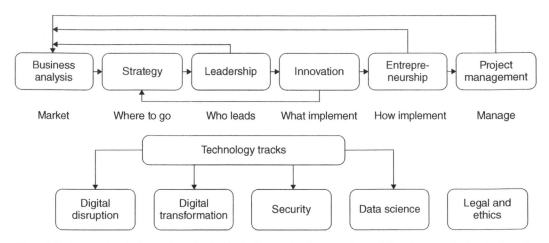

Figure I.1 The main sections of *TEMSBOK*, including technology tracks and showing a typical ordering of topics as addressed in innovation.

Section 1

Business Analysis

1

Profitability Analysis and Financial Evaluation of Projects

Mikael Collan[1,2] and Jyrki Savolainen[2]

[1] *VATT Institute for Economic Research, Helsinki, Finland*
[2] *School of Business and Management, LUT-University, Lappeenranta, Finland*

1.1   Profitability Analysis of Projects (Knowledge Area Fundamentals)

Financial analysis of projects typically can mean several different things; in this chapter, we talk about ex-ante profitability analysis of projects. Ex-ante profitability analysis means trying to figure out, before starting a project, whether the project will generate wealth – that is, whether or not the project will be able to pay back the costs invested into the project and give a return commensurate with the risk the project carries.

Profitability analysis is useful, when one is in the wealth creation business that is, when one is engaged in an activity with a profit motive. If a project is made for a purely "academic" reason and the success of the project, or whether it makes money or not, is of no consequence, then profitability analysis is typically a waste of time and other measures of goodness should be used. Also, if the project is of a trivial, very small size, then using common sense is more advisable than profitability analysis – after all, serious profitability analysis is time-consuming and requires work. Profitability analysis resembles budgeting in that it includes estimating costs and revenues (out- and in-cash-flows) and when they will take place – *estimation of the size and the timing of cash-flows is of importance*. In fact, profitability analysis is also called "capital budgeting."

Profitability analysis may also be useful when one must choose between projects. This means analyzing the profitability of alternatives and ranking them based (also) on the result of the profitability analysis. The ranking and selection of competing projects fall under what is known as project portfolio analysis and remains outside the focus of this chapter. Here, we introduce the main concepts that underlie classical profitability analysis, the most commonly used profitability analysis methods, and how they can be used in the context of evaluating projects.

1.1.1   General Introduction to the Main Underlying Concepts (Monetary Valuation of Inputs and Outputs, Opportunity Cost of Capital, Time Value of Money, and Where They Come From)

Profitability analysis is about money and more specifically about whether a project makes more money than it spends – an economist would ask whether a project creates or destroys wealth. The unit that is used to perform profitability analysis is money, dollars or euro, and as some inputs

IEEE Technology and Engineering Management Society Body of Knowledge (TEMSBOK), First Edition.
Edited by Gustavo Giannattasio, Elif Kongar, Marina Dabić, Celia Desmond, Michael Condry, Sudeendra Koushik, and Roberto Saracco.

(investments) into projects and outputs (proceeds, revenues) from projects are in a non-monetary form, the monetary value of these inputs and outputs must be determined for use in profitability analysis. For example, the cost of raw materials "already bought before for a previous project" is a cost that must be included in the profitability analysis, if the said raw material cannot be used again for another project – any significant use of resources in a project or a value created by the project should be reflected as a cost or revenue in monetary terms. The monetary value, the cash-flow, that is used for non-monetary costs and proceeds is always the best honest estimate, and it may be based on a market value of a similar good, estimated labor cost, expert estimate, or sometimes a best guess. The cash-flow estimation is most often than not an imprecise exercise. The level of aggregation typically used is thousands of dollars.

What also must be estimated is the timing of when resources are being spent and when revenues are collected from the project. Timing is of high importance, especially if proceeds from a project are estimated to take place in the faraway future – this is because a lot more uncertainty is associated with revenue that is estimated to be received faraway in the future. Determination of timing may be very difficult; most often, it is quite easy to estimate the timing of costs, but much harder to estimate the timing of revenues, for example, a single-lot sale of ready project results. Most often, the timing is estimated in terms of the year that a cash-flow takes place. The first year of analysis is considered year zero (0) and the cash-flows taking place within the first year are expected to be worth dollar for dollar the same as cash in hand at the time of initiating the project. By adding the cash inflows and outflows for each we get the net yearly cash-flow, also called the free cash-flow (FCF), see Table 1.1.

Table 1.1 is a simplification of reality; revenues and costs typically consist of multiple items added up to form the total cash in- and outflows.

For an investor, any given project is "only" one possible way to spend money, and there are also other investment opportunities around available to everyone, such as the stock markets. An investor is interested in maximizing wealth, and therefore the existence of alternatives is something that needs to be taken into consideration. In fact, the investor is asking the question: "If I invest in this project with a given level of risk, then how much return would I get from the best (highest return) alternative with the same risk-level that I forego by investing in this project?" Ideally, an investor will not invest in the project if they get more return elsewhere for taking the same amount of risk.

For example, if an investment on the stock market in Company A stock is the best investment alternative with the expected return of 12% for the same risk level that Project X has, then based on this information, the investor would want to use the 12% return as a benchmark for Project X. This highest alternative return from an investment with the same risk level available to us is the *opportunity cost* of investing in the project. It is important to note that the alternative that is used as a benchmark is a "real alternative" in the sense that it is realistically available to us – a fictitious alternative that is not really available should not be used as a benchmark.

Opportunity cost = the highest return (%) that a security or an investment realistically available to us with the same risk level offers

Table 1.1 Cash in- and outflows for years 0–8 and the sum of in- and outflows for each year.

Year	0	1	2	3	4	5	6	7	8
Cash inflow (revenue)	150	300	300	300	300	300	300	300	300
Cash outflow (cost)	1100	100	100	100	100	100	100	100	100
Revenue – costs (free cash-flow)	−950	200	200	200	200	200	200	200	200

One can see that it is important to be able to estimate the risk level that the project has. Estimating the risk level is less difficult for securities such as quoted stocks that have been "around" for years and have a clear performance history in terms of stock price volatility (a measure of financial risk) and recorded yearly returns. For projects about to be started, with no history it is a very different ballgame – past experience and common sense play a considerable role in estimation.

Because of this difficulty, especially in many larger companies, company-level proprietary opportunity cost rates have been set for different types of projects, and sometimes companies instruct that a single opportunity cost is used for all investments. Using a single opportunity cost for all investments may be dangerous, as such a practice will typically cause high-risk investments to be treated as having a lower risk than they actually have. Low-risk investments face the opposite dilemma when single rates are used.

When the risk level and corresponding returns are determined, the following estimates and comments can be used as guidelines/yardstick:

Liquid securities with very low risk have returns of 1–3%.

Very few if any commercial projects have a risk relative to this category, only projects that have to do with maintenance of existing equipment in an ongoing business or replacement investments for ongoing business may consider using such opportunity costs.

The stock markets on average have returns of 7–10%.

Projects that are a continuation of the ongoing business of an "average risk-level" company fall under this category. For example, a large production or energy-generation facility might have a risk level of this type.

Small publicly quoted companies have historical returns of 13–17%.

Clearly risky projects that are expansive and that take place in environments where markets have not yet fully evolved or the organization running the project is not "highly established" have a risk level that corresponds to this category or is higher.

Research and development projects typically have a high opportunity cost that depends on how revolutionary the envisioned project is. Enhancing existing equipment in ways that are known to create benefits may carry relatively low risk levels commensurate with "continuation of ongoing business" investment, while efforts to produce brand new products with untested technology may have risk levels that command alternative costs upwards of 40%. The point being that all R&D is not super high risk, but high-risk R&D is typically "higher" risk.

Time value of money is a central concept in analyzing the value and the profitability of projects. The main observation is that a present-day (time zero) dollar is worth more than a dollar in the future, and that is why today's present-day dollars are not comparable to future dollars.

The observation is based on the simple notion that today's dollar will earn a return if it is invested, and hence a dollar today will correspond to a dollar plus return in the future. As we have seen above, the return is commensurate with the risk taken to earn the return. The other side of the coin is that a promise of a dollar in the future does not mean that a dollar will with a certainty be obtained; therefore, the expected value of the promise is less than a dollar. How much less is again commensurate with the risk attached to the promise? To make today's dollars (cash-flows) comparable to future dollars (cash-flows), the value of future dollars (cash-flows) must be converted to what they are worth today, to *present value* (PV).

The mechanism of converting the value of future cash-flows into PV and hence making cash-flows additive is called *discounting*:

Discounting = conversion of the value of future cash-flows into present value (today's value)

Discounting is based on using the opportunity cost of capital as the *discount rate* in the *discount factor*, the term that is used to multiply future cash-flows, when they are converted to PV.

Table 1.2 Discount factors with a 12% opportunity cost of capital and the corresponding present values of free cash-flows from a project.

Project discount rate	12%								
Year	**0**	**1**	**2**	**3**	**4**	**5**	**6**	**7**	**8**
Cash inflow (revenue)	150	300	300	300	300	300	300	300	300
Cash outflow (cost)	1100	100	100	100	100	100	100	100	100
Revenue – costs (free cash-flow)	−950	200	200	200	200	200	200	200	200
Discount factor	1	0.893	0.797	0.712	0.636	0.567	0.507	0.452	0.404
PV of free cash-flows (FCF)	−950.00	178.57	159.44	142.36	127.10	113.49	101.33	90.47	80.78

The procedure is a simple one in the context of project profitability analysis, when yearly cash-flow estimates are used. The formula for the discount factor is:

$$\frac{1}{\left(1+r_d\right)^t}$$

where r_d is the discount rate and t is the timing of the cash-flow in years. For example, the PV of a 1000-dollar positive cash-flow on year five from a project with an alternative cost of 12% can be simply calculated as $1000 \times 1/(1+0.12)^5 = \$567.43$. The value of negative cash-flows can be calculated in the same way.

As discussed above, PV is additive, which means that all the cash-flows related to a project, negative (outgoing) and positive (incoming), for which the PV has been calculated can be added to understand what is the overall value of the project. Most often, a single discount rate for the costs and the revenues of the project is used to calculate the PV of FCFs for each year. Table 1.2 shows the PV of the FCF from a project calculated with a yearly discount factor with an opportunity cost (discount rate) of 12%.

It can be clearly seen from Table 1.2 how the discount factor becomes smaller the further into the future one goes. This is reflected in the PV of future cash-flows. Spreadsheet software is what is typically used in performing the calculations.

The following list sums up the main concepts/procedures that underlie profitability analysis:

1) Estimation of the *size* and *timing* of incoming and outgoing cash-flows
2) Estimation of the *opportunity cost of capital* (discount rate)
3) Calculating the *PV of the cash-flows* from a project

With the knowledge of these three "procedures," one can turn to the most commonly used profitability analysis methods. For more on the preliminaries, we suggest [1].

1.2 The Three Most Commonly Used Profitability Analysis Methods

According to published research [2], the most commonly used profitability analysis method in American corporations is the Net Present Value or the NPV method. The other two Top-3 methods are the "Payback method" and the "Internal Rate of Return (IRR) method." Each of these methods brings something to the table that the other two methods do not, and as such, they are complements to each other. When the profitability of large investments is contemplated, it makes sense to use them all simultaneously to gain a more holistic view of the investment.

1.2.1 Net Present Value

Net Present Value (NPV) method does exactly what the name of the method implies – the net of the PV of the cash inflows and outflows from a project is calculated, and the result is used to determine the financial goodness of the project. The "investment" rules connected to NPV are as follows:

If the NPV is lower than zero, the project does not create wealth in terms of PV. This signals that the project should not be started.

If the NPV is zero, the project does not create nor destroy wealth. The project pays back the investment and a risk-corrected return for the investment. In such cases, it is up to the decision-maker to decide whether it makes sense to go ahead with the project.

If the NPV is higher than zero, then the project creates wealth on top of paying back the investment and a risk-corrected return for the investment in terms of PV.

When the above "rules" for the NPV are interpreted, one must understand that the method is based on estimates, and thus the results are only as precise as the estimates used in creating it are. This is especially important when the NPV is close to zero.

Table 1.3 shows that the NPV of the project is 43.53. According to the NPV investment rule, this means that the project is estimated to pay back the invested capital, required compensation for taking the risk a yearly return of 12%, and 43.53 "money" over the invested capital and the required return.

This means the project creates wealth and as such "can be accepted" from the NPV point of view. The decision to start the project may also depend on many other issues, such as how good possible projects competing for the same money are and what the "technical" and "strategic" aspects of the project are.

1.2.2 Internal Rate of Return

The idea behind the internal rate of return (IRR) method is to answer the question: "What is the average yearly rate of return from this project?" To reach this information one uses the same information that is needed for the NPV calculation, sets the NPV to zero and solves the discount rate

Table 1.3 Net present value calculation for a project with a 12% discount rate.

Project discount rate 12%									
Year	0	1	2	3	4	5	6	7	8
Cash inflow (revenue)	150	300	300	300	300	300	300	300	300
Cash outflow (cost)	1100	100	100	100	100	100	100	100	100
Revenue – costs (free cash-flow)	−950	200	200	200	200	200	200	200	200
Discount factor	1	0.893	0.797	0.712	0.636	0.567	0.507	0.452	0.404
PV of free cash-flows (FCF)	−950.00	178.57	159.44	142.36	127.10	113.49	101.33	90.47	80.78
Net present value (NPV)	43.53								

Table 1.4 Calculating the IRR is setting the NPV to zero and finding the corresponding discount rate.

Project discount rate	???	Solve the discount rate							
Year	0	1	2	3	4	5	6	7	8
Cash inflow (revenue)	150	300	300	300	300	300	300	300	300
Cash outflow (cost)	1100	100	100	100	100	100	100	100	100
Revenue – costs (free cash-flow)	−950	200	200	200	200	200	200	200	200
Discount factor	1	???	???	???	???	???	???	???	???
PV of free cash-flows (FCF)	−950.00	???	???	???	???	???	???	???	???
Net present value (NPV)	0.00	Set to zero							

that returns NPV of zero. The discount rate that returns NPV of zero is the IRR. Finding the rate can be done by hand, by trial and error, but it can be simply accomplished with spreadsheet software that has a solver. Table 1.4 illustrates the procedure of calculating the IRR by setting NPV to zero and solving for the discount rate.

To understand whether an investment pays back the PV of the investment and at least compensates for the risk-level taken, one compares the IRR with the discount rate (opportunity cost of capital).

If the IRR is lower than the discount rate, the project is not as good as the alternative investment. This signals that if one simply considers the wealth creation point of view, the project should not be started.

If the IRR is the same as the discount rate, the project is exactly as good as the alternative and in theory the investor can be indifferent between the project and the alternative. In such cases, it is up to the decision-maker to decide whether it makes sense to go ahead with the project.

If the IRR is higher than the discount rate, then the project is better in terms of having a higher return than the alternative – there are grounds for making the investment and starting the project as it is expected to not only pay back the investment but also to pay a return that is more than enough to compensate for the risk-level taken.

The three "rules" are in line with the NPV investment rules, and the same caveats about the precision of the results remain.

There are some caveats connected to the IRR method. It must be noted that the IRR calculation works well only if the free cash-flow "schedule" is of the type: "Investment first, followed by positive cash-flows." If the cash-flow schedule is such that we first get revenues and make the investment payment later and we, in effect, borrow money and then pay it back – the IRR will reveal the lender's rate of return, not ours. Problems with the method returning multiple IRR rates (multiple solutions) occur, if the sign of the cash-flows $(-/+)$ changes more than once during the life of the project. Together these problems make IRR a delicate tool and "risky" in the sense that one must understand what one is doing.

One more thing to remember when using IRR, when projects are compared. As IRR returns a percentage the size of the project is not reflected in the result. A small project may be ranked higher by IRR, while a larger project "loses" and has a much higher profit in terms of dollars. To illustrate, consider that $1 million at a 10% return per year returns $100,000 per year, while $ thousand at a 20% return per year returns $200 per year. This points to IRR being most usable in investment comparisons, when the compared projects are of a similar enough size. In the attached material, you can find a spreadsheet workbook with an example IRR calculation template and detailed instructions on how to solve the IRR.

1.2.3 Payback Method

The speed at which a project is estimated to generate back the money invested can also be used as a benchmark for project goodness and an element with which competing projects can be compared. The speed of payback is the main point behind the payback method and the method gives an answer to the question: "How fast does the project pay back the nominal amount of money invested?" The simple decision rule is "The faster, the better." It is important to note that the classical method uses nominal, non-present value information of cash-flows. A more sophisticated version of the payback method that we can call Present Value Payback (PVP) method uses PVs of cash-flows.

The nominal value version is very simple and fast to use, because information about the opportunity cost of capital nor discounting is needed. This may be the main reason why the method is so popular. Table 1.5 shows the cumulative free cash-flows from which the payback time is calculated – one must count the first year (year 0) when determining the payback time.

The same project, if the payback time is calculated from the PV of the FCFs, has a payback time of eight over eight years. This is naturally due to the discounted FCFs being smaller than the nominal FCFs (Table 1.6).

Table 1.5 Nominal payback time for the project is over five years.

Year	0	1	2	3	4	5	6	7	8
Cash inflow (revenue)	150	300	300	300	300	300	300	300	300
Cash outflow (cost)	1100	100	100	100	100	100	100	100	100
Revenue – costs (free cash-flow)	−950	200	200	200	200	200	200	200	200
Cumulative FCF	−950	−750	−550	−350	−150	50	250	450	650

Table 1.6 Present value payback time for the project is over eight years.

Project discount rate	12%								
Year	0	1	2	3	4	5	6	7	8
Cash inflow (revenue)	150	300	300	300	300	300	300	300	300
Cash outflow (cost)	1100	100	100	100	100	100	100	100	100
Revenue – costs (free cash-flow)	−950	200	200	200	200	200	200	200	200
Discount factor	1	0.893	0.797	0.712	0.636	0.567	0.507	0.452	0.404
PV of free cash-flows (FCF)	−950.00	178.57	159.44	142.36	127.10	113.49	101.33	90.47	80.78
Cumulative PV of FCF	−950.00	−771.43	−611.99	−469.63	−342.53	−229.04	−127.72	−37.25	43.53

Put simply, the payback method is the calculation of the point of time, when the cumulative cash-inflows surpass the cumulative cash-outflows (including the initial investment). This is the only thing it does, which means that if there are (positive or negative) cash-flows that take place after the payback time, they do not have an effect on the answer from the method and the evaluation of "goodness" the method provides. Another important observation is that the nominal payback method does not necessarily rank investments in the same order than NPV does; for an example about this, see Appendix 1.A. For these reasons, we firmly suggest that the payback method be used only as a complement to the NPV method and not used alone to determine the financial goodness of projects.

1.2.4 Summary of the Three Methods

Now that we have visited the three most commonly used profitability analysis methods, also used for analyzing project profitability, we can shortly summarize them:

The NPV method takes into consideration the time value of money and gives a reliable picture of the absolute value added by a project. NPV should be used as the main method when project profitability is analyzed.

The IRR method is useful when one wants to understand how high the average yearly return from a project is. The method "is in line with" the NPV method; however, if the FCF schedule from the project is not "normal," translating the results from the method may become difficult. The method should be used as a complement to the NPV method.

The payback method is useful as a complementary method, when one wants to understand how fast a project pays back the investment. This is the case especially when investments with a very similar NPV are compared. The method is not suitable as a single measure of goodness or profitability for projects.

When one tries to determine whether a single project is good or profitable the main tool is NPV. When one is ranking competing projects, one will want to gather a more holistic view by using all three methods for the projects that are being compared and to make decisions based on the more holistic view. One must remember that any result from any of the three methods is only as reliable as the information used as the basis of the results that is, if the cash-flow size and timing estimates and the discount rates are imprecise, then the results are at best imprecise.

All the presented methods are based on looking at the world through one scenario meaning that they simplify reality greatly. Simplification is a way of coping with uncertainty. Let us keep this in mind, when using these methods and when results from them are being used to support decision-making. For more details on the methods, see [1].

In Section 1.4 of this chapter, a detailed project analysis is given to deepen the understanding of how the presented methods are used in practice.

1.3 Managerial Flexibility in Projects (Real Options)

Flexibility in projects is the ability of project owners and managers to steer the project if the circumstances surrounding the project change. The ability to execute changes comes from flexibilities embedded within or surrounding the project, and it can help the project save money (cut cost, protect against downside) or make more money (increase revenue, capture upside) depending on the circumstances. These flexibilities are commonly referred to as "real options."

The existence of managerial flexibility can be illustrated by a simplified comparison of two technologies employed in enriching ore in metal mining projects:

The first technology is mechanical enrichment, where ore is mechanically broken into small bits and then the metal is mechanically separated from the rest of the material. The process is done by machines that can be shut down and restarted at will. Relatively small costs are associated with shutdown and restarting the enrichment process.

The second technology is using a hydrometallurgical (biochemical) process to enrich the ore. The process takes a long time and the end result is that metal is separated from the rest of the material. Stopping and restarting can only be done with great costs.

If the market price and demand of the enriched metal drop dramatically, there are benefits to be captured by temporarily shutting down the enrichment and restarting it only when the market situation gets better. With mechanical enrichment this is possible; however, with biometallurgical enrichment this is not possible. The project with biometallurgical enrichment is stuck with having to run the project at a loss for an uncertain period of time, while the project with mechanical enrichment can cut losses and temporarily suspend enrichment.

The point made here is not that mechanical enrichment is the better choice but that when one is choosing an enrichment method, one should weigh the benefit brought about by the flexibility offered by mechanical enrichment against the possible cost disadvantages of mechanical enrichment. If the value of the flexibility outweighs the extra cost, then there is a case for choosing mechanical enrichment. This example illustrates what practical managerial flexibility, or a real option, is and shows that flexibility and real options are not always available. The choice of having or not having flexibility within a project can often be consciously made, if one understands that such a choice is at hand. Often, choosing to have flexibility (or optionality) within a project costs extra, which means that the flexibility has a price. If flexibility comes at no extra cost one typically should choose to take it. As flexibility is potentially worth money, it makes sense to know whether the cost is lower than what it is worth that is, whether building flexibility into a project is a good investment or not. This can be figured out by calculating the NPV of a project with flexibility and of the same project without flexibility – the difference is what the flexibility is worth.

Value of flexibility = Value of the project with flexibility – Value of the project without the flexibility

One has to be able to estimate the value of a project with and without flexibility, which may not be simple. When it can be done, this way of calculating the value of flexibility can be done practically with spreadsheet software. Performing calculations and helping the estimation of the value of flexibility can also be taken further, and the value of real options studied in more detail by modeling projects and using simulation analysis. The interested reader will find more information about the simulation-based approach in Appendix 1.B.

When there is very little risk of change and the world is expected to remain unchanged, flexibility is typically not worth a lot. On the other hand, if dramatic changes are possible (or even expected), then the value of having the option to change a project is typically high. This means that the value of real options (flexibility) moves in an opposite way to the NPV valuation of the project – as risks increase, discount rates increase and NPV decreases. But when risk increases, the value of flexibility and real options increase.

When the risk-level goes up, value of flexibility and real options go up

This is in line with the value of insurance contracts – in presence of high risks insurance is worth more than when there is low risk. Flexibility can, in fact, be seen as a sort of insurance against negative outcomes caused by change. Flexibility may come in many forms and previous literature has documented many of them, most often the types of flexibility (real options) are divided into four to six categories.

1.3.1 Different Types of Real Options

As discussed above, real options are a natural hedge against change that allows projects to either defend against negative outcomes or to capture positive outcomes. The flexibility is divided into different types – below we use a division into five categories.

1.3.1.1 Option to Temporarily Shut Down and Restart/(Completely) Abandon a Project

The above-presented mining project is a good example of what benefits an option to temporarily shutdown and restart a project may carry. This real option is typically connected to the technology or the business architecture used in the project, but also contracts may play a key role. For example, a contract of constant service provision may effectively hinder the use of a shutdown and restart option.

The option to abandon is a more "final" type of shutdown option, and here means that a project is completely and irreversibly abandoned. Such an option may have considerable value if the market conditions surrounding a project decline dramatically and render the project unprofitable. If the option to abandon exists, it can be exercised, and losses can be stopped. After the exercise of the option to abandon, remaining project assets can be sold to offset any incurred losses.

Option to abandon is often used in connection with managing capital-intensive industries like airlines that often terminate unprofitable routes as a matter of everyday management. Resources are in these cases allocated elsewhere and airplanes start servicing new, profitable routes. One must make sure one understands the value of the option to abandon when making long terms contracts!

Interestingly, not all managers seem to "accept" the existence of the option to abandon projects and stubbornly run money-losing projects to the bitter end, while it would make economic sense to terminate them before their planned ending time. This may also be sometimes due to not following up on project profitability and prospects (updating the NPV calculations with new information periodically).

1.3.1.2 Option to Switch Inputs or Outputs

The option to switch allows a project, typically a production system of some sort, to change either the input mix or the output mix of a machine, facility, or system. The decision to switch is typically triggered by market changes, more specifically price changes of inputs (raw materials) or outputs (end products).

A good real-world example of a project that has both the option to switch inputs and outputs is a combined heat and power (CHP) facility with the ability to burn many raw materials (oil, coal, biomaterials, combustible trash) and to produce power and heat. The switching decisions are guided by market prices of the inputs into the process and by the market prices of the output products. The flexibility allows a CHP facility to choose the optimal combination in terms of profitability if there are no restrictions by contracts that force the production to a set scheme.

Another real-world example is flex-fuel cars that allow their owners to utilize mixtures of alcohol and gasoline to power their cars. The flex-fuel car owner has the opportunity to fill up with the fuel that offers a better price-effect relationship. A flex-fuel car may cost more than a gasoline-powered car, but if the value of the flexibility is higher than the extra cost, then the investment into flexibility is a good investment.

The value of being able to switch between inputs or outputs may be considerable in cases where the volatility of raw materials prices used in the process is great.

1.3.1.3 Option to Stage Investment in Projects

Option to stage projects is a well-known way to control risks in research and development (R&D) projects that is used as a default mode of operation in pharmaceutical development. The idea is that a project receives funding only for one project stage at a time, after which the project results are typically examined and only if the results are in line with what is expected the project will get funding for the next stage. The alternative is to fund (or promise to fund) a project from the beginning to the end. This means up-front committing (large) resources to a project with an uncertain outcome.

This stage-gate process is in place to make sure that a (R&D) project continued only if it delivers the required and expected results that warrant spending more resources on it. If the project has not been delivered, it can be discontinued. In pharmaceutical R&D, there is a portfolio of multiple parallel drug-development projects ongoing simultaneously. If a project is abandoned, the resources (money, personnel) will start another project instead. No drama is attached to these decisions; they are business or "R&D as usual."

New business opportunities are often tested with a pilot (store, product, service), before "big money" is spent. Pilots are a way to "test the markets" and in effect to stage investments. Nowadays, every television series starts with a pilot that is screened with test audiences and the decision to proceed with an investment into the first season filming is triggered or abandoned based on the feedback received. Before the end of airing each season the decision to continue with another season is made, again based on the ratings the series has received. Staging investment has reached a status of "commonality" in some industries, while elsewhere similar practices have not been used a lot. Staging investment may carry high rewards in terms of lower risk and lower potential loss of capital. Interestingly staging the investment into a project may be a low-cost option for the project owner and therefore there is grounds to at least consider it with many types of projects.

1.3.1.4 Option to Change Scale in a Project or of a Project

Flexibility to change scale in a project refers to the ability of a project to adjust production according to demand within the limits given by the production process (machinery, staff) and thus refers to flexibility (optionality) that is internal to a project. Changing the scale of a whole project refers typically to the option to increase the overall size of the project (that then redefines the limits of internal flexibility). Such scale change is referred to in the literature as the option to grow.

The flexibility to change scale is not necessarily something that exists in the project setup unless it has been designed into the project. For example, a production machine that mass-produces in large batches only does not include full flexibility to scale down. Answering to demand of any size requires the production of a batch of products, which means overproduction in cases where the demand is smaller than the batch size. This brings costs of storage and inventory for the overproduced products or waste costs and may erode profitability. A production process that allows for producing exactly the demanded quantity of products (scalable production) typically does not face high storage costs or inventory costs. Most often there is a cost for flexibility and the cost should be compared to the benefit, when the production method is chosen.

Today, commercial spaces are often designed to be flexible, with the option to rent the same space (project) as one large space, or two or more smaller spaces. This option is created by designing removable walls and entrances into the space and by installing "extra" electrical outlets and other necessary utilities into the structures of the space that allow splitting it into multiple rentable wholes. Such extras cost in terms of investment, but they bring flexibility in return. The flexibility has value, because a flexible space is more likely to collect (at least partial) rent than a single large space. Transforming a single large space into multiple smaller spaces after it has been finished may cost considerably more than installing the extras, when the space is being constructed – the cost of the option is lower before completing construction and may be prohibitively large later on.

The option to change scale is typically very valuable in cyclical industries such as natural resource industries, commercial real estate, fashion, and the pulp and paper industry. The ultimate scaling down is temporary shutdown, already discussed above. Changing the scale of the (whole) project requires that the project exists in a "space" that allows for it. If we are looking at commercial real estate projects having the option to grow means, for example, having ownership and control over land adjacent to an existing building that allows for an expansion. Another possibility is to have the possibility to expand upwards that is, to have constructed a high-rise building in a way that allows for the construction of extra floors above the existing structure. Such projects have been realized already numerous times and are especially lucrative in locations, where potential for strong city growth exists. The cost of the option is to build in a way that allows construction to be restarted, and the value is in being able to grow in an area that experiences rising real estate prices.

On a more strategic scale, growth options include the possibility to grow a business by making corporate acquisitions. Also, patents and R&D are bearers of growth options, because without them a company may not be able to expand its business. Projects that are a "platform for growth" typically have their whole value based on the option to grow, which put in other words is "potential." Having potential is having the option to grow (revenue growth) and a volatile situation on the markets that may make the growth happen. The volatile situation alone does not carry value, the option with the volatility does. There is hardly any discussion about whether potential is valuable or not.

1.3.1.5 Option to Postpone a Project

Sometimes it makes sense to wait and see before acting. This is especially true in cases where the future is highly uncertain, there is no cost for waiting, only little value is estimated to be lost during the time that one waits, and it is expected that after the waiting time, some of the uncertainty (about the project outcome) has been removed. In other words, if there is a possibility to learn about how a project will do by waiting, then the additional information about the project may guide one to make a more informed (better) decision about whether to start the project or not. The cost/benefit between waiting and not waiting can be determined by estimating the costs of waiting (including the risk that a competitor's action will cause loss of value) and comparing them to the benefits from increased information gained during the waiting time. One can actively try to gain information during the waiting time by lunching, e.g. market studies or other probes. The pilot projects, discussed above, are a way to learn about the markets that can be used while waiting, thus making the option to postpone investment and the option to stage investment close relatives.

NPV analysis is based on the idea that we either make, or do not make, a decision to invest: "fire or forget." The investment decision is made now (time zero). The option to wait postpones the decision and offers a third alternative, "wait and then decide." This brings one to the world of "investment timing" that is, to discuss the optimal time of investment. Optimal timing is determined typically by the market situation (demand and competition). In cases where only one project can operate profitably, a competitor may invest and make follow-up investments unprofitable, in such cases exercising the option to postpone investment may destroy value. In other cases, it may be beneficial to be a follower on the markets. These "strategic games" are something that are always present in competitive markets and have also attracted a lot of attention in the investment and economics-related literature.

Option to postpone is the best-studied type of flexibility from the point of view of the academic literature because it resembles the type of options that exist behind financial call options contracts. Financial call options are contracts that give the holder of the contract (investor) the right to buy an underlying asset (usually a stock) for a pre-agreed price on a pre-agreed future date. This means that if the price of the underlying stock is higher than the price for which the option owner can buy it, buying the stock makes sense, and the investor makes money. On the other hand, if the market price of the stock is below the pre-agreed price, the investor will not exercise the right to buy, as the stock can be bought from the markets for cheaper. In the case of financial options, the uncertainty about the underlying stock price and about the decision to exercise the right to buy (or not) is completely resolved. Pricing of financial options contracts is typically based on valuing the option to postpone buying the underlying stock for the pre-agreed price and the well-known Black-Scholes formula [3] that elegantly does it received the Swedish Central Bank Prize for Economics (Nobel Prize).

The difference between financial options and project-related options to postpone decision-making is numerous, and real-world projects are typically much messier than simple contracts with a single pay-off. Directly using the original methods for valuing financial options to value project flexibility is not a safe bet. Newer methods for valuing real options, based on the logic behind financial option valuation exist, but they are left outside the scope of this discussion and refer the interested reader to, e.g. see [4–6] and Appendix 1.2 that presents and discusses simulation-based valuation.

1.3.2 Summary About Real Options (Flexibility Connected to Projects)

Flexibility found in projects carries value, and thus it should be taken into consideration when the profitability and feasibility of projects are estimated. Different types of real options may all be relevant to a project and that may hold value for the project. Real options may be valuable in running the projects (operational flexibility) or when decisions are made about the projects (strategy-level flexibility). The point is that unless the flexibility is understood to be there, it cannot be used to one's benefit: to improve the upside potential of projects and to limit the impacts of negative outcomes on project profitability. Sometimes the existence of real options and the ability to capture a very positive upside if the market conditions change for the better, may be worth a lot – if this value is not considered, when investment decisions are made, less than optimal decision-making may follow. The value of real options can be estimated by comparing a project with a real option with the same project without the real option – the difference is the value of the real option. If the cost of (creating) the real option is higher than the expected benefit, then creating the option is not a good investment.

A better understanding of the value created by having managerial flexibility to further company goals and to minimize losses and maximize profits allows decision-makers to make better investment decisions. This understanding also highlights the point that one must follow projects and the markets in which the project operates closely, combined with existing managerial flexibility the situational knowledge allows optimizing project profitability.

1.4 Additional Issues to Take Into Consideration in Profitability Analysis

Project cash-flows are a key issue in profitability analysis and if the estimation of cash-flows is done in a reliable (best possible) way the results are as good as they can be. Reliability includes completeness, which means that when estimating and identifying cash-flows that originate from the project, we must include also (positive and negative) "side-effects" of the project in terms of cash-flows. This means that the working capital that is needed to run the project must also be taken into consideration. Working capital is, after all, an investment into the project, and as such, it is a negative cash-flow when it is put into the project, and when it is taken out of the project (if it is), it is a positive cash-flow. It helps to think of the project as a box into which cash-flows go in and from which they come out. In the beginning, the box is empty and everything the project needs must be put in the box (and accounted for in cash-flow terms). Similarly, when the project ends the box is emptied and everything is taken out.

Sunk costs that have taken place before the project is started are irrelevant from the point of view of future project profitability. If a project starts using old equipment that can be sold and bought on the markets for $1000, then the money equivalent of inputting the equipment in the project is $1000 and not, e.g. $2000 that the company paid for the equipment three years ago. The value loss (sunk investment) is not the doing of the project, and therefore the project profitability should not reflect it. In other words, money lost has already been lost and does not matter anymore. One must not look backward!

When a single project that is a part of a large company is considered in isolation, consideration of taxes for the project and as a part of the profitability calculation may, in many cases, be left out. But when a project is the basis of a company taxes and depreciation play a more important overall role. Depreciation is an accounting procedure that allows for the deduction of the made investments from the taxable income and affects taxes. Depreciation is not a cash-flow and only affects profitability calculations via paid taxes (which are a cash-flow). If taxes and depreciation are taken into consideration, it is suggested that for simplicity and in an effort to avoid errors, the tax to be paid is calculated separately and included in the profitability analysis as a separate cash-flow.

The discussion on real options has shown that maximizing project profitability is a dynamic process that makes profitability analysis a part of project management also beyond making the (initial) decision to invest in a project. One should revisit and update profitability calculations when changes that remarkably affect the project cash-flows take place. It makes sense to be aware of whether a project is making or losing money, because that helps in determining the project's future.

1.5 Summary and Conclusions

Profitability analysis of projects concentrates on looking at the financial aspects, more specifically the cash-flows connected to a project. It is different from feasibility analysis that may include also other aspects, such as technological and strategic fit. The main underlying concepts behind profitability analysis are opportunity cost of capital and time value of money, which together allow us to understand how much future cash-flows are worth in today's money. This in turn allows us to determine whether the value going out to a project is compensated by the value coming in from a project – if more value is coming in from a project, then the project creates value and financially thinking we have a case to go ahead with it. This thinking underlies the most used profitability analysis method, the NPV method.

The two other Top-3 profitability analysis methods in terms of how much they are used are the IRR method and the payback method. These methods are complementary to the NPV method and using them gives more information about the project and how it is expected to fare financially. The IRR method tells one what is the average yearly return from the project (as a %) and the payback method reveals how fast the project takes to break even. The original payback method does not take into consideration the time value of money, which is a great handicap – however, using a variant that works with PV of the cash-flows corrects that problem. All in all, it makes sense to use all three methods to gain a holistic picture of project profitability and financial payback time. The analyses can be performed with spreadsheet-based tools, and their reliability is based on the accuracy of the cash-flow estimates (size and timing). The results from the methods are decision support, not "the truth."

The different types of flexibility connected to projects are called real options and they add value to the project. Real options allow project owners to change the project to escape costs and to capture extra revenue – they are a downside hedge and offer upside potential. Real Options are only valuable if the project owner understands that they exist. This may even sound stupid, but as real options must often be built into projects and they have a cost attached, stupid this observation is not. Being able to estimate the value of a real option and making the comparison of the value against the cost of the real option is sensible cost-benefit analysis. The three most commonly used profitability analysis methods do not take the value of real options (flexibility) into consideration and thus a project should be studied from the real options point of view in addition to traditional profitability analysis. Including the value of real options into the financial analysis of projects may help better understand the value of "platform projects," the value of which is to a large extent based on the potential included in the project. Even if the "calculated" paper value of real options may be high, one must not be lured into thinking that it will materialize into profit – a healthy dose of skepticism is a good companion to real option valuation. Again, profitability analysis and real option analysis are for decision support, they are not the truth, but done for better understanding – decision-making is left to the investor.

See also the enclosed spreadsheet workbook with a number of sheets including all the tables visible in the chapter, the information contained in the Appendices, and some extra material.

1.A   Appendix 1. Ranking of Projects by the Payback Method and the NPV Method

As discussed above the ranking of projects provided by the payback method does not always coincide with the ranking of the same projects by the NPV method. Table A1.1 shows the ranking of three projects with the payback method that results in Project 1 being the best. Ranking of the same projects with NPV (see Table A1.2) shows that Project 3 is the best by NPV. Furthermore, Project 2 that is considered second best by the payback method is considered clearly the least favorable by NPV, in fact the project is wealth destroying and thus unacceptable in that respect.

It is clear that the answer given by the payback method is different to that from the NPV method.

Table A1.1 Analyzing and ranking three projects with the payback method.

Project 1

Year	0	1	2	3	4	5
FCF	−100	10	30	70	20	10
Cum. FCF	−100	−90	−60	10	30	40
Payback				×		

Rank 1

Project 2

Year	0	1	2	3	4	5
FCF	−100	70	5	10	20	10
Cum. FCF	−100	−30	−25	−15	5	15
Payback					×	

Rank 2

Project 3

Year	0	1	2	3	4	5
FCF	−100	95	0	0	0	80
Cum. FCF	−100	−5	−5	−5	−5	75
Payback						×

Rank 3

Table A1.2 Analyzing the same three projects with the NPV method.

Project 1

Discount rate		12.00%				
Year	0	1	2	3	4	5
FCF	−100	10	30	70	20	10
PV FCF	−100.00	8.93	23.92	49.82	12.71	5.67
Cum. FCF	−100.00	−91.07	−67.16	−17.33	−4.62	1.05
NPV	1.05					

Rank 2

Project 2

Discount rate	12.00%					
Year	0	1	2	3	4	5
FCF	−100	70	5	10	20	10
PV FCF	−100.00	62.50	3.99	7.12	12.71	5.67
Cum. FCF	−100.00	−37.50	−33.51	−26.40	−13.69	−8.01
NPV	−8.01					

Rank 3, unacceptable

Project 3

Discount rate	12.00%					
Year	0	1	2	3	4	5
FCF	−100.00	95.00	0.00	0.00	0.00	80.00
PV FCF	−100.00	84.82	0.00	0.00	0.00	45.39
Cum. FCF	−100.00	−15.18	−15.18	−15.18	−15.18	30.22
NPV	30.22					

Rank 1

1.B   Appendix 2. On Simulation-Based Profitability Analysis

Simulation is dubbed as the "third mode of science" complementing traditional theoretical and empirical research efforts by way of offering the chance to create pseudo-empirical results that aid in understanding reality. Simulation analysis is based on a computer model (of a studied system) and on using simulation to create a large quantity of input–output pairs between inputs and outputs from the studied system. How closely the model resembles the studied real-world system (including in-model dependencies) determines how closely the results from the simulation resemble possible real-world outcomes. Simulation is used to study what effect different input combinations into the system have in terms of the outcome; in other words, simulation is used to study system output behavior in various situations (represented by the inputs).

Usually, the model used is static, meaning that the model used does not change with time/when changes take place in the (simulation) environment. Dynamic system models allow including the effects of systemic changes in the results. The simulation used in profitability analysis is typically based on static models, but lately, some system dynamic simulation-based profitability analysis has also emerged.

Simulation is a great way to explore possible future outcomes of a system. In the project profitability analysis context the studied system is the project, the inputs used are the determinants of the project cash-flows, and (main) studied output is the project profitability. Testing a wide array of input variable value combinations gives information about the distribution of profitability outcomes, as there may be a large number of input variables a pre-set number randomly chosen variable value-combinations (and the matching profitability outputs) is often used to act as the population, based on which the analysis is made. Using randomly chosen variable value-combinations is commonly called Monte Carlo simulation.

As a simulation returns a distribution of profitability outcomes (NPV distribution) under the many possible input-variable value-combinations that represent different states of the world, the simulation analysis is in effect scenario analysis that captures the uncertainty with regard to the project profitability outcomes.

The level of uncertainty present in a simulation can be specified by setting input-variable value ranges (as probability distributions of possible values) from within which the simulation (randomly) draws the values used. In the simple example underlying this appendix, only the revenue cash-flows are drawn from a probability distribution (simulated). In a complete simulation-based analysis, the values of all or many variables can be drawn from distributions in a similar way. The distribution form used can be chosen by the analyst. When values are drawn for each year (and yearly cost deducted) and the procedure is repeated multiple times (here 100 times), one can construct a set of cash-flow scenarios for the project (see Figure A2.1/left). The more scenarios are used the less robust the results are.

Simulation results are typically presented as a histogram (see Figure A2.1/right). From the output distribution, one can calculate descriptive numbers, such as the expected value of the distribution that serves as a single value representative for the simulated NPV of the project; in this case the E(NPV) of the distribution is 202 compared to the histogram in Figure A2.1. The E(NPV) is highly positive and signals that the project is possibly a good one. Seven out of the one hundred simulated scenarios returned a negative NPV outcome, and the great majority of scenarios were positive. This information is further confirmation of the expectation that the project is NPV positive.

If one wants to inspect the value of real optionality by using simulation, it is a common practice to run two simulations: first, with a model without the real option and, second, with a model that

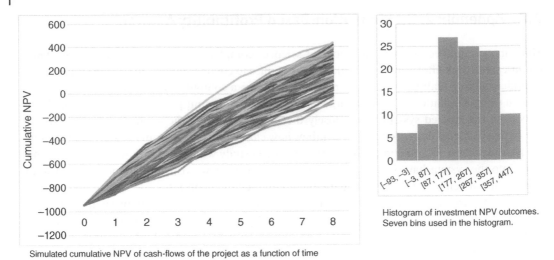

Histogram of investment NPV outcomes.
Seven bins used in the histogram.

Simulated cumulative NPV of cash-flows of the project as a function of time

Figure A2.1 Illustration of a profitability simulation. *Left:* 100 simulated scenarios of NPV for years 0–8; *right:* histogram of project NPV, from 100 simulated scenarios. The spreadsheet containing original values and the calculation is provided as supplementary material.

has the real optionality implemented. Two NPV histograms are produced and can be compared. To give a single number (crisp) representation for the value of the real option(ality), one may use the formula:

[RO-value] = [Simulated Average Project Value with RO] – [Simulated Average Project Value Without RO]

This is completely in line with what is presented in the chapter. However, one must take into consideration the risk level of the project with the real option and without the real option. If the real option changes the risk level, then this should be reflected in the discount rate used and thus will have (also) a risk-related effect on the project value, in case of the project with the real option.

References

1 Brealey, R.A. and Myers, S.C. (1996). *Principles of Corporate Finance*, 4e. McGraw-Hill Companies.
2 Ryan, P. and Ryan, G. (2002). Capital budgeting practices of the fortune 1000: how have things changed? *Journal of Business and Management* 8: 355–364.
3 Black, F. and Scholes, M. (1973). The pricing of options and corporate liabilities. *Journal of Political Economy* 81: 637–659.
4 Collan, M. (2012). *The Pay-Off Method: Re-Inventing Investment Analysis*. Charleston, NC: CreateSpace Inc.
5 Mun, J. (2002). *Real Options Analysis: Tools and Techniques for Valuing Strategic Investments and Decisions*. Hoboken, NJ: Wiley.
6 Cox, J., Ross, S., Rubinstein, M. et al. (1979). Option pricing: a simplified approach. *Journal of Financial Economics* 7: 229–263.

2

Fintech and Consumer Expectations: A Global Perspective

Reyna Virginia Barragán Quintero[1] and Fernando Barragán Quintero[2]

[1] *Facultad de Ciencias de la Ingeniería, Administrativas y Sociales, Universidad Autónoma de Baja California, Tecate, Baja California, Mexico*
[2] *Centro de Estudios de Asia y África, El Colegio de México, Mexico, Distrito Federal, Mexico*

2.1 Technology Fundamentals

In this age of globalization, the term technology seems to move forward embracing the ideas of digitalization, innovation, and, often, disruption, not only as the consequence of a multiplicity of new problems and new situations arising from new and old needs affected by new and changing industrial and financial paradigms but as a promoter of change, progress and of new levels of quality and competence. Yet, the phenomenon of globalization has also served to introduce this dynamic into people's everyday lives, never more so than now with the outbreak of the Covid-19 pandemic, the spread of which has also been facilitated by modern-day globalization dynamics.

Within these very same dynamics, the internet and digitization are two very important factors that have not only promoted, but altogether forced our society to transform, from the most ordinary and simple actions of acquiring electronic devices, and using mobile technology, to redefining the way we perceive and approach information, to changes in a consumer's mindset and his attitudes in reference to the market, to e-shopping, to paying, to convenience, speed, and so on; a phenomenon that has reached all industries and sectors of the economy bringing about new realities in business models, proposing new solutions to traditional problems, and even promoting a culture of new and constant expectations of further and more convenient advancements with better technology, better and faster solutions, at a smaller cost.

In the case of the financial industry, technology has simply changed the way traditional financial institutions are conceived today vis a vis financial institutions just 15 years ago; just to mention the most superficial but palpable example: the common view that financial institutions were mainly physical entities having several branches, where clients could seek financial assistance, services, and advice, is now a question of the past; nowadays, although clients can still go to their local branches, other options are available, digital options that are more convenient, speedy and that can save people a lot of time and even money. Traditional financial institutions have been forced not only to cope, but to evolve through a series of processes that have transformed the way in which financial services are offered: technology has brought about the development of financial institutions in the direction of technological innovation dynamics without precedents, by introducing, say, much more efficient financial products that are based on mobile technology. The financial-technological advances we are referring to are grounded in Big Data & Analytics, artificial intelligence, robotics, open platforms, apps (big data applications), mobile technology, APIS, legacy systems, machine learning, quantum computing, augmented reality, cloud computing, collaborative economics, IoT, Blockchain & cryptocurrency, etc. [1, 2].

Some areas in the financial world that have been directly affected in a very important way are payments and money transfers: these include the introduction of mobile applications, cryptocurrency transfers, remittances, P2P (peer-to-peer) transfers, digital loans, B2B (business to business loans), nano payments; as well as the technology applied to e-commerce (including retail shopping, on-line commerce,) and next-generation payments, mobile point of sale terminals, digital and mobile wallets (practically replacing debit and credit cards). Other technological advances in the financial industry include robo-advisors, chatbots, advanced analytics, crowdfunding, crowdlending and crowd investing (which have become very important to SMEs), blockchain and cryptocurrency, and so on: all of the above is only a part of what can be called financial technology (or fintech for short), and which has certainly become an international financial phenomenon encompassing all types of technology and bringing about new business models and more modern financial products [2].

The turn of the century has certainly been characterized by globalization and technology, the internet and digitalization, and the financial industry, has been one of its main beneficiaries: fintech or financial technology has become one of those groundbreaking achievements of our times: we might say that the financial services industry is the biggest contributor and the biggest consumer of technology [3] and though it is true that technological advancements in the financial industry are not new (take for instance the introduction of ATM's in the 1960s), the difference is that earlier advancements were, for the most part, sporadic, while nowadays, they have become constant, generalized and have decisively contributed to the way people approach to finance and financial institutions.

Unfortunately, the development of financial technology across the world has not been the same from one region to the next; it has been uneven at best; some may say that most advancements make their first appearance on the back of the biggest markets, and fintech is certainly the case, though its spreading has proven that a single phenomenon may experience multiple dynamics of adaptation and growth; indeed, now, more than ever before, with the Covid-19 pandemic, have the effects of globalization been felt in all corners of the world, not just because of the outbreak, but because of its effects to the economy, to the community and to the new dynamics with which we have lived our daily lives, the way we have interacted, the way we have shopped and paid, the way we have worked and been schooled: according to some estimates, our digitization processes evolved up to five years in a single year of lockdowns and social distancing. Yet fintech evolution implies a certain infrastructure, and that condition is not the same everywhere, so fintech evolution has spread across countries but not at the same level, and not for the same reasons. Fintech is now a reality in our daily lives, but certainly not in the same form everywhere, more so, it is continuously evolving and requires continuous revisions and updates, hand in hand with the latest technological advancements, as well as with the changing needs and expectations of ordinary consumers.

2.2 Antecedents and Context

Since fintech is a global phenomenon, it is difficult to speak of fintech in one region without taking into account what is happening in other parts of the world, risking, otherwise, a fragmentary view of the phenomenon; the reason is clear: the effects of fintech are experienced differently in different social, economic, and political contexts. Let us begin by saying that the first steps towards a new modernization are usually taken by the biggest economies, and fintech is no exception, so we will take a look at countries such as the United States, China, India, and Germany.

2.2.1.1 The United States

The widespread popularity of fintech services in the United States has a history of more than a decade, and it spread because of a question of trust; fintechs can be described as start-up companies that appeared in the American financial landscape especially between 2008 and 2010 [4] the 2008 financial crisis, was a time when people lost their money, their investments in insurance, wealth instruments, and real estate, they lost their jobs, and countries not only lost a great deal of wealth but had to spend taxpayers money (continuous massive injections of public money) to avoid the bankruptcy of firms and financial institutions firms and keep them afloat. So, consumers lost all trust in the government and especially, in the banking industry, a fact that has led to an increase in the adoption of fintech services [3, 5].

In 2008, large financial institutions defaulted on their financial obligations because of insufficient capital and high leverage; soon thereafter, a succession of huge defaults from related multiple entities followed, resulting in the bankruptcy of the financial services industry overall, including Central Banks. The government had to inject billions of dollars as capital into the financial system, followed by almost zero interest rates for the following years; economies worldwide entered a recession leading to alarming levels of unemployment, with millions of people unemployed in the United States alone, including executives, brokers, analysts, wealth advisors, bankers, etc. and other professionals from the financial services sector. Some of these professionals were employed by tech companies which started to have an interest in developing financial digital products, but other finance industry employees, in association with tech experts, had the initiative of starting their own businesses through venture capital funding; the infrastructure requirements were easy to come by since most transformative technologies such as Big Data and cloud were produced in the United States. Both processes ultimately resulted in the development of fintech start-ups, potentially creating disruptive technologies.

At the other end, the crisis resulted in low levels of capital availability; banks had to follow a series of regulatory and statutory requirements with supervision from a number of government agencies turning the banking institutions into a rigid industry, and loans and mortgages became unavailable or difficult to come by; it also made it more difficult for lenders to develop innovative lending business models; but Fintechs did not have these limitations, they were more agile and were able to introduce innovative lending products such as P2P lending, Payday lending, and Crowdfunding.

So, a direct effect of the crisis was the emergence of fintech start-ups, which were primarily tech companies (hence the name fintechs) associated with financial advisors that started disrupting the financial services sector, with the underlying promise of trust and transparency as a cornerstone for their fintech services. Another major factor in this process was the development of mobile phones which radically transformed the payment industry, as well as the health insurance industry and the digital economy in general; so, smartphones themselves are a very important source of technology disruption leading in a very real way to the rise of fintech services [3]. Fintechs have, in fact, been disrupting traditional financial services in a big way in areas such as payments and money transfers, wealth management and investments, and loans, but the biggest disruption to the financial industry caused by fintechs has been in automated financial advice with the development of robo-advisors.

Nowadays, the most used fintech services in the United States are payments and money transfers, lending, savings, and investments, and one of the most popular is robo-advisory; therefore, a large number of fintech companies are concentrated in those areas, as well as cybersecurity, artificial intelligence and analytics [3]:

- Mobile payments reached a half of a billion dollars last year; fintechs are disrupting the payments industry with products and models such as P2P payment, loyalty, digital wallets, payment banks, Bitcoin payments, via payment fintechs such as Paypal, Coinbase, Square and Stripe, and Apple pay.
- The online lending market, including the P2P model, will probably hit the trillion-dollar mark in the next 10 years, with companies like Lending Club and Prosper.
- Many fintech companies are focused on providing automated affordable services, such as financial advice, personal financial management, and robo-advising, for instance Wealthfront, Motif, Folio, and Betterment.
- A number of fintech startups seem to be disrupting the traditional insurance industry, by providing this service based on such factors as customer behavior, quick claim processing, donations, pay-per-use insurance, P2P insurance and using big data and analytics to detect fraud; some of these companies are Cyence, Bright Health, Lemonade, Metromile, Clover Health [3].

All in all, the United States maintains some of the highest levels of investment funding for fintechs, and so by 2020 the traditional financial services business had lost around a quarter of all their business to fintechs [3].

As could be expected though, the traditional financial industry is also scrambling to stay in tune with all technological advancements, investing billions of dollars to overhaul their systems and develop their own digital products through partnerships, mergers with fintech companies and even by establishing their own financial innovation labs. The benefits for the banks are many in spite of the costs: by resorting to digital technologies banks have an opportunity to open up a whole new market and acquire new customers by providing fintech services (Goldman Sachs), not to mention the need for the traditional bank industry to regain the lost levels of consumer trust.

Nowadays, banks that use online strategies tend to target new customers by appealing to their expectations, trying to foster trust and offering high rates, and keeping fees low; they provide user-friendly apps and tools that can make managing finances easier, and this is important because young consumers are especially attracted to digital technologies. Indeed, a 2015 research report from FIS (Financial Technology Company) found that 65% of all millennials between the ages of 18 and 26 had never used a branch at all, but had no qualms using fintech financial services [6]. By 2020 America's banking services seem to be widely recognized for their efficiency and consumer-friendly approach; bearing in mind that modern-day users seem to care more about the experience than names or brands, some of the best mobile banking apps of 2020 are based on consumer preferences, for example [4, 7, 8]:

- Ally Bank has been rated the best online bank overall for the year 2020, especially for best mobile banking app. With Ally the customer can perform all of the essential tasks that one would expect from a banking app, such as paying bills, money transfers, finding an ATM, mobile check deposits, reviewing transaction history. The customer can open up the app with a fingerprint instead of a password. Once logged in, the customer can track all his accounts directly from his device; stock market news can also be read directly from the app; the customer can even trade stocks and track investment performance; it can also help the customer keep tabs on his statements and tax forms. Ally also gives the customer free access to Zelle, which can allow the customer to transfer money between two Ally accounts or Ally accounts and an account from a different bank, using no more than an email address and a phone number, so there is no need to exchange personal account numbers or routing numbers. This app also allows the customer to download a built-in application called Card Control that allows him to customize his debit card.

- Discover Bank also rates very high with its Discover app. It has the best banking app for rewards. Discover offers all the essential mobile banking features anyone could expect from a big financial institution; however, it offers its clients Discover Deals: it is an in-app marketplace and offers deals on the latest brands. The customer can also redeem his credit card rewards through this app and check his FICO credit score.
- The Charles Schwab app is considered the best mobile banking app for managing multiple accounts; it is considered the best for frequent travelers and also helps the customer bring together all his finances.
- The PNC bank is noted for being the best banking app for cardless purchases.
- Alliant Credit Union is considered the best online bank for students.
- The first foundation bank is best for its high rates for savings.
- Chime has the best mobile banking app user experience.
- Chase has the best banking app for prepaid cards.
- Capital One is regarded as the best mobile banking app in consumer service.
- Simple is best for managing a customer's budget.
- Wells Fargo and Bank of America are recognized for best mobile banking app to monitor investments.

As we have seen, banks are trying to enhance a customer's digital experience according to the customer's specific interests. This fact stems from the customer's vision of fintech services and his expectations.

The Fintech Industry as such is a group of startups and companies that are introducing technological innovation into the financial services industry; some of those firms compete directly with banks, while others have negotiated a partnership with them, or simply provide goods and services to them. Whereas fintech, as a concept, is the application of technology to the financial sector, and it is not just limited to financing as it once was, but now covers any financial service, expanding to all innovations in technology, i.c. machine learning, cryptocurrencies, robo-advice, internet of things [8, 9]; fintech, in reality, means not just the action of rendering traditional financial services to their online versions, but rather providing new digital solutions for traditional financial services [2]. This fact was well understood when fintech companies started expanding; the original fintech companies were not limited to Silicon Valley, in California, they spread quickly to the East Coast, and then to Europe, Hong Kong, Singapore and much of Asia, and Australia; but the United States still accounts for more than half of the world's fintech companies [10].

2.2.1.2 Southeast Asia

Southeast Asia has traditionally been considered one of the most important regions for fintech investment and development. The reasons behind this region's special attention to fintech are the population's size and its much higher demand for financial digital services as compared to other parts of the world: the reason is that the overly complicated and sometimes even inaccessible service provided by traditional financial institutions such as banks in places like China has made the use of alternative banking options imperative, so a very large segment of the population, especially the unbanked group, turned to tech companies and fintech startups for financial services. By 2017 China had superseded the American fintech industry in terms of number of fintechs and overall fintech investment, and it is precisely the Chinese Internet companies that are currently dominating the global fintech transformation process [3].

It must be mentioned that China is now the largest fintech market in the world, with one of the most mature markets in fintech services, leading with 80% of the population using fintech digital products and services, most of them concentrated by the government in a single platform: WeChat [11, 12]. Additionally, the Chinese government has launched an "Internet plus" program to aid and fund local internet companies, and it has created an investment fund to promote the digital industry.

As to the Chinese banking industry, it soon realized the benefits of using financial technology and did not take long to follow suit [13].

- The case of the largest bank in China, the People's Bank of China and third-party payments may serve to illustrate Chinese fintech in action. The Bank of China acts as the Treasury and is in charge of issuing currency, of supervising money supply; it regulates monetary organizations, manages foreign exchange reserves, foreign exchange transactions, and formulates monetary policy [14].

A third-party payment is based on a licensed non-banking organization to facilitate the payment process: these are companies licensed by the People's Bank of China, which issues three types of licenses: online/mobile payment license, point of sale license, and prepaid card issuance license. As it turns out, a company can hold multiple licenses: Alipay has two and a half licenses enabling it to perform all three services, online payments, etc. The process is simple: when the buyer acquires an item online the money goes from the buyer to the third party who will in turn notify the seller to ship the goods. Only when the purchase is received and accepted by the buyer does the third party transfer the money to the seller.

Recently these companies have expanded to offline and P2P payments and now act as intermediaries in interbank payments. Payments done this way have a far higher rate of success than traditional interbank payments. Alipay is the most widely used payment method (Tenpay ranks second) with 62%, and also accounts for about 26% of all online mobile banking because it is convenient, secure, and low cost [15].

Other examples from the region are:

- **The China Construction Bank:** The CCB mobile banking service is a convenience that the bank offers in union with mobile operators, all based on a telecommunications data platform.
- **The Industrial and Commercial Bank of China:** Under the brand name Banking@home, the bank has an internet banking channel providing ICBC customers with online financial services including account information, online payments, transfer and remittance, investments and financing, donations, T-bond trading, gold trading, foreign exchange trading, financing products and more; it also allows for multi-account management. It is reputed to be safe and reliable in large part thanks to a rigorous standard digital certification system. The condition to use the application is an ID card and a registered ICBC local bank card or Passbook.

The bank also has a WeChat banking service and an ICBC messenger for internet banking clients that provides all kinds of financial information. For clients with special security needs, the bank can provide a USB shield, an E-banking code card, or ICBC e-password devices.

ICBC has moved to the next level in its fintech development by establishing seven innovation labs for internet research, big data, blockchains, AI, and increased capital and talent input in the innovation and tech areas [16].

Financial inclusion in China is still very low, since one in five adults remains unbanked; in any case, at least half the Chinese population uses new payment methods thanks to high Internet and mobile penetration, which has also sparked high levels of e-commerce transactions [3].

2.2.1.3 Hong Kong

A different case is represented by Hong Kong. Hong Kong is Asia's largest financial center [8] [6, 17, 18], (Hong Kong is the third largest financial center in the world), and this fact alone brings a natural attractiveness to the city for the introduction of fintech technologies and fintech growth, which has brought about Hong Kong's tremendous development in fintech in the region. Hong Kong's natural position as a super-connector in the region was very useful as well as Hong Kong's proximity to China, making it difficult for other locations to displace Hong Kong as a financial center and therefore as a fintech center. Hence, Hong Kong is becoming a strong fintech hub in Asia, backed by robust investment, and supported from the outset by the local government: in fact, the Hong Kong Monetary Authority (HKMA) has been very supportive of fintech, which has led to several initiatives to boost fintech projects. In 2016 the HKMA even established a Fintech Facilitation Office, and announced the designation of a fintech supervisory sandbox, although only for incumbent banks [3, 6, 17–19].

- A typical example of fintech adoption in Hong Kong, is the case of the Bank of China, Hong Kong (BOCHK) and its BOCHK Mortgage Loan Application. This DLT (Distributed Ledger Technologies) based application is the first of its type in Hong Kong: the goal of the application is to speed up property valuation and with it, the process of mortgage loan approval; it also increases the efficiency and reliability of the laborious paper-based process [6].
- Another example is the case of HSBC, Hong Kong (HSBCHK):

The HSBCHK's app allows mobile account opening, which is an efficient method of sending and receiving money; it also provides a mobile security key and ID, a faster payment system, etc., so most HSBC customers are comfortable using their phones for financial transactions; in fact, over 80% of all global retail banking transactions in the bank occur through digital means.

In April 2020, HSBCHK joined in partnership with Cainiao Network Technology, a logistics off-shoot of Alibaba, to offer quick trade finance approval to online merchants; and by using Alibaba's Tmall platform in Hong Kong, HSBC is able to offer seven-day approvals for trade financing loans of up to 500 thousand dollars; so, HSBC is using third party companies' data to approve trade finance loans [20].

2.2.1.4 Singapore

Singapore stands out as an example of unparalleled growth and development, ranking fourth in the largest financial centers of the world [14] and one of the most important fintech hubs, just behind London. It is a region known for a long tradition in capital investments, top of the line infrastructure, innovation investment, and for strong public support of fintech development. Some specialists even speculate that Singapore might even become the world fintech capitol, especially now that the United Kingdom has left the European Union (Mavadiya, 2017 cited by [1]). Singapore's strategic location gives it a huge potential to bring development to the region, including countries like Malaysia, Vietnam, Thailand, Indonesia, and the Philippines [14].

Government support for fintech in Singapore has been strong for the development of fintech projects. In 2016, the financial services regulator of the country, the Monetary Authority of Singapore (MAS) appointed a Chief Fintech Office, to be responsible for formulating regulations, policies and developing strategies to facilitate the use of technology and innovation, and to manage risk better, enhance efficiency and strengthen the financial sector. The Fintech Office, established in 2016 served as a single entity for all fintech-related matters and promoted Singapore as a fintech hub. In 2017 MAS also appointed a Chief Data officer, to supervise financial institutions [14, 17, 18].

The most widely used fintech products in the country include remittances and payments (MC Payment, PAYSEC YOLOLITE, PAYSWIFF, SMARTPESA, BILLPAY), crowdfunding, lending, asset management (CANOPY, SMARTKARMA, TRADERWAVE, BAMBU, STASHAWAY, GOINVEST) and Insurtech. Some of the key investors include about 500 fintech startups, such as Golden Gate Ventures, Sequoia Capital, SingTel INNOV8, and government entities such as GIK and Tamasek [1].

So, at this point of our study, we might say that the first phase in fintech evolution has been to improve financial services by making them more convenient, more efficient, more secure, easier, and faster, either through fintech startups or by applying mostly outside technology to traditional financial institution services.

In the second phase in fintech evolution, traditional financial institutions are seeking to depend less on outside tech, developing their own systems by collaborating closely with tech companies, including partnerships, mergers, or establishing their own fintech innovation labs [21].

The third phase in fintech evolution is much more radical: the introduction of virtual banks, meaning digital-only banks.

Coming back to Southeast Asia, the introduction of digital-only banks in Hong Kong has lagged behind other nations such as India, South Korea, and much of Europe.

2.2.1.5 India

The Indian fintech industry is basically divided into insurance, securities, and public and private banks. Many fintech companies are focused on financial services ranging from payment processing and trading solutions to banking [3].

The case of India is quite peculiar and though it resembles China's unbanked and underserved population situation somewhat, it is not the same: millions of people have been in practice, excluded from India's traditional financial institutions (about half the Indian population is unbanked) [3] for different reasons, mainly due to a poor financial infrastructure; that is to say that there are not enough local bank branches to satisfy such a huge population; so if a person were to need banking services in many cases he would have had to travel great distances. As can be expected, this fact alone accounts in part for the high levels of poverty in some regions of India, because of the absence of loans, investments, etc. At this juncture, fintech companies began to understand and take advantage of the tremendous opportunities that this situation offered and started to promote financial services through the use of mobile technology, specifically smartphones; needless to say, fintech services and the fintech industry as such have become a huge success in India, especially virtual banking. Virtual banking has increased exponentially along with the population's financial participation.

As we have just mentioned above, traditionally, many millions of Hindus have had no contact with traditional financial institutions, not even for the most basic financial services, but now, through the use of smartphones they no longer have to go personally to their nearest bank, they can go directly to digital services and virtual banks for financial services, for everything from payments and transfers to savings and investments. All of the above have made fintech and particularly virtual banking, especially meaningful to the country, so much so that there is no longer any need to improve or expand the financial institution's infrastructure, since the population is simply skipping (leapfrogging) this physical stage in financial evolution and moving directly to fintech services and virtual banks, for their own benefit, not just in regards financial services, but also financial education, financial inclusion and improvement in the quality of their lives [1]. A good example is MIaap; Mlaap is an online fundraising platform that allows people from around the world to fund Indian communities that are in great need of sanitation and access to drinking water; it also reviews loan applications and makes sure that the loan reaches the borrower [14].

As was the case for China, the Indian banking industry has also taken notice and is scrambling to modernize its systems. So, traditional financial institutions in India are going through a phase of firm adoption of automation processes and blockchain technology: in recent years, this sector has been changing, hand in hand with technology development; it is witnessing an increased adoption of robotic process automation hoping to improve productivity and efficiency in daily banking processes. Regarding blockchain, the Reserve bank of India has stated that it expects the introduction of blockchain technology to bring about cost savings, efficiency, and transparency to the Indian banking industry. In brief, supply chain automation, such as chatbots and blockchain technology, is transforming the banking sector in this country [17].

By contrast, a large proportion of the population resorts to cash payments, up to four-fifths of all payments (90% according to [14]), since Internet infrastructure is still being built, expecting to reach two-thirds of the population in a few years, and so Internet penetration and the use of mobile technology is still in its early stages in many regions; but the trend in cash payments is a matter of concern for government authorities because of cash hoarding, the absence of a money trail and outright corruption. This situation obviously represents an opportunity for fintech startups to offer alternative payment solutions, but it also represents opportunities for the government.

Indeed, India's government, being bent on accelerating financial digital processes, by means of a government-backed program called "Digital India" has now furthered India's fintech payment system. As a result, mobile payments and digital wallets have experienced an exponential growth; the use of cryptocurrencies in India is growing as well [17]. Other Indian public programs to hasten the fintech adoption in the country include India Stack (for startup growth), the Startup India Program, Jan Dan Jojana (a program for financial inclusion), Adhaar Adoption (for biometric authentication), National Payments Council of India initiatives, etc. [1]. Specifically, the Adhar Program issued 770 (UID) unique identities as key to uncomplicated mobile SIM card registration and mobile money, with the result that now 99.7% of urban households are connected to a bank account; but of a total of 1.12 billion Aadhar cards, only 399 million are numbers linked to Aadhar; there is still much to be done. [14, 22].

With all, the use of financial digital technology among the population is now widespread, second only to China; most consumers are young and middle-aged people, especially ages 24–35; surprisingly though, the adoption of these technologies is low in the 18–24 age group, lagging behind 45–54-year-old users, mainly due to unemployment [1, 11].

As we have witnessed for both the Chinese and Indian financial sectors, banks have gone to great lengths to be a part of this fintech phenomenon and so not lag behind and become anachronistic, which has also contributed for the opening of new markets and the inclusion of new clientele, of course, not just in these regions, but worldwide [13].

2.2.1.6 The European Scene

Moving on to the situation in Europe, we must begin by stating that factors such as online payments, regulations to support innovation via fintech startups, and digital banking are all spearheading the fintech transformation in Europe. Other emerging areas for fintech's in European countries include artificial intelligence, cybersecurity, and blockchain technology. Also, the increasing interest in adopting digital processes especially in identification and legitimation of customers is also motivating the emergence of new fintech companies [3].

The changes can be seen all across Europe, but a large part of fintech investment has been mostly concentrated in cities like London, Berlin, Barcelona, Paris, Stockholm, and Amsterdam, although one-third of those investments are centered around London [14].

2.2.1.7 **London**

London is the world's largest financial services center and is strong in areas such as wealth management, neobanking, and foreign exchange [20]; it is a leading force in insurance and trading, P2P lending, cross-border transactions, and banking; in addition, its tech sector is booming and has the highest mobile phone and Internet connectivity; these factors alone make London one of the most connected cities in the world, featured by digital success, high digital penetration, great levels of funding, capital talent, good regulatory environment, and government support along with demographic diversity, all of which establish London as an unquestionable fintech hub, as one of the most attractive fintech locations in the world [1, 17, 18]. London also has one of the world's most aggressive accelerator programs, which helps provide fintechs with the necessary infrastructure and helps them participate in the industry. The government has even established a public/private investment fund to secure London's place as a fintech world capital [1, 3]. Also, one of the main differences between the United Kingdom and other markets is the support fintechs are getting from the regulator and the Central Bank, in fact, it is the Bank of England itself who is the main promoter of fintechs within the United Kingdom [3].

The United Kingdom is doing well in this market, particularly London: the adoption rate of fintech services in London is significantly higher (25% vs. 14% for the rest of the United Kingdom), reflecting a younger and wealthier population that was enthusiastic about adopting fintech early on. fintech, especially in London, is growing faster than in any other major center, with more than a 65% growth rate in deal-making volume vs. an average of around 30% for the rest of the world, since at least 2015, and more than 51% of growth in investments over that same period; by 2017, this meant 224 operations and more than 800 million pounds. In fact, the fintech sector in London is reported to have created more job opportunities than in all of Europe, Hong Kong, Singapore, and Australia put together, with an estimated workforce of about a million people (direct or indirect employees) in the whole of the United Kingdom; and although California is still the largest global center, London's market size is bigger than New York and California put together [1, 3, 23].

The most widely used financial digital products in London tend to be money transfers (TransferWise is a fintech company that is now among the 20 largest in the world) and payments as the most common, reflecting the high use of e-commerce and payment apps such as PayPal. Following payments, the most widely used products are online services related to foreign exchange and investment (Algomi), and then lending, insurance, and crowdfunding (Funding Circle is a platform devoted to making funds available to SMEs that do not have the necessary requirements to approach traditional banks) [1, 18].

Some examples are the following:

The case of Lloyd's Banking Group: Lloyd's entered a partnership in July of 2020 with FORM3, a cloud-native technologies fintech hoping to simplify payments more efficiently, as well as provide support for enhanced data and new overlay services, while at the same time responding to the NPA New Payments Architecture initiative for UK's and Europe's banks and fintechs [24].

At the other end we might mention Monzo, a non-traditional financial institution: Monzo is the first fintech startup to receive a full banking license and has attracted more than 500,000 clients since 2015 [1].

The facts that most stand out in our review of the United Kingdom as boosters for fintech development are mainly London's modern infrastructure, its entrepreneurial environment, human capital, reputation, and a booming tech sector, but equally significant is its demographics, especially its young and wealthy consumers, as in many parts of the world. This feature will have special meaning when we deal with consumer's expectations.

In spite of everything, since Britain's decision to leave the European Union by means of the Brexit vote, many fintech companies in the United Kingdom have not had access to EU markets and resources, so other European actors are participating more aggressively in the European markets; Germany is one of those actors.

2.2.1.8 Germany

Germany and Sweden are the largest fintech destinations in mainland Europe, but they are both behind London's fintech industry. While Sweden is a leader in pursuing a cashless economy, the feature Germany has shown most interest in is AI and robotics especially applied to automated financial management, i.e. the robo-advisor market; other fintech areas of great interest to the Germans are payments and money transfers, lending and crowdfunding, but most fintech companies work in close association with banks in order to shape a more collaborative environment with traditional financial institutions [3].

As in other places, Germany is also a part of the trend to enhance the digital capabilities of its banking institutions through partnerships and mergers with tech companies or with fintech start-ups, hoping to develop in-house expertise.

The Deutsche Bank has spent more than 750 million during the past five years in Fintech technology. One of its newest features is the bank's establishment of a start-up factory called Breaking Wave, intended to attract tech staff that would otherwise not work in a bank environment. In July 2020 the bank took a step further, when it decided to partner up with Google, with the express intention of reimagining the scope of financial services for the present and for the future; that is to say, Deutsche Bank and Google hope to combine their resources in the exploration of new avenues of financial technology [25].

Additionally, through this partnership, Google will provide Deutsche Bank with its modern cloud capabilities and expedite its digital transformation by overhauling its traditional technical infrastructure: Deutsche Bank will then have access to Google's advancements in artificial intelligence, as well as its machine learning and data science capabilities, all in the hopes of making digital and online financial services faster and easier for their clients [25].

Through this merger, Deutsche Bank is also expecting to further develop cash-flow forecasting, increased risk analytics, and security for customer accounts, among other benefits [26, 27].

2.2.1.9 Latin America

As we mentioned earlier, modernization usually reaches all corners of the world by way of the biggest and more stable economies, yet the opportunities to do business with the greatest potential for growth and development can be found in solid economies that are still not saturated by financial technology initiatives; that is why it is very attractive for investors to turn to emerging markets that can yield profits and that have a good chance for growth, due in large part to the high levels of financial exclusion in Latin America that run between 45% and 65%, meaning no contact with traditional financial institutions [12, 28].

So, there seems to be a different type of transformation or evolution of Fintech activity in this region as compared to others; the reasons for this are mainly: unbanked population, a culture of overspending and debt, very low levels of financial literacy, for instance not having a proper understanding of interest rates, and especially, trust, because customers are aware that financial institutions offer products and services in their own interest not in the consumer's best interest; although fintechs are changing this by placing the customer at the center of their attention, people still do not trust financial institutions, traditional or otherwise, including start-ups.

The solution in this case appears to be to increase financial literacy as a previous step to fintech development [12].

In this region, Mexico is one of the countries competing to become a regional fintech center; its ecosystem is 2.5 times bigger than Colombia's, more than three times the size of Argentina's and four times the size of Chile's fintech industry, although Mexico's fintech industry is still second to the Brazilian ecosystem.

2.2.1.10 Mexico

Mexico's fintech ecosystem is comprised of a fintech market (dependent on a certain demand for financial digital products and fintech services), a group of investors, fintech companies, startups, incubators, start-up accelerators, fintech innovation labs, tech companies, and other tech providers, government, and regulators, as well as universities and research centers, human capital and consumers [29]. One of the main motivators in this country for the introduction of fintech technologies from the population's perspective is its convenience, as well as its speed and transparency; and one of its main features is its capacity to reach greater sectors of the population thanks to its rapid growth and development in the country (almost 100 new startups in one year) [1].

In Mexico, investments in fintech have been exponential: by 2018–2019, there were more than 273 fintech companies with a growth average of 52%; a few months later there were 334; fintech transactions have increased 42% a year, and the amounts involved have jumped from 3 billion dollars in 2017–2018 to 75 billion dollars the following year 2018–2019.

Most fintech startups in Mexico are located in Mexico City (more than 60%), around 10% are located in Guadalajara, and the rest are located either in Monterrey, Mérida, or Puebla [1].

Some of the most popular fintech services are loans, payments and transfers, remittances, crowdfunding, and management of assets (BID, 2018, cited by [29]). And among the most visible benefits fintech has had in the country we can mention: faster financial services, efficiency and competence, cost reduction, inclusion of large sectors of the population to the formal economy, the digitalization of the economy, the promotion of entrepreneurship and innovation in the country, financial education and the realization of the benefits technology can have in financial activity; also, the development of cybersecurity, financial inclusion, especially taking into account that approximately 60% of the adult population has no access to traditional financial institutions, but around 64% of the population does have a smartphone or electronic device with access to the internet: so, it is expected that up to 63% of the population might be in the process of bankerization and participating in financial services, in the short term [29].

In March 2018, one of the first integral laws in financial technology was published in Mexico. The law to Regulate Financial Technology Institutions (*Ley para regular las Instituciones de Tecnología Financiera*) or "The Fintech law" (for short) is the first of its kind in Latin America. The law was issued with the goal of reaching legal certainty and the preservation of financial stability vis a vis the changes that the Mexican financial institutions have undergone since the introduction of fintech technology; the law establishes the norms and fundamentals by which new actors may participate in the Mexican financial system; it also hopes to diversify the financial markets in Mexico, and allow access to the population to the benefits of innovation; it promotes financial inclusion, the protection of the rights of users-consumers of fintech services, and also contemplates crime-prevention (including money laundering and fraud) [1, 28, 29]. The idea for the law originally came from a recommendation by the Financial Action Task Force (FATF). According to the text, the legal framework of the "fintech law" is founded on a context of e-commerce, which involves the financial services offered by institutions of financial technology, as well as their organization, operations, and overall functions [28].

2.2.2 A Word on Financial Inclusion and Some General Expectations

In most economic regions of the world, there is still room for healthy fintech competition, though nowhere more than in emerging markets like many countries in Southeast Asia, China, India, and Latina America [20]. In countries where financial inclusion is still an issue to be dealt with some urgency. And this is important because financial inclusion has a direct impact on economic stability, but there are many people worldwide without the benefit of so much as a bank account, especially in regions such as Africa, Asia, and Latin America. These unbanked people do not have access to even the most basic banking services; according to some estimates, around half the world's adult population, that is, about two billion people, are unbanked or underbanked, mostly people living in developing economies, although this phenomenon is certainly not absent from more developed countries: just in the United States there is a 7% of the population that is still unbanked [30]. The main reason being that banks are usually absent in areas where economics makes it difficult to establish a bank branch, especially in the rural areas, and so people living in these areas are compelled to travel sometimes long distances, or wait in long lines, to receive financial services [31] In the opinion of some authors, financial inclusion per se cannot be a goal of fintechs [32], but rather a means for development; in practice though, both tend to go hand in hand.

Mobile technologies, mobile money, have helped tremendously in remedying the effects of financial exclusion and poverty [31]. Indeed, mobile financial services have been a very important instrument in boosting financial inclusion in a growing number of countries, bringing financial services to people who traditionally had no access to formal financial services, either due to the absence of branches in their area or the high cost of the service they required. One of the most important of the services fintech can offer in these countries is proving to be mobile money wallets, which can be used for transfers between subscribers, and all kinds of payments, to utility companies, schools, merchants, and so on [15]. Thus, mobile wallets are likely to become an important means of performing payments and remittances in the short term, as more and more people of emerging markets acquire a smartphone, especially in places like Asia [20]. In fact Africa has been a global leader in the adoption of mobile technologies for payments for unbanked people, mostly due to M-PESA, which is a simple money transfer and payments platform [30].

The problem is that the fast adoption of such services depends on how quickly an agent network can be established, or how fast an infrastructure can be built [33] since coverage and device penetration is not yet universal, particularly in regions where financial education is still under development, although financial education can be made accessible through mobile technologies as much as through fixed networks [31].

It is also expected that emerging market consumers will come up with their own innovations [32]; with two billion users currently influencing young fintech markets, it is still to be seen how those remaining five billion people, some living in countries characterized by conflict, instability, and autocracy, might influence fintech someday [32].

On another front, it is likely that many traditional financial institutions will not be able to cope very well with the new ecosystem since it requires constant efforts and resources to invest in modernization; so, it is more than likely that some banks will finally disappear or will radically be transformed: the concept of bank is changing, leaving behind such, now outdated models of branch-only banks [20].

So, banks must focus on a series of measures that will help them remain in the banking business [34]:

- Start by protecting a bank's bedrock services (storing, moving, and securing customers' finances).
- Training customers and employees in digital services.

- By moving from a keep-up budget to a step-up budget.
- By innovating and trying to deliver value added services.
- By investing in predictive analysis.
- By keeping pace with cybersecurity and new digital capabilities issues.
- By boosting digital transformation.

2.2.3 Regions and Expectations

1) Although China is still considered an emerging market in the financial sector, it is going through a phase of convergence without precedents between the financial services industry and technology. China has become the largest fintech market in the world, with a population of 1.3 billion, but it is also because of sheer necessity, since Chinese traditional state-based financial institutions are not in a position to satisfy the demand for capital for both consumers and businesses [5]. So, tech giants like Ant Financial, Tencent, and Alibaba are leading the fintech industry, not traditional financial institutions. Tech companies are focusing their efforts on the unmet needs of the population; and they are growing largely because they have mastered platform models, they have a favorable regulatory framework in their country, they are data-driven companies who are not afraid of partnerships, they are fast innovators, and they understand the adoption habits of consumers [12].

2) While the case of Singapore is different, the industry in Singapore is expecting to integrate digital solutions to sectors like health, Smart Cities, tech, and education. It is widely expected that Singapore will function as a springboard to the broader Asian region [14].

3) As to the Asian Pacific region, it is increasingly becoming more important by attracting huge amounts of capital: and because it has one of the largest unbanked populations of the world, it is seen as a land of opportunity; but this region's economies are growing at a firm and steady pace, and also have a solid private wealth sector; and equally important is the rapid development of Information and communication technology's role in transforming the entire industry geography by introducing a new trend in convergence services [5].

4) The newly formed fintech industry in India is having significant impact in the country, from job creation: more than 20 thousand jobs have been created in the banking, financial services, and Insurance sector; Infrastructure is currently being developed to take mobile technologies to unbanked populations. The country's regions needs are so diverse that cities are focusing on their particular interests and facing different challenges in the financial services sector [14]. It is expected that the country will show firm development in the near future in mobile technology infrastructure as well as digital banking and automated financial advisor services.

5) Latin America is experiencing a stage of digital evolution in many aspects of the business and financial sectors, by developing its digital infrastructure and adopting innovative technologies such as APIs and artificial intelligence. The greatest challenge that the region has is probably adopting and implementing technological advancements to the financial sector, and doing it coherently so as to avoid fragmenting the industry [35].

In practice, 2019 and 2020 have meant a huge step forward in the acceptance of technological and digital tools that were usually not popular prior to the Covid-19 pandemic in places like Mexico, referring for example to services such as digital payments, which have hugely increased vis a vis traditional payment methods, meaning that debit and credit cards have been diminished in their use. According to estimates by Mexico's Central Bank, Banxico, more than 30 million people (about a quarter of the total population) now prefer to use a cell phone for their banking services, including 12 million small businesses [36].

Both Latin America and the United States have moved on to the next stage of fintech evolution, i.e. virtual banking, though in the context of Latin America, it is been more the case of catching up; for Latin America, Brazil has Nubank, for Mexico Album, AcertumBank and Heybanco are virtual banks; in the United States, Bank Mobile, WeBank, CBD Now, Monzo, N26, Orange Bank, Standard Chartered, BankMobile, Hello Bank!, Digibank may be mentioned [1]. These banks offer services such as digital payments, money transfers and remittances, robo-advisory, application-based money management, loans, etc. [17].

Specifically in Mexico, in light of the pandemic, there is currently little support for fintech start-ups by the government; in addition, the so-called "Fintech law" is being criticized for not having regulated fintechs in their own right, because they have been regulated within the general framework for banks (Mario di Costanzo, ex-president of CONDUSEF cited by [37].

The changes in the payments systems are transforming industries in Mexico (and in all likelihood, in the rest of Latina America) such as the retail industry and telecommunications; but interestingly, technological innovation and data analysis have not been the sole factors to shape the new trends of accelerated growth in Mexico, consumer behavior and expectations have had a huge influence as well [38]. The revolution that Mexico has experienced especially in digital payments is expected to influence new concepts of money and money market and the way they will take shape in the future [38, 39].

2.2.4 New Consumers, New Expectations

As we have seen above, fintech, a phenomenon with worldwide reach, has been adopted in different regions for different motives, and it is these motives that usually define consumers' expectations in accordance to the type of fintech services people have been offered. In addition, there are always cultural differences, and culture is in itself a comprehensive concept that includes everything that influences a person's behavior, preferences, and decision-making. But the complexity of modern societies is such that culture alone does not determine behavior. A consumer's preferences are not static and can change either according to what an individual finds appealing or according to a society's relationship to its economic, technical, and physical environment [39]. So, there are factors that tend to unify preferences and expectations across borders.

Both in the United States and in China, there is a large population of savvy young people born at the beginning of the twenty-first century referred to as millennials or Generation Y, and the Zoomers, or Generation Z, born in an age of technological achievements and innovations, used to words like Google, Wi-Fi, the Internet, social networks, texting, all as part of everyday speech, as is their use of electronic devices, business applications, etc.: these digital natives are the younger generation who have grown up with the internet as an integral part of their lives, who see online, mobile and all other digital channels as being integrated into their lives; these individuals, more than the members of the earlier Generation X (digital aliens according to Prensky's definitions) tend to consider a good customer experience more important than any name or brand; one survey has shown that millennials in the US would switch banks if they did not receive a satisfactory experience; in China, millennials also prefer customer experience and the flexibility found in fintech companies vis a vis traditional banks that are rigid and state-oriented [3].

In addition, technology is constantly changing the way people approach their daily activities, like online shopping, so technology is raising people's expectations: for instance, the introduction of mobile points of sale that can be used on any smartphone to process payments, anytime, anywhere [20] so, in addition to the preferences and expectations of the younger generations, technology in ordinary life is also changing people's expectations. All within the framework of globalization which seems to be directing modern-day economic activity, and leading it to an international market integration, which will ultimately mean unity in consumer preferences and expectations [14].

There is a growing view that a single proposition financial services offering by a fintech company will come to replace traditional banking; according to one study, half the millennials in the United States are expecting fintech companies to disrupt traditional banking services, and about 75% of that consumer group would be more interested in receiving financial services from big companies such as Google, Amazon than they would from banks [14].

2.2.5 Recent Developments

An unexpected series of crisis, triggered by a modern-day epidemic, have been spreading all over the world with extreme expedition; the culprit is the new type of coronavirus designated 11 February 2020, as Severe Acute Respiratory Syndrome Coronavirus-2 or SARS-CoV-2 (for short) [40, 41], which is the virus responsible for the Covid-19 disease [42–44].

The impact of disease throughout history has represented at times, in more ways than one, a change in the direction of events [45–47]. Indeed, in the business sector, the Covid-19 pandemic has led to unprecedented changes, bringing companies to consider alternatives to the traditional methods they have used up to now. Hence, the current economic situation has allowed digital transformation to be a strategy for economic entities to renew themselves in order to remain competitive. In this new scenario, digital transformation has become a central issue for economic development. The pandemic has unexpectedly been responsible of the huge acceleration of the process of digitization: people have decisively turned to the internet and digitalization for everything from telework to schooling [48] from medical advice and reports [49] to e-shopping and all aspects of e-commerce [50].

But even if new solutions are still being outlined, what can be defined is the technology behind the strategy, meaning the use of mobile technologies, computers, the Internet, IP-based networking, wireless networking, online resources, blockchains, multimodal public reporting and notification, and many others [46].

So, needless to say, we are currently facing disruptive changes in many aspects of our society: the pandemic has been the engine of different processes that either, had already started and had to evolve more rapidly, or had to develop as new solutions to the unforeseen problems it caused. In this case, people have been achieving increasing access to the Internet around the world, interacting as individuals, sharing opinions, experiences, expectations, forming online communities and cultures, spreading to businesses, marketing, the world of finance, and of course other sectors, and finally, contributing to the development of the theory of Consumer Culture, which also deals with consumer expectations [31]. So, with greater digitalization in people's everyday lives, consumer expectations have evolved by hyper-personalization and superior interfaces that have been provided to them by digital companies such as Amazon, Google, and Netflix [12].

In conclusion, technological developments are truly spearheading new social and economic processes, but disruption is also being caused by changing consumer behavior and expectations, as much as by emerging younger and more agile competitors, the democratization of knowledge, and new up-to-date business models [12].

2.3 Digital Expectations for Consumers in the Financial Industry

As we have reviewed above, the unexpected consequences of the events stemming from the outbreak of the Covid-19 pandemic have greatly impacted people, markets, and financial institutions alike: with previous economic expectations completely challenged, consumers 'expectations have also shifted. In addition, technological advancements in areas such as financial services serve to raise consumers' expectations in regards the type of service they hope to receive; both factors converge, and it is left to the financial consumer to decide whether his expectations have been met by a bank or fintech company or whether he should look for the experience he expects elsewhere.

An expectation may be defined as a belief regarding the performance of a product or service. An expectation, in practice, is a desired outcome of the said product or service that derives from a pre-consumption belief regarding overall performance or attributes possessed by a product or service. And performance measures whether these expected outcomes have been achieved: either objectively, based on actual performance (fairly constant across customers), or subjectively based on an individual's own feelings, a factor that can vary from customer to customer [51].

There are several factors that can impact a consumer's preferences and behavior; these factors can be placed in three different categories [52]:

1) **External influences:** A consumer's culture has a direct impact on the internal decision-making process, and so do marketing, price, and information.
2) **Internal influences:** These include motivation, attitudes, decisions, perception, and exposure, all of which help predict the consumer's behavior, but may change over time as consumers gain access to new information.
3) **Post-decision processes (post-purchase behavior):** This depends by evaluating the outcome and the levels of satisfaction.

These three categories are dependent on certain decision-making factors, such as problem recognition (the identification of which needs to satisfy) and the products or services that are available: this process includes information, the search, a judgment, and decision-making [52].

Now that we have distinguished some general factors, we must now take into account that the consumer's behavior is a dynamic and fast-moving field, especially, if we refer to digital consumers. Some key features that define digital consumers are the following: [32]

1) **Digital consumers are increasingly satisfied with the medium:** The user demographic tends to be young people though it is, of course, not limited to this group. As consumers feel more comfortable with the medium, they tend to use it more frequently and more efficiently.
2) **They want it all and want it now:** Digital consumers have grown accustomed to getting their information from multiple sources simultaneously and on demand; the value here is time: Consumers scan for relevance before investing time. So, qualities such as "scannability" and "instant gratification" are desirable features to achieve positive results.
3) **The consumer is in control:** The web is certainly not a passive medium, the consumer knows he is in control, so the market has to be fashioned to be user/consumer-centric, as well as the trust-building, "permission-based" to appeal to the digital consumer; but still, the market has to be able to offer a real value proposition to have positive results.
4) **The digital consumer is fickle:** Factors such as transparency and immediacy of the internet tend to erode vendor loyalty, since the consumer has the capability to compare competing names and brands.
5) **Digital consumers are vocal:** Online consumers talk to each other frequently through peer reviews, blogs, social networks, and online forums; they communicate to each other their experiences online, both positive and negative [32]. The best fintech companies understand the information the customer can access and the ways in which these influence each other; so in a milieu where users are able to speak online about their experience as customers, companies can no longer believably exaggerate the quality of their service; with this, there seems to have been a shift in the balance of power between a bank and the customer, in favor of the customer [14].

For example, the case of mobile banking has shown at a global level what fintech startups and traditional financial institutions can achieve when they offer products and services focusing on the customer's experience [14].

2.3.1 Digital Customers' General Expectations

- Modern-day digital customers expect mobility, through such devices as smartphones, remote payments, remote automation (IoT), and telecommunications such as videoconferences and telework [10].
- Financial institutions have access to tremendous amounts of processing power that can be utilized to improve a customer's experience. For example, a bank's geolocation services can place a mobile banking customer at a car dealership and then send him alerts to extend a loan offer; this would allow the customer to compare and decide which loan offer is more competitive, the bank's or the dealer' [14].
- A financial services entity has to be aware that customers' experience more friction managing their financial interests than in any other area [14].
- Since financial consumers have been exposed to personalized shopping as a key component of the services they want: that is, they are becoming more and more accustomed to intelligent, personalized, online shopping, where their service providers know them and understand their wants and needs, and fulfill them, financial consumers expect banks to do the same with their personal and transaction data [34].
- Financial services institutions have to respond to such factors as the customer's needs, wants, expectations frustrations, so it will become easier to prioritize product and service offerings and development; but it is also a question of marketing and education; other important factors to be taken into account are awareness and trust-building [14].
- Financial customers' expectations are changing, expecting simplicity, ease, convenience, speed, transparency, reliability, and the possibility to receive financial services wherever and whenever the customer needs them [14].
- Financial services companies that adapt to the changing pace of consumer demand are able to solve the customer's friction with greater levels of satisfaction, increased adoption, and cost savings [14].
- In the wake of the pandemic, almost 90% of financial consumers expect companies to accelerate their digital transformation, while 68% have said that the pandemic has elevated their expectations of the digital capabilities of names and brands. It has also intensified the meaning of digital banking trends to financial consumers [53].
- Young consumers, like millennials, and the needs of customers are defining the trends in new Value-Added Services; the main problem is that the needs and expectations of customers tend to evolve more quickly as much as they are adapting even faster to new and more convenient services, which they then expect to receive in increasingly better versions [10].
- The concept of Value-Added Services is no longer what we understood it to be, and it has become the new core service. Customers are demanding new app-based services and features, and they expect certain *de facto* qualities, including streaming, booking services, etc. [10].

2.3.2 Value-Added Services that Banking Services' Consumers Expect

Protection: The great majority of financial consumers (91%) seek protection from money and personal data theft: consumers are increasingly aware that in the digital world, fraud and other cybercrimes are increasing, and so they demand bank protect their money and data [34, 53]. Consumers expect banks should protect them from Fraud and money theft (91%), bank failure (89%), identity theft (90%), data theft (89%), cyberattacks (86%), unsolicited product sales and services (76%), money lost due to market conditions (48%), money loss due to poor investment (46%). [34]

With the speed that digitalization is being applied to financial services, consumers of the financial sector are increasingly concerned about the security and protection of their finances, payments, and particularly their identity; they expect that the banks and fintech companies that offer them their financial services also guarantee the protection of their assets; and for that goal, customers (1 in 5) are even willing to pay premiums; with that in mind, It is expected that the banking industry promote their current protection services in order to turn protection into a revenue source. By extending their protection services banks could also form a valuable loyalty base that might give them a competitive edge over non-traditional financial institutions such as fintech companies. But less than half of banking services' consumers consider that banks deliver the services they value, as they expect they should be delivered [34].

So many customers protect their privacy and personal data by using digital personas and limiting the amount of online data they make available to the institution [34]: In addition, an average of 1 in 5 customers would pay fees for these and other services (in United Kingdom, 28%, in France 32%, in Germany 40%, in the United States 28%, in Sweden 34%): consumers would pay premiums for wealth building services, rewards services, cash management services, and others.

Receive rewards: Approximately 70% of financial customers across all countries, age groups, income levels, and bank types are in high demand for reward services. Some of the reasons that financial consumers expect rewards are the following: Length and volume of business, loyal years of membership 80%, number of transactions 57%, average balance (46%), number of products held (43%); other reasons financial consumers expect to receive rewards are: for meeting their financial goals, for referring new customers to the bank or company, for switching banks, for using financial products, because of their average cash balance, for their number of years as a customer, for money spent, and so on. Financial customers think that those rewards might be in the form of charitable endeavors, or community projects.

Advice on investments: Financial consumers are unanimous across all countries in their high demand for this type of service, although banks' success at offering these services varies geographically. These services include insight and updates on the world economy and how they affect my savings and investments; wealth transfers, advice on tax-efficient ways to build and transfer wealth, how to invest money for tax returns, loans for investing in wealth-building schemes, access to independent financial advisors, and access to global portfolios of investment options.

In a digital world, financial advice, updates, statements, and "what if" analysis can all be digitally provided using robo-advisors and personal finance digital tools. Some of the world's leading banks are already there, such as Bank of America's robo-advisory service called Erica. But new and innovative services are always being offered by non-banks, and if banks do not try to catch up quickly, they will no doubt loose customers.

"Know me as an individual" services: 63% of all financial customers expect a customized and consumer-centric service. Though Europeans are more likely to pay premiums for these services. This category includes [34]: Long- and short-term wealth-building goals, major life changes, location, the customers' other financial services providers, lifestyle and ambitions, the services the customer uses, and knowledge of the goods he buys. The digital economy is in the process of revolutionizing consumer-intensive industries: they deliver personalized services within all the financial activity they do online, and this has led banks to do the same. For example, financial consumers are asking banks to take a step beyond affordability checks and use advanced preference and goal-based insights combined with location indicators, to receive financial products "offers at the right time."

Financial consumers now expect to take care of their financial business with an institution or company that knows them, that understands their goals and aids them in achieving those goals: this is called my "digital best friend" available anytime, anywhere. So leading banks are embracing the concept of "my digital best friend" from more digitally advanced industries, such as retail communications, and combining it with smart phones and face to face support to deliver what financial consumers now expect.

Cash management services and advice: 60% of global financial consumers expect the bank to deliver cash management services. This service includes: Around the clock account access using multichannel means (phone, mobile device, the Internet), current account balances (over phone, mobile device, or online), easy and convenient access to money, easy access to information on payment transactions when they are coming in or going out, automatic savings mechanisms and alert features. Also receiving personal alerts before going over budgets and because of cash-flow problems, receiving alerts for new offers, automatically transferring passive money into savings accounts; this type of service also includes access to short-term loans, setting up and monitoring a monthly budget, choosing the best utility or insurance deals on the market; access to credit advice and support, information on what local and favorite stores have to offer.

Only 29% of global financial customers feel they have sufficient access to online planning tools from their current financial services provider, but 73% of consumers in the United States feel that their financial issues are resolved, higher compared to European countries such as the United Kingdom (59%) and Germany (37%). Some of the main problems financial consumers have with their financial service providers are the following: speaking to the right person, not being properly advised before things go wrong, sorting out charges customers do not agree with or did not foresee (including payment mistakes and fraud), problems using online and mobile tools, canceling products improperly sold to the client and getting their back money, getting advice on the best products at the right price: these are also the top reasons why a client would consider switching to another financial services provider.

To summarize, what financial customers expect when looking for financial services is that [34]:

- Payments should be made with few or no mistakes.
- Financial service providers keep customers' money safe.
- Financial service providers protect clients from fraud and safeguard their personal data.
- They deliver value for the money they have invested or saved.
- They understand when customers need additional liquidity or funding and offer loans.
- They help manage money on a monthly basis.
- They understand the customer, including their lifestyles and goals, and help those customers manage their money accordingly.
- They help customers build wealth.
- Tell customers what they are spending their money on and how they can save.
- Give customers anytime, anyplace access to their account balance.
- Provide customers great financial advice.
- Reward customers for their business.
- Give discounts on what they purchase.

2.3.3 Post-Pandemic Digital Expectations for Financial Service Providers and Consumers

- Expanding to the digital experience, transitioning to digital technologies [53, 54].
- Shifting to remote, thanks to platforms like zoom.
- Empowering an experience-based consumer culture by adapting quickly to digitization and integrating a consumer-centered thinking.
- Overcoming negative experiences, meaning the difference between a customer's expectations and the customer's experience in using a financial product or service, which can lead to decreased loyalty in a brand or simply switching providers. Consumers expect that a positive experience should be the norm, not the exception.
- Enhancing the financial customer's experience with an emotional connection, meaning sympathy, respect, and even kindness from brands and companies; in other words, focusing on the human element.
- Setting up experienced-based Key performance indicators to measure the satisfaction of the customer.
- Transitioning from an IQ perspective to emotional intelligence, EQ, referring to the emotional impact a product or service may have on a customer.
- Establishing consistency through the product ecosystem, by avoiding fragmentation in the financial consumer's experience.
- Consumers may expect contextual financial experience, including personalized experience, as well as simple, flexible payments.
- Collaboration between traditional financial institutions, fintechs, and tech companies.
- **Bringing a financial customer's experience to new levels:** 56% of all financial customers in the US switched brands due to the added value of the new products they experienced.

2.3.4 Financial Consumers' Expectations for Future Financial Services

- Financial services consumers expect that all banking will be done digitally, so branch-only banks will become anachronistic and disappear [34].
- Customers expect to use the phone for everything, including all their financial dealings.
- Some financial customers believe that banks in the future will charge high rates for their services, whereas others expect to see bank charges and fees dropped.
- Most financial customers fear that they may be at much higher risk for data ID theft but expect to be better protected by their financial services provider.
- Financial customers expect that smart tools will be available to them that can find the best deals.
- Most consumers expect to have access to a digital personal finance manager
- Many consumers expect that customer service will be worse, although others expect improved customer service.
- Customers expect to pay for things using their fingerprints and have all their financial data stored in their fingerprints, and that wearables and biometric devices will fast increase in use.
- Most financial customers believe that cash will not exist anymore in the future.

In conclusion, customers have been expecting banks to securely manage their finances, to understand their goals and preferences, and proactively deliver the right offers and services; the goal is to deliver a highly convenient, secure, and trusted digital experience that will set it apart from other banks and from non-bank entities such as fintech startups [34]. But banks have not been able to meet consumers' changing demands: many bank products and services lack customization. There are frequent complaints about the inefficiencies of the service and financial advice at bank branches or call centers. Banks are also charging high fees for overdrafts and other services [5]. Among the most important services are "see me as a person, but there is little improvement in delivery, so customers take their business to fintech companies [55]. Indeed, according to one report, in 2014 only 7% of customers went to the bank for financial services in the United States [34]. Understandably, between 2017 and 2021 almost 4500 bank branches have closed their doors in the United States [56].

It would seem that traditional banks have had trouble coping with consumers' expectations for reasons that go beyond just technological catching-up: there are problems in awareness and change management, in common platforms, in ecosystem partnership and connectivity, automation and legacy replacement, developing predictive analysis; then there is the resistance to change, the costs, and information disparities [5]. Digitalization requires learning new skills and a new set of competencies, it requires the introduction of new forms of leadership and new organizational capabilities, and depending on the degree entities achieve these requirements, they will be integrated into the new digital culture [57].

So, banks will have to pay more attention to investing in technological infrastructure, digitization, omni-channels and portals, big data, automation, the protection of customers, and legacy modernization, just to mention a few, if they hope to compete with younger and more agile non-bank financial service providers.

2.4 The Tech in Fintech

Fintech companies have founded their business models in the applicability of technology as an integral part of consumer's experience and services. Innovating technologies are in fact a cornerstone of fintech services [28].

In the next section, we will outline many of the technologies used by Fintech companies, and those technologies have been responsible for disrupting the financial industry through the introduction of new business models. And the introduction of these new business models is one of the principal factors that has resulted in a huge amount of losses by the traditional financial services industry; by contrast, traditional banks operate out-of-date business models "designed for old-style markets and consumers that in the meantime have changed their needs." Business models are continuously reconfigured, and that is why this classification may change in the future depending on the conditions of the market, new regulations, and changes in consumer behavior.

New business models [20, 28]:

- Digital payments and money transfers.
- Social media-based remittances.
- P2P, P2B, & B2B lending models.
- POS (point of sales) business model.
- Alternative financing: Crowdfunding business model, crowd investing.
- Wealthtech: robo-advisory.
- Business finance management.
- Digital banking.
- Insurtech. P2P insurance.
- Regtech.
- Proptech.
- Cryptocurrencies.

2.4.1 The Principal Enablers

- **4G and 5G networks and mobile technology:** Smartphones and mobile applications [3, 28]. They enable consumers' access to fintech services as digital channels and real-time virtual processes. Mobile devices such as smartphones, tablets, and laptops are consumers' preferred means to search for fintech products and services. The use of these devices has gained ground around the globe thanks to ever-increasing internet penetration and the lowering of costs; in practical terms, this means Internet access to more than four billion users worldwide, more than half the world population [28].

- **Human digital interfaces, security, and biometrics:** They refer to the means of how humans interact with computers, such as using voice to give commands instead of using a keyboard, measuring our emotional state after using a mobile device [1]. They are a means of authentication based on digital and facial recognition, as well as voice recognition in order to provide security and confidence to the consumer. In our growing digital economy, biometrics has reached a number of industries, including the retail industry and banking; biometrics is basically used by the industry as a means for authentication. In order to protect the integrity of banking operations and all kinds of transactions. The field of biometric authentication includes scanning fingerprints, retina scanners, voice recognition, heart rate reader, and API facial recognition.

- **Artificial intelligence and machine learning:** Is a technology based on algorithms that is used to eliminate human errors, automate processes, and reduce costs. Robotics and artificial intelligence technologies have evolved in their introduction to several industries; a significant part of the financial services industry has begun to use AI in order to add value, to cut costs, and save time. Both the traditional financial industry and fintech companies have discovered in AI, through its use of Big Data, the possibility to classify information, make decisions, and interact with clients in different contexts. AI and machine learning are expected to revolutionize financial risk management; in this sense, artificial intelligence can achieve a risk management-specific intelligence higher than ordinary humans; but conditioned to the availability of the appropriate data, and the availability of skilled technicians or staff to implement new techniques [58]. In this context, AI can be defined by the operations and general activities performed by a machine, showing the capability of imitating and simulating human reasoning patterns in functions such as planning, learning, perception, decision-making, as applied to various contexts. In other words, the term AI refers to the field of information science in charge of solving cognitive problems related to human intelligence parameters via pattern recognition, problem solving and learning. In practical terms, this means the capabilities of a machine to process huge quantities of complex, unstructured information, which allows a machine to see, talk, write, listen, and interpret information [12, 28]. The field of machine learning also derives from it, a field linked to robo-advisors and referred to the science by which computers show human traits, such as improving their performance and automated learning processes as they have access to information through the interaction with people and the observation of real-life situations [1]. Some of the main AI fintech technologies are chatbots, voice assistance, process automation, robo-advisory, fraud detection, and biometrics.

- **Application Programming Interface (API):** API refers to interfaces that allow the interconnection of platforms. It is a communications protocol that allows for the creation of new digital businesses and ecosystems through the means of integrating different platforms, processes, and data sources. This is a concept that focuses on the processing of consumer information. APIs allow institutions and fintech companies alike to use available information and to offer new products or transactions based on the knowledge of consumers. APIs are interfaces that can count on different degrees of exposure and are classified as private and public. The private type being available only to banks. The public type is divided into four subcategories: Associate: This is an open API available to the bank's selected members; it is also accessible to bank developers. The Member and Known types are available to bank developers and to members of the community. The Public type usually means a form of basic registration since it is open to all users. In practice, fintech companies that offer APIs are open to collaboration, and thus, larger products can be promoted when they connect different services [20]. And so, APIs may be used in banks and fintech companies, for instance if a bank offers its customers services on the PayPal platform through the bank's own mobile app, the customers are able to see the funds they have in the platform and would also be able to see details of their recent transactions [20]. So API is the answer to the question of how fintech startups are capable of coming up with faster and more efficient products [1].

- **Cloud computing:** Allow access to greater data storage space and remote access. This model is of great use to fintechs, for it enables these companies to focus on their offerings to their customers. Cloud computing encompasses several models: (i) The SAAS type, Software as a Service: It refers to the use of available applications allowing organizations to achieve solutions that have already been implemented with little help from IT. (ii) PaaS: Platform as a Service: It is based on the use of infrastructure on demand to operate a business. It is of low cost and easy access. (iii) IaaS: Infrastructure as a Service: It refers to the use of a platform on demand to develop new applications.

- **IoT:** It is the interconnection of ordinary things with the internet, incorporating sensory aspects through the measurement of physical events. The Internet of Things in the financial services industry offers opportunities for many industries and businesses to be transformed: it has a variety of uses in different industries: from retail, manufacturing, to health and hospitality to financial services. The Internet of Things network is building both thing-to-thing and person-to-thing relationships [20]. The IoT is where Internet Communication (both wired and wireless) is placed into everyday objects from cars to refrigerators, keys to watches: Basically, anything that can have a chip placed inside it. We will probably be wearing and watching and being monitored by chips in everything [55]. The Internet of Things will be the next wave of change, it could be the next industrial revolution; the next generations will probably take it for granted as naturally as smartphones today; mobile devices will be everywhere and will need management and security, along with robots and connected machines [20].

- **Legacy systems:** Including outdated systems and applications but that are still used by financial institutions because they are critical in their processes. The substitution of this type of system is expensive, so companies have to develop new systems or applications around them to allow a business to continue operating. Currently, there are many financial institutions that are interested in integrating fintech solutions to their chain of value, but one of their main challenges is to update their systems. Some solutions may be the development of APIs, cloud storage, migrating to new platforms, reconstructing, migrating or optimizing codes, and replacing systems.

- **Quantum Computing:** It refers to a technology that allows the superposition and linking of information, endowing computers with superior processing power. As a traditional computer system is formed by zeros and ones, quantum computing uses qubits, which allows a computer to go beyond two states to store an enormous amount of information while saving energy [1]. Quantum computing's objective is to solve extremely complex algorithms through fast data processing. Quantum computing aided by artificial intelligence is capable of extremely fast and precise fraud detection, thanks to its ability to detect patterns, self-learning, and the development of descriptive and prescriptive models. Quantum computers will not substitute traditional computers in the foreseeable future but will be used to solve difficult problems, such as maximizing investment profits based on a specific risk profile, dynamic portfolio optimization; Quantum Computing is expected to solve in just days problems that are normally expected to be solved in much longer periods of time [1].

- **Augmented reality:** It allows devices to include layers of visual information over reality; emulation is part of this technology: It offers optimized means to deliver information through digital frames that appeal to the user's senses. In other words, this technology is used to observe the real world through one or more technological devices that simultaneously add information through virtual reality. This type of technology has attracted a lot of interest and investment in the fintech industry. Fintechs use this technology to create innovative models where customers can interact with financial services, for example, they can visit a bank branch.

- **Big data:** This refers to large datasets that regular database software tools cannot store or analyze, they are large enough to require machine learning tools to be analyzed; this type of data is stored across many different machines; so, as technology advances, the size of big data will be dealt with as well. The definition of Big Data differs by industry, depending on the type of software used and the size of the datasets that need to be stored and analyzed; and it is the financial services industry that has the largest datasets due to the type of industry, taking into account data from customer experience, but also because of projections regarding the industry, trade activity, innovation, and so on [20]. So, the use of big data is expected to become a central factor for competition, bringing about innovation and growth. In addition, the growing volume and detail of information compiled by private companies along with social networks, and the Internet of things will bring an exponential growth to Big Data and as data grows, so will the tools to analyze the data improve [20].

- **Blockchain:** It is a registry of permanent transactions within a decentralized network which contains chronologically linked information using an encrypted code. Blockchain is considered the next revolution in the financial services industry [20].

2.5 Blockchain Technology

Blockchain is a technology that has been considered no less than revolutionary and can be applied not only to the financial services industry, but to a multiplicity of other sectors. Blockchain is one of the cornerstones of fintech and does not rely entirely on the web like Fintech 1.0 and 1.5; this system makes it possible to exchange valuables on the Internet [2].

It has shown itself to being a key factor in the digital transformation process of any entity or organization: the impact is concentrated in areas ranging from finance, insurance companies, banks, and international trade, to government, regulatory institutions, social networks, cloud storage, and represents practically endless possibilities; this is due to the tremendous capability that blockchain has to simplify a variety of processes in a large number of areas, as well as the capability it has to guarantee transparency in information, validation and security in operations. It is a technology that can potentially contribute in transforming a multiplicity of aspects in society, from the economic to the political [1].

Blockchain technology has become a veritable revolution in the way in which information is stored, in the role it plays in the development of a digital economy and in the way in which transactions are carried out [1].

2.5.1 Blockchain Procedures

The procedures within the blockchain can be described as follows [28, 59, 60]:

User 1 wants to send money to user 2, and both are members of a decentralized network (a blockchain in itself); the transaction is generated on the network and then published; the party that accepts the transaction proceeds to validate it digitally in order to authorize the amount of money, the information, the agreement or the digital assets that are the object of this transaction. Once this step has been completed, the transaction is sent to a node (a computer that is operating the applications) on the network with the data in order to proceed with its transmission, and then the information must be validated through a previously defined process.

If it has been accepted, the transaction is transmitted to all the nodes of the block by notification, in order to be validated again. Once the validation is done, it becomes part of a block (a block can contain thousands of transactions) and is published throughout the network (the blockchain is a chain or network of these blocks). Finally, it is registered in the blockchain, and its validity is confirmed by the rest of the blocks; with this, the transaction along with its time frame is permanently established in a highly detailed and decentralized ledger and is finally accepted by all the nodes in the blockchain.

So, there is no need to include intermediaries in this process (the intermediary is traditionally the party in charge of validating the accuracy and authenticity of a transaction as well as the information within the transaction): it is the network itself that seeks to replace the trust that was once placed in a transaction by the intervention of these intermediaries; but now, by virtue of the process itself, and without the need of a third-party intermediary such as a financial institution, trust in the operation should be high [28, 59, 60].

Every time a new block is generated (mined), an incentive is given to the person who was responsible for generating it; the reward can include a certain amount of money for each transaction incorporated into the new block, but it can also be a certain number of new cryptocurrencies as a reward for creating a new block. This process works as a capital gain mechanism for the block and, therefore, for the cryptocurrencies "added value" [28].

In short, the blockchain constitutes an extremely secure digital tool for the entire process of validation, handling, and storage of information, especially since one of its foundations is cryptography, thus, reducing risks, inefficiencies, and any cybercrime such as fraud. Blockchain is a system characterized by greater speeds in the operations it performs, with lower costs, greater privacy, and transparency and also represents equal access for all users. Blockchain helps boost productivity, as well as improvements in the quality of the operations and provides better overall results [28, 60, 61].

2.6 Benefits and Disadvantages of the Technology

The applications and benefits of Blockchain technology are many, but we can briefly mention the following [1, 59–61]:

- The blockchain has the potential to accelerate and simplify the process of transferring values in cross-border transactions, which can significantly reduce expenses. Hence it is simple and easy to use, and these features are strongly emphasized.
- The blockchain has been designed to work from peer to peer and is not based on a client-server network.
- With blockchain, it is possible to carry out transactions and agreements automatically, imposing the obligations of the parties contractually without the need for intermediaries; blockchain stock trading will facilitate accuracy and speed in this sector.

- Another feature of the blockchain is the removal of intermediaries such as traditional financial institutions. This feature is closely related to another very important aspect of the blockchain, which is decentralization.
- With the blockchain, users can choose the way in which they are going to identify themselves and to whom they will pass on their information: users must register their identity in the blockchain, but once this has been done, the same identification can be used in many other services. Cryptographic methods such as digital signatures operate here to demonstrate ownership and authenticity.
- The information registered in the blockchain is immutable, which means that it cannot be changed or altered, and although it is not without flaws, its immutability guarantees that the blockchain is safe and reliable.
- The blockchain also facilitates transparency in transactions, which benefits the client against corrupt institutions and entities, boosting confidence in the network and trust; at the same time, it benefits traditional banking institutions by renewing the trust of their longtime customers, and the commitment of their new clients, expressed for instance, in customer loyalty and the development of customer performance-based reward programs.

Current disadvantages of blockchain [60, 61]:

- Lack of privacy, which means that any user in the network can access all the information at any time.
- **Security problems:** Systems based on blockchain technology use advanced cryptography in order to encrypt the information contained in private passwords or personal access codes. But problems can arise for example in the case of cryptocurrency when the owner loses or does not remember the password: there is no way to access the money without it, unlike traditional banks where if the card is lost, or the passwords are lost, funds can be recovered if the client proves his identity.
- There is no centralized control, which means that if changes are to take place within the system, they have to be accepted by the majority of users, regardless of the disagreements of the minority.
- **Risk of 51% attacks:** when a malicious user control more than 50% of the computers in a particular blockchain, it is known as "51% attacks," which in theory could prevent new transactions or the much-dreaded double spending of cryptocurrencies.
- The blockchain has basically been used only with cryptocurrencies so far, so there is still no real experience of what would happen if this technology were to be used in other sectors, or what problems could come with it. It is expected that the system, in all likelihood, will mature in stages. In any case, further research in how blockchain changes over time is required [58].
- The "proof of work" algorithm implies that evidence of the electrical energy and the resources invested in creating a block is required before being accepted into the system: for example, a home ordinarily consumes 10–12 thousand kHz of electricity per year, but this power is only enough to generate four bitcoins worth one thousand dollars each: this implies that the system is more feasible and will be more easily developed in countries where electricity is low cost.
- Understanding the blockchain, its reputation, and trust in it: there is still little understanding about the way the blockchain system works, especially among the most common users of digital media. There is also the problem of blockchain and bitcoin, which in many parts of the world has a bad reputation, as it has been used by organized crime to launder money, undermining the trust of potential consumers [62].
- Virtual assets based on the blockchain system, cryptocurrencies, still have to face a long process of regulation and integration with pre-existing systems. It is no surprise that both government and banks are reluctant to accept this system; in fact, cryptocurrency is prohibited in some

countries. In Mexico, article 30 of the so-called Fintech Law, expressly discards cryptocurrency as legal tender; it cannot be considered foreign currency either (the government has even warned consumers about the risks of using virtual currency).

- There is also an unfounded belief among users that blockchain technology represents disruption, in the sense that it will alter the balance of governments and financial institutions, eliminate banks, and even change the world order. Users must remember that blockchain systems are only databases that store and manage digital information. However, blockchain has come to be called the "fifth disruptive paradigm": blockchain has the potential of reinventing certain concepts such as the money market, financial services, certain industries, and even some daily activities [60].

2.7 Real Examples and Use Cases

Examples of Blockchain in action in the financial industries:

In Mexico, Blockchain technology has been used in a variety of projects, however, some of them are still pilot projects and for this reason they should be considered as developing business models and not as fully developed projects [28]. In the public sector, the HACKMX Blockchain Network: This initiative was largely inspired by the use of blockchain for public services in other parts of the world; the Mexican initiative called "Smart transactions" is based on the Ethereum system [59, 63].

The program is to be developed in four stages: the initiative, based on blockchain technology, will generate a private network which is capable of using smart agreements with suitable mechanisms to authenticate all parties involved; the purpose of all of this is to generate a reliable public contract system in which there should be permanent and incorruptible records of the operations performed and the participants involved [28, 60].

In the field of insurance, the Mexican Insurance Institutions Association is studying the possibility of using the blockchain system to validate policies and optimize processes.

Another interesting initiative is the alliance between the Autonomous University of Southern Baja California and Bankcoin.global to start a pilot project to generate an intelligent contract system that should help foster e-commerce in Mexico.

In 2016, Everex, a Singapore-based startup, began working on platforms specializing in cross-border transactions based on blockchain technology for security; the platform is currently targeting around 3.5 million unbanked or underbanked people worldwide that need to send money abroad to their families: with that goal in mind, Everex has been able to create cryptocash assets using the Ethereum blockchain, which is now a resource used by expats, migrants, and international aid organizations to transfer money efficiently and at a very low cost to any country [30].

An estimated 20% of the world population does not have a legal identity, they have no proof of identity; this is a serious barrier for accessing government and financial services, especially for refugees. A platform called BanQu is being used to aid people such as small farmers in developing countries in creating an economic identity within the blockchain regardless of gender, land holdings, or income which not only allows these farmers to have access to credit, but introduces them to the local economy, and ultimately to the international markets.

With BanQu, these people are enabled to create a verifiable identity linked in the blockchain to their product, thus offering a solution to inequality and poverty [30]. BanQu has also piloted a blockchain-based scheme in Africa, helping Somali refugees create economic identities, and a long-term, secure economic profile intended to help these people access government services and financial aid [30]. BanQu is also providing small plot farmers, especially women farmers, land mapping in Latin America, where the absence of land rights is sometimes very conspicuous, and property registries are outdated, restricting access to financing [30].

2.8 Recommended Tool Box

The blockchain systems have many applications in the financial industry such as:

- The decentralization of markets through programmable money, which allows for the transfer of values, in addition to money via services and products such as loans, mortgages, titles, bonds, futures, etc., all of which are app-based [28].
- The system enables a context of fast and secure international commercial operations, without intermediaries, but contributing at the same time to the inclusion of new commercial actors in the ecosystem [28].
- This process cannot be applied to any situation though, but it is very attractive for various types of industries and sectors, for the financial services industry, the government, regulatory entities, and insurance companies. slockchain is also very attractive for carrying out cross-border business or international commerce, among many other possibilities; an example may be the following: the health sector, through the use of the blockchain network, intends to generate a medical record of patients, thereby seeking to increase efficiency in the diagnoses: providing doctors with a reliable, immediate and accurate record of all patient medical studies, where patients own their own information.
- In the case of the government, this system provides a decentralized and incorruptible record that provides transparency to the allocation of resources, public bidding, and to other public domains, such as voting [28].

For the financial sector, the impact and benefits have been numerous [28]:

- Among financial institutions, blockchain technology can help authenticate the identity and value of a transaction.
- In the transfer of assets, it can be useful in payment systems, money transfers, and transactions of goods and services.
- It has an impact on supply chains, on asset storage such as cryptocurrencies, all types of financial and digital assets, commodities, derivatives, and in government and corporate bonds.
- It has an impact on credit and debit card services, on asset titles and other types of credits, as much as in the exchange of securities, speculation, hedging, capital appreciation, interest, income, and dividends, among others.
- It also has an impact on guarantees, insurance, asset protection, property, life, health insurance, etc. [63].
- It can also have applications in payments abroad and in the administration of remittances.
- The blockchain is also useful in securing the identity of the clients, in platforms for the management and distribution of funds, in credit rating, in mortgage loans, in settlement platforms, compensation, trust funds, and pensions, among others (Don Tapscott cited by [28]).

For more information on Blockchain, please refer to the chapter "Section Security, Chapter Bitcoin, Blockchain, Smart Contracts and real use cases" in this same publication.

2.9 Summary and Conclusions

The process of Globalization for the past couple of decades has brought about dynamics that have no less than revolutionized many areas in technology, economics, communication, politics, education, and health, but probably nowhere more than in financial services. The full array of banking services has been completely shaken, not just by the arrival of new entrants into the financial system, namely fintech startups, fintech companies that have disrupted the entire financial services industry, but also customers' preferences and expectations. All have had a hand in reshaping the concept of modern-day banking and digital financial services. A very important element not to be taken lightly is the regional and cultural factors as well as the realities of the emerging markets that determine the particular fintech strategies that have most success in a particular community with specific needs, and which are responsible to the type of products and services that are available to the users, and it is this exposure that also shapes customers' preferences and expectations.

Financial consumers' expectations have changed overtime, they changed at the outset of globalization, they changed again with digitalization, and again with the outbreak of the Covid-19 pandemic; evidence indicates that consumer's preferences and expectations will keep changing as they are exposed to better financial products and services, or when they are simply exposed to any technological advancement that can be applied to the financial services industry.

Blockchain has raised a lot of expectations among financial consumers; although it already is one of fintech's technological foundations, it has a huge potential, along with quantum computing and robo-advisory, to bring the fintech industry to the next stage in its evolution, and lay the foundation for future financial services ecosystems [60]; another factor that will also have a voice in shaping fintech's future is the common everyday consumer, with his preferences and his behavior as a consumer, and ultimately, his expectations derived from his own digital experience, including human digital interface, on the one hand, and on the other, his expectations for a more human, more sensitive and friendlier digital interlocutor.

References

1 (2019). *Fintech en el mundo: la revolución digital de las finanzas ha llegado a México*. México: Bancomext, Pilar Madrazo, coordinadora editorial. 94 pp.

2 Kitao, Y. (2018). *Learning Practical Fintech from Successful Companies*, 163. New Jersey: Wiley.

3 Arjunwadkar, P.Y. (2018). *FinTech. The Technology Driving Disruption in the Financial Services Industry*, 261. CRC Press, Taylor and Francis Group.

4 Simon, J. (2020). 2020's best mobile banking apps. *Smartasset*.

5 Nicoletti, B. (2017). *The Future of Fintech. Integrating Finance and Technology in Financial Services*. Switzerland: Palgrave, MacMillan. 328 pp.

6 (2017). Whitepaper 2.0 on Distributed Ledger Technologies, Hong Kong Monetary Authority.

7 Delbridge, E. (2021). The five best banking apps of 2020. *The Balance* (August).

8 Pritchard, J. (2020). The best banks online. *The Balance*.

9 Greenspan, M. and Fox, S. The complete guide to Fintech trading and investments. In: *Building Wealth in the 21st Century*. eToro LTD. 261 pp.

10 Keyuan, Z. (2004). SARS and the rule of law in China. In: *SARS Epidemic. Challenges to China's Crisis Management* (ed. J. Wong and Z. Yongnian), 99–122. World Scientific. 226 pp.

11 Mills, K.G. (2018). *Fintech, Small Business and the American Dream*. Cham: Palgrave McMillan. 202 p.

12 The Wealthtech Book (2018). *The Fintech Handbook for Investors, Entrepreneurs and Finance Visionaries* (ed. S. Chishti and T. Puschmann). United Kingdom: Wiley. 318 pp.

13 High, M. (2020). *Fintech Profile: Marcus, by Goldman Sachs*. BizClik Media Limited.

14 Chishti, S. and Barberis, J. (2016). *The Fintech Book*. Wiley. 315 pp.

15 Digital Disruption (2016). Citi, GPS: global perspectives & solutions.

16 Hinchliff, R. (2020). *HSBC Joins Alibaba to Offer Hong Kong's Online Merchants Quick Loans*. Informaconnect.

17 India's Fintech landscape. 2016.

18 Fintech, U.K. (2016). *On the Cutting Edge*. London: HM Treasury. 130 pp.

19 Barbieri, J. and New Frontiers (2015). Fintech in Hong Kong. In: *The Fintech Capital*, 5e. London. 112 pp: Harrington Star.

20 Rubini, A. (2017). *Fintech in a Flash*. London: Simtac LTD. 235 pp.

21 Wong, P. (2020). For traditional banks, Fintech newcomers offer opportunities not threats. HSBC Group; August, 2019 World Economic Forum. How Fintech can help SMEs recover from the impact of Covid-19.

22 Agarwal, P., Chatterjee, S., and Agarwal, P. (2017). Digital Financial inclusion and consumer capabilities in India. In: *A Handbook for Financial Service Providers*. Institute for Financial Management and Research IFMR, Commissioned by J.P. Morgan. 131 pp.

23 The Fintech Scoping Review (2016). *Esatablishing Scotland as a FinTech Centre* (May 2016), Deloitte LLP, United Kingdom. 103 pp.

24 Lloyd's Banking Group (2020). *Lloyds Banking Group Enters Strategic Partnership with Fintech*. Lloyds Banking Group. Otto Benz, Director, payments technical services at Lloyd's Banking Group.

25 Girling, W. (2020). *Deutsche Bank Partners with Google to Transform Banking*. BizClik Media.

26 Bell, M. (2019). Fintech Raisin and Commerzbank launch savings platform for corporate clients. In: *Press Release Raisin and Commerzbank*. Berlin: Maggie Bell.

27 Nicholas Comfort (2019). Deutsche Bank wants in on the fintech start up scene, so it's going incognito. *Los Angeles Times* (December).

28 Bancomext (2018). *México, Nación Fintech. Coord. J. Santiago Rodríguez Suárez y Mariana V. Morales Rodríguez*, Bancomext, México. 188 pp.

29 Ernesto Ibarra, S. and Sánchez, P.L. (2019). *Fintech: regulación para la inclusión financiera en México*. México: Asociación Latinoamericana de Internet (ALAI).

30 Blackstad, S.e.A. (2018). *Fintech Revolution, Universal Inclusion in the New Financial Ecosystem*. Switzerland: Palgrave MacMillan. 406 pp.

31 Sánchez, A.y.B.C. (2017). Digital services in the 21st century. In: *A Strategic and Business Perspective* (ed. N.K. Cheung). New Jersey: Wiley, IEEE Press, and The Communications Society (ComSoc). 206 pp.

32 Ryan, D. (2014). *Understanding Digital Marketing*. London: Kogan Page. 410 pp.

33 Rory Macmillan and Bill & Melinda Gates Foundation (2016). Digital financial services: regulating for financial inclusion – an ICT perspective. In: *Regulatory & Market Environment*. Switzerland: ITU Development Communication Bureau. 65 pp.

34 (2015). Financial consumer demands for tomorrow's digital bank. In: *Financial Consumer Survey*. CGI Group INC. 42 pp.

35 Blanco, I. (2020). *La revolución de la tecnología*. en CIO.

36 Uniradio and Amigon, A. (2021). *Mexicanos optaron por pagos digitales en pandemia en vez de efectivo, revela Banxico*. NotiPress.

37 Albarrán, E. and Mor elos, Y.M. (2021). *La ley Fintech se atasca por burocracia a tres años de su promulgación*. El CEO.

38 Ikeda, F. (2020). *La revolución en el mundo de los pagos online ha llegado*. trendTIC.

39 Money 20/20 (2020). Payments innovation.

40 Zheng, J. (2020). SARS-CoV-2: an emerging Coronavirus that causes a global threat. *International Journal of Biological Sciences*. (Ilyspring International publishing 2020). Published online 2020 Mar 15. PMCID. PMC7098030. PMID. 32226285 10 (10): 1678–1685. https://doi.org/10.7150/ijbs.45053.

41 World Health Organization (2020). Coronavirus disease (COVID – 19. Situation Report – 138, 17 pp.

42 SARS (2006). *How a Global Epidemic was Stopped*. World Health Organization. 307 pp.

43 Sahut, J.M., Dana, L.P., and Laroche, M. (2020). Digital innovations, impacts on marketing, value chain and business models: an introduction. *Canadian Journal Administrative Sciences* 37: 671–667.

44 World Health Organization (2020). Novel Coronavirus (2019-nCoV) Situation Report – 1, 5 pp.

45 Beltz, L.A. (2011). Emerging infectious diseases. In: *A Guide to Diseases, Causative Agents, and Surveillance*, 734. San Francisco, CA: Jossey Bass, A Wiley Imprint.

46 National Research Council (2001). *Improving Disaster Management*. Washington DC: The National Akademies Press. 277 pp.

47 Smil, V. (2008). *Global Catastrophes and Trends. The Next Fifty Years*. Cambridge, MA: The MIT Press. 307 pp.

48 Bogdandy, B., Tamas, J., and Toth, Z. (2020). Digital transformation in education during Covid-19. *A Case Study in the 11th IEEE International Conference on Cognitive Infocommunications-CogInfoCom2020,* Mariehamn, Finland (23–25 September 2020).

49 Carbajo-Martín, L. (2020). Covid-19 y su oportunidad de transformación digital. *Revista Clínica de Medicina de Familia* 3: 177–179.

50 Kim, R.Y. (2020). The impact of Covid-19 on consumers: preparing for digital sales. *IEEE Engineering Management Review* 48 (3). Third Quarter: 212–218.

51 Hoyer, W.D. and MacClinnis, D.J. (2008). *Consumer Behavior*. South-Western CENGAGE Learning. 493 pp.

52 Noel, H. (2009). *Consumer Behavior*. United Kingdom: Ava Publishing S.A. 167 pp.

53 Alex Kreger (2021). Post-Covid digital strategy; banking customer experience trends of 2021. *Finextra* (January).

54 Levertov, A. (2021). Consumer expectations for financial services have changed-make sure you can meet them. *Forbes*. (September).

55 Skinner, C. and Digital Bank (2014). *Strategies to Launch or Become a Digital Bank*. MC Marshall Cavendish Business. 315 pp.

56 Ghosh, P. (2021). Three major banks plan more branch closings as thousands shutter – in the US and UK- amid Covid, digital growth. *Forbes*. (August).

57 Shaping the Digital Enterprise (2017). *Trends and Use Cases in Digital Innovation and Transformation*. Switzerland: Springer. 335 pp.

58 Lynn, T., Mooney, J.G., Rosati, P., and Cummins, M. Disrupting finance. In: *Fintech and Strategy in the 21st Century*. Switzerland: Palgrave MacMillan. 175 pp.

59 Lewis, A. (2018). *The Basics of Bitcoins and Blockchains*. Florida: Mango Publishing. 399 pp.

60 Swan, M. (2015). Blockchain, O'Reilly, California, 129 pp.

61 Gates, M. (2017). *Blockchain. Ultimate guide to Understanding Blockchain, Bitcoin, Cryptocurrencies, Smart Contracts, and the Future of Money*. Wise Fox Publishing. 60 pp.

62 Soze, K. (2017). *Blockchain, Novice to Expert*, vol. 1. Sabi Shepherd Ltd. 171 pp.

63 Dhillon, V., Metcalf, D., Hopper, M. et al. (2017). *Blockchain Enabled Applications*. Apress. 218 pp.

Section 2

Strategy

3

Strategic Roadmapping for Technological Change

Eduardo Ahumada-Tello[1], Oscar Omar Ovalle-Osuna[2], and Richard D. Evans[3]

[1] Faculty of Accounting and Management, Autonomous University of Baja California, Tijuana, Baja California, Mexico
[2] Faculty of Sciences in Engineering, Management and Social Studies, Autonomous University of Baja California, Tecate, Baja California, Mexico
[3] Faculty of Computer Science, Dalhousie University, Halifax, Nova Scotia, Canada

3.1 Fundamentals of Technology Management, Innovation, and Strategy

3.1.1 Business Strategy

As markets become more competitive, firms must create mechanisms and guidelines that allow them to clarify and predict their future. As firms attempt to respond to the changes that arise in competitive markets with a high number of competitors, they must create a culture of innovation that is rooted in their operations. In such turbulent environments, firms must take an innovation and technology-based approach to competition; similarly, they must show intent and direction toward the products they are developing. By doing so, firms can create a positive change and, it is by adopting such an approach, that the goods and services produced by firms can become accessible to the largest number of customers and end users. In food production, for example, many firms have innovated their production processes to allow them to increase productivity [1].

Numerous industries, including medical, transportation, and construction, have benefited from technological innovation and change. However, any initiative related to these factors is bound to have a positive effect on a firm's competitiveness. For firms to benefit the most requires them to plan, analyze, and implement differentiated strategies from their competitors. In doing so, technology and innovation can integrate with firms' management processes [2].

Many firms have ill-defined strategies – this means that they do not adequately know or understand the opportunities or threats that may arise when carrying out process changes derived from innovation and technology [3]. Much of the time, firms may not even know what skills, abilities, knowledge, and internal resources they possess [4]. It is challenging, therefore, to set goals that are compatible with a firm's vision and to manage the innovation process while, at the same time, improving the technology used for creating products [1]. As a result, it is necessary for firms to plan the foundations of their strategy; this is required to develop a complete view of the philosophical foundations of the firm, such as their mission, vision, and corporate values. Subsequently,

IEEE Technology and Engineering Management Society Body of Knowledge (TEMSBOK), First Edition.
Edited by Gustavo Giannattasio, Elif Kongar, Marina Dabić, Celia Desmond, Michael Condry, Sudeendra Koushik, and Roberto Saracco.

firms must conduct an internal and external analysis using appropriate tools to understand fully the state of their operation. In the following sections, we introduce (i) the elements needed to analyze the requirements and (ii) the actions necessary to complete the process of analyzing the firm and generating key activities based on its resources, capacities, competencies, and abilities [5–7].

3.1.2 Strategy Formulation

Nowadays, more than ever, firms must be flexible and adapt quickly to changes caused by technological advancement. Concepts such as strategic positioning, which were previously considered a fundamental approach to competitive strategy, are now losing relevance in the face of technological advancement and digital transformation. Firms must, therefore, benchmark with competitors to detect necessary changes which can affect their long-term viability and, to the best of their ability, should focus on activities that provide them with value and resign those that do not. Ultimately, they should focus on creating competitive advantage, which allows them to be superior to competitors [8].

We define a strategy as the decision process that a firm uses to achieve its vision and objectives. Firms must create and maintain a strategic direction that integrates their functional areas; these include management, marketing, finance, accounting, production, operations, research and development, and information systems [9]. The responsibility of maintaining an aligned strategy that involves technology and innovation in the other functional areas is part of the firm's actions to improve performance. From this, firms can create a sustained competitive advantage that is difficult for competitors to assimilate.

Scholars provide varying definitions to the term planning, but one of the most significant definitions is: "It is the design of the desired future and the effective means of realizing it" [10]. Once the strategic direction of a firm is defined, it can proceed to develop its strategic planning process. This process evaluates the firm from an internal and external perspective to propose an action plan which is monitored and supervised by the firm's management. This involves the process of planning to compete, followed by the formulation, implementation, and evaluation of the strategy used to manage the firm's value. If the strategy is successful, it continues with implementation; otherwise, it can make adjustments to respond to market demands [11, 12]. Figure 3.1 illustrates the steps required to develop a strategy, including its formulation, implementation, evaluation, and feedback.

The raison d'être of a firm's strategy is to improve its competitive capacity. In industries such as manufacturing, firms must carefully plan their operations to achieve success; for example, by not planning, firms may overorder on stock, spend too much on resources, or not achieve an expected return on investment. It is for these reasons that we propose a conceptual model to analyze the competitiveness of firms in the following sections.

3.1.3 Competitiveness Models for Firms

The core goal of a firm is to achieve its corporate vision and successfully compete against others in the marketplace. To do this, firms must be able to satisfy the needs and expectations of the customers and end users of their products or services. Competitiveness is the ability of firms to produce goods that are of better quality and/or lower cost than those of their competitors or have unique features that are highly valued by consumers [5]. As such, competitiveness can be defined as generating the required amount of production for market demand at the lowest possible cost.

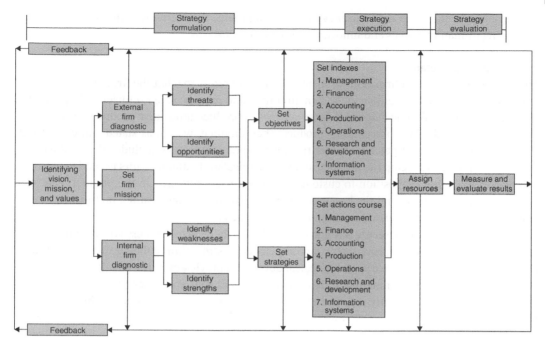

Figure 3.1 Strategic process development.

Four fundamental attributes determine a firm's competitiveness, namely factor conditions, demand conditions, related and support industries, and the strategy, structure, and rivalry of competitors. Such attributes and their interactions explain how firms innovate and remain competitive [6]. To innovate, however, often involves social costs. In Mexico, for example, competitiveness has existed in the maquiladora sector since the 1980s due to low employee wages and government infrastructure for industries [13]. China, on the other hand, creates greater industrial production, mainly due to its labor force and low wages too and a political environment that creates opportunities for firms. To analyze this situation, we must study the educational system, technological and communications infrastructure, labor laws, public institutions, financial system, cultural uses and customs, among other determining aspects [14].

Ultimately, competitiveness refers to the ability of firms to produce products or services efficiently by reducing costs and increasing quality and attractiveness, both inside and outside of the country of production. To achieve this, firms must increase productivity so that profitability also increases. A necessary condition for this is the existence, in each region, of a stable macroeconomic institutional environment that transmits trust, attracts capital and technology, and generates a productive and humane environment that allows firms to absorb, transform, and reproduce knowledge and technology, as well as have the ability to adapt to changes in the international context to export products efficiently. Countries and governments that provide such environments have proven to be the most dynamic in world markets [15].

The presence of resources (e.g. human, economic, or natural) is what defines the competitiveness of a region. To conclude our understanding of competitiveness, we must consider a region's or country's prosperity as a reference. The quality of life of citizens has an essential relationship with the productivity of their economy. It relates to the value of goods and services produced concerning human, economic, and natural resources. Productivity also depends on the value of a region's products and the efficiency of producing them. Productivity, therefore, depends on mobilizing human resources and innovation as a culture [16].

3.1.3.1 Core Competence

The core competence of a firm is in its business units and refers to what the firm produces better. Business units can be competitive in their environment and must be able to efficiently improve the skills, techniques, and knowledge of their employees to produce innovative products. In so doing, firms must know the best processes or products they produce, which, as stated, are considered their primary or core competence. Under this focus, core competence includes the product, the processes required to develop the product, and the technical and human factors needed to improve and deliver the value proposition to customers and end users reflected in the product. It is also essential that they understand the best sector in which to grow. If they cannot do this, they must redefine their objectives to create a competitive advantage [17].

The Japanese electronics sector is a good example of where firms possess strong core competence. Firms in the sector improve their core competence by streamlining their growth strategy and focusing on creating excellence in technological development. However, the creation of core competence is subject to environmental variations and changes that arise over time. It is necessary, therefore, that firms constantly redefine and strengthen their core competence; otherwise, competitors can identify, imitate, or nullify firms' competence. Effective management of knowledge requires data about the core competence that the firm generates [18]. A firm that uses strategies that are similar to their competitors will enter a state of vulnerability and, therefore, should develop unique core competence. We identify five concepts that shape the central competence of firms [19]:

- The basis of competitiveness is competition between companies;
- Structure the competition in a portfolio of competencies, including their products and core business units;
- Business unit is the repository of core competencies;
- Allocation of resources, businesses, and skills is the unit of analysis: senior management allocates capital and talent;
- The added value of senior management, architecture strategy, and the core competence of a firm secures its future success.

In conclusion, core competence can range from the products and services you create and sell to consumers, to the people who make up the firm and represent that difference in the market and mark a competitive advantage in the sector [20, 21].

3.1.4 Introduction to Technology Roadmapping

Technology Roadmapping (TRM) is an important tool that has, in recent years, become an integral part of creating and implementing strategies for technology development, research, and innovation in many firms. The graphical, descriptive, analytical, and collaborative nature of technology roadmaps becomes a supporting tool for strategic alignment and dialogue between functions in a firm and between organizations, such as multi-tier supply chains. A complete technology roadmap addresses the identification, selection, acquisition, development, exploitation, and protection of developed technologies (i.e. products, processes, and infrastructure) necessary to achieve, maintain, and grow market position and business performance commensurate with the strategic objectives of the organization [1–4]. Technology roadmapping is used extensively nowadays by firms to chart technology issues that are important to their future success. Just as a map shows us our starting point and where we want to go, as well as critical choices in-between, so do technology roadmaps [22, 23].

The underlying principles of a technology roadmap generally relate to three characteristics as they provide firms with an illustration of the current state, a desirable future state, and strategies to achieve the future state [24]. We provide a set of key benefits of technology roadmapping as follows:

- Establishes alignment of commercial and technical strategies;
- Improves communication between teams and the wider organization;
- Provides the ability to examine potential competitive strategies and ways to implement those strategies;
- Enables efficient time management and planning;
- Identifies gaps between technology, market, and product intelligence;
- Allows for the prioritization of investments;
- Establishes competitive and rational objectives;
- Guides and leads project teams; and
- Allows for the visualization of results, including objectives, processes, and the progress of projects.

From the identified benefits, we can understand that technology roadmapping allows firms to define a path for increasing the relevance of technology in all the processes and their infrastructure [25]. These changes must have at least four approaches to integrate a complete solution:

1) Use emerging Information Technologies (IT) in every corporate process;
2) Focus on innovation, research, and development;
3) Analyze, using different methodologies, the strategy, competencies, competitiveness, and human factors that drive improved firm performance; and
4) Analyze all business units or departments of firms (e.g. production, marketing, logistics) to include specific goals.

3.1.5 Technology Roadmapping

To develop sustainable competitive advantage, firms must deploy innovation and technology projects, which is not a simple process. Good practices indicate that to reach a successful conclusion, it is important to have technology-based strategies to help coordinate technologies effectively, their use within the firm, and the maturity of the firm's processes to complete the management strategic plan [26]. It is noticeable how new projections in companies affect their strategic propositions. Industry 4.0 and the digital transformation of an organization show the importance of developing a technology approach to a firm's strategy. This will affect every functional area of the organization, including technical and human factors assets. All of them must be included, as much as possible, in the technology roadmapping.

Technology roadmapping, in its simplest form, refers to one of the strategic objectives usually derived from the general strategic plan of a firm. It is also related to the future of the departments in firms and is generally built on a defined timeline containing milestones, tasks, activities, and resources that search to obtain an acceptable competitive performance result in a firm or company being prepared with the knowledge and tools required to manage and adapt to changes [25].

Scholars have provided several definitions of technology roadmaps that include the following:

1) Prior research contemplates the visions of possible future technological developments, products, or contexts;
2) Technology roadmaps are a fundamental tool for planning the technological and scientific resources of an organization;
3) Technology roadmaps act as a high-level planning tool that helps adopt technology management and planning in the firm;
4) Roadmapping involves a planning process aimed at the future needs of the market. It helps firms identify, select, and develop technology options to meet future services, products, or operational needs;
5) Technology roadmaps are management tools that help plan and forecast the steps necessary to achieve one or more technology goals;
6) Roadmaps act as a planning tool that facilitates the science environment, technology priorities, and the integration of new technologies with business, policy, and the needs of the market.

Considering the above definitions, we can broadly define technology roadmapping as a technological and strategic planning tool that helps firms identify, select, and develop technological alternatives to satisfy future market needs and achieve their goals [8, 27]. To create efficient technology roadmaps, diagrams can be used to illustrate how to control actions while improving the technological conditions of the firm. The main and most representative uses and benefits of TRM are listed as follows:

- Develops an internal consensus about the technological needs and requirements to satisfy market demands;
- Represents a mechanism that helps firms forecast technology development in desired areas (goal/objective);
- Provides a structure to coordinate technological development with the firm;
- Defines the technological goals that refer to the desired future;
- Helps firms predict future market technologies;
- Defines the "path" that firms take to compete successfully;
- Guides technological research and development decisions;
- Increases collaboration and allows for the sharing of stakeholder knowledge; and
- Helps firms take advantage of future market opportunities.

3.1.5.1 Phases Required to Develop a Technology Roadmap

To develop a technology roadmap requires an action plan [26]. Typically, a roadmap is defined as a three-phase process, namely preliminary activities, development of the plan, and monitoring of its effectiveness. Roadmaps are often divided into six central phases, which are incremental allowing firms to improve the structure of the process, as well as maintaining the development of the strategy [28, 29]. The suggested phases to develop these actions are as follows:

3.1.5.1.1 *Phase 1: Identify Objectives and Goals* The first phase in producing your technology roadmap is to identify your objectives and goals. This phase involves planning, scheduling, organizing, and controlling changes in your technological environment. To do this effectively, you must have a clearly defined purpose. With your objectives defined, you can more easily chart the most suitable path to achieve your vision.

3.1.5.1.2 *Phase 2: Focus on Innovation and Research and Development* As your firm develops its roadmap, competing firms will not stop. Suppose you go too far in planning and defining the action plan or ignore the industry's innovation as your plan's central approach; in this scenario, you will

not achieve the expected results due to technological developments that occur every day. Keeping an eye on the actions of opinion leaders in your industry is important as you try to develop an idea of where your industry or technology is heading in the future. Even as speculative as it is, any change in the industry is possible, and when it happens, the entire proposal should be reassessed with changes made.

3.1.5.1.3 Phase 3: Identify Stages and Organize Priorities In this phase, the development of your roadmap is divided into incremental stages with small tasks being completed to achieve a larger goal. These incremental stages should be identified and arranged in order of priority. Some tasks will be necessary before starting other projects, while other tasks will provide immediate business value. Securing the hierarchy to create benefits in places of obstacles is part of this phase.

3.1.5.1.4 Phase 4: Cost Analysis During the process of creating your roadmap, it is crucial that you complete a cost analysis using real costs; otherwise, you may artificially limit your scope or strategy for fear of going over budget. In this phase, you must calculate your estimated costs and review the different stages and priorities. If necessary, you can adjust due to exceeding costs.

3.1.5.1.5 Phase 5: Project Timelines The phases of your roadmap are likely to be successful if you place them on a timeline. Timelines may not go according to plan, but a timeline is required to improve your chances of success. The deadlines set must be flexible, and permanent supervision is necessary.

3.1.5.1.6 Phase 6: Assign Staff The last phase in developing your technology roadmap is to assign personnel to the various stages described in Phase 3, who become responsible for the roadmap's execution. You should divide this responsibility among employees. If your firm does not possess the appropriate skills to complete a stage, you must devise another plan. If necessary, you can consider adding new stages or seeking third-party help to complete the necessary stage(s).

3.1.5.2 Business Strategy and Technology Roadmap Alignment

To guarantee the successful implementation of a technology roadmap, firms must make it part of their business strategy. Firms should consider their roadmap when developing strategic actions and when analyzing resources and collaboration with other strategies [30].

3.1.6 Examples of Technology Roadmap Development and Implementation

In recent years, technology roadmapping has been widely adopted across a variety of industries. Similarly, extant research has reported successful implementations of roadmaps in Table 3.1, and we provide some examples with references to the corresponding study.

Table 3.1 Examples of implementing technology roadmaps.

Industry or approach for implementing TRM	Year	Reference
Methodology	2004	[24]
United Kingdom	2004	[23]
Energy services sector	2008	[22]
Motorola technology roadmap process	1987	[31]
Strategic management	2017	[27]
Semiconductor applications	1997	[32]

3.2 Recommended Toolbox and Benefits

3.2.1 Drafting a Technology Roadmap

Every technology roadmap must include a thorough analysis of where the firm's internal and external environment currently is and a projection of where it would like it to be at the end of the implementation of the roadmap. The more thorough these analyses are, the more organized the roadmap will inevitably become [33]. A firm's analysis should cover the following:

- Performance and age of firms assets;
- The business expectations of its value proposition to customers;
- System dependencies;
- Application dependencies;
- Current costs and return on expenses; and
- Operational procedures.

By completing this analysis, it allows firms to contrast between where they are and where they hope to be. Firms must identify where in the environment they can find a lack of steps and the actions required to move where they want to be as a successful organization.

3.2.2 Diagrams

The use of diagrams for technology roadmapping has become widespread across industries. Diagrams are used to help firms carry out and visualize the development of their strategy. It is important that firms understand how to propose technological improvements, and it is necessary that they appreciate the interrelationships between different activities, tasks, goals, and dates. For this, it is convenient to analyze each activity, define contact points with other processes, and identify the sub-processes in defined activities. By doing this, it allows firms to reveal existing technological problems, which encourage improvements. Process diagrams or maps are the methods used to represent these processes [26]. There are, of course, some disadvantages pointed out in this development of graphical processes or diagrams, including:

- They can be distracting; firms can become more concerned with map integrity than speeding up their actions;
- They may lose relevance for employees if not implemented correctly;
- Inadequate preparation may cause them not to be used appropriately for communication with senior management; and
- Sometimes, if taken as rigid elements, management may not perceive the possibility of change.

However, benefits do exist, which include:

- They are useful for explaining key project goals to stakeholders;
- They show the processes involved in production clearer than using words;
- By making diagrams, workers can more easily understand the tasks and problems being experienced by the firm; and
- They are useful tools for the redesign of corporate procedures.

3.2.3 Why Should I Draft a Technology Roadmap?

3.2.3.1 Staying Competitive

A firm can perform better than its competitors only if it can differentiate itself. For firms to stay competitive, they must use their knowledge, human and technological resources, and tangible and intangible assets. The intended use of technology roadmaps is to align the firm's strategy with the technology strategy described in them. The analysis and selection of the strategy mainly involve making subjective decisions based on objective information to stay competitive. In this process, organizational aspects (e.g. political, cultural, ethical, and social responsibility) are included when formulating the competitive strategy of a firm, and this proposal is based on technology and knowledge [34].

3.2.3.2 Predictability of Expense
The expenses generated by the firm departments of a firm can be difficult to predict [9, 27, 30]. Hardware may contain parts that, when failing, cause considerable cost to replace and possibly affect the production or manufacturing process. Technology roadmaps provide firms with an approximation of the budget. All preparation is subject to events that can alter the budget, but this is the basis of good planning.

3.2.3.3 Ensuring Firms Meet Regulatory and Compliance Needs
Technology roadmaps allow firms to be prepared for any regulation, compliance, or requirement requested in the operation of the technological strategy.

3.2.3.4 Attract Better Talent
When the vision of a firm is clear and stakeholders know what is expected, firms must hire and retain qualified personnel to carry out the business activities. Technology roadmaps clarify these human requirements by identifying the functional departments and/or employees responsible for each activity. Being clear about the current and future requirements of the firm provides the possibility for professional growth; a strategy that makes working for the firm attractive will source talented workers more efficiently. In this regard, firms should aim to become a learning organization, continuously developing and transforming their employees so that they can remain competitive. Similarly, from a leadership perspective, technology roadmaps should encourage buy-in from all stakeholders and, therefore, their simplicity in understanding is paramount to their success.

3.2.3.5 Using Employee Time Wisely
For firms, it is easy to lose their way due to unplanned problems arising during business-as-usual. These problems can lead to timewasting if problems are not appropriately addressed. Technology roadmaps assign long-term objectives for the department, which guarantee that employees maintain an organized approach, ensuring problems are delivered on time and budget. If roadmaps are too ambitious or make false promises, employees' morale will reduce and other stakeholders, such as customers, will become less confident in the firm.

3.2.3.6 Improving the Relationship Between IT and the Firm
The IT department is sometimes seen as a barrier to organizational development when it should be a success factor. The accumulation of fault repair activities can become an impediment to meet the requirements of other departments. This happens when updates arrive in systems and the IT department cannot fulfill the requests. Technology roadmaps guarantee that IT departments meet the firm's needs in a stable, secure, and profitable manner. The improvement of communication between departments increases capabilities and reduces limitations. By creating a roadmap, firms can prioritize activities and assign schedules to employees, allowing them to increase firm performance and chances of success.

3.3   Realistic Examples and Use Cases

Firms must develop a strategy that can be integrated with the technology and innovation activities they carry out. At this juncture, the use of tools, such as technology roadmaps, helps them to adopt new technologies. Faced with new ways of working, the fundamental role of technology and innovation is recognized. When firms make decisions, they do so based on information processed from the use of mechanisms that manage data and seek to improve the strategic decision-making process in the medium and long term (see Figure 3.2).

Firms often use roadmaps to plan and manage their technology-related projects and other non-related projects like human resources and cultural interrelations. Technology roadmaps can come in a variety of forms, such as word-processed documents, slide decks, or project timelines depicted in spreadsheets. Some companies choose to develop one-off visual representations of their roadmap, while others choose several documents that constitute their roadmap; essentially, the design of the roadmap document(s) is far less important than its ability to communicate the value you intend to deliver to consumers and the way in which you intend to achieve your vision. For example, diagrams and guidelines are used by large technology-based companies, including Google, Amazon, Microsoft, Mayo Clinic, Apple, and Walmart, which have used them to achieve success. From these examples, we can see the importance of aligning corporate strategies and taking advantage of technology in a planned manner. A general template for developing a technology roadmap is provided in Figure 3.2.

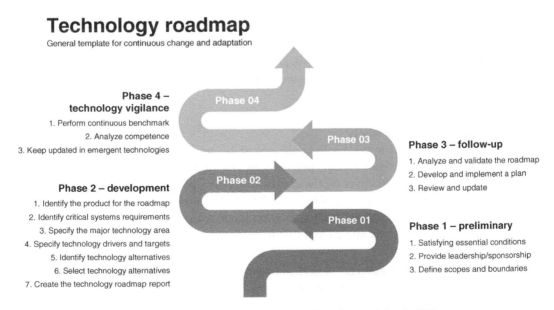

Technology roadmap
General template for continuous change and adaptation

**Phase 4 –
technology vigilance**
1. Perform continuous benchmark
2. Analyze competence
3. Keep updated in emergent technologies

Phase 2 – development
1. Identify the product for the roadmap
2. Identify critical systems requirements
3. Specify the major technology area
4. Specify technology drivers and targets
5. Identify technology alternatives
6. Select technology alternatives
7. Create the technology roadmap report

Phase 04
Phase 03
Phase 02
Phase 01

Phase 3 – follow-up
1. Analyze and validate the roadmap
2. Develop and implement a plan
3. Review and update

Phase 1 – preliminary
1. Satisfying essential conditions
2. Provide leadership/sponsorship
3. Define scopes and boundaries

Figure 3.2   Technology roadmap template. *Source:* Adapted from Bray and Garcia [32].

3.4   Summary and Conclusions

This chapter describes the importance of strategy and competitiveness as essential foundations in the development of a firm. Technology roadmaps are used as a strategic communication tool to improve firms' performance and alignment of their products with achieving their strategic goals. Technology has, therefore, become a fundamental part of all strategic planning and any technology-based firm. For plans to function correctly, firms must be open-minded and provide support to all involved in planning. Technological applications (e.g. cloud servers) that base firms' infrastructure outside the corporate boundaries are becoming increasingly common. However, technology roadmaps allow us to assess and plan access to these technologies while minimizing risk.

Technology roadmaps show, in a detailed manner, the route to producing the best possible results with limited impact and interruptions. Process diagrams must be kept under permanent supervision to avoid losing the timeliness of the strategies and to remain aligned with the firm's business strategy. The important thing is that there must always be a plan that keeps the strategy updated, evaluated, and following the environment, the firm's objectives, and the business strategy. Ultimately, technology roadmaps are used as a strategic guide that unfolds in an industry that increasingly depends on technology and where the selection of new ones is often chaotic and not always well planned. Although there are enormous amounts of technological solutions available for different business activities, technology roadmaps keep the business focused on its objectives and attentive to the changing nature of the industry [26].

References

1 Peteraf, M.A. (1993). The cornerstones of competitive advantage: a resource-based view. *Strategic Management Journal* https://doi.org/10.1002/smj.4250140303.

2 Barragan-Quintero, R.V., Ovalle-Osuna, O.O., Ahumada-Tello, E., and Evans, R.D. (2019). Measuring the Effects of Innovation in Wine Companies in Baja California, Atlanta, GA, USA (11–14 June 2019). http://doi.org/10.1109/TEMSCON.2019.8813746.

3 Ahumada Tello, E. and Perusquia Velasco, J.M.A. (2016). Business intelligence: strategy for competitiveness development in technology-based firms. *Contaduria y Administracion* 61 (1): https://doi.org/10.1016/j.cya.2015.09.006.

4 Tjemkes, B. and Mihalache, O. (ed.) (2021). Strategizing the unknown and transformative strategizing. In: *Transformative Strategies: Strategic Thinking in the Age of Globalization, Disruption, Collaboration and Responsibility*. Routledge. https://doi.org/10.4324/9780429274381.

5 Delgado, M., Ketels, C., Porter, M., and Stern, S. (2012). *The Determinants of National Competitiveness*. National Bureau of Economic Research. https://doi.org/10.3386/w18249.

6 Porter, M.E. (1990). *The Competitive Advantage of Firms in Global Industries*, 33–68. UK: Palgrave Macmillan https://doi.org/10.1007/978-1-349-11336-1_2.

7 Ahumada-Tello, E., Castanon-Puga, M., Evans, R. D., and Gaxiola-Pacheco, C. (2018). *Contributions of Knowledge Management to Firm Competitiveness from a Complexity Approach*, Chicago, IL, USA (28–30 June 2018). https://doi.org/10.1109/TEMSCON.2018.8488416.

8 Chen, C. and Zhou, X. (2022). Impact of enterprise strategic mode on technological innovation under information technology. *Lecture Notes on Data Engineering and Communications Technologies* 84: 290–299. https://doi.org/10.1007/978-981-16-5857-0_37.

9 Sehwa, W., Songquan, P., and Wenzhe, L. (2020). The development of strategic management and its relationship with innovation studies. *Towards the Digital World and Industry X.0 – Proceedings of the 29th International Conference of the International Association for Management of Technology, IAMOT 2020*, Cairo, Egypt (13–17 September 2020), pp. 451–465.

10 Ackoff, R. (2010). *El paradigma de Ackoff: una administración sistemática*. México: Limusa Wiley.

11 Hart, S.L., Milstein, M.B., and Caggiano, J. (2003). Creating sustainable value. *Academy of Management Executive* 17 (2): 56–69. https://doi.org/10.5465/ame.2003.10025194.

12 Nelson, R.S. III and Sandborn, P. (2012). Strategic management of component obsolescence using constraint-driven design refresh planning. *International Journal of Product Lifecycle Management* 6 (2): 99–120. https://doi.org/10.1504/IJPLM.2012.052656.

13 Bernardi, C. (2021). The maquiladora between neoliberalism and Mexican nationalism (1964–1988) [La maquiladora tra neoliberismo e nazionalismo messicano (1964–1988)]. *Confluenze* 13 (1): 270–298. https://doi.org/10.6092/issn.2036-0967/13105.

14 Flores, V. and Avendaño, M. (2016). Theoretical models of knowledge management: descriptors, conceptualizations and approaches. *c* 4 (10): 201–227.

15 Villavicencio, D. and Casalet, M. (2002). *{VIII}. Desarrollo tecnológico en las pequeñas y medianas empresas. Aproximaciones al caso de México*, 191–217. Centro de estudios mexicanos y centroamericanos. https://doi.org/10.4000/books.cemca.2668.

16 Kayal, A.A. (2008). National innovation systems. *International Journal of Entrepreneurship and Innovation Management* 8 (1): 74. https://doi.org/10.1504/IJEIM.2008.018615.

17 Ergün, C. (1997). *From Know-How to Know-Why*, 89–93. Netherlands: Springer https://doi.org/10.1007/978-94-015-8810-2_10.

18 Paoloni, M., Coluccia, D., Fontana, S., and Solimene, S. (2020). Knowledge management, intellectual capital and entrepreneurship: a structured literature review. *Journal of Knowledge Management* 24 (8): 1797–1818. https://doi.org/10.1108/JKM-01-2020-0052.

19 Prahalad, C.K. and Hamel, G. (1990). *The Core Competence of the Corporation*, 275–292. Springer-Verlag https://doi.org/10.1007/3-540-30763-x_14.

20 Barney, J. (1991). Firm resources and sustained competitive advantage. *Journal of Management* 17 (1): 99–120. https://doi.org/10.1177/014920639101700108.

21 Toro-Jarrín, M.A., Ponce-Jaramillo, I.E., and Güemes-Castorena, D. (2016). Methodology for the of building process integration of Business Model Canvas and Technological Roadmap. *Technological Forecasting and Social Change* 110: 213–225. https://doi.org/10.1016/j.techfore.2016.01.009.

22 Daim, T.U. and Oliver, T. (2008). Implementing technology roadmap process in the energy services sector: a case study of a government agency. *Technological Forecasting and Social Change* 75 (5): 687–720. https://doi.org/10.1016/j.techfore.2007.04.006.

23 Saritas, O. and Oner, M.A. (2004). Systemic analysis of UK foresight results. *Technological Forecasting and Social Change* 71 (1–2): 27–65. https://doi.org/10.1016/S0040-1625(03)00067-2.

24 Phaal, R., Farrukh, C.J.P., and Probert, D.R. (2004). Technology roadmapping – a planning framework for evolution and revolution. *Technological Forecasting and Social Change* 71 (1–2): 5–26. https://doi.org/10.1016/S0040-1625(03)00072-6.

25 Schiele, H., Bos-Nehles, A., Delke, V. et al. (2021). Interpreting the Industry 4.0 future: technology, business, society and people. *Journal of Business Strategy*. https://doi.org/10.1108/JBS-08-2020-0181.

26 Baez, C.A., Otto, R.B., da Silva, F.P., and Dechechi, E.C. (2019). Technology roadmap applied in projects. *Advances in Transdisciplinary Engineering* 10: 279–288. https://doi.org/10.3233/ATDE190133.

27 Arianto, B.G. and Surendro, K. (2017). Implementation of building process integration of business model canvas and technology roadmap for strategic management: case study: PT. XYZ. *2017 International Conference on Information Technology Systems and Innovation, ICITSI 2017 – Proceedings*, Bandung, Indonesia (23–24 October 2017), Vol. 2018, pp. 61–66. https://doi.org/10.1109/ICITSI.2017.8267919.

28 Brown, P., Von Daniels, C., Bocken, N.M.P., and Balkenende, A.R. (2021). A process model for collaboration in circular oriented innovation. *Journal of Cleaner Production* 286. https://doi.org/10.1016/j.jclepro.2020.125499.

29 Braun, A.-T., Schöllhammer, O., and Rosenkranz, B. (2021). Adaptation of the business model canvas template to develop business models for the circular economy. *Procedia CIRP* 99: 698–702. https://doi.org/10.1016/j.procir.2021.03.093.

30 Cho, Y., Yoon, S.-P., Kim, K.-S., and Chang, B. (2014). Industrial technology roadmap as a decision making tool to support public planning. *PICMET 2014 – Portland International Center for Management of Engineering and Technology, Proceedings: Infrastructure and Service Integration*, Kanazawa, Japan (27–31 July 2014), pp. 2986–2995.

31 Willyard, C.H. and McClees, C.W. (1987). Motorola's technology roadmap process. *Research Management* 30 (5): 13–19. https://doi.org/10.1080/00345334.1987.11757057.

32 Bray, O.H. and Garcia, M.L. (1997). Technology roadmapping: the integration of strategic and technology planning for competitiveness. *Innovation in Technology Management. The Key to Global Leadership. PICMET'97*, Portland, OR, USA (27–31 July 1997), pp. 25–28. http://doi.org/10.1109/PICMET.1997.653238.

33 Ahumada-Tello, E., Castañón-Puga, M., Gaxiola-Pacheco, C., and Evans, R.D. (2019). Applied decision making in design innovation management. In: *Applied Decision-Making*, vol. 209. https://doi.org/10.1007/978-3-030-17985-4_5.

34 North, K. and Babakhanlou, R. (2016). Knowledge management tools for SMES. In: *Competitive Strategies for Small and Medium Enterprises: Increasing Crisis Resilience, Agility and Innovation in Turbulent Times* (ed. K. North and G. Varvakis), 211–222. Cham: Springer International Publishing. https://doi.org/10.1007/978-3-319-27303-7_14.

Tutorials, reports, and websites

Technology Roadmapping

https://roadmunk.com/guides/how-to-create-a-technology-roadmap
https://www.productplan.com/learn/three-example-technology-roadmaps
https://www.ifm.eng.cam.ac.uk/uploads/Research/CTM/Roadmapping/roadmapping_overview.pdf

Strategic Planning

https://www.hcc.edu/Documents/About/Strategic%20Planning/Strategic%20Planning_%20A%20Ten%20Step%20Guide.pdf
https://www.gartner.com/en/insights/strategic-planning
https://ocw.mit.edu/courses/sloan-school-of-management/15-902-strategic-management-i-fall-2006/index.htm

Section 3

Leadership

4

Guided Continuous Improvement (GCI): The Key to Becoming a Disciplined Agile Enterprise

Scott Ambler

School of Computing, University of Leeds, Leeds, UK

4.1 Agile Methods/Frameworks Are Only a Good Start

Many teams, and sometimes entire organizations, start their agile journey by adopting agile methods such as Scrum [1, 2], Extreme Programming (XP) [3], or Dynamic Systems Development Method (DSDM)-Atern [4]. Large teams dealing with "scale" may choose to adopt frameworks such as SAFe® [5], LeSS [6], or Nexus® [7], to name a few. Throughout this paper, I will use the term framework to be inclusive of both methods and frameworks. These frameworks each address a specific class of problem(s) that agile teams face, and regardless of claims to the contrary, they are rather prescriptive in that they do not provide you with many choices. Being prescriptive makes them relatively easy to adopt, but as I show below, it makes long-term process improvement difficult.

Sometimes, particularly when frameworks are applied to contexts where they aren't an ideal fit, teams often find that they need to invest significant time "descaling" them to remove techniques that do not apply to their situation, then add back in other techniques that do. Having said that, when frameworks are applied in the appropriate context, they can work quite well in practice. When you successfully adopt one of these prescriptive frameworks, your team effectiveness tends to follow the Satir change curve shown in Figure 4.1 [8]. At first, there is a drop in effectiveness because the team is learning a new way of working (WoW), it's investing in training, and people are often learning new techniques. In time, effectiveness rises, going above what it originally was, but eventually plateaus as the team falls into its new WoW [9]. Things have gotten better, but without a concerted effort to improve, you discover that team effectiveness plateaus because they have hit the limits of the framework and find themselves in "method prison" [10].

Figure 4.1 differs from many of the claims of the agile community, such as the promise that you can do twice the work in half the time with Scrum [11], let alone the marketing rhetoric of 10× or greater improvements to be gained from other agile frameworks. Sadly, this claim of four times productivity improvement does not seem to hold water in practice. Although the adoption of an agile framework, or several frameworks, may prove to be positive this clearly should not be your end goal. What you really want is to become a Disciplined Agile® Enterprise (DAE), a learning

IEEE Technology and Engineering Management Society Body of Knowledge (TEMSBOK), First Edition.
Edited by Gustavo Giannattasio, Elif Kongar, Marina Dabić, Celia Desmond, Michael Condry, Sudeendra Koushik, and Roberto Saracco.

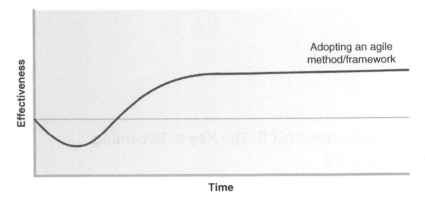

Figure 4.1 Team effectiveness when successfully adopting an agile framework.

organization that can sense and respond swiftly to changes in its environment. A learning organization is one that is skilled at creating, acquiring, and transferring knowledge, and at modifying its behavior to reflect new knowledge and insights. Agile enterprises – such as Amazon, Apple, Google, Tesla, and more – choose to continually improve which enables them to steadily increase their competitiveness.

4.2 Organizational Transformation: Context Counts

Context counts when it comes to organizational transformations. An organization that is relatively new to agile will need a different transformation strategy than an organization that has already adopted one or more agile frameworks. Adopting agile WoW [9] in your information technology (IT) department is a different proposition than adopting it within your Finance department, which is different yet again than adopting it within Marketing. But regardless of your initial transformation path, your end goal is always the same, to become a DAE.

Figure 4.2 summarizes a three-stage transformation strategy:

1) **Align:** Although it sounds obvious, you will start your transformation journey where you are – but do you know where that is? You must honestly assess your situation and identify your

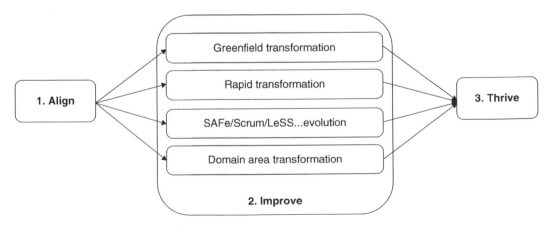

Figure 4.2 A contextualized transformation strategy.

strengths, your challenges, and the outcomes you hope to achieve (all of which will evolve over time). This in turn enables you to identify the best-fit improvement strategy for you. This may be a greenfield transformation if you are relatively inexperienced with agile, a high-risk rapid transformation strategy if you are in trouble and need to transform fast, an evolutionary strategy where you learn to improve upon existing frameworks such as Scrum or SAFe, or perhaps a tailored transformation strategy for one or more domain areas such as Finance, Marketing, or Procurement.

2) **Improve:** The focus of this stage is to improve in place via a fit-for-purpose strategy. You will train, educate, and coach your staff to give them the skills to follow new techniques. During this stage you want to improve, to get better, and you want to learn how to learn – in other words, you want to learn how to continuously improve.

3) **Thrive:** During this stage you have become a DAE that improves continually. You have normalized continuous improvement, and your teams have accepted responsibility for improving their WoW.

4.3 Continuous Improvement

A fundamental philosophy of agile is that teams should own their own process, or as we like to say in Disciplined Agile (DA) teams should choose their WoW [9]. This is easier said than done in practice. The challenge is that every team is unique and faces a unique situation – in other words, context counts. Furthermore, there are no "best practices," rather, every practice has tradeoffs and works well in some situations and poorly in others. Worse yet, you really do not know how well a technique will work for you until you try it in your environment. Given all of this, how can a team choose its WoW?

A common strategy is for teams to take an experimental approach to improvement. This is referred to as kaizen, where a team experiments with a small change in their WoW, adopting the change if it works in their given context and abandoning it if it does not. The goal of kaizen is often to reduce or better yet eliminate waste (muda) or to eliminate overly hard work (muri). Continuous improvement is the act of applying a series of small improvements to improve your WoW over time. Arguably the most famous strategy to do so is W. Edward Deming's plan-do-study-act (PDSA) improvement loop [12], sometimes called a "kaizen loop."

Figure 4.3 depicts the logic for a continuous improvement loop, expanding on Deming's PDSA steps to provide overall context. This approach is typically implemented by a team to improve their WoW but is also viable for two or more collaborating teams (more on this later) to improve their shared WoW.

Let us consider each step of Figure 4.3 in greater detail:

1) **Identify an issue:** A potential issue, perhaps an inefficiency in your current WoW or an unanticipated side effect of your WoW, becomes apparent to you. This may have occurred as an insight during a reflective session such as a retrospective, as the result of a customer complaint, as an observation, or simply as an insight that someone has. Regardless of the source, you have identified an issue that you hope to address.

2) **Identify potential technique(s):** The team identifies a technique, either a practice or strategy, that they believe will work for them. This may be something someone on the team has done before, something they have read about, or even an idea that they have identified on their own. An important consideration is to try a technique where it is "safe to fail" at it – you should not be putting your team at risk by experimenting with a new WoW. If an experiment

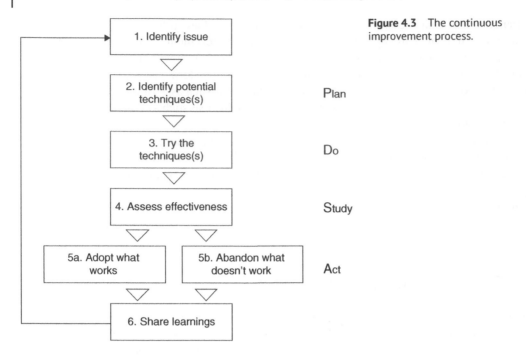

Figure 4.3 The continuous improvement process.

is deemed too risky to try, attempt to break it down into a series of smaller experiments that are less risky. It is also critical to identify the criteria against which you are going to assess the effectiveness of the technique. Ideally, most of the criteria is quantifiable in nature, although do not ignore qualitative measures too. Remember that your team can, and perhaps should when they become sufficiently skilled to do so, run multiple experiments in parallel, particularly when the potential improvements are on different areas of your process and therefore will not affect each other (if they affect each other, it makes it difficult to determine the effectiveness of each experiment).

3) **Try the technique(s):** The team needs to see how well the technique works for them in their environment. Do they have the skills to perform the technique? How well does it work given their technology platform? How well does it work given their organization and team culture? The team needs to give the experiment sufficient time to determine how well it works in practice. This could several anywhere from several days to several months, although in most cases several weeks is sufficient.

4) **Assess effectiveness:** After you have run the experiment, you should assess how well it worked for you. This assessment should be against the criteria that you agreed to at the beginning.

5) **Adopt or abandon the technique(s):** Sometimes you'll find that a new technique works incredibly well for you, and other times you'll experience the other extreme and discover that a technique is a failure for you. But in most cases, there will be some aspects of the technique that work well for you and other aspects that do not (an indication that maybe you need to consider running another experiment to identify a solution). Our advice is to adopt what works well for you and to abandon or better yet improve upon what does not.

6) **Share learnings:** When you learn something about a technique you should share it with others.

4.4   Guided Continuous Improvement

While the continuous improvement strategy works, it is possible to do better. The linchpin is the second step in the process of Figure 4.3, the identification of a technique to experiment with. The idea is to get better at identifying techniques that are more likely to work for you in the situation that you face. You will still have failed experiments, you are only human, but you'll have fewer of them because you are making better decisions. This increases your rate of process improvement – you succeed fast rather than fail fast. This is called "guided continuous improvement (CGI)" [9, 13]. GCI improves the kaizen strategy to use proven guidance to help teams identify techniques that are likely to work in their context.

There are three ways that improve how you identify techniques to experiment with:

1) **Hire an experienced coach (and listen to them):** Although it can be hard to find an experienced agile coach they do exist and if you are lucky enough to have them then follow their guidance.
2) **Apply the DA tool kit:** There are several ways you can apply the DA tool kit to help you to make better process decisions. Instead of prescribing the "best practices" you must adopt, DA instead advises you on process-related issues you need to think about and the options available to you. Such issues include what you should consider when choosing a life cycle, what you should consider when forming your team, and what you should consider when addressing changing stakeholder needs (to name a few important things). The DA tool kit then presents you with potential options and the tradeoffs associated with those options. This gives you an idea as to what techniques you might want to experiment with and what is likely to happen if you do, enabling you to make better process decisions. Later in this paper, several examples of doing exactly this are presented.
3) **A combination of both:** Good coaches have the humility to recognize that they do not know everything. As a result, they will apply the DA tool kit to help your team to make better decisions about new WoW to experiment with.

The improvement curve for (unguided) continuous improvement strategies is shown in Figure 4.4 as a dashed line. You can see that there is still a bit of a productivity dip at first as teams learn how to identify potential improvements and then run the experiments, but this is small and short-lived. The full line depicts the curve for GCI in context; teams are more likely to identify options that will work for them, resulting in a higher rate of positive experiments and, thereby, a

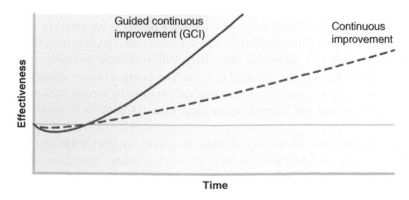

Figure 4.4   Comparing CI and GCI.

faster rate of improvement. In short, better decisions lead to better outcomes. Of course, neither of the lines in Figure 4.4 is perfectly smooth. A team will have ups and downs, with some failed experiments (downs) where they learn what does not work in their situation and some successful experiences (ups) where they discover a technique that improves their effectiveness as a team. The full line, representing GCI, will be smoother than the dashed line because teams will have a higher percentage of ups.

4.5 Applying Guided Continuous Improvement at Multiple Levels

Let us explore how to apply GCI-based improvement at multiple levels of complexity within your overall process improvement effort:

1) **Improve at the team level:** The challenge here is that without guidance teams will flounder and "fail fast" rather than "succeed fast." Team improvement is the domain of agile team coaches.
2) **Improve across disparate teams:** Many agile transformations have hit an "organizational glass barrier" in that they struggle to apply agile outside of IT. Agile got its start in the software development realm and as a result many agile coaches and experts come from that background, and unfortunately struggle with other areas of your organization. To be successful, you need to understand and respect the backgrounds and priorities of people working within these other areas. Improvement across disparate teams is the domain of senior agile coaches.
3) **Improve across a value stream:** Value streams, described below, require people from a wide range of organizational domains to work together effectively. To engender improvement across a value stream requires you to work with many disparate teams, increasing the chance that "organizational antibodies" will attack your efforts. Value stream improvement is the domain of value stream consultants and enterprise coaches.
4) **Improve across the organization:** This is similar to improving across a value stream, yet can potentially be harder because the customer focus that exists within a value stream may not exist across teams in different parts of your organization. This level of improvement is the domain of enterprise coaches and change management professionals.

4.5.1 Guided Continuous Improvement at the Team Level

To explore how to apply GCI at the team level, we will work through a scenario based on an actual experience. In this scenario, our team produces and supports a product that exists in the marketplace. Our existing customers love it and appreciate new releases, and it is clear to us that our team's product owner understands our customers intimately. Unfortunately, our customer base is not growing even though our team has committed to do so. The product owner struggles to identify functionality that would attract new customers. Originally trained in Scrum, the team has recently received training in DA and has learned how to apply the DA tool kit to improve their WoW.

The DA tool kit organizes several hundred techniques that potentially support team agility, organizing them into 24 process goals. One of these process goals, Explore Scope, aggregates several dozen techniques for requirements elicitation, exploration, and capture [14]. The Explore Scope goal diagram is depicted in Figure 4.5. Where some agile methods will simply advise you to initially populate a product backlog with a collection of user stories and epics, the goal diagram

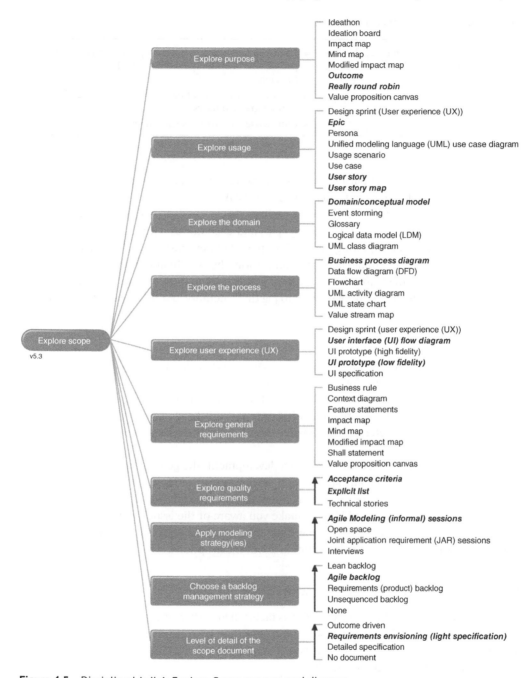

Figure 4.5 Disciplined Agile's Explore Scope process goal diagram.

makes it clear that it is possible to be more sophisticated in your approach. There are several issues, called "decision points," that you should consider when exploring scope. What level of detail should you capture, if any? How are you going to explore potential usage of the system? Or the UI requirements? Or the business process(es) supported by the solution? For each of these decision points, there are several techniques that may be applicable for your situation. Suggested starting

Table 4.1 Summarizing the persona technique.

Detailed descriptions of fictional people who fill roles as stakeholders of the solution being developed.	• Used as a technique to build empathy for users as real people and to understand the optimal user experiences for each. • Useful when we do not have access to actual end users or potential end users. • Can be used as an excuse not to work with actual users.

points, strategies that are often applicable in straightforward situations, are shown in bold italics to guide the initial process decisions of teams new to agile WoW.

To solve the problem that we face, the team realizes that it needs to improve its approach to exploring usage. The product owner is creating epics and user stories with existing customers to capture their needs but is not exploring the needs of potential new customers. We need to adopt a technique to either replace or supplement epics and stories, but we do not know what that would be. If we had someone on the team who is an expert in business analysis techniques, then the process goal diagram is likely to have been enough to jog their memory, but this is not the case, so we need more advice.

The team dives down into the details using the DA Browser, a free tool developed by the Project Management Institute (PMI) [15]. They use the DA Browser to browse the supporting information behind the goal diagrams, and eventually identify personas as a technique that will likely work for them given their current situation. Table 4.1 shows the information captured in DA Browser about personas – a brief description of the technique and its associated tradeoffs to guide your decision regarding whether you want to experiment with it or not. If this level of information is still insufficient, the browser provides references to detailed descriptions of the techniques, in this case to a blog [16] and a book [17].

Although this scenario was focused on product development, the general strategy is applicable to all types of areas. The DA tool kit captures options applicable to finance, vendor management (procurement), legal, marketing, people management (human resources), security, data management, and more. The aim of the tool kit is to make you aware of the issues you need to consider, potential options to address those issues, the tradeoffs associated with those options, and then provide references to great sources of information about those options.

4.5.2 Guided Continuous Improvement Across Disparate Teams

Most agile coaches operate at the team level, but as mentioned earlier struggle to break through the glass barrier of applying agile outside of software teams. Let us work through an actual example where two very different teams, in this case, your Finance team and a Product Delivery team need to find a way to work together more effectively:

- **Finance:** "Your funding runs out next month. We need you to create a business case, a project plan with ±10% estimates, and have it reviewed by the PMO, before we can extend your funding for another six months"
- **Delivery team:** "But that means we will need to stop for two weeks to do this planning! We will be producing no value during this time! Why do you insist on this?"
- **Finance:** "This is our standard process that we adopted to avoid all the problems we had in the past with time and materials (T&M) funding."

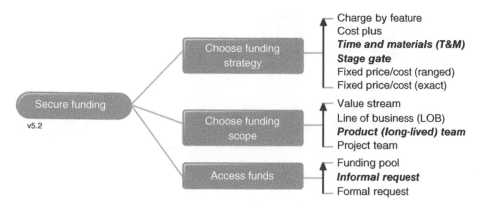

Figure 4.6 Disciplined Agile's Secure Funding process goal diagram.

The challenge here is that the delivery team would like to have a less onerous, and likely more flexible, approach to funding their initiative. Looking at the Secure Funding process goal diagram of Figure 4.6, you see that the funding strategy is "fixed price/cost ranged" and the approach to access the funds is a formal request [18]. The diagram itself indicates that these two options are relatively ineffective, the arrow beside an option list indicates that the list is ordered in that the generally more effective strategies appear at the top of the list and the less effective ones at the bottom. Exploring the details via the DA Browser, you discover that you are in fact suffering from many of the downsides of these practices, yet Finance is justifying their "best practice" by focusing on the advantages. Other agile teams have tried to convince Finance to adopt T&M, that's the most common funding strategy for agile teams, to no avail. Your organization has failed several times with T&M, for the most part, because the Finance team did not know how to govern such an approach and now has no appetite to try again.

The delivery team once again uses the DA Browser to explore their options, learning more about each one and getting a better understanding of the implications of changing the approach to funding their initiative. This better understanding enables them to have a more productive conversation with Finance, pitching the idea that a stage gate strategy is better aligned with an agile WoW and is something that Finance could likely govern successfully. Choosing to experiment with this approach has a greater chance of success in this context, and if successful may prove to be a step toward experimenting with T&M in the future.

4.5.3 Guided Continuous Improvement Across Value Streams

Getting two or three teams to agree on a new WoW to experiment is hard enough; imagine trying to get everyone involved in an entire value stream to work together effectively. A value stream is the set of actions that take place to add value for customers from the initial request through the realization of value by the customers. In the case of a government agency, the "customer" would be citizens. A value stream begins, ends, and hopefully continues with a customer. Figure 4.7 depicts the DA FLEX lifecycle, overviewing the high-level workflow for a value stream [19]. As you can see, a value stream begins with the initial concept, moves through various stages for one or more development teams, and on through final delivery into business operations. Along the outside edge is the primary workflow, the hexes representing process blades/areas that must work together to serve your customers. Inside the oval are supporting process blades that may also be involved as appropriate.

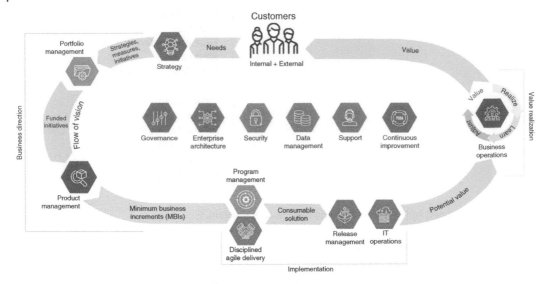

Figure 4.7 The Disciplined Agile FLEX workflow for value streams.

Let us consider the following conversation between four managers who are part of one of the value streams within your organization:

- **Portfolio manager:** "We need to release new offerings to market faster."
- **Product manager:** "We can do that if we develop minimum business increments (MBIs) rather than big product releases."
- **Portfolio manager:** "We fund projects, not new features."
- **Release manager:** "We would love to support a quicker release cadence, but any database change takes three months."
- **Data manager:** "Data is different. We need to think through everything up front, agile doesn't apply to us."
- **Portfolio manager:** "We need to release new offerings to market faster."

They need to negotiate a new WoW to experiment with that they all believe they can perform. They can enable that conversation by collaboratively navigating the goal diagrams for their four respective process blades [20–23], looking for a combination of techniques that they believe they could successfully experiment to improve their overall WoW. Very likely they would identify several experiments that they would then prioritize and run appropriately. These four process blades would just be a start; other aspects of their WoW may also need improving.

4.5.4 Guided Continuous Improvement Across the Organization

An organization offers one or more value streams to their customers, and these value streams are supported by other enterprise process blades/areas not shown in Figure 4.7. These enterprise blades include finance, people management (human resources), vendor management/procurement, marketing, sales, legal, and many others. The strategies required to improve across enterprise blades are the same as those improving across the various aspects of a value stream.

4.6   Summary and Conclusions

The true goal is to become a DAE, a learning organization that improves continually, not to adopt an agile framework. Adopting an agile framework may be a good starting point for a team or organization, but it is not an ending point. Start where you are, do the best that you can in the situation that face, and always strive to get better. The GCI strategy, supported by the DA tool kit, enables you to get better at getting better.

References

1 Sutherland, J. and Schwaber, K. (2020). The scrum guide. http://Scrum.org/resources/scrum-guide (accessed 21 March 2023).

2 Schwaber, K. and Beedle, M. (2001). *Agile Software Development with SCRUM*. Pearson.

3 Beck, K. and Andres, C. (2004). *Extreme Programming Explained: Embrace Change*, 2e. Addison-Wesley Publishing.

4 Stapleton, J. (1997). *Dynamic Systems Development Method (DSDM)*. Addison-Wesley Professional.

5 Knaster, R. and Leffingwell, D. (2018). *SAFe 4.5 Distilled: Applying the Scaled Agile Framework for Lean Enterprises*, 2e. Addison-Wesley Professional.

6 The LeSS framework. LeSS.works. https://less.works/ (accessed 21 March 2023).

7 (2020). The nexus guide. http://Scrum.org/resources/nexus-guide (accessed 21 March 2023).

8 Satir, V. (1991). *The Satir Model: Family Therapy and Beyond*. Science & Behavior Books.

9 Ambler, S. and Lines, M. (2020). *Choose Your WoW! A Disciplined Agile Delivery Handbook for Optimizing Your Way of Working*. Project Management Institute.

10 Jacobson, I. and Stimson, R. (2019). *Tear Down the Method Prisons! Set Free the Practices!* ACM Queue.

11 Sutherland, J. and Sutherland, J.J. (2014). *Scrum: The Art of Doing Twice the Work in Half the Time*. Currency.

12 Deming, W.E. (2002). *The New Economics for Industry*. Government, Education, MIT Press.

13 Ambler, S. (2017). Guided Continuous Improvement (GCI): Speeding Up the Agile Kaizen Loop. http://PMI.org/disciplined-agile/gci (accessed 21 March 2023).

14 Explore Scope (2020). http://PMI.org/disciplined-agile/inception-goals/explore-initial-scope (accessed 21 March 2023).

15 The Disciplined Agile Browser (2021). DA Browser. http://PMI.org/disciplined-agile/da-browser (accessed 21 March 2023).

16 Pichler, R. (2013). 10 tips for creating Agile Personas. http://RomanPichler.com/blog/10-tips-agile-personas.

17 Cooper, A. (2004). *The Inmates Are Running the Asylum: Why High Tech Products Drive us Crazy and How to Restore the Sanity*. SAMS-Pearson Education.

18 Secure Funding (2020). http://PMI.org/disciplined-agile/inception-goals/secure-funding (accessed 21 March 2023).

19 Value Streams (2020). http://PMI.org/disciplined-agile/process/value-streams (accessed 21 March 2023).

20 Portfolio Management Practices (2021). http://PMI.org/disciplined-agile/process/portfolio-management/portfolio-management-practices (accessed 21 March 2023).

21 Product Management Practices (2022). http://PMI.org/disciplined-agile/process/product-management/product-management-practices (accessed 21 March 2023).

22 Release Management Practices (2021). http://PMI.org/disciplined-agile/process/release-management/release-management-practices (accessed 21 March 2023).

23 Data Management Practices (2020). http://PMI.org/disciplined-agile/process/data-management/data-management-practices (accessed 21 March 2023).

5

Leading and Managing Optimal Working Groups, the Belbin Methodology

Marcelo Da Costa Porto

Consultant at Avanza Consulting and Teacher at School of Economic Science and Administration (UdelaR), Montevideo, Uruguay

5.1 The Academic Foundations of the Belbin® Methodology

Why do some teams succeed and others fail? This question posed by the management team of the Administrative Staff College of Henley (now known as Henley Business School) was the baseline of the research conducted by Dr. Meredith Belbin in 1969. This research on team functioning was carried out for a decade and consisted of a business simulation game in which managers competed in teams. They sought to be high-performance teams [1].

5.1.1 Research at Henley Business School, University of Reading UK

The research consisted of three business simulation games per year. These simulations gathered the main variables that characterize the difficulties related to the decision-making process in a business environment. The experiment was designed rigorously by following scientific precepts and making careful measurements at each stage.

The contestants completed a series of psychometric tests and a critical thinking assessment questionnaire. Based on the results, the teams were made up. Duly trained observers recorded each person's contributions every 30 seconds and classified them into different categories. At the end of the year, the financial results of each team (operating as a company) were presented, which allowed the most effective and the least effective "companies" to be compared [2].

As the research progressed, it became apparent that the difference between the teams' success and failure did not depend on elements such as intellect or academic record, but on behavior. The research team began to identify different patterns of behavior, each of which constituted a distinctive contribution or "Team Role." A Team Role was defined as: "A tendency to behave, contribute and interact with others in a certain way." It was found that various people displayed different Team Roles in varying degrees of intensity. Different people were found to display different Team Roles and with different degrees of intensity [3].

The main discovery was that specific individuals assumed particular roles, and the Team Roles pattern played a fundamental role in their results. Unbalanced groups performed poorly. Teams composed of valid people did not necessarily get good results as they might not be balanced. On the

IEEE Technology and Engineering Management Society Body of Knowledge (TEMSBOK), First Edition.
Edited by Gustavo Giannattasio, Elif Kongar, Marina Dabić, Celia Desmond, Michael Condry, Sudeendra Koushik, and Roberto Saracco.

other hand, a team needs valid people in order to be successful. Therefore, the composition of the team, almost entirely neglected at the time, turned out to be of crucial importance. The types of behavior that people adopt are infinite, but the ones that make an effective contribution to the team performance are not [4].

Most people have a series of "Preferred Team Roles" or behaviors which they execute more frequently and naturally. There are also "roles capable of assuming," those that could be taken on by the individuals if required in a given situation, even though said behavior does not come naturally to them. Finally, "Less Preferred Roles" are defined as the ones that should not be taken on by the contestant since they go against that contestant's nature. In the latter case, the effort to be made is great and the result is poor. If the job requires Team Roles apart from those available, it is a much better bet to find and work with other people who have roles that are complementary to the existing ones [5].

The methodology indicates that by recognizing Team Roles, it is possible to obtain higher performance of strengths and manage weaknesses in a better way. Sometimes this means being aware of difficulties and making an effort to avoid them.

There are several scientific studies on the reliability and validity of the methodology. A 2005 investigation verifies the convergent validity of the Team Roles Self-Perception Inventory (TRSPI) with the Kirton Adaption-Innovation Inventory (KAI) for 109 managers and 114 students [6]. In 2007, the self-perception inventory was compared with 43 empirical studies to evaluate the probable behavior of a person, concluding that, together, the model and the inventory that accompany it have an adequate convergent validity [7]. In 2003, the University of the Basque Country held a study consisting of 34 teams during the course of four months. The study showed that balance in the Team Roles favored effectiveness [8]. In 2017, research from the Autonomous University of Yucatan proved that teams formed based on Belbin's Role Theory achieved significantly better results than those who were grouped together randomly [9].

5.1.2 The 9 Team Roles

5.1.2.1 Resource Investigator Role

Strengths: They are natural communicators who are quickly able to build a rapport with other people and expand the range of contacts and useful collaborations of the team [10].[1]

They are highly trained to get out of the environment and discover new possibilities to nurture the team with. Without them, the team takes the risk of stagnating – looking only at itself and losing track of its market.

Potential weaknesses: As they move on to the "next big thing," they can leave other team members dealing with the consequences, which can cause resentment, or even worse, lead to a missed opportunity.

Resource Investigators focus on potential and may leave out important details. This forces other members to make efforts that were not initially foreseen.

---

1 All role icons are under copyright of Belbin – Belbin Associates, UK. 2015 – www.belbin.com.

5.1.2.2 Teamworker Role

Strengths: They are the most helpful team members, helping to maintain a positive team atmosphere.

They are experts in responding to people and reacting in a sensitive, correct, and empathetic way. They play a key role in defusing conflict and helping all team members to contribute effectively.

Potential weaknesses: They do not like conflicts, especially interpersonal conflicts, and they will try to avoid them. This can be a valuable addition; however, sometimes high-pitched feedback is healthy and necessary for the team to progress.

They are people who can hesitate when making important decisions in crucial situations, which can cause annoyance to other team members.

5.1.2.3 Coordinator Role

Strengths: They are responsible for pooling the group's efforts toward a shared goal.

They are very good people at facilitating meetings, ensuring that all important contributions are put on the table, and encouraging consensus to be reached for the team to move forward.

Potential weaknesses: They can create a negative atmosphere through entanglement and intrigue in order to achieve their personal gain.

They are highly competent at delegating tasks so they may end up doing rather little and letting other team members take on their *share of the work*.

5.1.2.4 Plant Role

Strengths: Open-minded, they are the most creative and lateral-thinking people, recognized as the generators of the initial spark during research at Henley.

Those who have this role often find a different way of facing and solving problems, even complex ones. The challenge is to generate the place and the moment so that they can develop their maximum potential.

Potential weaknesses: When they are taken up with whatever they are doing, they can disconnect from what is happening in other parts of the team.

Due to their original way of thinking, the ideas of the Brains can be radical and lacking in practicality. Also, they can focus on an idea that interests them rather than the one required by the team. For this reason, they depend on other team members to evaluate their ideas and put them into practice.

5.1.2.5 Monitor Evaluator Role

Strengths: The Monitor Evaluators are people with the critical eye that every organization or project needs. This working style with which they face each situation allows them to guide their logical rigor away from emotionality.

In meetings, they tend to stay away from the rest of the team, that allows them to maintain their objectivity, which can be important in high-tension situations.

Potential weaknesses: They can be understood as negative, extremely critical, and even unsociable. Consequently, it would not seem appropriate to entrust them with tasks that require motivating others.

The Monitor Evaluators are likely to take their time to deliberate when making a decision, which can be frustrating for other team members who want things to move faster.

5.1.2.6 Specialist Role

Strengths: Committed to a specific interest or area, specialists pride themselves on acquiring knowledge in their particular field of study and bringing exceptional skills that are often essential to successful team performance.

Specialists often enjoy talking about their topic with others, particularly those who share their interests. They are likely willing to attend courses in their subject area, not just out of a sense of duty but out of love for their subject.

Potential weaknesses: Due to their limited approach, they may have a hard time seeing the bigger picture.

Since they work in a particular area, specialists may isolate themselves from other team members and gain a distorted view of the team's role or purpose based on their own interests and priorities.

5.1.2.7 Shaper Role

Strengths: They are dynamic people, highly motivated, with a lot of nervous energy and a great need to achieve goals. They are mainly interested in getting things done as quickly as possible.

Boosters like to lead and push others to action. They grow under pressure and are well-suited to managing change. They are good at tackling inertia and complacency in teams or intervening when the team is straying too far from its original commission.

Potential weaknesses: In their rush to get things done, boosters can ignore other people's feelings and upset their moods.

Too many boosters on a team can also cause problems as they tend to be confrontational and get involved in heated matches.

5.1.2.8 Implementer Role

Strengths: Implementers focus on what is feasible and then move on to the task. They tend to approach work systematically and methodically, ensuring maximum efficiency.

Implementers are reliable and disciplined, and they are helpful in organizations because of their inclination and ability to do whatever needs to be done.

Potential weaknesses: Once plans have been put in place, Implementers may be reluctant to receive further suggestions in case the resulting outage threatens productivity and leads to inefficiency.

Implementers are more effective when they achieve a balance between accommodating important changes and maintaining productivity.

5.1.2.9 Completer Finisher Role

Strengths: They are ideal for working in areas that require careful monitoring, deep concentration, and a high degree of precision, such as proofreading or verifying figures.

In teams, they can be relied on as "quality controllers" and detectors for errors that could otherwise jeopardize or ruin the final product or the result.

Potential weaknesses: Completer finishers "finish" in the sense of polishing the work rather than actually finishing it.

While they may be called in the latter phase of a task to put the finishing touches on work, they may struggle to meet deadlines because they place more importance on doing things right than on time.

5.1.3 Differences with Other Methodologies

There are other methodologies that pursue similar objectives, in terms of identifying traits that allow understanding the behavior of individuals in organizations. However, the Belbin® methodology has some distinctive attributes that make it considered a worldwide reference methodology in the management and development of people and teams, with reports available in 22 languages.

Belbin® does not seek to define the contestants' personality, but the behavior groups within the teams in order for them to be successful, identified as a consequence of nine years of intensive research by Dr. Meredith Belbin. For this methodology, it is considered key to measure what really influences work, that is, what we do (behavior) and not what we are (personality).

The other difference, and it is one of the essential points of this methodology, is that it focuses on the results. Meredith Belbin identified the inputs that are required to be used in team processes in order to achieve exceptional team results. The inputs he was able to identify were the skill sets and behaviors that would become the nine Belbin Team roles.

Perhaps the most relevant difference is that it is a methodology with strong academic foundations and arises as a consequence of serious research. Despite being a solid methodology in its foundations, it is easy to apply and effortlessly accepted and understood not only by people who work in the field but also by every person who works in an organization.

5.2 Operation of the Methodology and Benefits of Its Application

Although the methodology has an academic origin, with validity tests and periodic updates, the true value of the Belbin methodology has not been lost sight of: its practical application.

5.2.1 How Belbin® Works

The methodology is based on the preparation of individual self-perception and observer reports under the 360-evaluation format.

The Belbin Self-Perception Questionnaire was designed by Dr. Meredith Belbin and his team in the 1980s, after nearly a decade of research at Henley Management College (UK). It is a behavioral test (not psychometric) designed for its application in the workplace and is internationally recognized as one of the most influential in the field of industrial psychology.

Belbin reports are a powerful, reliable, and validated tool that allows you to gain a better understanding of how a person actually behaves at work. The reports contain personalized guidance and helpful advice for both the individual and managers and middle managers. All this information provides a very wider perspective of people's strengths and weaknesses and how they will contribute at the individual, relational, and team level, which is extremely advantageous for any organization [11].

One of the distinctive values is the possibility of adding the Observer Assessment to the Belbin Self-Perception Inventory. Observer evaluations (OEs) offer an independent point of view on the Team Roles of a given person. The validity of a self-perception test will depend on the realistic vision that a person has about themself. Some people answer this type of test based on how they would like to contribute rather than how they actually behave [12].

5.2.2 Benefits of Applying the Belbin® Methodology

Applying Belbin® can be understood as an effective tool for the identification of talents, minimizing the failure generated by the particularities of a job. In other words, the performance of a worker in a position that does not take advantage of their preferred competencies can lead to errors of judgment when evaluating their performance [13].

It is a communication tool that enables cooperation and teamwork. When people work as a team, they are able to raise quality standards and achieve much more ambitious goals. In fact, not only do performance and productivity improve but there is also an increase in motivation and commitment and, therefore, an improvement in the work environment [14].

As a team leader, applying the Belbin® methodology enables the improvement of personal and team effectiveness, through the identification and development of talent. Provides substantial elements for the preparation of individual and team development plans.

Ultimately, the Belbin® methodology adds value to individuals, teams, and organizations.

Belbin® promotes transparency and trust between people through common language, constructive feedback, and recognition of potential failures, even when teams cannot meet in person. Enables teams to respond to their environment and adapt as they go [15].

5.3 Real Examples

5.3.1 RCD Recycling

RCD Recycling arose in 2011 as a project that was carried out at the School of Architecture – *Universidad de la Republica* (Uruguay), where a prototype of a floor tile was made from crushed rubble. In 2017, sponsored by Incubadora Sinergia, a "Seed Capital" was obtained as an innovative project in the National Agency for Research and Innovation (*ANII*). This allowed the company to settle inside the municipal dump in Montevideo during the first year of support, to carry out our first tests and trials of the material processed on site [16].

In 2018, thanks to the results obtained, RCD was declared of Municipal Interest. In 2019 it obtained the support of *ANDE* – National Development Agency – and Biovalor within the framework of the Circular Opportunities program. In turn, it was distinguished as an inspiring project by DERES.

Today, RCD has the first Construction Waste Management plant in Uruguay, where it manages construction waste and offers different types of products from processed rubble such as recycled aggregates of different granulometries, and new products from recycled concrete, in order to avoid the final disposal of this waste and to be able to reinsert it once again in the construction cycle.

In 2020, the RCD Board of Directors considered it necessary to review the composition of the team of partners and the roles assigned to each one, taking in an agile vision for their management. Within this framework, the Belbin Self-Perception Test (IAP) and Observer Assessment tools were implemented. With the inputs obtained, the individual and the team reports were generated.

As a result, development plans could be defined for each partner, and a role reassignment, based on a new form of management, moving from a predictable framework to an agile format. The arrival of the pandemic caused by Sars-Cov-2 served as the basis for dedicating time to reflection and preparing the company for a new reality.

In addition to the role reassignment, new dynamics were generated for the management of board meetings, fostering the exchange of ideas, the recognition of the differentiated contributions of each partner, and the development of the team based on strengths.

Currently RCD recycling, which continues to grow, has decided to review its current structure, assigning new roles and responsibilities to its team of collaborators based on the Team Roles of the Belbin® methodology.

5.3.2 British Hospital

The British Hospital was built in Uruguay in 1857. Its history proudly unfolded alongside that of a thriving country, which grew and became one of the most beautiful places to live in Latin America [17].

The British Hospital was a pioneer in health insurance issues, offering, more than 45 years ago, the British Hospital Scheme, a health coverage of the highest quality that has been transformed to become one of the most prestigious health products in the country.

At the same time, the development process of the hospital's infrastructure was always following international standards with a team of professionals and technical staff that filled it with talent, service, and professionalism.

In 2020, the British Hospital considered it important to implement several Patient Safety solutions proposed by the WHO. Among them, progress in the improvement of the Patient Identification process, for which it implements several optimizations from the technological and communicational point of view, seeking to ensure its application in an optimal and universal way.

Within the framework of this, it hires the consulting service of Avanza Consulting to carry out a project whose objective is to ensure the effective identification of the hospitalized patient through the optimization and correct execution of the process that goes from admission to leaving the hospital, incorporating practices for effective self-management and direction of transverse processes, based on the current strengths of the working teams [18].

Avanza Consulting decides to emphasize the correct selection of the counterpart team, prioritizing it to be multidisciplinary and diverse in its composition. The Belbin® methodology is used to determine individual roles and for the formation of two balanced teams that contribute to achieve the objectives of the three phases in which the project is designed.

The tools IAP and Observer Evaluation (EO) were used, the individual and the team reports were generated, and work was done on the development of the optimal roles in each team.

As a result, individual capacities were developed in each member of the teams, while at the same time, innovative, practical, adequate, and feasible solutions were obtained for the fulfillment of the specific objectives of the project.

This approach for designing work teams contributed to the fact that the team members' supervisors could have a different view of the capacities and preferred roles at work. This introduced a new kind of management system for future projects, positively valuing the distinctive contribution that each individual has to achieve the organizational objectives.

5.4   Some Recommended Uses for the Belbin® Methodology

The application of the methodology requires a deep knowledge of the culture, structure, regular practices, process policies, and talent management within the organization and management of human talent. Taking the above-mentioned into consideration, some recommended applications and a brief analysis on the consistency between Belbin® and agile methodologies are described.

5.4.1   Belbin® Applications

5.4.1.1   Identification of the Right People for a Role

Belbin® provides a solution to the question of identifying the right people to fulfill a given role.

Given the freedom to choose the members of a team and the need to form a high-powered group, how should we proceed?

Job assignment can be a straightforward task when job requirements are fully predictable, can be precisely specified, and individuals can also attend training courses that equip them with the required skills. However, there is a drawback: it is a very volatile script. Everything mentioned above has nothing to do with the world executives inhabit. It is not unusual for the specific requirements of a position to change even before the chosen person has settled in or the person does not know in detail what the position entails. This is why assigning jobs can be so volatile [19].

Once the profile of a person has been established in terms of Team Roles, through self-perception and evaluations from observers, we have to store the information and retrieve it when necessary. Thus, job requirements can be assessed, and it is technically possible to turn them into the profile of an ideal person [19].

5.4.1.2   Design Teams

The compatibility of team members is essential for its effectiveness [19].

This is not less important than whether the members of a team are good accountants, production engineers, or salespeople. The problem is that human compatibility is more difficult to assess than technical competence.

According to Belbin®, every team needs an optimal balance between functional roles and Team Roles. A team can make the most of its technical resources only when it has the necessary Team Roles to ensure effective teamwork. Furthermore, the effectiveness of a team will be fostered to the extent that its members recognize and correctly adapt to the strengths of the team, both in knowledge and in the ability to perform certain Team Roles [19].

5.4.2   Belbin® Compatibility with Agile Methodologies

Recently, research has been carried out that seeks to identify the degree of coherence between Belbin® and agile methodologies for project management.

Agile empowers cross-functional teams, enabling them to come together, self-manage, and produce complex products in challenging environments. It is credited with greater employee engagement, reducing the time it takes to get a product to market, and greater customer satisfaction [20].

A Scrum team does not have sub-teams, roles, or titles. The team can have specialists such as experts in testing, development, architecture, etc., and this could also be good for the team.

But these specialists contribute so that the team can achieve the objective of Sprint and the product as a team, complementing their skills. In this team, a member does not own the tasks of their specialty; as a team, everyone is responsible for doing the necessary work for achieving the Sprint objective and helping each other in the work to be done. Creating an identity as a team, helping each other, growing and learning from others by making commitments beyond their area of expertise allows the team to maximize its effort to achieve goals as well as feeling motivated and confident in others [20].

As a philosophy, Agile is about identifying uncertainties within your work environment and figuring out how to adapt on the fly. With its principles of prioritizing people over process and responding to change rather than rigidly sticking to a plan, it is not hard to see why it might be a popular approach for our current circumstances [20].

But of course, to work quickly with cross-functional teams, which are now primarily virtual teams as well, is a great question. Fortunately, we can affirm that the Belbin® methodology is convenient when implementing the Agile principles related to teamwork, which will be mentioned below [20].

5.4.2.1 Business Managers and Developers Work Together on a Daily Basis Throughout the Project

For the Belbin® methodology, to work as a team is more than working together. Belbin® understands teams as "a limited number of people selected to work together toward a shared goal and in a way that allows each person to make a distinctive contribution."

Teams are typically more productive when creative solutions are required that are beyond the capabilities of a single person, or in any case, a team is likely to provide better solutions than an individual person.

Similar to Agile, Belbin® argues that "the essence of a team lies on its players acting together and engaging dynamically" [21].

5.4.2.2 Projects Are Developed Around Motivated Individuals. You Have to Give Them the Environment and Support They Need and Entrust Them with the Work Execution

A key principle of Agile is that teams must feel safe to fail, and fail quickly. This means fostering an environment where failures are not only recognized but celebrated as part of the learning process. The reason for this is that embracing and fostering failure allows teams to become more adventurous, open, and confident, learn from past mistakes, and avoid wasting time with non-starters and blaming culture.

However, fostering failure, and the kind of trust that allows teams to fail without fear of individual or group recriminations, is easier said than done. For many organizations, failure is not part of the business speech.

The language of Belbin Team Roles shares the same honesty. We believe in focusing on strengths but also giving time to weaknesses, areas where we might struggle, and asking others for help. The process of doing so identifies individual failures (and how they can obstruct the team) and allows the team to formulate strategies for compensating, building interdependence, trust among team members, and psychological security in the team as a whole.

5.4.2.3 The Most Efficient and Effective Method of Communicating Information to the Development Team and Among Its Members Is Face-to-Face Conversation

In a virtual environment, it is important to generate and maintain transparency and trust. Ongoing training and delivery feedback is key to creating a "culture of listening" in which agile teams thrive.

While ideas, hopes, and concerns can be conveyed informally and ad hoc, the Belbin® Report helps consolidate feedback on individual and team behaviors into a key metric and an excellent starting point for individual and team discussions.

5.4.2.4 The Best Architectures, Requirements, and Designs Emerge from Self-Organizing Teams

Agile teams are democratic (there is no leader), and so is Belbin®. While some measures are based on a person's self-knowledge, Belbin® uses observer feedback, giving each team member the opportunity to express how individual behaviors affect them and the team as a whole. This creates a consistent picture of how the computer (and each of its components) works. Moreover, this process can be repeated whenever things change on the computer, to help diagnose problems that arise or to point out problems that may pose problems later.

5.4.2.5 At Regular Intervals, the Team Reflects on How to be More Effective and Then Adjust and Refine Their Behavior Accordingly

For Scrum leaders, understanding how diversity of behavior develops in a team is crucial to building trust among team members. For a cross-functional Agile team that has never shared office space, different work styles require some adjustments. It can be worrisome for a strong Implementer (who prides themself on updating the Kanban board by the hour) that someone with "brainy" behaviors spends a lot of time off the grid only to come back with a new idea. Of course, it is up to the team to decide what is acceptable in their work styles, but understanding behaviors and motivations can help build trust as long as everyone is complying.

5.5   Conclusions and Takeaway

Currently, multiple organizations throughout the world use the Team Roles identified by Meredith Belbin.

It is a methodology that breaks paradigms linked to leadership and provides solutions for strategic management, project management, and innovation management.

The Belbin® methodology contributes to the debate on the degree of effectiveness of academic credits and IQ as predictors of good performance.

If self-knowledge of people's preferences while working, translated into obvious behaviors, are the key factor for better individual and team performance, the leaders of organizations should take this methodology into account to design and implement their Human Resources processes.

It is favorable, feasible, and advisable to find common points that facilitate consistent and coherent work between both work methodologies.

Regarding Agile methodologies, Belbin® could become a simple and objective application tool for processes to generate agile business cultures. It is important to note that approaches to the relationship of synergy between Belbin® and Agile are currently being investigated around the world.

References

1 https://www.belbin.com/about/our-story.
2 Belbin Associates. https://www.belbin.com/resources/blogs/development-of-the-dream-team-in-change-management.
3 https://www.belbin.com/about/belbin-team-roles.

4 Belbin Associates. https://www.belbin.com/resources/blogs/a-team-role-guide-to-suncream.

5 https://www.belbin.com/resources/articles-directory/team-role-communication-styles.

6 Aritzeta, A., Swailes, S., and Senior, B. (2005). Belbin team-role preference and cognitive styles: a convergent validity study. https://journals.sagepub.com/doi/10.1177/1046496404273742.

7 Aritzeta, A., Swailes, S., and Senior, B. (2007). Belbin's team role model: development, validity and applications for team building. https://onlinelibrary.wiley.com/doi/abs/10.1111/j.1467-6486.2007.00666.x.

8 Aritzeta, A. and Ayestarán, S. (2003). Aplicabilityof belbin roles. https://journals.sagepub.com/doi/abs/10.1177/1059601114562000.

9 Aguileta-Guemez, A. and Ucán-Pech, J. (2017). Exploring the influence of belbin roles. https://www.springerprofessional.de/en/exploring-the-influence-of-belbin-s-roles-in-software-measuremen/16154498.

10 https://www.belbin.com/about/belbin-team-roles.

11 https://www.belbin.com/about/belbin-reports.

12 https://belbin-dacostaporto-com.translate.goog/?_x_tr_sl=auto&_x_tr_tl=en&_x_tr_hl=en&_x_tr_pto=wapp.

13 https://www.belbin.com/resources/videos.

14 https://www.belbin.com/.

15 https://www.belbin.com/resources/belbin-and-agile-building-responsive-teams-and-failing-fast.

16 https://www.rcdreciclaje.com/#/about.

17 https://www.hospitalbritanico.org.uy/english/.

18 https://www.avanzaconsulting.biz/.

19 Belbin, M. Team roles at work. https://books.google.com.uy/books/about/Team_Roles_at_Work.html?id=hF2yJzYfUBAC&redir_esc=y.

20 https://www.belbin.com.au/post/belbin-and-agile-building-responsive-teams-and-failing-fast.

21 Belbin, M. (2013). *Roles de equipo en el trabajo*, 2da ediciône, 96.

Section 4

Innovation

6

Managing Innovation

Joseph Tidd

Science Policy Research Unit (SPRU), University of Sussex, Brighton, UK

6.1 Knowledge Area Fundamentals

6.1.1 Managing Innovation: A Process Approach

Most of the fundamental interdisciplinary research and knowledge base on technology and innovation management was conducted in the 1970–1990s, but since then studies by management disciplines and business school functional groups have fragmented the field into more narrow disciplinary questions and solutions, to the detriment of innovation research and practice.

Early models of the innovation process were simply a linear sequence of functional activities [1]. Some began with technical developments, which eventually found their way to the marketplace (so-called "technology push" models), or alternatively, started with market opportunities or users' inputs and then applied technology to develop solutions (so-called "market pull" models). However, empirical studies challenged this simple dichotomy. In practice, innovation is a coupling process where interaction between the technical and commercial is critical. Depending on the maturity of the technologies and markets, sometimes the "push" will dominate, but other times the "pull" will. Successful innovation results from the interaction between the two.

Whilst there remain fundamental uncertainties in managing innovation, there are also clear patterns or regularities in good management practices. These practices will, of course, need to be adopted selectively and adapted for different challenges, in particular, the degree and type of innovation and the resources available (Figure 6.1).

6.1.2 Developing an Innovation Strategy

The aim of developing an innovation strategy is to guide the degree and direction of innovation. The degree of innovation includes decisions regarding the balance between incremental and more radical innovation, and the direction of innovation influences the types of innovation, for example, product vs. process or technological vs. commercial. The outcome is more a portfolio of projects and investments rather than a definitive plan, as innovation is inherently uncertain.

IEEE Technology and Engineering Management Society Body of Knowledge (TEMSBOK), First Edition.
Edited by Gustavo Giannattasio, Elif Kongar, Marina Dabić, Celia Desmond, Michael Condry, Sudeendra Koushik, and Roberto Saracco.

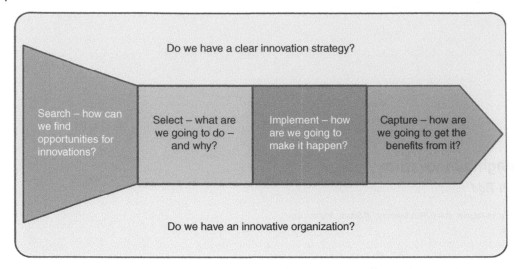

Figure 6.1 A process to manage innovation. *Source:* Reproduced with permission from Tidd and Bessant [2]/John Wiley & Sons.

The relationship between strategic planning and innovation is complex. At the project and product levels, planning and innovation tend to have a positive relationship, but at the corporate level, the relationship is more antagonistic. There is a tension between the so-called "rational planning" approach of strategic management, with the more incremental, iterative "resource-based view" (RBV) of innovation management. The empirical evidence shows that the RBV is a better explanation and guide to innovation strategy [3]. This has two practical implications for managers. Firstly, that innovation strategy begins with the assessment of organizational-specific capabilities rather than a generic analysis of markets and competitors. Secondly, that innovation strategy is more a form of iterative organizational learning rather than simply planning.

6.1.3 Building an Innovative Organization

The core innovation process cannot function without being embedded and supported by the wider organization. The organizational factors that may encourage or constrain innovation include leadership, other key individuals, culture and climate, teams and an open, external focus (Table 6.1). It is critical here to understand that there is no single "best" recipe for innovation, but rather that

Table 6.1 Components of the innovative organization.

Component	Key features
Shared vision, leadership and the will to innovate	Clearly articulated and shared sense of purpose. Stretching strategic intent "Top management commitment." Significant LMX – leader-members exchange.
Key individuals	Promoters, champions, gatekeepers, and other roles that energize or facilitate innovation.
Effective team working	Appropriate use of teams, cross-functional and inter-organizational levels. Requires investment in team selection and building.
Creative climate	Environment that stimulates and supports the development of creative problem-solving.
External focus	Open innovation, external orientation, extensive networking.

Table 6.2 Climate factors influencing innovation.

Climate factor	Most innovative (score)	Least innovative (score)	Difference
Trust and openness	253	88	165
Challenge and involvement	260	100	160
Support and space for ideas	218	70	148
Conflict and debate	231	83	148
Risk-taking	210	65	145
Freedom	202	110	92

Source: Derived from Isaksen and Tidd [5].

different combinations or configurations will better fit the nature of the project and organizational environment, the so-called "contingency view" [4].

Organizational culture refers to the deeper and more enduring values, norms, and beliefs within the organization and is therefore difficult to measure or manage. Climate is distinct from culture in that it is more observable at a surface level of an organization, and therefore more able to be measured, managed, and improved [5]. Research shows that organizations have larger differences in practices than in values, for example, the levels of autonomy or risk. Therefore, it is useful to breakdown the overall organizational climate into dimensions that can be measured and managed (Table 6.2). The purpose is to identify which dimensions most need to be improved.

6.1.4 Searching for Sources of Innovation

One of the major challenges of managing innovation is to expand the search space. Typically, when seeking opportunities or solutions, we favor local knowledge drawn from our own discipline, experience, and organization. The aim of actively and systematically searching is to increase the scope of opportunities and solutions (Figure 6.2).

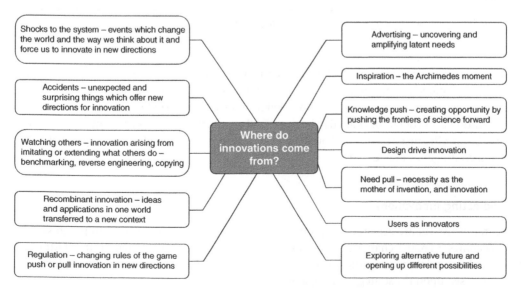

Figure 6.2 Sources and triggers of innovation. *Source:* Reproduced with permission from Tidd and Bessant [2]/John Wiley & Sons.

Table 6.3 Benefits and challenges of applying open innovation.

Six principles of open innovation	Potential benefits	Challenges to apply
Tap into external knowledge	Increase the pool of knowledge Reduce reliance on limited internal knowledge	How to search for and identify relevant knowledge sources How to share or transfer such knowledge, especially tacit and systemic
External R&D has significant value	Can reduce the cost and uncertainty associated with internal R&D and increase depth and breadth of R&D	Less likely to lead to distinctive capabilities and more difficult to differentiate External R&D also available to competitors
Do not have to originate research to profit from it	Reduce costs of internal R&D, more resources on external search strategies and relationships	Need sufficient R&D capability in order to identify, evaluate, and adapt external R&D
Building a better business model is superior to being first to market	Greater emphasis on capturing rather than creating value	First-mover advantages depend on technology and market context Developing a business model demands time-consuming negotiation with other actors
Best *use* of internal and external ideas, not *generation* of ideas	Better balance of resources to search and identify ideas rather than generate	Generating ideas is only a small part of the innovation process Most ideas are unproven or no value, so cost of evaluation and development high
Profit from others' intellectual property (inbound OI) and others' use of our intellectual property (outbound IP)	Value of IP is very sensitive to complementary capabilities such as brand, sales network, production, logistics, and complementary products and services	Conflicts of commercial interest or strategic direction Negotiation of acceptable forms and terms of IP licenses

Source: Reproduced with permission from Tidd [7]/John Wiley & Sons.

The concept of open innovation emphasizes that firms should acquire valuable resources from external sources and share internal resources for innovation [6]. However, the question of when and how an organization sources external knowledge and shares internal knowledge is less clear. Open innovation is too often considered to be universally applicable and positive, but research suggests that the specific mechanisms and outcomes of open innovation models are very sensitive to context and contingency (Table 6.3).

6.1.5 Selecting Innovations

If the search activities are expanded as proposed, selecting potential innovation to progress and support becomes a greater challenge as the number of opportunities is increased. In addition to conventional financial and quantitative assessments, innovations require qualitative judgments to be made based upon the strategy and capabilities of the organization.

Table 6.4 Use and usefulness of criteria project screening and selection.

	High novelty		Low novelty	
	Usage (%)	Usefulness	Usage (%)	Usefulness
Probability of technical success	100	4.37	100	4.32
Probability of commercial success	100	4.68	95	4.50
Market share[a]	100	3.63	84	4.00
Capabilities[a]	95	3.61	79	3.00
Degree of internal commitment	89	3.82	79	3.67
Market size	89	3.76	84	3.94
Competition	89	3.76	84	3.81
NPV/IRR	79	3.47	68	3.92
Payback period/break-even[a]	79	3.20	58	4.27

Usefulness score: 5 = critical; 0 = irrelevant.
[a] Difference in usefulness rating is statistically significant at 5% level.
Source: Adapted from Tidd and Bodley [8].

The choice of criteria for screening and selecting innovations to implement will depend on the novelty of the proposal (Table 6.4).

Clearly, financial assessments and probabilistic estimates of technical and commercial success are the most common criteria, and considered to be of critical importance in all types of projects. For more routine projects, competition and the market size and share are important, but for more novel projects the fit with organization-specific capabilities is more relevant, consistent with the earlier discussion of innovation strategy.

The purpose of such assessments is to build a balanced portfolio of innovation projects, which are characterized by different degree on risk and reward.

6.1.6 Implementing Innovation

The implementation of innovation is not simply good project management. It requires the reduction of uncertainty over time, through the acquisition of more technological and market knowledge, and by mobilizing internal and external resources throughout development. This iterative and emergent process is closer to experimentation and learning than conventional project management (Figure 6.3).

For more novel innovation projects, which demand the sharing and exchange of diverse knowledge, cross-functional teams are a proven, but resource-intensive method to support the development of innovations. Quality function deployment (QFD) is a structured tool to help support such cross-functional collaboration [9]. In QFD, customer-required characteristics are translated or "deployed" by means of a matrix into language that engineers can understand. The construction of a relationship matrix – also known as "the house of quality" – requires a significant amount of technical and market research input, including technical, market, and user data to identify potential design trade-offs and to achieve the most appropriate balance between cost, quality, and performance.

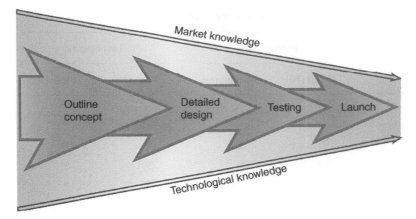

Figure 6.3 Reducing market and technological uncertainty during implementation. *Source:* Reproduced with permission from Tidd and Bessant [2]/John Wiley & Sons.

6.1.7 Creating and Capturing Value

Managing innovation is more than simply commercializing the technology. It involves the creation of value for a target market or user and the capture or appropriation of some of this value. By definition, innovation must create some value, not simply in a narrow financial sense, but in a broader business or social sense. Business model innovation does not sufficiently capture these broader benefits to the organization or users.

Typically, the development of a business model will include consideration of the value proposition, mechanisms for revenue generation, capabilities and processes, and position in the value network or ecosystem (Figure 6.4):

- **Value proposition:** How does the innovation or venture create value and for whom? The value created will be specific to target market segments and customer groups, and different types of innovation will contribute in different ways.

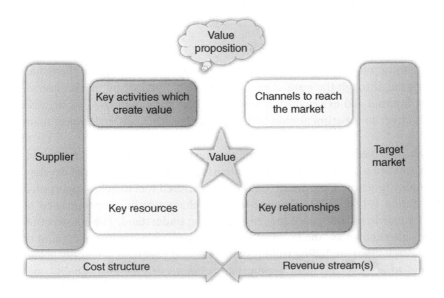

Figure 6.4 Business Model Canvas. *Source:* Adapted from Tidd and Bessant [2] and Osterwalder and Pigneur [10].

- **Revenue generation:** How does the enterprise capture and appropriate the benefits (or "rents" as economists call them)? In the case of public and social ventures, capture and revenues are less important than demonstrating value, and ensuring that resources, human and financial are sustainable.
- **Capabilities and processes:** How can the innovation or venture deliver? This is much more than access to financial and other resources. It requires a (rare) combination of resources, knowledge, and capabilities. A common mistake made by entrepreneurs is to focus too much on the initial creation of value and not to pay sufficient attention to how value will be captured in the longer term.

Innovation can change any, and all, of these core components of a business model, so-called "Business Model Innovation." In many cases, this features the exploitation of novel combinations of tangible and intangible assets as complementary building blocks to create value in new ways, for different applications and users, for both commercial and social benefit.

6.2 Managing Approaches, Techniques, and Benefits to Using Them

Each of the four core components of the innovation process has a range of methods and tools to support good practice. Some tools are useful for the diagnosis and analysis end of things, helping to clarify and focus information on what needs to be done; other tools are more concerned with implementation, helping to make monitor and manage to ensure things happen. Many tools exist to support search strategies (Table 6.5) and implementation (Table 6.6).

Table 6.5 Search strategies for wider exploration.

Search strategy	Mode of operation
Sending out scouts	Dispatch idea hunters to track down new innovation triggers.
Exploring multiple futures	Use futures techniques to explore alternative possible futures, and develop innovation options from that.
Using the web	Harness the power of the Internet, through online communities, and virtual worlds, for example, to detect new trends.
Working with active users	Team up with product and service users to see the ways in which they change and develop existing offerings.
Deep diving	Most market research has become adept at hearing the "voice of the customer" via interviews, focus groups, panels, etc. But sometimes what people say and what they actually do is different. In recent years, there has been an upsurge in the use of anthropological style techniques to get closer to what people need/want in the context in which they operate. "Deep dive" is one of many terms used to describe the approach – "empathic design" and "ethnographic methods" are others.
Probe and learn	Use prototyping as mechanism to explore emergent phenomena and act as boundary object to bring key stakeholders into the innovation process.
Mobilize the mainstream	Bring mainstream actors into the product and service development process.

(Continued)

Table 6.5 (Continued)

Search strategy	Mode of operation
Corporate venturing	Setting up of special units with the remit – and more importantly the budget – to explore new diversification options. Loosely termed "corporate venture" (CV) units, they actually cover a spectrum ranging from simple venture capital funds (for internally and externally generated ideas) through to active search and implementation teams, acquisition and spin-out specialists, etc.
Use brokers and bridges	Increasingly, organizations are looking outside their "normal" knowledge zones as they begin to pursue "open-innovation" strategies. But sending out scouts or mobilizing the Internet can result simply in a vast increase in the amount of information coming at the firm – without necessarily making new or helpful connections. Increasingly, organizations are making use of social networking tools and techniques to map their networks and spot where and how bridges might be built – and this is a source of a growing professional service sector activity. Brokering and facilitation across the open innovation landscape has become a growth area with players like 100% Open, a spin-off from Procter and Gamble's Connect and Develop program, online patent matching agencies, brokers, and idea connectors like NineSigma, and design houses like IDEO all playing a role.
Deliberate diversity	Create diverse teams and a diverse workforce.
Idea generators	Use creativity tools.

Source: Reproduced with permission from Tidd and Bessant [2]/John Wiley & Sons.

Table 6.6 Tools for development and implementation.

	High novelty		Low novelty	
	Usage (%)	Usefulness	Usage (%)	Usefulness
Segmentation[a]	89	3.42	42	4.50
Prototyping[a]	79	4.33	63	4.08
Market experimentation	63	4.00	53	3.70
Industry experts[a]	63	3.83	37	3.71
Surveys/focus groups[a]	52	4.50	37	4.00
Trend extrapolation	47	4.00	47	3.44
Latent needs analysis[a]	47	3.89	32	3.67
User-practice observation	47	3.67	42	3.50
Partnering customers	37	4.43	58	3.67
User-developers	32	4.33	37	3.57
Scenario development	21	3.75	26	2.80
Role-playing	5	4.00	11	1.00

[a] Difference in usefulness rating is statistically significant at 5% level ($n = 50$).
Source: Adapted from Tidd and Bodley [8].

6.3 Real Examples

The innovation process and tools presented in this paper are based on systematic research and codified experience and practice, but there are numerous real case examples of their application and impact. Here discuss three examples of organizational innovation based on methods we have used.

Case Study 6.1   Increasing Challenge and Involvement

The organization was a division of a large, global electrical power and product supply company [5]. When this division was merged with the parent company, it was losing about $8 million a year. A new general manager was bought in to turn the division around and make it profitable.

An assessment of the organization's climate identified that it was strongest on the debate dimension but was very close to the stagnated norms when it came to challenge and involvement, playfulness and humor, and conflict. The quantitative and qualitative assessment results were consistent with their own impressions that the division could be characterized as conflict-driven, uncommitted to producing results, and people were generally despondent. The leadership decided, after some debate, that they should target challenge and involvement, which was consistent with their strategic emphasis on a global initiative on employee commitment. They committed to increase communication by holding monthly all-employee meetings, sharing quarterly reviews on performance, and using cross-functional strategy review sessions. They implemented mandatory "skip level" meetings to allow more direct interaction between senior managers and all levels of employees. All employee suggestions and recommendations were invited, and feedback and recognition were immediately given. A new monthly recognition and rewards program was launched across the division for both managers and employees that was based on peer nomination. The management team formed employee review teams to challenge and craft the statements in the hopes of encouraging more ownership and involvement in the overall strategic direction of the business. In 18 months, the division showed a $7 million turnaround, and in 2003 won a worldwide innovation award. The general manager was promoted to a national position.

Case Study 6.2   Developing a Creative Climate in a Medical Technology Company

A Finnish-based global healthcare organization had 55,000 employees and $50 billion revenue. The senior management team of one division conducted an assessment and found that they had been doing well on quality and operational excellence initiatives in manufacturing and had improved their sales and marketing results, but were still concerned that there were many other areas on which they could improve, in particular, technological and product innovation [5].

The team targeted challenge and involvement, freedom, idea time, and idea support as critical dimensions to improve to enable them to meet their strategic objectives. The organization was facing increasing competition in its markets and significant advances in technology. Although a major progress had been made in the manufacturing area, they needed to improve their product development and marketing efforts by broadening involvement internally and cross-functionally and externally by obtaining deep consumer insight.

Key personnel in new product development and marketing were provided training in creative problem solving, and follow-up projects were launched to apply the learning to existing and new projects. One project was a major investment in reengineering their main product line. Clinicians were challenged with the current design of the equipment. The project leader decided to use a number of tools to go out and clarify the problem with the end users, involving project team members from research and development as well as marketing. The result was a redefinition of the challenge and the decision to save the millions of dollars involved in the reengineering effort. Since the professionals in the research and development lab were also directly involved in obtaining and interpreting the consumer insight data, they understood the needs of the end users and displayed an unusually high degree of energy and commitment to the project.

Case Study 6.3   Customer Integration in Action

A supplier of technologies for monitoring and controlling energy for welding machinery, solar converters, and battery charging systems [11]. When developing a new generation of traction battery charging systems, the company applied QFD with the goal to develop features that were a clear unique selling proposition and would simultaneously decrease the charging systems' production costs by 10%. In a first step, they gathered its immediate customers' and endusers' requirements.

The realization of the QFD was challenging because it required several QFD team workshop meetings, which took place over a period of several weeks instead of days, due to the team members' busy agenda. This required time-consuming updates at the beginning of each workshop. Unclear formulations of the requirements and the differing degrees of their abstraction levels led to time-consuming discussions that were sometimes enlightening and sometimes unproductive. Overall, the QFD improved the information exchange between the departments, the R&D employees' understanding of the customer, and the inter-functional information sharing of the expected future product features, which allowed the employees involved in later NPD stages to prepare accordingly.

6.4   Recommended Best Use

As we discussed earlier, the standard innovation process needs to be adapted according to the specific management challenges, in particular the industry sector, degree of innovation, and resources available. However, it is possible to identify general good practices. For example, we assessed the use of different groups of innovation management practices (IMPs) within and across sectors, based on a sample of 300 firms [12].

This annual survey consists of 69 questions sent to senior innovation executives with responsibility for innovation management, in large firms with significant innovation investments and outcomes. The responses are grouped into eight functional types of IMPs by task or phase in the innovation process, each with a number of specific supporting tools (Table 6.7). Innovation performance is assessed by a composite index of product, process, service, and business model innovation. Case studies, workshops, and reports to respondents are used to validate the results.

Table 6.7 Groups of innovation management practices (IMPs).

Innovation phase or task	Examples of specific innovation management practices (IMPs)
Innovation strategy	Innovation goals and targets, use of cross-functional innovation steering groups
External business intelligence	Use of benchmarking and scenario, use of internal and external sources of data, customer segmentation
Idea management	Idea generation and assessment processes, use of ideation support, use of Lead-users, guidance from strategic priorities and customer needs
Product portfolio management	Process to align corporate strategy with portfolio, Integration of portfolio management into annual strategy setting process, Multi-criteria Analysis
Technology portfolio management	Use of technology Roadmaps, prioritization of clear technology focus areas
Development and launch	Use of milestone-based project reviews, use of stage-gate development, organization of development projects
Post-launch	Project learning, use of continuous product/service improvement teams, separation of product improvement budget from R&D budget
Resource and competence management	Capability assessment, use of competency strategy, use of programs to access skill within and outside the company

This analysis of the relationship between the use of IMPs and innovation performance reveals a strong, positive relationship across all sectors. We find that IMPs that support the following functional goals to be significantly associated with innovation performance, which can be mapped onto the process model:

1) **Strategy and selection:** To support technology/product/service portfolio reviews.
2) **Value creation and capture:** Technologies are evaluated in terms of their contribution to corporate goals.
3) **Organization:** Mobilizing the entire organization for innovation.
4) **Sources and search:** External sources (e.g. clients, suppliers, experts, academia, etc.) are exploited in a structured way.

The use and effectiveness of most other IMPs vary by industry and type of project. This suggests there is significant potential for the more widespread application of IMPs but that managers must be highly selective.

6.5 Summary and Conclusions

Innovation is more than simply applying technology, and can, and should, be managed to increase the likelihood of successful outcomes. There is a significant interdisciplinary body of knowledge that is distinct from business and management studies [13]. This knowledge can be best applied by adopting a process approach to managing innovation. It is possible to identify different innovation management challenges – strategy, organization, search, select, implementation and value creation and capture, and to apply proven methods and tools to support good practice. However, the standard process and practices need to be adapted for different specific contexts, in particular: the industry sector, which influences the sources of innovation, search strategies, and how value is created and captured; degree of innovation, which influences the selection criteria, organization and methods of implementation; and size of the organization, which will determine the optimal balance between internal and external resources and capabilities.

References

1 Rothwell, R. (1977). The characteristics of successful innovators and technically progressive firms (with some comments on innovation research). *R&D Management* 7: 191–206.

2 Tidd, J. and Bessant, J. (2020). *Managing Innovation: Integrating Technological, Market and Organizational Change*, 7e. New Jersey: Wiley.

3 Teece, D.J., Pisano, G., and Shuen, A. (1977). Dynamic capabilities and strategic management. *Strategic Management Journal* 18 (7): 509–533.

4 Tidd, J. (2001). Innovation management in context: environment, organization and performance. *International Journal of Management Reviews* 3 (3): 169–183.

5 Isaksen, S. and Tidd, J. (2006). *Meeting the Innovation Challenge*. Chichester: Wiley.

6 Chesbrough, H.W. (2003). *Open Innovation: The New Imperative for Creating and Profiting from Technology*. Boston, MA: Harvard Business School Publishing.

7 Tidd, J. (2013). *Open Innovation Research, Management and Practice*. London: Imperial College Press.

8 Tidd, J. and Bodley, K. (2002). Effect of novelty on new product development processes and tools. *R&D Management* 32 (2): 127–138.

9 Terniko, J. (2018). *Step-by-Step QFD: Customer-Driven Product Design*, 2e. New York: CRC Press.

10 Osterwalder, A. and Pigneur, Y. (2010). *Business Model Generation: A Handbook for Visionaries, Game Changers, and Challengers*. New Jersey: Wiley.

11 Schweitzer, F. and Tidd, J. (2018). *Innovation Heroes: Understanding Customers as a Valuable Innovation Resource*. London: World Scientific.

12 Tidd, J. and Thuriaux-Alemán, B. (2016). Innovation management practices: cross-sectorial adoption, variation and effectiveness. *R&D Management* 46 (3): 1024–1043.

13 Fagerberg, J., Fosaas, M., and Sapprasert, K. (2012). Innovation: exploring the knowledge base. *Research Policy* 41 (7): 1132–1153.

7

Innovation and R&D
Sudeendra Koushik

IEEE, Bangalore, Karnataka, India

7.1 Introduction

Every business today, more than ever, needs to innovate to sustain in their businesses and grow their business. As they say, either you are growing, or you are dying as a business. The need for new products, services, processes, user experiences, business models, and more is imperative for the market presence and growth of organizations, small or big alike. Innovation can deliver these innovations and help organizations thrive in their businesses. Hence, the role of innovation is extremely significant. Every business today is under pressure to innovate, and the pressure is increasing due to the speed at which technology is changing. Also, the added pressure is due to the fact that customers are evolving and so are their preferences, needs, and aspirations. The number of choices that are available to customers, specifically due to global network and exposure due to internet is mind-boggling. Due to globalization and ease of doing business, competition is very intense. To make matters worse, the competition is no longer coming from the expected or even from the same industry. Today competition to a business can come from virtually anywhere, even from another industry. The disruptors can be smaller organizations, from seemingly unrelated, unconnected businesses and locations. All these factors imply that, every organization faces the risk of being a slow innovator. Being slow is not an option and every industry and firm needs to innovate really fast. Regular innovations can keep the organization abreast of competition.

Innovation is at the heart of business success in today's economy, where every organization needs to be innovative in its offerings. This holds good for any industry, any location, or any size of the organization. According to the consulting firm two out of three organizations will disappear without innovation (www.paconsulting.com). Several companies which grew or disappeared in the recent 20 years, have clearly pointed out the link between change and adaptability. The link that bridges these two is innovation, which makes it so important it to be embedded in any organization. As Jack Welch, the prolific leader of GE once said, "if the rate of change outside is more than the rate of change inside an organisation, the end is near."

Innovation is one of the United Nations Sustainable Development Goals, SDG 9. According to UNESCO Institute for Statistics (http://uis.unesco.org/apps/visualisations/research-and-development-spending) Global spending on R&D has reached a record high of almost

IEEE Technology and Engineering Management Society Body of Knowledge (TEMSBOK), First Edition.
Edited by Gustavo Giannattasio, Elif Kongar, Marina Dabić, Celia Desmond, Michael Condry, Sudeendra Koushik, and Roberto Saracco.

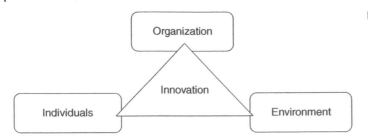

Figure 7.1 Innovation triad.

US$1.7 trillion. About 10 countries account for 80% of spending. As part of the Sustainable Development Goals (SDGs), countries have pledged to substantially increase public and private R&D spending as well as the number of researchers by 2030.

We will focus on innovation relevant for businesses, led by research and/or technology in the scope of this discussion.

7.2 The Innovation Triad

Innovation can be broadly seen applicable for B2C or B2B. Any organization whose growth strategy is led by research or technology innovation can be defined using the innovation triad. The innovation triad involves three elements (see Figure 7.1):

- The organization.
- The people in the organization.
- The environment in which the organization operates.

7.3 The Organization

The organization is the entity which is serving its stakeholders. Since the changes outside the organization are continuously evolving, the organization has to continuously adapt. Whether the change is in its core areas of operation or in the so-called "support functions," the organization has to adapt. Interestingly many times, the key areas to adapt are not in the core areas of the organization but in other parts of the organization. A bank has its core business in financial transactions, but new ways and means of mobile banking might force it to offer mobility solutions for how a customer deals and interacts with a bank, while the transactions themselves, be it transferring money, investments etc., might not have changed. Similarly, organizations might need a new way of procurement, a different way of hiring people, digital transactions and cyber security, and so on.

In the end, every organization depends on its people to achieve its goals. In order to achieve the expectations around innovation set by the organization, the individuals in the organization need to be supported with skills and tools, not to mention the processes to go with it. This will equip and empower the people to play their roles. The organization has to develop a solid strategy to pursue innovation for its growth. This is a top-down approach and is a must for the organization to be successful, especially when it is starting its innovation journey or when the organization is a large one, spread across the globe. This innovation strategy is crucial to deploy, develop and sustain a culture of innovation that many organizations aspire for. The expectations are to be set by the organization leadership to its workforce. This is not limited to its own workforce but also toward its partners. Many organizations are known to co-innovate with its partners and suppliers. This sets up the motive and motivation for the organization and the people to pursue innovation.

7.4 The People

Just because the motive is set and the expectation of innovation is set by the leadership, the people will not be able to deliver on that expectation. Expectation needs to be articulated across the organization, so that everybody understands what the organization is trying to achieve. The key reason is not so much about the fact that people do not appreciate it but more often they do not feel connected to this goal or objective, or strategy of the organization. As a leader one needs to understand that people need to connect to the message that the leader is putting out. That happens effectively when (i) they know clearly what the message is and of course they buy into that strategy or goal, but (ii) more importantly that they know their "role" in that journey. Many leaders fail to get commitment for missing out on communicating what role people have to play in the organization's journey. This message of expectations and role they need to play can be strengthened by providing adequate support to deliver on those expectations. Individuals come into the picture here to deliver on those expectations.

7.5 The Environment

The environment is beyond the organizations and the individuals. While the individuals discussed above were within the organization, the environment is outside the organization. This also implies the control of the organization was more within the organization and would be significantly less on the outside environment. The environment is critical for the success of the innovation journey ensued by organizations. This environment, which is outside the organization needs to be understood by the organization very well. Also, the environment should be conducive to develop, receive, and consume the innovation produced by the organization.

Businesses exist to make money. Many people we come across irrespective of their line of business are either unclear or uncomfortable to say so. In our experience, the bigger the organization the higher the probability of not hearing that businesses exist to make money.

7.6 Case in Point

In one of our engagements with a fortune 500 global MNC technology center, we posed this question, why do businesses exist? What was expected to be a quick discussion lasted almost 45 minutes. The group consisted mainly of senior leaders in both technology and business side of the organization. So there was a very good representation of the various functions of the organization; the answers were all over and each was defending their own view. To converge the discussion, we brought it down to two options. Option one, the reason businesses exist, in this case, their business which was mainly about computers and storage business, is to make customers happy. Option two similarly, was to make money. We also made it explicit that the second option of making money was assumed to be using ethical means, just to avoid stray conversations and assumptions. Those who said the purpose of their business was to make customers happy were majority. The fact was that they sold computers as one of their business lines. We suggested, especially to those who chose option one, that for instance, if they gave the computers they manufactured for free, there would be very many happy customers and their goal could be reached. Perhaps these happy customers would bring more customers, who would also turn out to be happy, and so on. There was another round of arguments and debates. To summarize this discussion, the leaders in the room were unable to separate the "what" part of the goal and the "how" part of the goal of their business, the vision, and the mission of their company. When we sum this up and state, the goal of our business is to make money through happy and satisfied customers it appears simpler and more apt.

7.7 Practical Tip

Lack of clarity of goal, vision, or strategy impacts decision-making and makes it a huge problem. People in the organization, at almost every level, makes hundreds of decisions all the time. The more senior you are in your role, the more significant decisions you make which will impact the business. Articulating the goal, vision, or strategy with clarity is one of the key skills for any leader. Purpose of the business should be clear, specifically the what and the how as well. Because when the purpose is unclear, one cannot ensure the right decisions are made. When the decisions are not well conceived, the organization goals becomes elusive not to mention managing or achieving them. If innovation is chosen and communicated to be the growth driver for the organization, then the day-to-day decision-making should include innovation. Decision precedes action; one will execute what one decides.

> In business, the competition will bite if you keep running, If you stand still they will swallow you.
>
> —*Victor Kiam, CEO of Remington*

7.8 Innovation, Research, and Business

It is important to understand how businesses and innovation are connected (see Figure 7.2).

Innovation helps a business to deliver value through new offerings and hence monetize the innovation. The innovation offering can be new products, new services, new processes, new

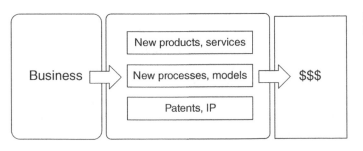

Figure 7.2 Innovation and business.

business models, new use cases, and more. Innovation could also help generate revenue for a business through intellectual property. This is typically through licensing the intellectual property owned by the organization.

It is also important to understand and appreciate the difference between research and innovation in the context of a business. Dr. Nicholson of 3M has articulated this difference beautifully. Dr. Nicholson opines, that research converts money into knowledge, while innovation converts knowledge into money (see Figure 7.3).

Usually, research projects are funded to generate new knowledge, new insights, new understanding, and so on. While the input to the research process is funding, the output of research is new discovery, insights and is published as a report. A good research always adds to the existing body of the knowledge. The outputs of research projects, which could be new knowledge, new insights, new materials, new process, and so on.

Through innovation, these outcomes of research could be used to create new products and services, business models, experiences, and so on. These could be sold to customers who would like to consume these products, services, experiences, etc. (see Figure 7.4).

Businesses scale innovation to deliver value to customers in the form of a product or a service, and in the process create wealth and well-being of its stakeholders (see Figure 7.5).

Organizations spend money to create new knowledge, about new processes, materials, etc. From the new knowledge innovation helps creates products and services, the main currencies of business and consumption. A business can then take that to masses and markets to enhance the quality of life of humanity. Businesses are being built and are being tested on sustainability (see Figure 7.6).

A business or an entrepreneur would then pick up this idea and convert it into a profitable and sustainable business. The startups do try to incubate a business potential before they build factories and warehouses, develop marketing and distribution channels and reach consumers for a sustainable business. It is important to understand that whatever be the innovation, monetising the innovation is key to reap the benefits of research and also enhance the standard of living through new products, services, experiences, and so on.

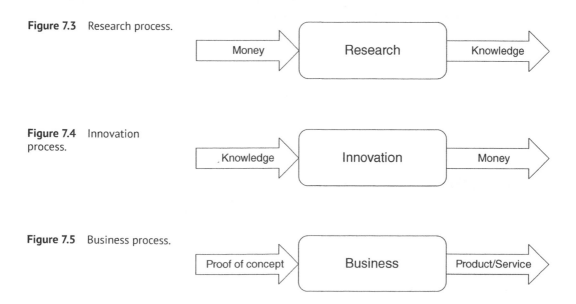

Figure 7.3 Research process.

Figure 7.4 Innovation process.

Figure 7.5 Business process.

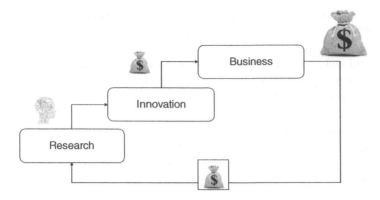

Figure 7.6 Research, innovation, and business.

7.9 Monetizing Innovation

There are three important ways to monetize innovation at an organization level as follows:

- New products.
- New services.
- Intellectual property.

These three important ways to monetize innovation can be applied mostly to "Top-Line" innovation which is described later in this chapter (see Figure 7.7).

As an example, suppose a research organization comes up with a new material which is much stronger and lighter than regular material used in making luggage or carry-on bags. This organization has sponsored or funded this project and hence has converted the funding, i.e. money into new knowledge, i.e. the new material. This is the research part, where a detailed report about the test, characteristics, etc. would be published, and possibly a patent be filed or intellectual property protection be sought.

To take this example further, now that the new knowledge is available, an innovation would enable the conversion of this knowledge into money. This can be done by designing and developing a new product using this newly discovered material. This new product, in this case, could be a

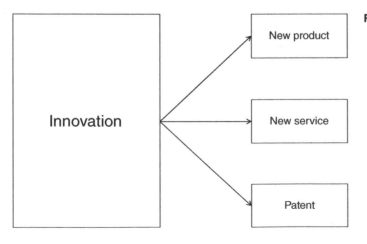

Figure 7.7 Monetizing innovation.

new carry-on bag generally used by air travelers to take it with them on board a flight. This new product would be different from the regular one, in that the new luggage would be much lighter than the conventional carry-on luggage. This lightweight bag was possible due to the knowledge of the new material by research. This "new product" could be marketed to potential customers, frequent travelers, who would need lightweight luggage. The potential customers could be specifically air travelers. Why it could be air travelers because there is a restriction of both size i.e. form factor and also the weight of the carry-on luggage. This new product, the "lighter bag" could enable frequent travelers to carry more of their own stuff compared to the heavier, conventional bag. Hence innovation has enabled a business-worthy product. Whether this would be a sustainable business which is scalable and profitable is another point.

7.10 Case in Point

Aircrafts have improved vastly since the time of wright brothers. And one of the aspects that has made a huge difference is the weight of aircraft. Ideally, the payload of aircraft compared to its own weight should increase for a profitable business. From the model of research, innovation, and business we can clearly see the link in this example too. Research developed new materials called "composites," which are lighter but stronger than conventional materials. Innovation made it possible that these new lightweight composites can be used in aircrafts and Boeing built Dreamliner. According to Todd Johnson, The Dreamliner has an airframe comprising nearly 50% carbon fiber-reinforced plastic and other composites. This approach offers weight savings on average of 20% compared to more conventional (and outdated) aluminum designs [https://www.thoughtco.com/boeings-787-dreamliner-820385].

> The value of an idea lies in the using of it.
>
> —*Thomas Edison, co-founder of General Electric*

7.11 Value in Innovation

Innovation is generally describing as problem-solving. In fact, innovation in the context of business, is much more than problem-solving and providing a solution. Innovation in a business context does not succeed if it just solves a problem. Innovation succeeds if it converts the solution into a viable and sustainable business, by adding "value" to a user and to the society (see Figure 7.8).

Innovation is about delivering "value" by organizations through a product or a service, but the key is value. The customers pay for a product or service, but they actually consume a value offering in the form of a product or as a service. Innovation in business becomes successful innovation when it adds value to a consumer and that consumer would be the one who would pay for it. The value in the example of the carry-on luggage was that the new lighter luggage helps air travelers to pack more in their carry-on baggage. Not only this but if it happens that these air travelers are looking a faster way in and out of an airport and perhaps avoid check-in baggage altogether, this solution could be very "valuable" to them. Then how does it create value to the luggage manufacturer? This new product could extend the range of products that a Samsonite or American Tourister as a company is offering. This product could help build the brand of the company as a user-friendly, technologically advanced company. Nike or Adidas among others, always markets its products including the new materials that it uses, be they recycled material or user-friendly materials.

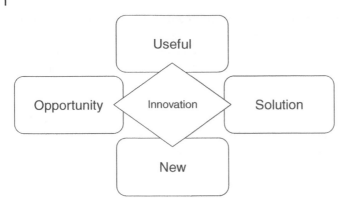

Figure 7.8 Innovation in business.

7.12 Value Alignment

Focus on how the end-users, customers will perceive the impact of your innovation – rather than on how you, the innovators, perceive it.

—*Thomas Alva Edison*

Value in the context of innovation is a combination of things. The value is also a matter of perspective. If we classify the parts of innovation in business, as the "maker" who develops this innovation, and as the "user" of this innovation, the perspective for value becomes clearer. The value should be for the maker or the user? Should the value be for the organization who is the maker or to the consumer who uses it? Traditionally we would hear people answer that it should be valuable to the consumer who uses it. Absolutely true, but that is not sufficient for a successful innovation in a business context. The innovation should be equally if not more, valuable to the maker as well. When both of these conditions are met, we have a possibility of a sustainable business led by innovation. If the innovation is only valuable to consumers, then the business will suffer. This situation will make it unattractive to the business as there may not be any substantial benefits to the maker, the organization. If the Innovation is only valuable to the maker, the organization then also the business will suffer. This is because, if there is no value to the customers, there would not be any demand for this innovation, and hence the profitability and sustainability will be challenged. Majority of the startups end up as unsuccessful because of the latter, where the value to the consumer was undermined and the value to the maker, the startup was accentuated. To make it a win-win, where both the users as well as the makers are benefited, value has to be understood and appreciated from both the perspectives. Apple does not make phones because it is beneficial to them only. People want to own and experience apple devices too. Since Apple as an organization want to deliver those experiences and people are willing to buy those experiences, the value is mutual and beneficial from both sides. This mutually perceived value sets up a sustainable platform for business.

An ax without a shaft is no threat to the forest

—*Bulgarian proverb*

Users pay money for buying a product or a service of their choice. They do so as they expect a value in return for themselves through this product or service. The organization who delivers this product or service, in other words, the maker, delivers the value that the customer is looking for

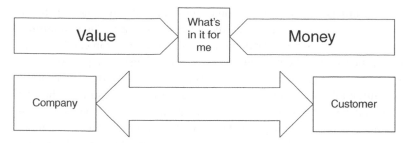

Figure 7.9 Value in innovation.

through this product or service. When these two "values" matches, the value proposed by the maker and the value perceived by the user, a potentially successful innovation led business becomes possible. The value, in other words, is also known as "what is in it for me." The focus of innovation in business is the meeting point of these two into one common point. This happens when both the user's and maker's "what's in it for me" are in sync.

Many organizations and startups struggle to achieve the "synchronization" of value or "what's in it for me." The cases where there is no synchronization are not rare but can be found regularly in larger as well as smaller organizations. When the two what's in it for me do not align or coincide or synchronize, the organization delivers innovation that users do not perceive as valuable. This is recipe for failure. If the organization continues to do so regularly, delivering products and services to users (remember the users perception of the value proposition is important) there will be a longer term damage to the organization in terms of track record but also erosion of brand equity (see Figure 7.9).

7.13 Top-Line and Bottom-Line Innovation

Innovation for a business can take many forms. From a business perspective, it can be categorized as follows:

- Top line innovation.
- Bottom line innovation.

While, in top-line innovation the focus of innovation is to "increase the revenue," in bottom-line innovation the focus of innovation is to "decrease the cost." Typically, bottom-line innovation is more internally focused. External facing innovation is ideally pursued under top-line innovation (see Figure 7.10).

Figure 7.10 Top-line and bottom-line innovation.

What the customer can see

Top line

Bottom line

What the company can see

Innovation

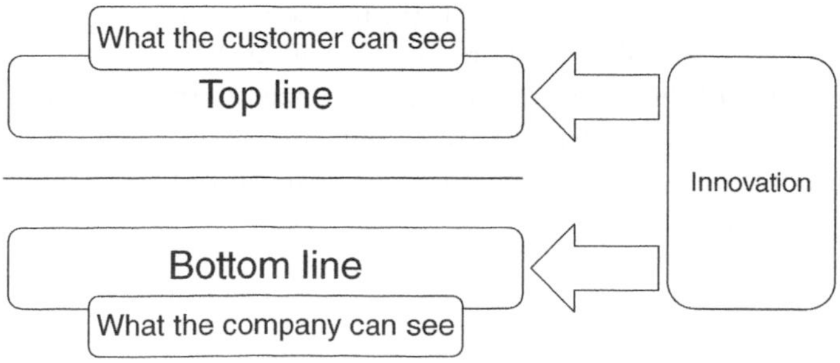

Top-line innovations are those innovations that lead to an experience that can be demonstrated to the end user. This innovation is meant to enhance the experience of the user. When an innovation that enhances the experience of the user, can be demonstrated, the innovation can be potentially monetized too, assuming the value of the innovation exists from the user perspective. As an example, in an automobile, if the automaker comes up with an innovation where the wipers of the front windshield switches on automatically, it can be demonstrated to the user. Assuming the user appreciates the value and would like to have that feature, it can be monetized too. Note all innovations need not be monetized at a product or service level. Some features are not even fully demonstrable. Again, in the example of an automobile an Anti-skid Brake System (ABS) or a Secondary Restraint Airbag system (SRS) are innovations that are explainable but not demonstrable per se. But the value can be articulated using simulations or with the assistance of videos, etc. The key in top-line innovation is how the innovation helps the customer. When top-line innovations are successful, they usually allow the organization to sell more products with these innovation or allow the organization to charge more for the products with these innovations. This results in the growth of total revenue for the organization.

An eyewear, such as a reading glass, is a product that is sold on multiple innovations. All the innovations can be chosen separately and are also charged separately. Some of these features, top-line innovations are that the lens can be acrylic or glass, the lens can be having a oleophobic coating, the lens can have the blue-light protection coating, the lens can be photochromatic, and so on. All these are clearly the value that the users can get and are also in this case easily demonstrable. Since each of them are mutually exclusive and to be paid individually, the user chooses what features are of "value" and which they could pay for. For example, for someone who does not use laptop or mobile often, might not find blue-light protection valuable. For someone who spends more time in the sunlight, the photochromatic feature might be valuable. This eyewear company can sell more products as they have multiple innovations, serving multiple audiences. This would result in a higher revenue for the company.

Contrary to top-line innovation, in bottom-line innovation, the organization innovates for the benefit of the company. What could be beneficial to a company or organization? Most obvious and common benefit would be in reducing costs – cost of operations, cost of procurement, cost of hiring, etc.

A typical business needs usually both top-line and bottom-line innovation to sustain. The ratio of efforts into top-line or bottom-line innovation can vary based on the strategy of the organization. It is possible that the two types of innovations are interconnected. A top-line innovation can lead to bottom-line benefits and the other way around too.

In top-line innovation, the innovations have a direct impact on the user. The various benefits of the innovation are directly linked to the user. This helps the business to create the link between the innovation and how it benefits the user, and monetization is possible while demonstrating the benefits of the innovation to the users. The top-line innovation is what the user can see. This, in turn, will help (see Figure 7.11) the business generate "revenue."

In bottom-line innovation, the innovations have a direct impact on the business. The various benefits of the innovation is directly linked to the company or business doing it. This type of innovation helps the business to improve what happens inside the company. The bottom-line innovation is what the company or business can see. The innovation helps in improving the bottom-line.

Figure 7.11 Top-line and bottom-line, revenue vs. cost.

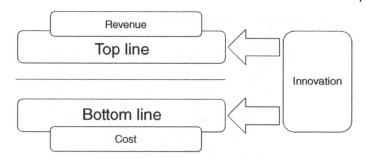

7.14 Success Factors for Innovation in Business

In innovation, four factors have a direct impact on the success of innovation and business. The four factors are

1) Value of the problem is a key element, as this is the situation that people want to get rid of. This problem could exist in the way people live or work. Not every problem is a problem to everyone. What is a problem to one may not be to another.
2) Effectiveness of the solution is how well the solution solves the problem, for whom it is a problem as described above. This also includes the solution is available at what cost, convenience, experience, etc.
3) User value is how the user perceives the situation with the solution as compared to the situation without the solution being offered. Many a times, a solution creates additional problems or may not solve the problem effectively. Remember the user gets to decide the value of any innovation, and this perception is the user's perceived value.
4) Business value is the profitability, sustainability, and strategic importance for the organization pursuing this innovation. Business value includes the business case of pursuing an innovation for the organization. It is important to note (see Figure 7.12) that the benefit or the return on Investment (RoI) need not always be money.

Figure 7.12 Components of successful innovation.

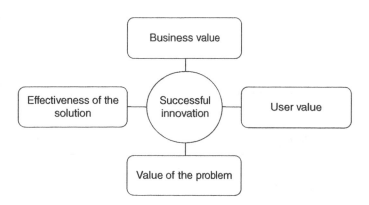

7.15 Scope for Innovation

In top-line innovation, there are no limitations, in which areas to innovate, like introducing new features, new products, new services, new experiences, and so on.

Similarly in bottom-line innovation every aspect is an opportunity to innovate, including new business models, new processes, new organization structures, new materials, new supply chain, and so on.

Every function of an organization is an opportunity to innovate (see Figure 7.13), as the respective experts are in those groups and departments.

Whether it is top-line or bottom-line innovation, it need not be restricted or meant to be in certain areas or few departments of an organization. It is a common myth that the R&D or design department of an organization is responsible for innovation. As a matter of fact in many cases, the R&D department of a large organization will be a single digit percentage of the entire organization headcount. Every function and every interaction that an organization engages in, is a potential area for innovation.

A purchase department is equally equipped to deliver innovation in how it interacts with supplier, exchanges documents, manages approvals of designs and parts, and so on. The human resource department is equally expected to Innovate in how effectively and efficiently it manages its people, or organizes reimbursements, or conducts programs, and so on. That means every person in every function is an opportunity (see Figure 7.14) to be an Innovator, in principle.

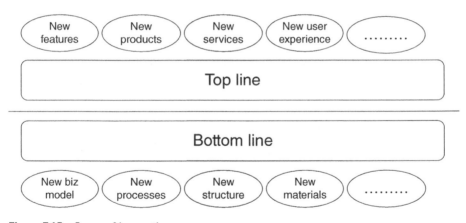

Figure 7.13 Scope of innovation.

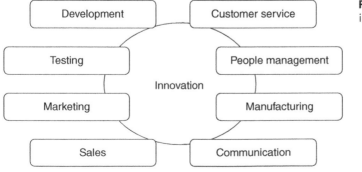

Figure 7.14 Functions and innovation.

7.16   Case in Point

A company was situated in the suburbs of a major city. Being a major city, it was prone to its share of traffic owes, which meant long travel times from home to work and back. This was starting to result in attrition where people started choosing a different company to work for which would be better for traveling. The exit interviews by the human resources (HR) department showed that many of those who left the organization, did so not by choice but were sort of forced to do so. Though they liked the company, the work they were doing, and the work culture, their travel time skewed the work-life balance, and they also remarked that the travel was too strenuous and time-consuming. Moving closer to the company could have helped, but they felt the accommodation and housing in and around that suburb were on the expensive side. This presented an opportunity for the HR team to innovate. Instead of finding new people all the time, which would cost the company a lot of time and money, they went about innovating how to fix the attrition due to this reason. They combined the facts that people wanted to work for this organization and the other fact that housing nearby was expensive and introduced a new scheme called "nearness allowance," to encourage people to stay closer to their offices, which would in turn reduce traffic congestion and transit time for its employees. This organization looked at the scheme Innovatively and incentivized the same.

7.17   Internal and External Innovation

A successful Innovative organization will realize that the internal and external innovation value chains are rather different processes. However, the various parameters discussed earlier are applicable to both trajectories. The processes are rather different, but the goals are same – to create value to the customer through an innovation.

Internal innovation has been more an "in-control-of" process various executives. This is due to the fact that the teams are inside; the execution, the budget, and change management are all in the direct control of the organization. Having said this many organizations report the internal process of innovation can be a slow process.

External innovation is more "giving away control" process. Those who are familiar with internal innovation processes struggle with external engagement as it makes them behave in a relatively opposite way, the known way. In external innovation executives have to work with external parties such as startups or other companies with innovation, even academia and subject-matter experts where information or input has to be provided, and the results are to be awaited. It is also possible that the already available solutions or innovations with the external parties may not be ready to be used by the organizations. In such cases a joint development can happen where there is significant contribution from both sides. Both parties can come up with intellectual property and an agreement upfront helps manage the outcomes of such joint development. The executive and the execution team of engineers of large organizations may behave rather differently compared to say a startup here. Speed for one can be a variable wherein the startup tries to move fast and the organization due to its very nature of size can be slow. Engineers often have to overcome a "not-invented-here" syndrome as they feel challenged by the solutions on the table provided by the external parties. Control of the large organization is not as much as they have during internal innovation process. This control is not just about the joint development project but also who all the startup or partner is working or wants to work with. This is where a good agreement helps to manage this efficiently.

So, the goals and what makes an innovation successful would not change much between external and internal – however, the way of working between them will be anything but same.

7.18 Conclusion – The Innovation Conundrum

A successful innovation is one that gets implemented and delivers the results as planned and expected. Though there seems to be so much of opportunities and possibilities, we often hear "Very few of our ideas turn out as good innovations leading to new products and services." This is a very often heard complaint from senior leaders across organizations. The word "often" stems from the fact that many of the innovations seems to be of very poor quality and do not make it to the end. Interestingly this conundrum is a problem faced by companies who have the best of the talents available on the planet and by organizations who are very competent and experienced in their domains. These organizations still do not deliver on innovation even as per their own expectations.

Further Reading

1 Dyer, J. and Christensen, C. (2009). The Innovator's DNA. *Harvard Business Review* 87 (12): 60–67.

2 Piva, M. and Vivarelli, M. (2009). The role of skills as a major driver of corporate R&D. *International Journal of Manpower* 30 (8): 835–852.

3 Leiponen, A. (2005). Skills and innovation. *International Journal of Industrial Organization* 23 (5/6): 303–323.

4 Spithoven, A. and Teirlinck, P. (2010). External R&D: exploring the functions and qualifications of R&D personnel. *International Journal of Innovation Management* 14 (6): 967–987.

5 Katzenbach, J. and Vlak, G. (2008). Finding and grooming breakthrough innovators. *Harvard Business Review* 86 (12): 62–69.

6 Barsky, N.P. and Catanach, A.H. Jr. (2011). Every manager can be an innovator. *Strategic Finance* 93: 22–29.

7 Chell, E. and Athayde, R. (2009). *The Identification and Measurement of Innovative Characteristics of Young People*. NESTA (National Endowment for Science, Technology and Arts.

8 Peterson, F., Kerrin, D.M., Gatto-Roissard, G., and Coan, P. (2009). *Everyday Innovation – How to Enhance Innovative Working in Employees and Organisations*. NESTA (National Endowment for Science, Technology and Arts).

9 Gaynor, G.H. (2001). Innovator: what does it take to be one? *Antennas and Propagation Magazine, IEEE* 43 (3): 126–130.

10 Martins, E.C. and Terblanche, F. (2003). Building organisational culture that stimulates creativity and innovation. *European Journal of Innovation Management* 6 (1): 64–74. https://doi.org/10.1108/14601060310456337.

11 Scarbrough, H. (2003). Knowledge management, HRM and the innovation process. *International Journal of Manpower* 24 (5): 501–516. https://doi.org/10.1108/01437720310491053.

12 Johannessen, J.-A., Olsen, B., and Lumpkin, G.T. (2001). Innovation as newness: what is new, how new, and new to whom? *European Journal of Innovation Management* 4 (1): 20–31. https://doi.org/10.1108/14601060110365547.

13 Amabile, T.M. (1988). From individual creativity to organisational innovation, Chapter 5. In: *Innovation: A Cross-Disciplinary Perspective*. Oslo: Norwegian University Press.

8

Women in Technology and Innovation

Elif Kongar[1] and Tarek Sobh[2]

[1] Economics & Business Analytics Department, Pompea College of Business, University of New Haven, West Haven, CT, USA
[2] Lawrence Technological University, Southfield, MI, USA

8.1 Introduction

The role of underrepresented and underprivileged populations in the ever-emerging science, technology, engineering, and mathematics (STEM) fields, in addition to other interdisciplinary technologically-dependent professions, has always been a concern with unsatisfactory results despite several well-planned initiatives. Specifically, the role of women in technology and innovation has been an ongoing discussion, almost equal to the efforts of increasing the contribution of the female workforce to this exciting field. Despite the growing numbers in engineering education, for example, the yield to academia and industry has always been relatively low compared to men.

The literature focusing on the same issue fell short in addressing the lack of female workforce in technological fields, even though many other professions have witnessed growth in the entry of women into fields where rigorous education and effort were needed to better understand the discrepancy between fields where women are underrepresented, gaining insight regarding the defining characteristics of environments where academic preparation and professional practice take place [1].

Focusing on the first aspect, the preparation for technical careers, Etzkowitz et al. [1] argued that there was no evidence that showed lack of encouragement by personal caregivers, academic mentors, or from the media. They, however, found evidence that would imply exclusion from the profession either before or after entry. Barriers to entry and success were specifically directed to female workforce two decades ago. This chapter also claims that these barriers still do exist, however, veiled and well disguised as they may be, today.

Heavy utilization of technology by underprivileged and underrepresented groups is promising. In particular, social media technologies allowed several groups to gain power and have a voice. However, the participation in creation of these technologies has historically been male-dominated. Perhaps, data science and other fields in exponential technologies will remain to be one of the highly demanded professions where the need will never exceed the ready workforce.

IEEE Technology and Engineering Management Society Body of Knowledge (TEMSBOK), First Edition.
Edited by Gustavo Giannattasio, Elif Kongar, Marina Dabić, Celia Desmond, Michael Condry, Sudeendra Koushik, and Roberto Saracco.

This changing landscape is also the main driver of the need for fundamental restructuring in STEM education and practice. Today's global and highly interactive, consumer and society-driven business ecosystem require unique approaches to training future STEM workforce. There are many reasons for this. First and foremost is the increased availability and accessibility of information. Directing the future workforce to value-adding data and information is one of the key elements of training capable employees. In this regard, mentoring underrepresented women is as significant as equipping them with skills. In addition, with rapidly advancing technologies, fostering a culture of continuous learning for groups that are historically underrepresented is also important. Another, perhaps more important driver, is the interdisciplinary nature of the current technological advancements. Today, tearing down the walls between STEM and other fields and creating well-rounded individuals who possess technical communication, data visualization, entrepreneurship, project management, team-building, and leadership skills is key to success for especially talent-scarce small-and-medium-sized businesses and industries. Therefore, it is important to create awareness of focusing on technology and innovation from a wider lens, not limiting the discussion to STEM-related disciples, which can serve as a deterrent.

Women in STEM also suffer from lack of networking opportunities. In this regard, building long-term collaborations with industry via women-led mentorship opportunities would establish an effective bridge between academia and industry, thus streamlining the female workforce interested in STEM careers. The mentorship structure would potentially generate internships that are known to help graduate career-ready students with an existing established network. Provided that the educational community is now training students for job titles that do not exist yet, skill development and experiential learning are paramount when technology- and innovation-related positions are concerned [2]. Industry, which is now a driver of emerging technologies, would also benefit from internships via gaining close proximity to the future workforce and allowing them to contribute to the curriculum.

8.2 The Changing Landscape

The rapid enhancements and deployment of Internet of Things (IoT), artificial intelligence (AI), robotics, machine learning (ML), and other technology-driven advancements are predicted to continue to transform professional and personal lives. Compatible with this trend, and as an example, within the United States, STEM occupations are projected to grow over two times faster than the total for all occupations in the next decade reaching 7.5–10% growth [3]. The growth of the digital economy along with the increasing need for risk management, and changing rules and regulations, increase the need for management occupations groups over the next decade.

Ending all forms of discrimination against women, including wage discrimination, is one of the 17 Sustainable Development Goals (SDGs) of the United Nations. Project Drawdown also highlights the importance of women's political, social, and economic empowerment expansion [4]. Figure 8.1 depicts the importance of wages to achieving the SDGs. It should not be surprising that the sustainable growth of the new business landscape and, more generally, of local, national, and global economies also rely on the empowerment of women, which is integral to all dimensions of inclusive development.

The need for professionals who are capable of managing new technology and innovation-driven businesses is also expected to grow, given that their managerial positions depict resistance to automation effects due to their essential role in navigating technological change [5]. Advancements in innovation and technology, while rendering some of the workforce redundant, will also increase

Figure 8.1 Wages as an accelerator for achieving the UN Sustainable Development Goals. *Source:* Adapted from the UN Department of Economic and Social Affairs site: http://sdgs.un.org/.

the need for some unique skill sets which are inherently innate to women. For example, consensus building, conflict resolution, and creative decision-making are three skills that the next-generation industry calls for.

It is not a secret anymore that females are exposed to a set of different and, in many cases, more strictures than their male counterparts. Among these, personal, societal, and financial considerations are three factors that steer women away from pursuing demanding careers in technology and innovation. For example, in the United States, female high school students made up only 15% of engineering technologies concentrations, 8.5% of manufacturing, 14.5% in computer and information sciences, and 9.6% in construction and architecture [6].

At this moment, women still remain underrepresented in the STEM academic workforce due to systemic gender inequalities including numeric underrepresentation, lower social capital, and threatening academic climates [7]. Many of the laws, legislations, regulations, or directives are decidedly inadequate and lack sufficient incentives or mandates. The majority of professional women repudiate taking part in active platforms often due to lack of time, compensation, and/or support.

There are predictions that by 2030, the middle classes will be the most important social and economic sectors in the globe [8]. Efforts that do not target long-term change serve more as a conversation-starter, but are not effective in challenging the status quo for the female workforce of the future. Despite the existing efforts, a deeper understanding of the factors that motivate or discourage women from choosing a career in STEM field is needed, and following are some of these factors [9]:

- Women have the ability and drive to succeed in science and engineering.
- Women who are interested in science and engineering careers are lost at every educational transition.
- The problem is not simply the pipeline.
- Women are very likely to face discrimination in every field of science and engineering.
- A substantial body of evidence establishes that most people – men and women – hold implicit biases.
- Evaluation criteria contain arbitrary and subjective components that disadvantage women.
- Academic organizational structures and rules contribute significantly to the underuse of women in academic science and engineering.
- The consequences of not acting will be detrimental to the nation's competitiveness.

Having more women in technology and innovation fields would not only bring the unique female perspective and thinking into the creation of new products and services, but it would also increase the number of available technologies that are specifically designed to address the needs of women. The efforts in identifying and eliminating the impediments women face on their journey to these fields would also strengthen their role in society. Historically growing numbers of females in STEM disciplines do not lend themselves directly into professions as technologists and innovators. Invidious compensation, promotion, networking, and recognition practices set invisible and visible barriers for females.

Unless the society inculcates the value of female contribution into technology and innovation, and embraces an inclusive culture that cultivates and promotes equal representation, underrepresentation of women will continue to be an unsolved problem. Milgram [6] argues that frequent exposure to women in technology occupations would aid in eradicating the common misperceptions about the female workforce and help in associating the image of females with technology. Bringing exposure to females in technology and innovation via effective community outreach activities while eradicating the wage gap, creating incentivizing recruitment strategies, and shifting the entry and success barriers paradigm to support and award effective efforts would ultimately lead to higher numbers joining the professional pipeline.

8.3 Methodology and Recommendations

Challenges specific to women in technology and innovation fields are best addressed by creating preventive and proactive responses [10] to known obstacles and barriers. Feedback from the workforce professionals who hold positions in these fields gain significance to create a platform where these otherwise hidden impediments transform into visible problems that need to be addressed and resolved. Unless these problems can be carried into a mutual shared platform, common issues continue to be solved by individuals as they appear, and does allow a body of knowledge and common wisdom to be generated and nurtured.

As a prefatory to the discrepancy between female and male workforce in technology and innovation fields, it is customary to observe the historical data. Indicators of such discrepancies are reflected in the number of positions held, compensation levels, and average duration of employment by gender, in addition to other criteria. The problem gains complexity when the data is not visible to outside researchers. When the underlying issues are investigated, many problems remain hidden, often unsolved, or solved by individuals without being shared with other parties, making in-depth analysis a challenge. Furthermore, the compensation of female workforce has historically been below men's salaries.

To this day, there is no proof of concept to cement women's established position in related fields. Majority of the efforts offer no permanent solutions and fail to establish a systematic action plan to address and resolve the impediments exclusive to the female workforce. As well-intentioned as they are, overall, they are a farrago of disconnected activities with no clear goal plan and hence fail to be impactful. Despite the number of studies that focus on increasing the contribution and involvement of female workforce, the number of women in technology-driven fields continues to be outnumbered by men. Despite the growing number of women in STEM, Hill et al. [11] also indicate that the difference grows larger as we move into the upper levels of these professions.

Perhaps the solution lies in taking gender-conscious actions that actively attempt to increase female participation. Creating groups where all genders are equally represented would increase the knowledge base and overall synergy of the project teams regardless of the task. Setting the tone carries weight when it comes to sustaining such efforts. For the supporters of the cause, it is often required to act like a fugleman so that the message to parties that are in the same network would be conveyed.

The role of women in technology is not field specific but rather universal. Unless it is well understood that regardless of the field, inclusion of women in the workplace has benefits to the national economy, the grassroots movement will lack the support of clear messaging. According to McKinsey, 12 trillion additional GDP can be generated if the gender gap is narrowed in the next five years [12]. Again, regardless of the industry, it is crucial that companies invest in addressing the challenges women are facing [13].

Women in the workplace, especially women in leadership positions, are perceived differently than men even though their actions might be identical. This Gorgon-like stigma not only hinders opportunities for career advancement but also shortens their durations within the positions they hold. There is also significant bias and prejudice against women in industry and academia. It becomes almost quotidian for the women in workplace to be vilified, especially in male-dominated fields and in positions of power. The role of females as mentors often involves edifying the young female workforce, in addition to mentoring and even tutoring. The typical first instinct is often to overlook the insinuations or undermining remarks even though the solution lies in communicating clearly and not condoning any hostile comments, behaviors, decisions, or actions.

Despite the overwhelming nisus, studies, disquisitions, initiatives, and awareness regarding the inadequate participation of women in technology, a systematic approach to help ratiocinate the impact of current academic and industry landscape would aid in creating meaningful and impactful solutions. The twists and turns in the diegesis of women's journey are almost puzzling and often highlight the success stories and personas that are exceedingly rare. Women testimonials are rare and typically never provide ample evidence for bias and discrimination. They, therefore, fall short in depicting the true picture and in offering effective solutions to overcome obstacles that are exclusive to the female workforce. Women are often forced to create a tougher version of themselves exhibiting a gender neutral, almost removed superficies in the workplace, often avoiding any mention of personal and family-related issues.

Increasing and sustaining women enrollment in STEM can only be possible by a collaborative effort rather than independent individual institutional projects. In this regard, universities, businesses, and government agencies should work together to support research, create programs, and develop and implement regulations to promote women in engineering, technology, and computing. Industry also has a responsibility to accelerate the readiness of the female workforce, which will help alleviate skills availability issues. Honorary societies, funding agencies including federal agencies and foundations, along with the Congress should take the necessary steps to encourage adequate enforcement of female workforce development. There is also no doubt that family-friendly policies help women to combine paid jobs with family work [9]. Emphasizing in the media and other agencies' communiqués the suitability of the technologically-focused jobs of the twenty-first century with their unique requirements such as interdisciplinarity, need for teamwork, great communication skills, sustainability, presentations skills, and need for direct client/customer interaction and feedback for the female workforce, students, and mentees would be very helpful.

8.4 Conclusions

As a unique lagniappe to organizations, women create a workplace that is outcome oriented. Deputing the proper response and corrective actions to others is rarely the solution. Unless there is a common and collective wisdom that would hold against the common misperceptions and mistreatments which have penetrated into the daily vocabulary and activities of individuals, a fundamental change will be a farfetched utopia. Being more demiurgic is innate to the female workforce, especially when innovative projects are concerned.

There is a gap between the number of women entering STEM fields and the number who should be contributing in these areas. The types of skill sets needed have changed along with changes in the economy to focus more heavily on skills that women more typically possess such as organization, communication, and flexibility. There are several corrective actions that can be considered to attract women to the field and more importantly to get them to stay in the field.

Women are well-positioned to make major advances in interdisciplinary research. They like to integrate across various academic fields and use multi-pronged, multi-disciplinary approaches. They perform research well in teams and are committed to connecting their research with societal concerns. Using interdisciplinarity to attract women to the technology jobs of the future as well as other underrepresented minority groups into science and engineering is only practical and ethical if it leads to stable and secure pathways through scientific, technological, and academic careers [14]. We look forward to collective action, policies, and legislation to be developed as a collaborative effort between academia, industry, the media, and government to affect the required change.

References

1 Etzkowitz, H., Kemelgor, C., and Uzzi, B. (2004). *Athena Unbound: The Advancement of Women in Science and Technology*. Cambridge University Press.
2 Galbraith, D. and Mondal, S. (1997). The potential power of internships and the impact on career preparation. *Research in Higher Education Journal* 38: 9. https://files.eric.ed.gov/fulltext/EJ1263677.pdf.

3 U.S. Bureau of Labor Statistics (2021). *Employment in STEM Occupations: U.S. Bureau of Labor Statistics*. U.S. Bureau of Labor Statistics. https://www.bls.gov/emp/tables/stem-employment.htm (accessed 10 December 2022).

4 Hawken, P. (2021). *Regeneration: Ending the Climate Crisis in One Generation*. Penguin Books.

5 Rieley, M. (2021). *We Can Manage: Employment Trends for Management Occupations: Beyond the Numbers: U.S. Bureau of Labor Statistics*. U.S. Bureau of Labor Statistics. https://www.bls.gov/opub/btn/volume-10/we-can-manage-employment-trends.htm (accessed 10 December 2022).

6 Milgram, D. (2011). *How to Recruit Women and Girls to the Science, Technology, Engineering, and Math (STEM) Classroom*. Technology and Engineering Teacher.

7 Casad, B.J., Franks, J.E., Garasky, C.E. et al. (2020). Gender inequality in academia: problems and solutions for women faculty in STEM. *Journal of Neuroscience Research* 99 (1): 13–23. https://doi.org/10.1002/jnr.24631.

8 National Intelligence Council (2012). *Global Trends 2030: Alternative Worlds (Global Trends Reports)*, 2e. Center for the Study of Intelligence.

9 Kongar, E., Russo, N., and Sobh, T. (2009). Women in science, engineering and technology: changing roles and perceptions within the technical services industries. *Journal of Engineering and Applied Sciences* 4 (1): 46–50.

10 Orser, B.J., Elliott, C., and Leck, J. (2011). Feminist attributes and entrepreneurial identity. *Gender in Management: An International Journal* 26 (8): 561–589.

11 Hill, C., Corbett, C., and St Rose, A. (2010). *Why So Few? Women in Science, Technology, Engineering, and Mathematics*. Washington, DC: American Association of University Women.

12 McKinsey & Company (2021). *Women in the Workplace 2021*. McKinsey & Company. https://www.mckinsey.com/featured-insights/diversity-and-inclusion/women-in-the-workplace.

13 McKinsey and Company & http://leanin.org (2020). *Women in the Workplace 2020*. McKinsey and Company. Published. https://wiw-report.s3.amazonaws.com/Women_in_the_Workplace_2020.pdf.

14 Rhoten, D. and Pfirman, S. (2007). Women in interdisciplinary science: exploring preferences and consequences. *Research Policy* 36 (1): 56–75.

Section 5

Entrepreneurship

9

Roadmap for Entrepreneurial Success

Arun Tanksali

Nearex Pte Ltd, Bengaluru, India

9.1 Introduction

Entrepreneurship is best defined by building on Schumpeter's [1] description [2] as a "process that harnesses innovation to create an organization and deliver economic value." This definition contains the three key elements that are at the heart of entrepreneurship – Innovation, Economic Value, and Organization.

Innovation, as has been covered earlier in this *Book of Knowledge*, takes the form of new products or processes, new markets or business models, new sources of resources, or any other aspect that provides a differentiating factor.

Economic value is a reflection of value derived by customers of the enterprise, measured in revenue, profit, or valuation. Today, economic value is no longer considered the only measure of value. Social entrepreneurs and impact entrepreneurs consider [3] value delivered to society, beyond the economics, as equally important.

The organization represents the formal structure that entrepreneurship takes, covering the 6Ms – Manpower, Method, Machine, Material, Milieu, and Measurement [4]. This, depending on the jurisdiction, could be any kind of a limited liability company, a partnership, a limited liability partnership, a proprietorship, or any other form that's permitted in the jurisdiction.

Entrepreneurship, by its very nature and as is evident by its definition above, defies attempts to be captured as a single definitive source of truth. However, there are many aspects of the entrepreneurial journey that have become well-understood, and there are regular courses that are designed to teach how to become an entrepreneur. This section attempts to distill that knowledge into a set of pointers that will allow a prospective or active entrepreneur to understand the subject.

This section is structured to reflect the typical life-cycle of an entrepreneurial journey that one might go through. The five stages of the life-cycle described below are a convenience to help group related aspects together. It is, of course, to be clarified that not every entrepreneur will go through every one of these stages nor when they do, will it be in this exact sequence (see Figure 9.1).

There are other definitions [5, 6] with different structures, and these will be instructive to consider as well. Each of these stages in Figure 9.1 and the associated aspects that would be most

IEEE Technology and Engineering Management Society Body of Knowledge (TEMSBOK), First Edition.
Edited by Gustavo Giannattasio, Elif Kongar, Marina Dabić, Celia Desmond, Michael Condry, Sudeendra Koushik, and Roberto Saracco.

Figure 9.1 Five stages of an entrepreneurial journey.

prominent in them is described next. It is to be noted that many aspects are relevant at multiple or all stages, and the description in any given stage is where it is likely to be most critical. Risk assessment, for example, would be present in every stage in different contexts and with different magnitudes. It is, however, described in the first stage, Preparation, as the most critical assessment would need to be made there.

9.2 Preparation

This is the first stage of an entrepreneurial journey. This stage may repeat many times before moving on to the next one. Many ideas and opportunities may be considered and discarded before one is assessed to be worthy of taking to the next stage.

9.2.1 Opportunity Identification

The first step in the entrepreneurial journey is to identify a candidate opportunity. Often, an opportunity might seem intuitive, and it is important to understand the motivations behind that intuition correctly. Opportunities can originate from first principles from two primary sources

- from within the entrepreneur (The Knowledge domain) or
- from external sources (the Problem domain).
  1) **Knowledge:** The capabilities of the entrepreneur, or those that an entrepreneur has access to or becomes aware of, can be the source of opportunity [7]. An opportunity emanating from knowledge might be stated in the form of "____ was not possible earlier and knowledge/ skill/technology/process _____ now allows us to do it. This opens an opportunity for an offering ____."
  2) **Problem:** The unmet need of a certain segment of the population is considered to be the most powerful source for an opportunity. It implies a ready market might be available if a solution to the need can be delivered. Opportunities from this source might be stated in the form "Users (who share a certain attribute) are unable to do _____ (well). This opens an opportunity to help them by creating _____." Airbnb famously "scratched their own itch" by identifying a problem they saw as an opportunity [8, 9].
     There is a third source of inspiration to identify opportunities beyond the two primary sources described above.
  3) **Competition:** The success (or potentially, failure) of an entity in delivering a particular product or service offering may be a trigger for similar offerings in detached markets or domains. This is a form of opportunity identification following someone who has used one of the earlier two sources to launch their own product. This opportunity statement might be of the form "____ has succeeded (or failed) in market/region/domain _____. This points to an opportunity to do the same in _____."

It is very rare for an opportunity to belong to just one or the other domain, and it is often a mix of both or all three. A clear understanding of the motivations behind the opportunity will help articulate the opportunity more clearly and help communicate its value in simpler terms to all future stakeholders. In all cases, though, moving to the next step of assessment will require bringing in elements of all of them.

9.2.2 Opportunity Assessment

An opportunity, once identified, will require fleshing out the details to see if it has the potential to become successful. This is also known as hypothesis validation or testing, with the opportunity statement representing the hypothesis. This assessment is usually different from that of an idea funnel as the intent is not to filter a large number of ideas to a few but to assess if a given idea is worth pursuing. This can be approached in different ways, and there are a lot of resources for this critical step. A great place to begin is with the book *Start with Why by Simon Sinek* [10]. There are different frameworks, for example [11, 12], that can be used to carry out this assessment. This process helps the entrepreneur state the assumptions and risks and carry out further due diligence or research. The Lean Canvas [13] is an excellent framework to use to brainstorm and think through all the critical aspects and thereby also carry out the assessment. There are many excellent resources like [14], and some imagined models for well-known companies at [15] on how to apply the Lean Canvas framework (see Figure 9.2), and hence this is not described in detail here.

It will be readily apparent that completing the nine sections in this framework will require thinking of all the key aspects involved in converting an opportunity into a startup. This, or any of the other frameworks including the classic Porter's Five forces model [16], assist assessment by prompting entrepreneurs to look for insights and arrive at their own conclusions.

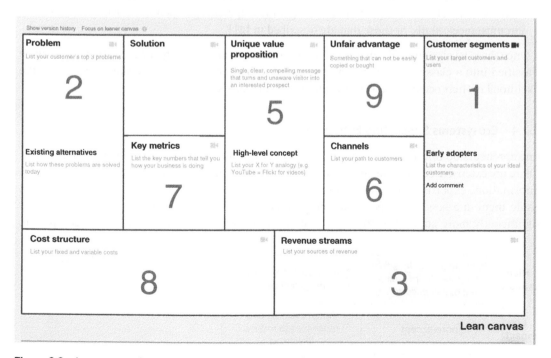

Figure 9.2   Lean canvas framework © http://Leanstack.com.

9.2.3 Understanding Risk

Risk is a constant companion in an entrepreneurial journey, at every stage. The earlier the stage, the greater the unknowns and therefore greater the risk in taking any particular path. Hence, this topic is in the Section 9.2, but it should be seen as an activity that needs constant re-assessment as some of the initial uncertainties get resolved or new ones arise. The Lean Canvas framework was intended by its creator, Ash Maurya, to capture the riskiest, that which was most uncertain, and hence provides a great place to begin [17].

When modeling risks, it is worth considering them as belonging to four distinct levels.

1) **Universal risks:** These are risks that would apply to any entrepreneur trying to create any product. Examples would include general market conditions, easy availability of finance, presence of supporting ecosystems, etc. Broad risk analysis is covered in many resources like [18, 19].
2) **Industry risks:** These are risks that are specific to the industry being considered and would include regulatory as well as aspects of the Universal risks that may take a specific form. This would include risks like licensing, long gestation periods, the presence of large incumbents, etc. In addition to this, there may be additional risks specific to certain domains – regulatory risk. Finance, health, aviation, and many others are heavily regulated industries for good reason. Opportunities in this space come with regulatory risk in that the regulations can act as an additional barrier protecting incumbents, or new barriers may arise any time due to regulatory action.
3) **Enterprise risks:** These are risks that are specific to the entrepreneur's organization. Since this may still be in the future, the anticipated organization can be considered. These would include risks like management team alignment, ability to hire team members, access to resources, etc. [20].
4) **Personal risks:** These are risks that are completely based on the entrepreneur's own circumstances themselves. These could include planned upcoming expenses, existing debt, family circumstances, skills, or those that are described in [21].

Once these levels and the Lean Canvas framework are used to identify the risks, they can then be classified into a classic assessment matrix based on the impact, if the risk materializes, and the likelihood of their occurrence as shown in Figure 9.3; see also [22].

9.2.4 Ecosystems Supporting Entrepreneurs

Entrepreneurs do not have to do this all independently and on their own, in many regions today. There are extensive ecosystems that have been created to support entrepreneurship. Availability of information, expertise, and mentors can help entrepreneurs think through the opportunity, guide them in assessing the potential and in understanding the risks. Many ecosystems provide significantly more support in the form of co-working spaces, access to hard-to-get equipment,

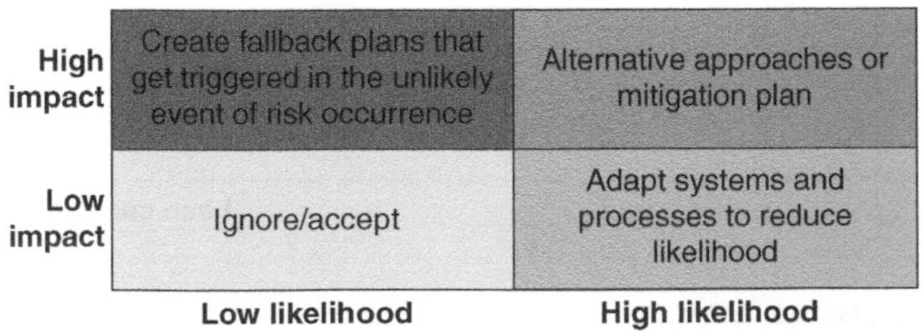

Figure 9.3 Risk classification.

networking opportunities, and more. These are commonly in form of incubators, accelerators, or entrepreneurship labs. These can be run by local universities, local government administrations, large corporations in the area, or groups of entrepreneurs themselves. Many of these also provide significant benefits in the form of credits for key online services like cloud computing or commonly used office software.

In addition, there are plenty of online ecosystems, like that from IEEE Entrepreneurship [23] and AngelList [24], and others that provide virtual access to the same resources on a global basis.

All these facilities are, at best, intended to help the entrepreneur learn, understand and plan well. In all cases, the eventual decisions, and the risks, will still need to be taken by the entrepreneur.

9.3 Initiation

9.3.1 Team

It is very often the case that the pursuit of a selected opportunity is carried out by a team of more than one person, though there is nothing to prevent a single person from doing it too. The skills and competencies of the team, their motivations, their shared vision and objectives play an outsized role in their success [25]. The quality of the team is, anecdotally, the primary parameter that venture capital investors look for with the idea itself having lower priority. A team may form naturally or be explicitly created or a combination of both. Key aspects that affect the functioning of a founding team:

- **Cohesion** through shared values, vision, and goals on a foundation of mutual trust.
- **Decisiveness** with one of the team being first among equals.
- **Work-ethic** to get things done, with skills and know-how required to execute.
- **Adaptive** with the ability to listen and respond to each other, market feedback, and changing circumstances.
- **Complementary and Diversity** of skills, characteristics, and aptitudes.
- **Defined** roles and responsibilities and equity participation.

There is an enormous amount of research on the impact of the team on the success of the enterprise with diverse perspectives. Wasserman's 2013 book [26] is a particularly good reference.

9.3.2 Legal Entity

Once an opportunity has seemed worthwhile enough to pursue, most often, the entrepreneur will need a legal entity to hold within and manage this opportunity. The most important reason to create such a legal entity is to separate the finances of the entrepreneur from the enterprise. A second reason is to allow multiple sources of funds to be used, independently and transparently, to fund the pursuit of the opportunity. This could be from multiple founders or investors, and in both cases, it serves to protect the interests of all parties. It protects the founders from the liability that may arise and protects the other founders and investors from commingling personal expenses with that of the organization. The exact form of such a separation of concerns may vary across jurisdictions but are often variations on the names of limited liability companies or limited liability partnerships. Other structures exist and selecting the right structure needs careful consideration of the rights and obligations associated with each including aspects like compliance requirements and taxation. This varies considerably by jurisdiction and, indeed, even with time.

9.3.3 Business Models

A business model represents the monetization model for the startup. There are a variety of possible business modes that a startup can adopt. Established market practices may dictate the models in certain industries, particularly if they are also heavily regulated. The model also depends on whether the startup is positioning itself as a B2B or a B2C player. Quite often, in B2B models, the buying organization may have established purchasing policies that would need to be met to do business with them, thereby dictating the business models that need to be followed. B2C players have more flexibility in defining their own business models including creating innovative new ones.

At a fundamental level, there are just two models that operate. These are Capex and Opex oriented and represent how the payment for the product or service reflects in the books of the buying organization. Capex, derived from Capital Expenditure, would mean that the payment would be considered as primarily a one-time expense. This is also the ownership model where the ownership rights transfer to the buyer. An Opex, derived from Operating Expenditure, would imply a periodic payment model and represents a rental-style model where the buyer would merely be using the product or service without owning it. Many B2B business models would have a mix of both elements.

A consumption-oriented model is used where the product/service requires resources to be expended when used. It is rarely the case that the raw resource consumption is directly exposed to the users. Such consumption gets packaged into friendlier proxies like the number of users, the number of uses, or a similar user domain metric. These metrics are modeled to represent a utilization factor.

Another way to consider the business model is from the perspective of who pays.

1) **Two-sided business models:** These are also known as platforms and tend to have at least two types of users who generate revenue. These are platforms that bring users of one type to service providers of another type and charge both sides for the facilitation [27]. Examples of this are Uber, eBay, Airbnb, and similar marketplaces.

2) **Ad funded business models:** In this model, the revenue is generated from those who wish to advertise to the users of the service with the users themselves not being charged. This is very common in the media world with many other services also considering this as a viable option. YouTube, many newspapers, Google search, and other services are examples.

3) **Freemium:** In this model, individual customers or small businesses tend to not pay for a product and use it freely at no cost but with associated functionality restrictions. Larger or paying customers may get access to a rich set of services. Many products particularly in the open-source community, follow this to get a large user base and convert some into paying customers, particularly when there are low consumption-linked costs. Evernote, etc. are good examples.

This is not an exhaustive list of business models and many references describe many more [28, 29].

9.3.4 Minimum Viable Product

The concept of a Minimum Viable Product (MVP) [30] has taken a very strong hold in the startup environment today. Fundamentally, this acknowledges that it may not be realistic to have a completed definition of a product when one starts building it. The MVP model implies that user and market feedback is critical in refining the product definition. The MVP represents the smallest definition of the vision behind the product or service that users can use and achieve the essence of what the full-blown service will be.

For an MVP to achieve its purpose, it is important that the MVP [31]:

1) Deliver an actual usable service and tangible value to the users.
2) Be stripped of all features that are not necessary for the functionality to be used.
3) Have a path open to initial users to get their feedback.
4) Have short cycle times to understand the feedback, derive actions, and provide updates.

There are certain areas where an MVP has to be relatively more complete, with every t crossed and i dotted. These are in the highly regulated areas where the possibility of user harm is immense. Any healthcare or financial product, for example, would need to comply with all applicable regulations on day one of the launch.

Coupled with the concept of an MVP is the idea of an initial target audience. This goes back to the classic Crossing the Chasm by Geoffrey Moore [32], which formalizes the early-adopter definition. An MVP is designed to target early adopters, who may be more eager to use the product or service, maybe more forgiving of issues they may face, or are willing to try a new and unknown product.

Getting traction with these early adopters is the goal of the MVP. These users are characterized by their willingness to spread the word on something they love while being equally and vocally critical of something they do not. This helps improve the product while gathering more users, as long as they are listened to and their feedback is acted upon.

This cycle of user feedback and improvements is considered the path to reach a product definition that the larger market, beyond the early adopters, will accept.

9.4 Traction

9.4.1 Product-Market-Fit

The idea of a Product Market Fit (PMF) has become a cornerstone of a successful product or service (the term Product is used in a broad sense and refers to all offerings). This is considered to be such a critical parameter that achieving PMF is considered equivalent to assured success.

At its heart, PMF is not new. Ensuring customers exist and are happy with their product or service has always been critical. What is new is the kind of data-driven approach that is being used to assess whether one has achieved PMF [33].

Key metrics that are most often used include [34]:

- **Adoption rate/rate of sign-up of new users:** Is it increasing in every period of measurement? Usually a daily or weekly or, rarely, a monthly cadence for this metric that is expected to show a sustained positive trend.
- **Activity rate:** This is the percentage of signed-up users who actively use the product. Often measured in Daily Active Users or their weekly/monthly counterparts, this is a metric that is unlikely to be monotonic as some users sign up and realize this is not a product for them. This is also a very critical metric when the users are capable of being segmented, as this may be the best source to identify the segment that finds it most valuable, measured by their activity level.
- **Attachment:** This was a metric popularized by Rahul Vohra, the founder of the Superhuman mail app [35]. If less than 40% of a certain active user segment says that they will be disappointed if the product is no longer available/accessible, then PMF is yet to be achieved according to this model.

Metrics across different domains, products, or user segments are unlikely to show the same trend or scale and have to be assessed on their own past and, where available, directly comparable products. Concluding that an offering has achieved PMF is, ultimately, a subjective decision. These data metrics help in providing evidence that might support such a conclusion.

It is important to note that the idea of PMF is very much linked to the ability to influence those metrics in relatively short cycles. Changing the product on a weekly/monthly basis and observing the change in metrics leads to better quality decision-making and faster PMF.

9.4.2 Valuation

Among all the aspects of entrepreneurship that this section is covering, with all its attendant uncertainties, there is none more esoteric and less exact than the valuation of startups. Yet it does have an outsize influence on how a startup behaves. The challenges in valuation stem from the fact that many startups at an early (or even mature!) stage are unprofitable, have negative cash-flows, and are growing or have expectations of growing at an exponential rate. The second reason is that an acceptable valuation is really one that two (groups of) people agree on and has very little to do with the broader market or what might be a consensus value, as in a publicly listed company.

Several models are commonly considered when valuing startups and are described in [36, 37].

While valuations are always supposed to reflect future earnings, for early-stage startups with rapidly changing metrics, this is not a particularly easy exercise. Key factors that can influence valuation are:

- **Traction:** This can be measured in new customer sign-ups, usage activity, rate of incoming enquiries, conversion rates for ads, or any other metric that can serve as a proxy for revenue particularly when revenue itself is very small or non-existent.
- **Team:** The ability of the team to execute their plan, as perceived by potential investors.
- **Moat:** The presence of barriers to entry for others that naturally exist or have been created. This could be technological assets, via business contracts, preferential partnerships that can be leveraged to significantly elevate one from the competition.
- **Market:** Multiples used in the market over any metric (revenue, customers, etc.) set a benchmark that tends to become the guiding multiple for others in similar or related domains, particularly when the benchmark is set by what is considered a savvy investor.

Valuation, using any or all of the metrics, is almost always needed in the context of raising funds.

9.4.3 Funding

Startups do not always need to raise funds to succeed. There are many examples of those who do not do so and still succeed. These go by the term *bootstrapped* startups [38]. Nevertheless, funds get raised by startups more often than not, and these can be classified under four different categories depending on the stage of the startup.

- Angel or pre-seed investments represent the funds raised at the earliest stage of maturity of a startup. Often, before there are customers and perhaps even when it is just an idea. Many times this investment may be from informal investors like friends and family.
- Seed or Startup investment is usually the first formal investment from a professional venture capital firm. This may still be in the absence of customers though there is probably market research or a prototype product in place. Usually, this is intended to allow a team to be put in place to create the MVP and take it to the market.

- Expansion stage investments are those raised to help propel the startup into higher orbits. These can be labeled Series A, B, C, or Stage-1, 2, 3 but shorn of the nomenclature, they are all add-on investments to the original investments, either by the same investors or new ones.
- Growth stage investments are often intended to support organizations that are otherwise self-sustaining and can use this investment to pursue aggressive growth via acquisitions, new markets, or products.

All funding needs a valuation and these are often what are known as pre-money valuations (prior to the incoming funds being considered). The investment results in a higher post-money valuation to the extent of the investment with the previously existing shareholders getting diluted by the proportion of the equity traded for the investment.

There are other, more conventional, sources of funding that have existed in the past and are still relevant today. These include:

- Debt financing from banks and similar lending institutions.
- Grants from Government, philanthropic organizations, and others particularly when the social impact from the startup is high.
- In-kind by trading equity for access to goods or services like ads or physical facilities or access to equipment.

9.5 Growth

9.5.1 Organization

Growth requires an organization structure that is very different than at the early stage [39]. There is an anecdotal model that says that the organization structure would need to evolve when the team size crosses breakpoints of 10, 20, and 50 and their scaled-up versions with zeros added, 100, 200, 500, 1000, and further.

This change is often through the need to introduce layers, either horizontally or vertically. These become necessary to manage the increasing complexity of the operations that growth brings in, the specialized skills that need to be brought in to strengthen the team, the increase in the number of people requiring new management layers to be introduced, the potential geographic spread of the customers and other factors.

Vertical layering: This is often the easiest to understand and takes the form of additional management layers between the founders and the junior-most members of the organization. While lean organizations and flat hierarchies remain the desired goal in many organizations, the introduction of intermediary layers is inevitable in most cases. This is neither a bad thing nor something to avoid and is a necessity to bring in the ability to let different layers focus on different levels of abstraction of the company. The highest layers, usually including the founders, focus on aspects like strategy or investor management, while layers lower down take care of converting organizational imperatives into goals and actions and further down take care of routine day-to-day operations. This creates new roles, designations, and responsibilities. This also, inevitably, brings in the need to define responsibilities and goals better, the need to monitor and track achievements against defined goals, and matching accountability to responsibilities. The greater distance between founders and those at the bottom of the vertical layers requires fresh articulation of the vision, the objectives, and the plans so that they can be percolated to everyone in the organization.

Horizontal layering: The need for multiple people with distinct specialized skills results in the formation of departments or the horizontal layering described here. People with similar skills need access to each other, have similar professional concerns, possibly access to a common set of tools and a department focused on that becomes a natural evolution. The consequences of such horizontal layering are the potential formation of silos of information and the creation of hurdles in communication across departments.

An additional layer can get overlaid on this due to the need for physically distributed teams, due to the availability of customers, ease of market access, or other factors.

Each organization may have varying levels of scaling in the vertical, horizontal, or geographic axes, but the outcome of all of them is an increased need for formality in the way the organization functions. This need for formality results in the processes that an organization will need to adopt. While these structures may evoke classic organizational models, now considered outdated, it may be worth keeping in mind that the essence of what organization structures need to achieve has not changed much over time [40].

This stage is tricky as the natural, informal, approach of the early-stage startup needs to evolve into a more structured and formal approach [41]. An inability to navigate through this change is a leading cause of failure in startups at this stage. Equally challenging would be if this is done too early resulting in a premature rise in complexity and costs.

9.5.2 Processes

The increased complexity of the organization results in new processes being needed in nearly every area of the company. Well-defined processes play a primary role in fostering growth from this stage forward. These processes can be classified into a few buckets based on the time frame they operate upon.

1) **Annual:** This operates on a full-year horizon, with parts of it covering multi-year scope. This would take the form of annual plans, structured budgets, org-level goals, and KPIs that would support the annual plan [42].
2) **Monthly:** These would be around setting short-term goals, monitoring and tracking progress, reviews, reporting/dashboards, corrective actions, and process improvements. This last bit is important in a young organization with new processes as adapting them to reality and what works would be important.
3) **Routine:** These would cover structured methods for the dozens, or maybe even hundreds, of small activities that get done on a daily or weekly basis. Handling inquiries, customer complaints, shipping, recruitment, staff absences, website updates, emails, document templates, product releases, purchases, facilities, celebrations, and the myriad activities that constitute life in any organization all become candidates for formalizing. These are the ones that end users most often encounter directly and require the most thought to deliver a consistent experience.
4) **Strategic:** These are processes that are not necessarily periodic and operate on a varying time-scale of several quarters to several years and provide the bedrock on which everything else is built. These are usually around compliance, quality, security, investor interactions, and similar important aspects which may operate from milestone to milestone.

Following the organization structure, some of these processes will be at a department level or even lower layers. Or could be defined by individual departments for use across the entire organization. The processes may (and often, should) evolve regularly, particularly where the organization is growing rapidly as they may get outdated very quickly.

Processes take time to define as well as adopt. Conventionally, they are designed for a time in the future where they will be necessary rather than based on the need of the day. It is important that the future state of the organization be identified correctly in order to not over-emphasize the processes involved and make it unwieldy for the potentially leaner organization that will begin implementing it.

9.6 Exit

Many startups, having gone through the growth phase, find that they may be able to exploit the opportunities or the market better if they had access to significantly more resources. Often, this may be also needed to deliver on the expectations of investors as well. This calls for founders to consider what direction their organization should take. The options can be summarized as follows:

1) **Natural growth:** Letting it grow naturally, investing its own generated returns, is a perfectly valid possibility though not taken as often.
2) **Initial Public Offering (IPO):** This is often considered the most desirable outcome for a startup. The shares get listed on a public stock exchange with retail investors becoming shareholders of the company.
3) **Merger or Acquisition:** These are variants where the startup joins hands with another organization to form a larger one. A merger usually involves similarly sized entities with equity being mutually exchanged while an acquisition usually has the acquiring entity much larger than the acquired. There are also aspects of branding, product portfolio, people, and what happens to all of them that may vary between them. The startup could be acquired by a larger entity or could acquire others smaller than itself to strengthen its portfolio.
4) **Winding-up:** A significant number of startups do not get a chance to take one of the first three options. This is inherently an outcome of the entrepreneurial risk that was discussed earlier. For those startups, closing is a very rational option as it can allow the entrepreneurs to take the lessons learned and apply them to a new opportunity.

9.7 Conclusion

Entrepreneurship is one of the greatest paths available to anyone, but it is not necessarily the easiest or most obvious path. It is inherently riskier than many other career paths. It is also significantly more rewarding when the risks pay off and, often, even when they do not.

Persistence and adaptability are considered two of the primary characteristics that allow some entrepreneurs to succeed while others do not. Entrepreneurship is as much art as science and there remain no proven ways to succeed – just ways to avoid failures. There are excellent sources of education that can serve an entrepreneur very well, like [43]. The landscape remains largely unmapped and following best practices, as outlined in this chapter, very much improve the chances of success by avoiding the most common failure modes. This chapter lays out a roadmap that should help prospective and early-stage entrepreneurs identify the aspects they should consider learning more about and get a sense of what comes next.

References

1 Henderson, D.R. and Schumpeter, J.A. (2018). https://www.econlib.org/library/Enc/bios/Schumpeter.html (accessed December 2021).

2 Schumpeter, J.A. (1942). *Capitalism, Socialism and Democracy*, 5e. New York: Harper and Brothers.

3 Roger, S.O. and Martin, L. (2007). Social entrepreneurship: the case for definition. https://ssir.org/articles/entry/social_entrepreneurship_the_case_for_definition (accessed December 2021).

4 Kaufman Global. 6ms of production (man, machine, material, method, mother nature and measurement. https://www.kaufmanglobal.com/glossary/6ms-production-man-machine-material-method-mother-nature-measurement.

5 Goel, N. (2017). The 7 stages of entrepreneurship. https://www.entrepreneur.com/article/293463 (accessed December 2021).

6 Clark, C. (2019). The 5 stages of the entrepreneurial journey. https://greyjournal.net/work/work-business/the-5-stages-of-the-entrepreneurial-journey (accessed December 2021).

7 Arentz, J., Sautet, F., and Storr, V. (2013). Prior-knowledge and opportunity identification. *Small Business Economics* 41 (2): 461–478.

8 Wharton School, University of Pennsylvania (2017). The inside story behind the unlikely rise of Airbnb. Airbnb example/https://www.youtube.com/watch?v=axqh6SJ0O0c/https://knowledge.wharton.upenn.edu/article/the-inside-story-behind-the-unlikely-rise-of-airbnb/ (accessed December 2021).

9 Fundersandfounders (2014). How Airbnb started – founding story. https://www.youtube.com/watch?v=axqh6SJ0O0c (accessed December 2021).

10 Sinek, S. (2011). *Start with Why: How Great Leaders Inspire Everyone to Take Action*. Portfolio.

11 Fuqua School of Business, Duke University. (2019). The duke entrepreneurship manual. https://sites.fuqua.duke.edu/dukeven/new-venture-guidelines/evaluating-an-opporunity/.

12 Cagan, M. (2006). Assessing product opportunities. https://svpg.com/assessing-product-opportunities/ (accessed December 2021).

13 Maurya, A. (2012). Lean canvas. Leanstack. https://blog.leanstack.com/why-lean-canvas-vs-business-model-canvas/ (accessed December 2021).

14 Pereira, D. (2022). What is lean canvas? https://businessmodelanalyst.com/lean-canvas/.

15 Korolev, S. (2019). Lean canvas examples of multi-billion startups. https://railsware.com/blog/5-lean-canvas-examples/ (accessed December 2021).

16 Institute for Strategy & Competitiveness, Harvard Business School. The five forces. https://www.isc.hbs.edu/strategy/business-strategy/Pages/the-five-forces.aspx.

17 Ceren Çubukçu, S.G.S.T.T. (2017). Risk management in startups: how using lean principles helps startups manage their risks more efficiently. *International Conference on Engineering Technologies (ICENTE'17)*, Konya, Turkey (7–9 December 2017).

18 Lettich, T. (2021). 4 startup risks entrepreneurs should consider when launching their startup. https://fundingsage.com/14-startup-risks-entrepreneurs-should-consider/ (accessed December 2021).

19 Enterprise Risk Management Academy. What is it that kills a startup? https://www2.erm-academy.org/publication/risk-management-article/what-it-kills-startup/.

20 Hayes, A. (2021). What risks does an entrepreneur face? https://www.investopedia.com/ask/answers/040615/what-risks-does-entrepreneur-face.asp (accessed December 2021).

21 Entrepreneur India (2014). 7 risks every entrepreneur must take. https://www.entrepreneur.com/article/238319 (accessed December 2021).

22 Niedbala, C. (2021). Risk management: avoid, reduce, transfer or accept? https://foundershield.com/blog/risk-management/ (accessed December 2021).

23 IEEE Entrepreneurship. IEEE Entrepreneurship. https://entrepreneurship.ieee.org/.

24 AL Advisors Management Inc. AngelList ventures. https://www.angellist.com/.

25 Davda, A. (2014). How to build a fabulous founding team. *Wharton Magazine* (8 December).

26 Wasserman, N. (2013). *The Founder's Dilemmas: Anticipating and Avoiding the Pitfalls That Can Sink a Startup*, 1ste. Princeton University Press.

27 Reason Street. Two-sided marketplace. https://reasonstreet.co/business-model-two-sided-marketplace/.

28 Cudmore, D. (2021). 20 business model examples (and how to pick the right one). https://digital.com/how-to-start-an-online-business/business-model/ (accessed December 2021).

29 WhatAVenture GmbH. 10 business models that will inspire you. https://www.whataventure.com/blog/10-inspiring-business-models/.

30 ProductPlan (2021). Minimum viable product (MVP). https://www.productplan.com/glossary/minimum-viable-product/ (accessed December 2021).

31 Brikman, J. A minimum viable product is not a product, it's a process. https://www.ycombinator.com/library/4Q-a-minimum-viable-product-is-not-a-product-it-s-a-process.

32 Moore, G.A. (2006). *Crossing the Chasm: Marketing and Selling High-Tech Products to Mainstream Customers*. HarperBusiness;. Revised edition.

33 Chang, J. The best metric for determining quantitative product market fit. https://www.growthengblog.com/blog/the-best-metric-for-determining-quantitative-product-market-fit.

34 Link, J. (2021). What is product-market fit? how do you best measure it? https://builtin.com/product/product-market-fit.

35 Vohra, R. How superhuman built an engine to find product market fit. https://review.firstround.com/how-superhuman-built-an-engine-to-find-product-market-fit.

36 CFI Education Inc. Startup valuation methods. https://corporatefinanceinstitute.com/resources/knowledge/valuation/startup-valuation-methods/.

37 McGowan, E. (2018). 10 real-world startup valuation methods. https://www.startups.com/library/expert-advice/startup-valuation-methods.

38 Pahwa, A. (2021). Startup bootstrapping: a detailed guide. https://www.feedough.com/startup-bootstrapping/.

39 Toolshero. Greiner growth model. https://www.toolshero.com/strategy/greiner-growth-model/.

40 Clarke, S. (1990). What in the F-'s name is fordism. *British Sociological Association Conference*, Surrey.

41 Henshall, A. (2017). How 4 top startups are reinventing organizational structure. https://www.process.st/organizational-structure/ (accessed December 2021).

42 GoCardless Ltd. How to develop an effective annual plan. https://gocardless.com/guides/posts/how-to-develop-an-effective-annual-plan/.

43 Combinator, Y. Startup School. https://www.startupschool.org/.

10

Toward Smart Manufacturing and Supply Chain Logistics

Eldon Glen Caldwell Marin

Industrial Engineering Department, University of Costa Rica, San José, Costa Rica, Central America

10.1   Introduction

The production and logistics systems of goods and services have been the basis of world economic growth in what has been called "modernity" since the mid-nineteenth century. But technological advance has always determined the acceleration of the productivity of resources and money, as well as the reduction of delivery times and the achievement of standardization and the breaking down of commercial borders.

The needs for diversification, massification, security, speed, standardization, customization, flexibility, differentiation, integration, predictability, automation, autonomization, sustainability, and profitability of manufacturing and supply chain systems have inspired the configuration of different schools of manufacturing and logistics theoretical thinking and also best practices of excellence. Within the most prevalent in the last 200 years the craft manufacturing, mass production ("Fordism"), Japanese production (or "Toyota Production System"), and the "Theory of Constraints" (TOC System) can be mentioned. In addition, Concentrated Manufacturing, Modular Manufacturing, Flexible Manufacturing Systems, Automated and Autonomous Manufacturing, Lean Manufacturing and Lean Logistics based on "Lean Thinking," Cyberphysical Manufacturing and Logistics Systems and, emerging strongly in the last 20 years, Smart Manufacturing and Logistics Systems (SMLS).

These schools of theoretical thinking and practices of excellence have responded to the needs of the competitive environment at different times in history and many of its principles have been prevalent. They evolve as a technological change into a constant source of innovation in all industrial sectors, from agriculture and aquaculture to outer space to our planet; from the physical and biological world to the virtual world; and from the world of nanosystems to the quantum world.

This chapter delves into SMLS and its relationship with the different thinking schools, recognizing the dynamism in which the application of its principles has been successful or has provided lessons learned. In addition, conceptual key rocks are established that clearly differentiate, but not mutually exclusive, SMLS from cyber-physical systems (CP Systems) that have been widely developed as the spearhead of the so-called Fourth Industrial Revolution (Industry 4.0), as well as from systems based in flow development such as those based on "Lean Thinking" or the "Theory of Constraints."

IEEE Technology and Engineering Management Society Body of Knowledge (TEMSBOK), First Edition.
Edited by Gustavo Giannattasio, Elif Kongar, Marina Dabić, Celia Desmond, Michael Condry, Sudeendra Koushik, and Roberto Saracco.

Next, the basis of the technology and technical descriptions that are associated with SMLS development is exposed. This technological approach has different edges in terms of the possibilities that are currently most marked as a trend and does not have a sense of an updated toolbox but rather of a reference framework of the business architecture of an SMLS, based on its impact on competitiveness and the achievement of operational excellence results.

It must be remarked that the technological base for SMLS has areas of convergence in relation to cyber-physical systems that seek the implementation of the "cyber-physical factory or supply network," for which a taxonomic base is used in three essential dimensions: interconnectivity, interoperability, and virtualization. However, it is critical to emphasize that an SMLS is not necessarily a cyber-physical system nor is the cyber-physical supply network necessarily a smart manufacturing and logistics system. However, it is clear that most organizations that seek operational excellence make decisions so that their operations are both cyber-physical and intelligent systems.

Each technological application space refers to benefits linked to the achievement of the SMLS goals and their positive impact on the business model and specifically the value proposition. Strategic, operational, economic, and ecosystem benefits are described for each technology, but without losing sight of the fact that SMLS are synergistic, so the applied technologies also become sources of holistic impact, especially in supply network management.

Subsequently, this chapter addresses real cases of application of the principles and technologies of SMLS in the American continent and Europe, as a way to expose different frameworks or emphasis on the design of intelligent systems in competitive environments with different characteristics.

Each case has its peculiarities and lessons learned and they do not necessarily show the application of all the "toolbox" available in the implementation of SMLS, however, they provide analytical dimensions that refers to frequent configurations of manufacturing and delivery of goods and services.

It is clear that there are marked trends such as the industrial application of co-bots, assistance robots, digital twins, digital phenotypes, and data analytics, as well as applications based on Industrial Internet of Things (IIoT), computer vision, smart sensors, and blockchain. However, the cases show that what is essential is not the technological tool as a core goal in itself but the ability to manage innovation and provide an advantage to the system in terms of the value proposition for customers.

Finally, before presenting the conclusions and recommendations, methodological approaches and technological tools to manage implementation projects are going to be described, as well as a generic roadmap that can guide management decisions. Likewise, risk factors are provided that must be taken into account according to the characteristics of the environment or execution ecosystem and possible ways of solving possible contingencies.

10.2 Smart Manufacturing and Logistics: Knowledge Area Fundamentals

Throughout history, humanity has developed knowledge, methods, tools, and techniques to obtain products that enhance its well-being. However, it is not until the end of the nineteenth century and the beginning of the twentieth century that what has been known as "manufacturing and supply chain systems" has come to establish itself as a formal science with dynamism and complexity without referents.

A simple way of understanding manufacturing and logistics can be found in [1] conceived as the creation of goods and services or as defined in the dictionary of the American Production and

Inventory Control Society (APICS), "the conversion of inputs into finished products" (goods and services) [2].

The first thinking school in manufacturing is craft production, which has the following essential characteristics [3]:

1) Relies on a highly skilled workforce in design, manufacturing operations, and assembly. In addition, it generally presents an extremely deconcentrated organization, although commonly located in a single city or locality.
2) The administrative and operational system is coordinated by an owner/entrepreneur who is in direct contact with all those involved in the substantive transformation activities of the company: customers, employers, and suppliers. It also emphasizes the use of general-purpose machine tools and it is not common to observe artisan workshops with machines or tools dedicated to a single function.
3) It is commonly operated with a very small production volume. Although in theory, an artisanal process could have thousands of people and thus produce large volumes, so this topic is a matter of capacity rather than of the system. However, most of the production is for very specific market segments.

With craft manufacturing, in general, it is possible to obtain a greater orientation toward flexibility, rapid product development, and of course, a great satisfaction of the customer (and the craftsman) in terms of the unique characteristics of the product. Table 10.1 outlines the advantages and disadvantages of craft manufacturing.

Every production system must meet three specific objectives to optimize its overall effectiveness: flow, balance, and synchronization [1]. The concept of flow refers to getting materials to move seamlessly through the entire process. That is, the materials "flow like water," which is why balance and synchronization are necessary [3]. The concept of balance refers to keeping the workload balanced in each phase of the process, operating the lines at a stable rate of production while synchronization refers to starting operations at the exact moment it is required. For example, without unnecessary delays or material supply problems.

In the late nineteenth and early twentieth centuries, F.W. Taylor's scientific management reached a paradigmatic pinnacle, and mass production was born with the first industrial revolution. The goals focused on the development of flow, balance, and synchronization were approached from the perspective of the division of labor, line balancing, and feeding at the earliest possible moment and at maximum capacity [4].

Table 10.1 Characterization of craft production system.

Craft production system		
Characteristics	**Advantages**	**Disadvantages**
1) Highly skilled workforce in design, manufacturing, and assembly operations.	Customer orientation. Workforce flexibility.	Higher cost of labor. Greater difficulty in reducing variability.
2) Extremely decentralized corporate organization although commonly concentrated in a single location.	Cooperative relationships with subcontractors to reduce costs and quality problems.	Less control of operations and greater dependence on subcontractors.

Table 10.2 Impact of the Ford system on production cycle times.

Mass production vs. craft manufacturing (1914)			
Assembly minutes	**Craft manufacturing (1913)**	**Mass production (1914)**	**%**
Engine	594	226	62
Magnet	20	5	75
Axle	150	26.5	83

Source: Adapted from Womack and Jones [6].

Henry Ford was the benchmark for the success of mass production and for many years the River Rouge plant in Detroit became the greatest example of global competitiveness. In 1924, the car's manufacturing time was 48 hours from the extraction of raw materials to its final assembly [5]. To achieve this, Ford engineers used team technology dedicated to exclusive operations. Such as the machines capable of working previously hardened metals, and in addition, equipment was developed in the plant that would be used in specific operations of its process, in order to take advantage of the specialization and division of labor and in this way, radically increase productivity.

In those years, for the system to work perfectly, there was one fundamental requirement: to minimize variability in product design. Ford system achieved productivity optimization as long as the plant was dedicated exclusively to making black and Model T cars. Table 10.2 establishes a comparison of the results obtained by Ford with its manufacturing system, in terms of cycle time, compared to craft manufacturing [6].

Ford's mass production dominated the automotive industry for more than half a century and was embraced by almost every industrial activity in North America and the world. Table 10.3 summarizes the principles of mass production, which until today is the most widely used production system [6].

10.2.1 Toyota Production System and Theory of Constraints

In 1950s, representatives of the Toyota Company visited Ford's Rouge plant, located in Detroit, assuming that it was the best expression of mass production and therefore, a school of process design that Toyota would surely take advantage of to increase its competitiveness. However, after carefully analyzing the principles applied there, they were disappointed to conclude that the system, as it was proposed, would not work in Japan [6].

First, the Japanese economy lacked capital and foreign exchange, which meant that massive purchases were almost impossible. Large vehicle producers were eager to start operations in Japan and ready to defend the markets they gained against Japanese exports. Furthermore, the domestic market was small and demanded a wide range of vehicles; the resources for rapid transfer of computer technology from the western world did not exist. On the other hand, the native Japanese workforce had particular conditions after World War II, with very strong and legally consolidated union movements [7].

Taiichi Ohno, when analyzing the existing limitations to apply the mass production of the Ford system, decided to create a production system that would rescue both the advantages of artisanal production, as well as those of mass production. This is how what has been called the "Toyota Production System" (TPS) [8] emerged, a fundamental basis of Japanese manufacturing techniques, which is characterized by the search for the maximum possible elimination of waste.

Table 10.3 Characterization of mass production.

Mass production

Characteristics	Advantages	Disadvantages
1) Total and consistent interchangeability of parts and simplicity in assembly.	Higher levels of productivity.	Greater bureaucracy.
	Fewer adjustments (changes) and lower cost of labor.	Lower systemic vision.
2) Maximum division of work both in operations and in the organization.	More accessible prices.	Systems fragmentation.
	Simplicity of activities.	Low labor flexibility.
3) Standard measurement system.	Greater delimitation of functions.	Lack of systemic vision and agility in decision-making due to verticalization.
4) Extreme specialization.		
5) Little variety of products.		
6) Centralized and vertical organization.	Greater knowledge development.	Higher investment.
	Less variability in quality.	Greater complexity in operations.
7) High vertical integration.		
	Fewer tuning interruptions.	Creation of business exit barriers.
	Development of specific skills.	
	Repeatability of the execution.	
	Economies for having less variability, greater volume, fixed production rate, discounted and predictable purchases, and adaptation of machines for dedicated use.	
	Defined authority and better delimitation of functions.	
	Creation of entry barriers to potential competitors.	
	Cost savings and control in the value stream.	

In the West, Just in Time (JIT) is the term by which the system led by Taiichi Ohno was known and refers to a philosophy of planning, direction, and control of manufacturing and logistics and is closely related not only to the execution within the plant but also to the relations of the company with its suppliers.

The fundamental principles of TPS can be synthesized as follows [8]:

1) **Minimum work in process**

Reducing inventories brings financial benefits: better use of working capital, lower costs caused by inventories (obsolescence, personnel, insurance, etc.), lower opportunity cost of money, higher turnover of money (which is relevant in terms of return on investment), and of course, lower operating expenses associated with production control. However, the major benefits of reducing inventories are more operational than financial [3]. In general, high inventories represent protection for production and by increasing that protection, alluding to that "inevitable variability of processes," problems simply begin to hide [6].

2) "Pull" production (balanced and synchronized) vs. "push" production

The traditional way of scheduling production is known worldwide [4] under the name of "PUSH" and it is so called because this system works with a material feed at the earliest possible moment and at the maximum capacity of each process throughout of the supply network. However, TPS seeks to process production batches with the smallest possible transfer runs and through a "front-to-back" work order placement system, adjusting production transfer runs according to demand, in such a way that the dependent stages within the process request the material from the preceding ones as needed and at the time it is needed [6].

The ideal feeding system is to transfer the pieces one by one in a synchronized way (one-piece flow). However, since this is not always possible, Ohno devised a system in which he used cards (Kan Ban in Japanese) to achieve that the flow is kept to the minimum depending on the variation existing between the capacity of the different processes, the setup times and a coefficient of variations in the queue.

The number of Kan Ban cards is calculated as follows [1]:

$$y = D \times L \times (1 + b)/a$$

where,

y: Number of sets of Kanban cards
D: Demand per period of time
L: Lead time
a: Capacity of the container or transport mechanism.
b: Coefficient of variation, or safety stock.

The "Kan Ban" feeding system consists of controlling the flow of materials between one work center and another (or between one operation and another) so that each one of them is only authorized to produce or transport material when it has been indicated. For optimal operation, it is necessary to have the following operating characteristics:

1) Reduction of set up times.
2) Work standardization.
3) Flexible workers.
4) Automatic error-proofing devices (poka joke).
5) Rational use of automation and autonomization.
6) Collaborative logistics and distributed systems for the manufacture of parts and assembly.
7) Kaizen or Kaikaku process (continuous or disruptive-radical improvement).
8) Intelligent use of technologies under the principle of minimum investment and focus on the value proposition for customers.

Taiichi Ohno's work, in developing TPS, influenced the manufacturing world as much as Ford did with Mass Production. However, from the mid-1970s to the early 1990s, a theory was developed that has impacted the schools of thought in manufacturing and logistics especially in the West and that became known as TOC by its creator, Eliyahu M. Goldratt [9].

This theory has been widely applied in manufacturing and logistics, although it has also been extended to the decision-making process in marketing, information systems, distribution systems, finance, and many other areas. In short, it seeks the integration of the manufacturing and delivery process with the entire value stream of the company, so that it works as a global system whose goal is the sustainable increase of the economic value of the company. In simple words, "earn as much money as possible today as well as in the future" [9].

The TOC proposes that operational decisions should be evaluated in terms of three variables and not only by means of unit cost, as is the usual practice:

1) **Throughput (economic throughput):** It is money that the production system generates through sales. The throughput per unit is calculated in a similar way to the contribution margin per unit, subtracting from the price per unit the unit cost of raw materials (components, parts, materials, inputs from external suppliers, considering customs costs, broker commissions, costs transportation and freight, when the resources or systems do not belong to the company's operating system). These items are subtracted because they are not money generated by the company's system but are throughput generated by other external systems.

2) **Inventory (*I*):** All the money that the system has invested in the purchase of resources that it intends to sell.

3) **Operating expenses (OE):** It is the minimum money necessary that the system spends to be able to convert inventory into throughput. Under TOC thinking, it is not relevant (for operational decision making) that operating expenses are allocated to products. The costing system associated with TOC thinking is variable costing, with the only difference that direct labor becomes an operating expense, as long as it is not paid piecework.

Throughput (*T*), inventory (*I*), and operating expense (OE) are operational measurements or decision criteria, which according to TOC thinking, better evaluate actions in terms of achieving a company's goal. Thus, the best decisions are not those that minimize costs, but rather those that increase throughput while simultaneously reducing inventory and operating expenses. So, using these three variables, return on investment (ROI) can be calculated and analyzed as follows:

$$\text{ROI} = \left(T - \text{OE} \right) / I$$

T: throughput
OE: operational expenses
I: inventory (investment)

Figure 10.1 shows the material feeding system proposed in the TOC, which has been called "DBR" (Drum, Buffer, Rope) and seeks to improve the results of the PUSH system and the Kanban. The buffer works as an inventory in process (WIP) that allows to protect the constraint from possible variations in the material feed and the arrow ROPE, refers to a pull signal or reorder point to push more material from operation 1 to the bottleneck (OP 2).

This system establishes that the throughput of the plant and the investment (inventory) are determined by the restrictions of the system (bottlenecks). If a constraint is a machine and it is stopped for an hour, the overall effect is as if the entire system was stopped for that time. TOC, therefore, states that one hour saved in the constraint is equivalent to one hour saved throughout the system.

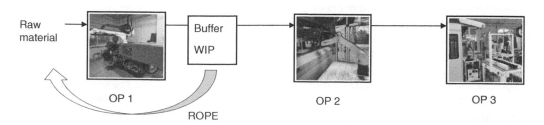

Figure 10.1 DBR system.

For this reason, it is vitally important to "protect" the operation from the restriction or bottle-neck and this implies preventing interruptions caused by lack of material, maintenance problems, or of any other nature. According to TOC, it is necessary to recognize that there are uncontrolled factors, fortuitous or eventual, that can endanger the throughput of the plant if the bottleneck stops. Until such events are completely eliminated, it is necessary to place a safety inventory or time buffer at the entrance of the restrictive resource, to protect its operation.

Based on these concepts, TOC proposes a series of principles that contrast with the traditional way of managing manufacturing in the West, which is presented in Table 10.4. The DBR system is a particular case of five postulates that define TOC [9]:

1) **Identify the constraints of the system:** These can be physical or political. The first ones are structural, like the capacity of a machine, a process, or a market. The second ones are philosophical impositions of the decision-makers.
2) **Decide how to exploit the constraint (s) of the system:** Exploit means to get the most out of the constraint. In the DBR system, it is clear that the existence of the time buffer is to prevent the bottleneck from stopping because it has no material to process.
3) **Subordinate everything else to the previous decision:** The overall performance of the system is determined by the constraint. If the system is visualized as a chain, that is, as a set of interlocking links, its strength depends on the weakest link.
4) If necessary, eliminate bottleneck from the system.
5) If, in the previous steps, a constraint is broken, go back to step 1, but do not allow inertia to cause a system constraint.

Table 10.4 shows a comparison of fundamentals about the TOC way vs. Fordism and TPS. The TOC questions whether the Japanese approach or mass manufacturing aims to achieve the global

Table 10.4 TOC way vs. fordism and TPS.

Fordism and TPS	TOC way
Balance the capacity, then try to keep the flow going.	Balance flow, not capacity.
The level of utilization of any workplace is determined by its own potential.	The level of use of an unrestricted resource is not determined by its own potential but by some constraint in the system.
Resource utilization and activation are the same.	Use and activation of a resource are not the same.
One hour lost on a constraint is just one hour lost on that resource.	One hour lost in a constraint is one hour lost throughout the system.
An hour saved on a constraint is just one hour saved on that resource.	One hour saved on a constraint is one hour saved across the system.
Temporarily restricted resources limit throughput but have a low impact on inventory.	Restricted resources govern both throughput and inventory.
The process batch must remain constant over time and throughout its path or process.	The process batch can be variable. The process batch may differ from the batch size to be processed.
Paradigm: the way to achieve the global optimum is by adding the local optimum.	**Paradigm:** the sum of the local optimum does not lead to the global optimum of the system.

optimum by adding the local optimum. For example, Kaizen promotes continuous improvement in all areas of the company; TOC promotes continuous improvement, starting from constraints.

In addition, TOC specifies that even in cases where the Kan Ban system is more effective than the traditional mass production system (Push), throughput is constantly at risk. This is because, no matter how much perfection is sought, there will always be sources of variability that affect the use of restricted resources.

10.2.2 Lean Thinking, Smart Manufacturing/Logistics, and the Cyber-Physical Systems Era

Japan's rebound in competitiveness in the US market prompted the development of operations engineering approaches and material feeding, sequencing, and synchronization systems to achieve flow development in the value stream. In this way, names such as Concentrated Manufacturing, Modular Manufacturing and World Class Manufacturing emerged, promoting the idea of eliminating waste, increasing flexibility, synchronization, and meeting market demands.

The term Lean Production (or Lean Manufacturing) was coined by MIT researcher John Krafcik [10], referring to a strategic production system adjusted to the client and to the objectives. An organization adjusted to the nature of the worker, to his mentality, to his training, to the tools at his disposal; stocks adjusted to the market situation; supplies tailored to needs; research and development adjusted to the objectives and market conditions and use of technology-adjusted to the prevailing pattern of competition and, essentially, processes without waste [6]. So, the school of lean thinking was started by Taiichi Ohno, when he realized that Henry Ford's ideas could be radically improved. He developed a model that results in lean thinking, using less of everything compared to the classic mass production used at the beginning of this century [7].

Lean Manufacturing is an approach to dramatically increase global productivity, that is, "do more and more with less and less" or more simply, it is a "systematic approach to identify and eliminate waste" [3]. "Lean Thinking" is a philosophical effort to clarify essential theoretical concepts about lean production and its implementation [6]; applicable to manufacturing and service environments and that guides the organization toward achieving the comparative advantages of craft manufacturing at the same time as those offered by mass production. In addition, it highlights the sustained creation of value for both internal and external clients and the increase in global profitability through the elimination of "muda" (everything that is far from the minimum necessary to produce and deliver value to the client).

The objectives of increasing productivity, flexibility, quality, and service, while reducing variability, response times, and waste, are the thematic axes of "lean thinking." These axes are developed by applying three strategic premises: development of the flow in the value stream pulling production from the needs of consumers, having intelligent and self-directed personnel, and the intelligent use of technology. This last principle has given a great boost to the concept of "Smart Manufacturing" and "Smart Logistics."

Logistics is an art and at the same time a science that allows obtaining, producing, and distributing materials and products in the required locations and quantities [2]. Furthermore, logistics is intrinsically linked to manufacturing and to the concept of "value stream" that refers to the processes of creating, producing, and delivering a good or service to the market. For a good, the value stream encompasses the raw material supplier, the manufacture and assembly of the good, and the distribution network. For a service, "the value stream consists of suppliers, support personnel and technology, the service producer and the distribution channel" [2]. Table 10.5 presents an approximation of what lean manufacturing promotes, compared to mass and artisanal manufacturing.

Table 10.5 Lean production vs. mass production and craft production.

Comparison of manufacturing systems		
Craft manufacturing	**Lean manufacturing**	**Mass manufacturing**
Qualified workforce in design, manufacturing, and assembly operations.	Qualified and multifunctional workforce. Emphasis on teamwork.	Maximum division of labor.
Extremely decentralized corporate organization.	Decentralized and horizontal organization based on processes oriented to the development of the value stream.	Centralized, vertical, and typically departmental organization.
Use of general-purpose machines and tools.	Combination of general purpose machines and specialized use machines.	Machines and tools for specialized use.
Typically low production volume with low capacity equipment.	High or low production volume with equipment that has capacity adjusted to the conditions of demand.	Typically high volume production with high capacity equipment.
Lesser normalization and high variability.	High normalization and standardization but with flexibility for change.	High standardization and standardization of operations and quality.
Highly flexible design and with a wide range of models.	Flexible design with high standardization of parts and components used in a wide range of processes.	Standardized design and with a reduced range of models.

Source: Adapted from Womack and Jones [6].

The term "smart manufacturing" comes from "intelligent manufacturing" that apparently was coined around 1990 by Kusiak [11]. When the term "smart" is used as an attribute of manufacturing or logistics, it is necessary to refer to the following characteristics of the system:

1) **Orientation to operational performance:** The design, management, and resources for the execution and control are focused on the value proposition for customers, shareholders, or owners of the corporation; value for the workers, suppliers, and distributors and, in addition, the social-environmental value proposition for stakeholders.
2) **Cause-Effect:** The systems that support managerial, operational, and continuous improvement decisions are based on cause-and-effect analysis through the collection and management of data and objective evidence.
3) **Prediction and anticipation of undesirable effects:** There are business intelligence, market intelligence, and interconnected distribution systems that allow the anticipation of undesirable effects in the value stream, implementing collaborative models to make strategic, operational, financial, and legal risks visible and preventively managed. In addition, the organization learns to solve problems by attacking their root causes, avoiding making a tremendous effort in what is not important or essential.
4) **"Core Competencies/Core Processes" vs. "Nice to Have":** Decisions on optimization or continuous improvement in the value stream or delivery logistics system are based on "core competencies/core processes" avoiding the fundamental changes in the enterprise architecture that do not impact on the value proposition and the competitive position in the medium or short term ("nice to have" decisions). This is especially important when making strategic decisions on "automation/autonomization" with special emphasis on "not making large investments in these

fields on what is wrong," that is, "paving old roads that lead nowhere." So, the premise of a "Smart system" is the development of technological architecture applying the cycle "**analyze-eliminate** muda (waste) and **simplify** processes (keep it simple)" and not fall into cycles of radical change due to fashion or a "me too" behavior following pioneering industry leaders.

Since the 1950s of the previous century, advances in computing, automation, and telecommunications gave way to solutions that "transforming" the competitive pattern in all sectors.

Great advances were deployed, such as MRP (Materials Requirements Planning), CRP (Capacity Requirements Planning), MRP II (Manufacturing Resources Planning) systems until reaching ERP (Enterprise Requirements Planning) systems and Flexible Manufacturing Systems (FMS) based in CAD (computer aided design), CAM (computer aided manufacturing), CIM (computer integrated manufacturing) applications; without neglecting IR (industrial robotics) applications in controlled and highly standardized environments. This technological landscape laid the foundations for what is now known as Smart Manufacturing and Smart Logistics, but until the first decade of the twenty-first century, architecture was still not intensive in cyber-physical applications.

As of 2011, the term Industry 4.0 was coined at the Hannover-Messe Technology Fair in Germany. From that moment on, the topic grew in popularity and collective construction of knowledge until it reached the 2016 edition of the Davos Forum organized by the World Economic Forum (WEF), where the consolidation and deployment of the Fourth Industrial Revolution were announced [12].

The first industrial revolution is the one that transformed manufacturing and transportation through hydraulics and the steam engine discovered by James Watt at the end of the eighteenth century. It allowed the development of the concept of "industrial factory" that with the great contribution of the Scientific Management of F.W. Taylor and Henry Ford's mass production consolidated the second industrial revolution that brought with it the combustion engine, the control of electricity, the development of aviation, the automobile, the telephone, radio, and television. Since the Neolithic, humanity has not known such a momentous change, including the economic, social, environmental, and political dimensions.

The third industrial revolution began approximately at the end of the 1960s and the beginning of the 1970s. Technological change was based on electronics and information and communication technologies. In addition, microprocessors, the internet (html language), and the miniaturization of components were the protagonists of a new architecture that accelerated computing, smartphones and mobile devices, cloud computing, and the internet of things (IoT), creating a bridge for the development of cyber-physical systems (CP Systems) and which are precisely the foundation of Industry 4.0 [12]. Figure 10.2 presents some frequently mentioned disruptive technologies when developing the concept of Industry 4.0 [13].

Cyber-physical systems are those that interconnect objects in the physical world with cybernetic environments in virtual worlds. These systems apply sensors, computer vision, the construction of virtual scenarios, digital twins, and algorithms or procedures with different degrees of autonomy for the control, operation, teleoperation, simulation, and assisted emulation of objects and subsystems [14].

In this way, the concept of a cyber-physical factory and a cyber-physical supply network was born. Technological applications related to IIoT, interconnected robotic platforms, "cloud computing" and mobile devices have intensified, impacting teleoperated and autonomous control and the possibilities of reducing delivery times and integrating the flow of the value stream. This new platform is known as "Interconnected Manufacturing and Logistics" or "IIoT Manufacturing" given its technological application emphasis. If the application of autonomous objects is intensive, these systems are known as "Autonomous Manufacturing and Logistics Systems" (AMLS) [15].

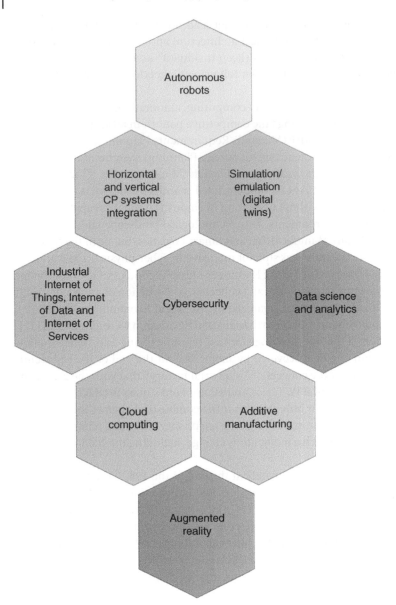

Figure 10.2 Disruptive technologies. *Source:* Adapted from Rüßmann et al. [13].

10.2.3 Design and Management Guidelines for Smart Systems

Next, managing approaches, specific techniques and its benefits are presented, giving emphasis in the strategic role of the smart manufacturing and supply chain systems in the Fourth Industrial Revolution. As already mentioned, there are different theoretical schools for the configuration of production and supply systems.

From these schools, variants emerge according to the dominant technologies in enterprise architecture. Figure 10.3 shows the main approaches and their variants for systems of production and supply of goods and services.

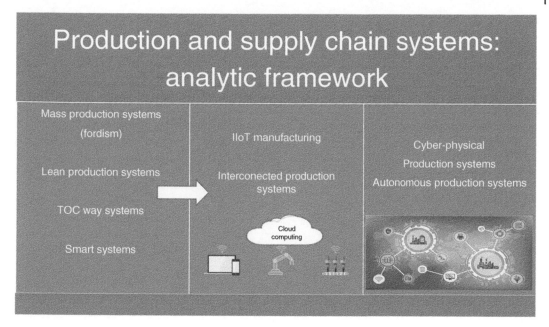

Figure 10.3 Production and supply systems (finished goods and services).

First, based on the explanation in the previous section, it is clear that a "lean" system is not necessarily a "Smart" system or a "CP" system. A "Smart" system can be designed without cyber-physical applications or "lean" tools, and although the Fourth Industrial Revolution is based on technologies such as artificial intelligence, robotics, interconnectivity, and cyber-physics, this does not mean that these applications become "Smart" or "lean" to the design of the manufacturing and supply system. Therefore, in the implementation of SMLS as a competitive weapon, the clarity that the technology is at the service of the competitive strategy is essential to win the preference of customers, and this covers the entire supply chain system.

The analysis, design, installation, and evaluation of applications focused on disruptive technologies to obtain a "value stream" with the characteristics of a smart system can have at least three "approaches": customer (client) orientation (CO), product or service orientation (PSO), and process orientation (PO). Table 10.6 presents an analytical framework to implement a Smart Manufacturing and Supply Chain System referring to these orientations.

In customer orientation, the starting point for the implementation of an SMLS is to understand the needs and expectations of customers when defining their preference for a product or brand. Figure 10.4 shows examples of trends in consumer decision criteria regarding preference toward brand categories in both, production of goods and services.

Companies that implement smart systems with a customer orientation relate their value proposition to operational objectives and indicators to positively meet the preference criteria, be it economically, environmentally, or socially [16]. For example, artificial intelligence applications based on emotion analysis, highly flexible and customizable additive manufacturing or also IIoT, IoD (Internet of Data), IoS (Internet of Services), and IoRT (Internet of Robotic Things) most often have an impact emphasis on customer satisfaction or distribution or delivery channels [17].

On the other hand, the orientation toward the product or service refers to the tendency of the company to carry out transformations in its business architecture and production and supply system with SML principles so that its offer acquires new attributes or differentiating quality

Table 10.6 Analytical framework of the smart manufacturing and supply chain systems.

Smart manufacturing and supply chain systems

Corporate goals	Dominant orientations	Schools of production systems thinking	Linked 4.0 disruptive technologies
• Recruitment, retention, and profitability of consumers • Brand preference, top of mind • Consumer satisfaction • Customer experience rating • Usabilit, functionality, customization	1) Customer Orientation 2) Product/Service Orientation	1) Smart and CP Systems 2) Interconnected Systems 3) Autonomous Systems	• IIoT, IIoD, IIoS • IIoRT/Computer Vision • Cloud • Social Robotics/Assistive Robotics • IA/Data Analytics • Cybersecurity • Block Chain • Additive Manufacturing • Digital Twins/ Augmented Reality • Autonomous Robots and Objects
• Product/Service Cost • User/product interaction • Expand service range. • Remote attention, assistance or collaboration services (omnichannel).	1) Product/Service Orientation 2) Process Orientation	1) Smart and CP Systems. 2) Interconnected Systems. 3) Autonomous Systems. 4) Customized Production (Craft Production Systems)	• Domotics • IIoT, IIoD, IIoS • IIoRT/Computer Vision • Cloud • Social Robotics/Assistive Robotics • IA/Data Analytics • Cybersecurity • Block Chain • Additive Manufacturing • Digital Twins/ Augmented Reality
• Value Stream Profitabiity; Global Throughput • Dock to dock time • Lead Time/Takt time • Make Span, Scrap, Yield, Variability, MUDA Savings • Inventory Management Cost, Warehousing Costs, Distribution Costs, transportation costs. • Stock levels/Fill rate	1) Process Orientation 2) Product/Service Orientation 3) Customer Orientation.	1) Lean Manufacturing and Logistics 2) Lean Supply Chain Systems 3) Smart and CP Systems. 4) Interconneted Systems. 5) Autonomous Systems. 6) Mass Production Systems	• IIoT, IIoD, IIoS • IIoRT/Computer Vision • Cloud • Social Robotics/Assistive Robotics • IA /Big Data Analytics • Cybersecurity • Block Chain • Additive Manufacturing • Digital Twins/ Augmented Reality • Horizontal and vertical systems integration

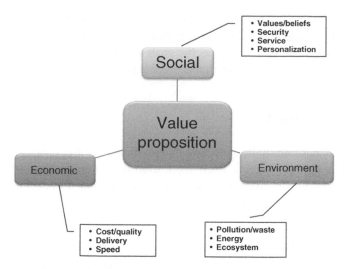

Figure 10.4 Customer preference criteria.

specifications and that in turn it, must once impact competitiveness or operational excellence [18]. For example, the design of Smart Systems but also cyberphysicists with IoRT and computer vision systems (CVS) applications that allow distribution channels to control inventories in the supplier's warehouse in real-time and with visualization; or also the industrial and residential domotics applications [19].

Companies that seek to operate with process-oriented SML systems are those that have identified their greatest opportunities for impact in the continuous or radical improvement of what they have defined as "core competencies or core processes."

Then, goals related with flow development in the value stream are highlighted, seeking increased throughput and reduction of cycle times, waiting and queuing times, inventory levels, and buffers [20]. In addition, companies aim to do better prediction of incidents with remote monitoring to improve "tracking" and reduce transportation costs, picking, container consolidation, "dock to dock time," loss of merchandize in transit, "scrap" or unproductive times, and costs due to unplanned downtime of manufacturing and logistics equipment. Also, typically seek the reduction of costs due to non-remote corrective and preventive maintenance or set-up times and other activities that do not add value and that affect the "overall equipment effectiveness" (OEE) through the supply network [21].

Technologies are also implemented that allow the reduction of occupational, operational, environmental, and legal risks, including the waste of clean energies and the optimization of sales and operations planning (S&OP) to take advantage of capacity and buffer management (TOC techniques) [22]. This impacts the entire supply chain and integrated enterprise resource planning systems (ERP 4.0), which today are designed with applications of artificial intelligence, machine learning, intelligent sensing, computer vision, robotics, and multi-agent cybersecurity, as well as others disruptive 4.0 technologies, such as IoT/IoS systems, digital human–machine interaction, cyberphysical simulation, and teleoperated control based on digital twins and digital phenotypes.

Obviously, the implementation of smart manufacturing and smart supply networks are in practice a mixture of the three orientations (product, process or customer) but it is foreseeable that some of them will be dominant according to the competitive position of the company. For example, companies that compete in mature markets tend to be more process-oriented while those in emerging or pioneer markets tend to be more product/service or customer oriented. So the design, installation, and optimization of smart factories and smart supply networks relate both, schools of theoretical thinking and strategic orientations, to aim for specific goals in order to sustain and increase competitiveness, as well as disruptive technologies that achieve agile operations.

10.3 Case Analysis: Lessons Learned

This section presents a "quick overview" of some application cases that exemplify the implementation of SMLS with a sense of "Lean" and "CP" Systems. Each application has an emphasis that does not necessarily fall on a "totally pure" perspective with respect to the theory, since the adaptation of the design to the competitive goals of the organization prevails.

10.3.1 Infineon Technologies AG Dresden

Infineon Technologies AG is a company that was founded in 1994 and was initially part of Siemens. Today, it employs more than 46,700 people worldwide from more than 100 countries and is one of the most recognized companies in the semiconductor industry. Its main production plant is located in Dresden, Germany and is a mandatory benchmark in facilities and operations in Europe. Infineon acquire the US company Cypress Semiconductor Corporation in April 2020 and now is a global top 10 semiconductor company.

The company has as part of its mission, to collaborate with its clients in the digital and cyber-physical transformation of their operations and applications under a sustainability and circular economy approach. Its factories stand out for their high level of automation and the interconnectivity of their stages in the value chain for their production lines in Dresden. In recent years, they have grown to become one of the most automated factories in the world; however, they have the objective of moving toward a "smart" and "cyber-physical" operation to increase their operational and financial performance.

This case study begins with the goal of designing and implementing a cyber-physical system for the production of semiconductors [23]. The objective is to speed up decision-making by reducing response times and the strategy is to replace a centralized system with a decentralized one, especially in areas where high autonomy and automation are not yet available, for example in "wafer test" facilities. This will improve production flexibility to increase product customization and reduce lead times.

To achieve the company's goal, a system was designed based on the concept of "self-control" (which includes the search for decentralization and autonomy) to increase flexibility, in terms of system changeability; robustness, transferring the decision-making capability to individual logistics objects; and data availability [23].

The required flexibility and robustness are sought through a software platform based on multi-agent technology (Multi-agent System, MAS) was developed and it allows establishing negotiation and interaction routines between different objects or agents (machines, people, materials, or tools). In this case, a "Process for Agent Societies Specification and Implementation" (PASSI) was established which includes five models: systems requirements model, agent society model, agent implementation model, code model, and deployment model.

The Hierarchical Agent Society (HAS) is based on smart sensors, data acquisition system, processing system, mobile devices for human interfaces, and cloud computing. In addition, a hierarchical model of interactions and response decisions contains a human interface and planning/controlling and execution process.

This case shows an SMLS application with a dominant PO. But the design aims to improve the company's competitiveness with faster deliveries and a more personalized product offering. Therefore, a customer orientation is evidenced.

The system has been developed and is implemented at a prototype level [23]. The technical validation has included the design and testing of the radio network in the semiconductor factory. Furthermore, the integration "of the carrier with the new electronic components" and the integration of MAS in the IT landscape have been tested. The expected result in the medium term is the reduction of changeover times to produce smaller and personalized batches, but clearly, this means an increase in economic throughput and a decrease in cycle times, which brings the system closer to a "Lean-Smart and Cyber-physical" operation.

10.3.2 Stanley Black & Decker, Inc.

This company, headquartered in New Britain, Connecticut, US, was formerly known as The Stanley Works and changed its name to Stanley Black & Decker, Inc. in March 2010. It was founded in 1843 and engages in the tools and storage, industrial, and security businesses worldwide [24].

The Tools & Storage segment offers tools and equipment and sells its products through retailers, distributors, and a direct sales force to professional end users, distributors, and industrial customers. Industrial segment provides fastening systems and products in several industries like automotive, manufacturing, electronics, construction, and aerospace as well as provides pipeline inspection services; and sells hydraulic tools, attachments, and accessories. Finally, its Security segment designs, supplies, and installs commercial electronic security systems and provides electronic security services to healthcare, educational, and financial institutions as well as commercial.

Stanley Black & Decker started its transformation strategy five years ago; the initial idea was to take advantage of the tools, technologies, and strategies of Industry 4.0 to help its clients. But this effort made them rethink their own business and its operation in all countries through technology. The goal of the corporation is to generate two to three dollars per dollar invested in this 4.0 strategy in all its operations in the world [24].

The first finding of their initial analysis was that they should work on the optimization of their processes but with higher levels of efficiency. So the first step has been to work on a digital transformation program to modify the processes that impact service and delivery so that they are more data-driven and automated throughout the supply chain. This has enabled the vision of creating an intelligent connected enterprise that will act as the backbone for discovering and unlocking data.

This involves building a holistic suite of business applications and control towers to connect operations in more than 175 countries. An example of this effort is the plant located in Reynosa Mexico, which since 2005 manufactures a wide range of products for the DeWALT brand with a volume of millions of power tools per year that come out of its 40 production lines.

In this plant, as in many others of the corporation, the manufacturing process was connected with smart sensors and the Real-Time Location System (RTLS), made up of a wireless network developed by Cisco and business visibility solutions from AeroScout Industrial. Wi-Fi RFID tags have been used that adhere to virtually any material and provide assembly operators, shift supervisors, and plant managers with the real-time location and status of each asset. In addition, Cisco connection points are leveraged to offer mobile access to production line information through the tablets and smartphones of plant managers. This allows the production of each line to be known as well as the performance and effectiveness indicators both globally and in detail.

Another factor of change and success at the same time for Stanley Black & Decker is the focus on creating value and opportunities for employees. Through the comprehensive and technical skills upgrade of more than 60,000 employees, the company created professional growth programs and development paths increasing retention. This has been achieved through teleoperation platforms, digital twins, and data-driven systems and digital tools. Increasing training and teamwork through staff with high-value digital skills also enables them to make more strategic business-driven decisions, generating new ideas and better execution of innovation initiatives [24].

A crucial element of the implemented strategy is that Stanley Black & Decker has been able to articulate the ROI and the value of the digital transformation. They have developed a clear strategy to improve key process indicators (KPIs), including performance, asset and labor utilization, quality, and profitability. This has made it possible to align efforts around the world and also keep investment focused on the value proposition, avoiding "nice to have" applications [24].

10.4 Summary and Conclusions

The last decade of this century expresses the ultimate meaning of rapid change in the competitive world. Change factors and trends introduce new criteria of brand preference by consumers and new competitive patterns to achieve or maintain leadership in a sustainable way. For example, climate change and the explosion of disruptive technologies clustered participation in emerging industries such as biotechnology, space industry (space tourism, resource exploitation, satellite applications, among others), genomic bioinformatics, nanotechnology, and high-speed telecommunications (5G and data transmission by laser communication); as well as the dynamization of the digital transformation of businesses

The first and second industrial revolutions marked an era of competitiveness focused on productivity and the reach of mass markets. On the other hand, the third industrial revolution was characterized by the phenomenon of globalization that pressured the development of "mass customization" models to manage differentiation (design, quality, brand positioning), diversification, quick response with high inventory buffers, sales margin protection (low costs and elimination of waste).

The response was the development of thinking schools like "Lean Thinking," TOC, and Smart Manufacturing, among others. With the fourth industrial revolution (Industry 4.0), the acceptance of a new technological platform for production and supply chain systems is in the way of consolidation: cyber-physical systems. These systems, which integrate the physical world with the virtual world, seek the optimization of processes, quick response, the reduction of operational costs and energy use, as well as the entire value stream agility through robotics, AI applications, and Data Analytics; as well as the implementation of other disruptive technologies.

However, as we have seen in this chapter, a cyber-physical system is not necessarily smart, just as a smart system is not necessarily cyber-physical or managed in a "lean" or "TOC" or "Mass Production" way. Following this same logic, what is more common to find is that "Lean," "TOC" or "Mass Production" systems are not "Smart" or "Cyber-physical."

SMLS are a clear global trend to respond to competitive needs and implementations that take advantage of the technological development of cyber-physical systems of the Fourth Industrial Revolution. The analysis, design, installation, and evaluation of applications focused on disruptive technologies to obtain a "value stream" with the characteristics of an SMLS can have at least three "approaches" or a combination of them: consumer orientation, product orientation, or service and PO. In practice, the success study cases allow us to conclude that combinations between consumer orientation and PO predominate, taking into account the transformation of practices to involve the development of people from operator levels to senior management levels.

The best practices demonstrate that the implementations that articulate goals aligned with performance are successful, from an impact point of view on the value proposition for customers, the profitability of the operation, collaboration, and optimization in the industrial ecosystem as well as seizing opportunities in terms of innovation at the right time.

Finally, in the near future, some trends can be highlighted and will surely guide scientific research and the development of innovations; among them, is the autonomization of assistance platforms for the remote recovery of equipment in the event of breakdown risks; for example, in the case of teleoperated or fully autonomous sea or air transport, the introduction of multi-agent systems with machine learning will be an operational and strategic demand. Likewise, the investment in cybersecurity of all systems will continue to grow, and even more so with the growth of the development of digital twins and digital phenotypes in virtual sensory environments or metaverses.

References

1 Hopp, W.J. and Spearman, M.L. (2008). *Factory Physics*, 3e. USA: Waveland Press Inc.

2 Pittman, P.H. and Atwater, J.B. (ed.) (2020). *APICS Dictionary*, 16e. USA: APICS.

3 Conner, G. (2019). *Becoming a Lean Savant*. USA: BookBaby Press.

4 Stein, R.E. (2013). *Re-Engineering the Manufacturing System: Applying the Theory of Constraints*. USA: CRC Press.

5 Hobbs, D.P. (2004). *Lean Manufacturing Implementation: A Complete Execution Manual for any Size Manufacturer*. USA: J. Ross Publishing-APICS.

6 Womack, J.P. and Jones, D.T. (2003). *Lean Thinking*, 2e. USA: Simon & Schuster, Inc.

7 Hobbs, D.P. (2011). *Applied Lean Business Transformation: A Complete Project Management Approach*. USA: J. Ross Publishing.

8 Ohno, T. (1988). *Toyota Production System: Beyond Large-Scale Production*. USA: CRC Press.

9 Goldratt, E.M. and Cox, J. (2003). *Production the TOC Way*. Great Barrington, MA: North River Press.

10 Womack, J.P., Roos, D., and Jones, D.T. (2007). *The Machine that Changed the World: The Story of Lean Production – Toyota's Secret Weapon in the Global Car Wars that Is Now Revolutionizing World Industry*, 2e. USA: Simon & Schuster, Inc.

11 Kusiak, A. (2017). Smart manufacturing must embrace big data. *Nature* 544 (7648): 23–35.

12 Schwab, K. and Davis, N. (2018). *Shaping the Future of the Fourth Industrial Revolution*. USA: Currency.

13 Rüßmann, M., Lorenz, M., Gerbert, P. et al. (2015). *Industry 4.0: The Future of Productivity and Growth in Manufacturing Industries*. USA: Boston Conulting Group.

14 Ghobakhloo, M. (2020). Determinants of information and digital technology implementation for smart manufacturing. *International Journal of Production Research* 58 (8): 2384–2405.

15 Kusiak, A. (2018). Smart manufacturing. *International Journal of Production Research* 56 (2): 508–517.

16 Kusiak, A. (2019). Fundamentals of smart manufacturing: a multi-thread perspective. *Annual Reviews in Control* 47 (1): 214–220.

17 Qu, Y.J., Ming, X.G., Liu, Z.W. et al. (2019). Smart manufacturing systems: state of the art and future trends. *The International Journal of Advanced Manufacturing Technology* 103 (9): 3751–3768.

18 Lu, Y., Morris, K.C., and Frechette, S. (2016). Current standards landscape for smart manufacturing systems. *National Institute of Standards and Technology, NISTIR* 8107 (39). https://doi.org/10.6028/NIST.IR.8107.

19 Zhegn, P., Wang, H., Sang, Z. et al. (2018). Smart manufacturing systems for Industry 4.0: conceptual framework, scenarios, and future perspectives. *Frontiers of Mechanical Engineering* 13 (2): 137–150.

20 Tao, F., Qi, Q., Liu, A., and Kusiak, A. (2018). Data-driven smart manufacturing. *Journal of Manufacturing Systems* 48: 157–169.

21 Lu, Y., Xu, X., and Wang, L. (2020). Smart manufacturing process and system automation – a critical review of the standards and envisioned scenarios. *Journal of Manufacturing Systems* 56: 312–325.

22 Abubakr, M., Abbas, M.A.T., Tomaz, I. et al. (2020). Sustainable and smart manufacturing: an integrated approach. *Sustainability* 12 (6): 2280.

23 Keil, S. (2017). Design of a cyber-physical production system for semiconductor manufacturing. *Proceedings of the 2017 Hamburg International Conference of Logistics (HICL)*, No. 23, ISBN: 978–3–7450-4328-0, epubli GmbH, Berlin. http://dx.doi.org/10.15480/882.1442.

24 Rockwell Inc (2020). Stanley Black & Decker's digital transformation journey accelerates innovation. *Automation Today* 67: 12–13.

Section 6

Project Management

11

Project Management

Celia Desmond

World Class – Telecommunications, Mississauga, ON, Canada

11.1 Introduction

The accepted base of Project Management processes and process areas used in this chapter is defined by an organization called the "Project Management Institute" (PMI). This organization defines project management processes categorized into different process areas. The processes, areas, and techniques vary over time which fit the evolving technology industries.

The process areas to be covered are Project Integration Management, Project Scope Management, Project Schedule Management, Project Cost Management, Project Quality Management, Project Resource Management, Project Communications Management, Project Risk Management, Project Procurement Management, and Project Stakeholder Management. Although most project managers are selected from the engineering ranks, project teams are composed of people from many different departments. The Project Manager (PM) relies on the skills and knowledge of all of the team members, so project aspects will be considered from multiple perspectives. It is important that the PM and the team members can communicate well with all members of the team, and with the customer.

A few examples of the different types of technology projects which one might encounter include, but are certainly not limited to:

- Develop a new product or service.
- Analyze the introduction of another company's competing product or service enabling the provider to determine the best competitive response.
- Manage a serious problem such as a cable cut or pole destruction in a remote area.
- Equip the current company with a new ordering or billing system which is more flexible that the current one, allowing new rating models to be adopted when desired.
- Introduce a new culture to the employees that is more conducive to determining customer requirements clearly before initiating the design of a new service, and provides them with the tools to be able to do this.

We will view projects from various perspectives, as it is important to keep the business picture in mind while focusing on the details of the results produced via the project.

IEEE Technology and Engineering Management Society Body of Knowledge (TEMSBOK), First Edition.
Edited by Gustavo Giannattasio, Elif Kongar, Marina Dabić, Celia Desmond, Michael Condry, Sudeendra Koushik, and Roberto Saracco.

This chapter will be useful to engineers working in any technology area, and it will be understandable and meaningful to technical teams. This material is at an introductory level. Each of the references for this chapter provides a wealth of additional material and more in-depth discussion, for those wishing to pursue any area more fully.

11.2 Content

There are many definitions of Projects and Project Management. There are also many recognized organizations worldwide which focus on project management, providing guidance on processes, tools, techniques, and methodologies. The base for those provided in this document is those recommended by PMI. Recommendations from all organizations address knowledge and processes which can be applied in projects across all technology areas. When the recommendations are followed there is no guarantee of perfection, given the complexity and evolving nature of the system and its components, but the end results produced will be much closer to the goals and objectives for the project.

Among the many methodologies recommended for managing projects, two of the most popular today are Critical Path Method (CPM) and Agile. CPM applies well to projects which have clear objectives, and for which a high-level path can be foreseen, while Agile applies well to projects such as software development and some research, where direction can easily change based on the results of ongoing developments. We will discuss Agile briefly in this chapter, with a main focus on CPM, since Agile methodologies are covered also in other areas of this Body of Knowledge.

Let us start with definitions of a project and of project management.

A project is a temporary endeavor to produce a unique product or service with limited resources within a certain timeframe. Project management is the application of knowledge, skills, tools, and techniques to project activities to meet the project requirements [1].

In CPM the process areas to be covered are Project Integration Management, Project Scope Management, Project Schedule Management, Project Cost Management, Project Quality Management, Project Resource Management, Project Communications Management, Project Risk Management, Project Procurement Management, and Project Stakeholder Management. PMI also divides these into five process groups reflecting the flow of projects – Initiation, Planning, Execution, Monitoring and Control, and Closing.

Using Agile the team plans for releases, allowing the provision of an initial feature or set thereof, upon which the team subsequently builds. The flow of the releases is defined by a roadmap, and each release can be built via a series of iterations. Each iteration plan schedules a series of features and these are then tracked daily, usually with short standup meetings.

The environments in which technology projects occur can be quite varied. Consider the industrial supply chain shown in Figure 11.1. Each enterprise in this chain will have multiple customers, so the number of supply paths will grow quite large: imagine how many end users might be at the end of a chain started by a company which mines and refines iron ore! However, any final

Figure 11.1 Technology life cycle.

product in the hands of an end user will have a unique path going all the way back to that mining company. A final product will usually comprise many different parts, hence many different supply chains coming together to make the final product. A project approach to doing business may be found in any of the companies in the supply chain. These projects are typically confined within one company, but in some rare instances, multiple links in the chain may collaborate in a project of importance to all of them. A modern example of such collaboration is the development of new battery technology in conjunction with the release of an electric vehicle.

Let us examine the electric vehicle example a bit further. A project approach is to be found at each stage of the supply chain.

- The resource supplier is a mining company producing raw Lithium metal. It has implemented a project to introduce a revised process which will reduce power consumption in the refining of the ore.
- The component manufacturer produces highly purified Lithium metal strip for use in the production of Lithium-ion batteries. It has implemented a project to improve a clean room environment necessary to prevent contamination in their product.
- The equipment vendor is a manufacturer of Lithium Ion cells for automotive use. There are many emerging candidates for greatly improved battery chemistry, and the company has implemented a project to examine these to determine the design for their next-generation product.
- The equipment/service provider is an electric vehicle manufacturer. They have implemented a project to roll out a new model based on experience with existing models and opportunities arising from the availability of more advanced batteries.
- The end user is a large metropolitan taxi company. They have already made the decision to go with electric vehicles, and have implemented a project to determine the feasibility of creating a dedicated network of charging stations, vs. the use of existing public charging infrastructure.

With the multitude of technologies existing today, and the interdependence in many areas of multiple technologies to provide one product or service, we can see that a very wide range of highly technical projects may exist. Project management methodologies which can apply across the board in all of these areas will be needed.

A few examples of different types of technology projects which one might encounter could include:

- Develop a new product or service. This might happen at any stage in the industry value chain and might involve a single technology or the integration of multiple ones. Consider self-driving cars that communicate with each other and other devices.
- Analyze the introduction of another company's competing product or service enabling the provider to determine the best competitive response.
- Manage a serious problem such as a cable cut in a remote area, which could be an extremely complex project with severe time constraints.
- Research a new technology – maybe a more efficient battery for electric vehicles.
- Equip the current company with a new ordering or billing system which is more flexible than the current one, allowing new rating models to be adopted when desired.
- Introduce a new culture to the employees that is more conducive to determining customer requirements clearly before initiating the design of a new service, and provide them with the tools to be able to do this.

All of these are run as projects, and similar projects could happen in any technology area.

11.3 Knowledge Area Fundamentals and Examples

Effective project management requires the use of processes in all knowledge areas that are relevant to each specific project. Here we briefly outline the recommendations in each of the process areas listed above. Describing all 10 Process areas in detail cannot be done within this chapter, so those interested in additional information should access the reference books.

11.3.1 Project Integration Management

In this section, we separate the various components of project management, but none of these operate independently of the others, so it is important to consider the integration of all of the elements. Integration starts at the planning stage with the development of the Project Charter and Project Management Plan. The Charter is a short, high-level document that overviews the entire project, with information about the objectives, scope, timing, risks, critical team members, assumptions, and any other information that could be relevant to the success of the project. The Project Plan is a document outlining how the project team will ensure that the goals for each of the areas, and for the overall project, can be met. It does not contain the plans for the other areas, just the information on how these can be implemented. Integration continues with the management of the work and also the knowledge of the project and its product, specifying how the work will be directed and controlled. At the closing stage of the project all aspects must be confirmed. Once the product has been accepted and all actions closed integration completes.

11.3.2 Project Scope Management

Project Scope is defined at a high level in the Project Charter, and then in considerable detail in the Scope Description.

11.3.2.1 Project Charter
The Project Charter is a high-level description of the project, which is used to recruit the PM, and subsequently to communicate the project information to those who need to know it. The signing of the Charter officially kicks off the project, providing authorization to begin work and charge to the project.

This sample Charter shows a very short-term project which is highly critical. The company has a structure for a Charter for such projects to be used when needed in these emergencies.

Keeping Connected – Field OPS Project Charter

Cable Repair Project Charter
Project Manager: Jessica Fletcher
Sponsor: VP Operations
Project Customer: Keeping Connected
Date: 21 September 2021

Project Purpose: A serious cable cut in a location which is remote from any major city, in difficult terrain is preventing communications for major companies and residents in and between two major cities. The location of the cut must be found, and the connectivity restored.

High-Level Project Description: Keeping Connected provides service to 70% of the major companies and 85% of the residents in and between Detroit Lakes Minnesota and Minneapolis. Competing companies are actively advertising to convince customers of connected to switch. This service discontinuity has the potential to lose a significant portion of Connected's business in this area. The location of the break must be quickly determined, repair materials and equipment deployed, media continually informed, and customers kept abreast of the status.

Key Deliverables: The major deliverable is restoration of service, but in addition the customer loyalty must be preserved. High-level deliverables are:

- Location determined
- Cause of break assessed for future planning
- Deployment of restoration team and equipment
- Media coverage every three hours
- Customer communications
- Compensation plan and delivery

High-Level Requirements:

- Facility testers must locate the problem area within two hours
- Network design must be informed to facilitate the determination of other potential risk areas
- Media must be informed of progress every hour. No employee other than corporate communications to provide any information to the media
- Restore all communications within 20 hours
- Management must be informed hourly, round the clock
- Develop a compensation package to be delivered to the customers within 56 hours

11.3.2.2 Project Scope Statement

This document might also be called by another name, such as "Scope Description." In this document the planners describe all aspects of the scope, using any text, standards, diagrams, photos, brand names, charts, or pictures required to ensure that all readers will have the same understanding of what the project will produce. Needless to say, these documents can become quite long.

11.3.2.3 Project and Product

Before discussing more details, we need to make the distinction between the project and the product. The project is in place to deliver a product. The "product" might actually be a service or a process, but we can call it a product for the purposes of this discussion. The product will have a description. This product description will be one of the project information documents. The "project" in this context is the work that must be completed in order to produce the product.

11.3.2.4 Work Breakdown Structure

Based on the above documents the team will then create a structure which shows the full project at the top, broken into high-level deliverables at the next level, and continue breaking till eventually each branch ends with activities that must be performed to provide that deliverable. Each box is broken into its components in such a way that the items below completely add up to the one above, with no gaps and no overlaps anywhere in the structure. For example, in Figure 11.2 you see a high-level work breakdown structure for creating a conference on wireless communications. A few of the deliverables are broken down to show the complexity that is needed to fully describe all the deliverables and project work.

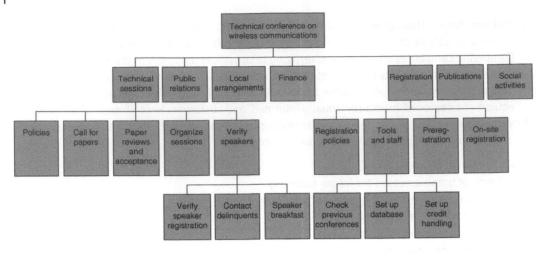

Figure 11.2 Work breakdown structure.

The creation of the WBS starts with the high-level deliverables which are listed in charter. These are then broken into smaller deliverables, gradually working down to smaller pieces, until the "bottom level" is reached. The bottom-level elements are actually activities. Instead of expressing them as deliverables (the what), it is possible to include a verb, expressing them as activities. These activities must meet certain criteria. They must be

- assignable
- independent
- measurable
- schedulable
- budgetable
- suitable size

Once the WBS is available, the team can use it to determine the resource assignments, the project costs, and the timelines.

11.3.3 Project Cost Management

Prior to the project approval a high-level cost would have been determined as part of the business plan, to be used in a GO/NOGO decision for the project. If this estimate was approved via the Charter, then this is the total amount available for the project. The team must now estimate the costs for the work and the deliverables to determine how well this fits into the total budget available. Using the WBS it is best to start with estimating the cost for work and materials for every bottom-level item. Since the bottom level items in each chain fully add up to the one above, complete the bottom level costs first, add these to find the totals for the level above, and continue adding till the top is reached. This shows the full cost for completing the project, as long as all costs were considered in the formation of the item estimates.

Figure 11.3 shows the costs at the top level of a WBS for the creation of a new product. One of the top-level boxes is missing in this figure – the project management. Those costs are usually hours of labor for the PM. As can be seen from the figure, there are many types of costs to consider for new product development. One great advantage of using a project management program such as

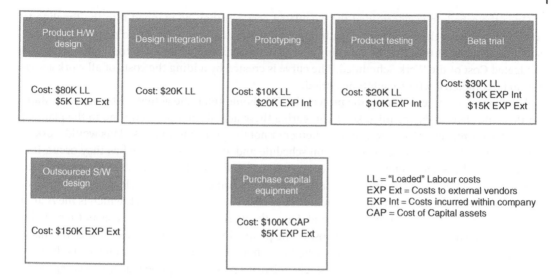

Figure 11.3 Development cycle (deliverables) costs.

Microsoft Project is that once you enter the costs for time unit for each of the people, and the material costs, then assign the work to the people, and the materials to the activities, the program calculates the costs at each level. Then, as the work progresses, the actual costs for the work can be entered, and graphs can be produced to show the planned value, the actual cost, and the earned value.

There is one additional cost that must be considered. This is the project contingency. Included somewhere in the costs we should always have contingency, extra cost to be included to cover for any of the risks that are considered to possibly occur during the project. These risks are called "known unknowns" because we do not know whether any individual one will occur. Some will and others will not. We need to have enough funding to cover the ones that do occur. This is the contingency cost that must be added to the cost of the project. The methodology for determining the amount of contingency to include is shown in the Risk Management section to follow. The riskier the project, the more contingency will be required to cover all the costs, as many things are likely to change from the initial plan and some of the impacts of the changes could be significant. It is expected that the entire amount will be needed in order to complete the project. The issue is only what the money will be used for, and this is something that the PM should manage closely.

If this bottom-up total including contingency cost does not fit within the available budget, the costs must be reviewed to ensure that each item is covered properly. If the budget is exceeded, the team should first consider trade-offs in work or materials or scope reductions which could allow the budget to be met along with all of the project objectives. If this proves impossible, perhaps additional funding can be requested, requiring a clear analysis of why it is needed. The final option could be that the project should not be undertaken, but this should be considered a last resort if there is no workable solution.

As a rule the outcome of the project must recover the total project cost including all direct and indirect costs, while showing a profit or surplus. Most organizations have guidelines for the PMs to follow to ensure project viability. If the project team is involved at early stages, perhaps when a bid is being prepared to be forwarded to a client, it is important to ensure that this bid includes the direct costs, all overhead, contingency, and the required margin; otherwise, the project will result in a loss to the company. That should happen only with the full knowledge and intent of the management.

Once the schedule is determined the calculated cost can then be mapped into the timeline to allow planning for the amount of spending planned for each time unit of the project. The curve showing the map of the planned spending over time is called the "Planned Value," or by some, the Budgeted Cost of the Work Scheduled. The curve is created by adding the costs of all work activities to be completed within each time period.

As the work completes during the project the PM should track the actual costs incurred. Many do this, and then make the mistake of comparing these actuals at any given time to the planned cost for that time, in order to determine whether or not the project is on track. This would work as long as all of the work proceeds exactly on schedule and all costs are entered as they occur. It is critical that all costs are entered as they occur; otherwise, the total actual cost will not be correct. Figures 11.4–11.7 follow to illustrate the showing of Planned Value, which is the cumulative Budgeted Cost of Work Scheduled for each work period, then the Actual Cost, which is the Actual Cost of the Work Performed in each period. Since the work performed often does not match the work scheduled, comparing the actual cost to the planned value is usually misleading. Yet this is what many PMs do, thinking it will show whether or not they are under or over budget. Only if the work performed exactly matches the work scheduled at a point in the project does this comparison show the true performance against the budget.

First, we see a simple project with only nine bottom-level activities. For simplicity we show that each box has a planned cost of $1000. Thus, we can see the cumulative planned spending for the project over time.

Next view the project after the first week: we are slightly ahead, as some work scheduled for week 2 has been accomplished. The areas shaded show the work that has been completed in the first week, and the AC figures show the actual cost of this completed work. Everything clear must still be completed, and costs for it will be incurred later as the work happens. The AC curve shows the actual cost incurred in the first week. Note that there is another curve, a shorter solid line. This curve is the one that shows the actual real progress of the project. It is called "Earned Value," calculated from the Budgeted Cost of the Work Performed. At the end of the first week, we can see

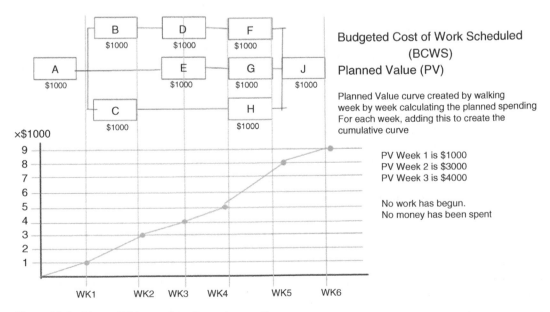

Figure 11.4 Planned Value – plans for costs over time.

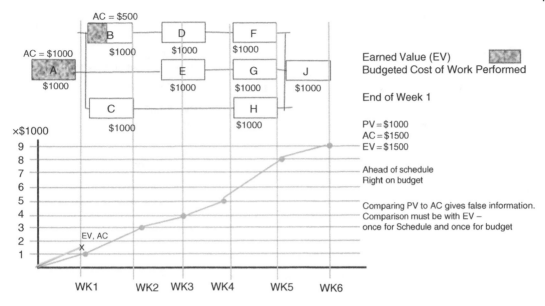

Figure 11.5 Earned Value Week 1.

that we have completed one package planned at $1000 and ½ package planned at $1000, and that the cost of this work is $1500. Thus, we are right on track for budget. Comparing PV to AC would have shown that we are over budget – which is not the case. To determine the status as per the budget, Subtract AC from EV. That compares the amount spent for the work performed with the amount planned for that same work. In this case, we find that we are right on track for the spending. However, looking at the green in the boxes, we can see that we are actually ahead of schedule with the work, because some of activity B has been done even though it was not scheduled till week 2. We can use the same numbers to determine how the work stacks up against the schedule. Subtract PV from EV to see this. In our case, we get a positive value of $500, showing that we are ahead of schedule by this amount. A positive answer for the schedule or cost is good. A negative answer is bad.

Figures 11.6 and 11.7 track the work and the costs through the following weeks. These can be used by the reader to determine the actual status of the project, by doing the subtractions from Earned Value.

11.3.4 Project Quality Management

In a project, "quality" essentially means meeting the expectations of all stakeholders. This is a challenging task. PMI defines three processes for this.

- plan quality management
- manage quality
- control quality

The quality management plan must specify the quality expectations – objectives and goals. This requires the definition of the quality objectives to be met, so that the team can focus on these as a priority. Whether the project is producing software, producing hardware, or essentially research, some mix of these, or possibly something that is non-technical, such as an office move or culture

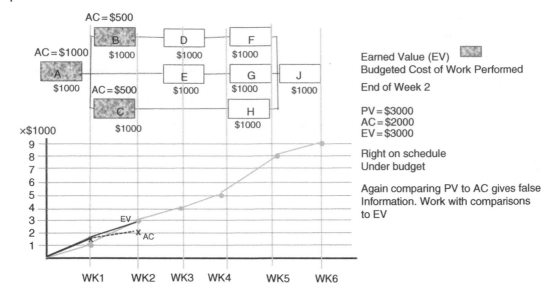

Figure 11.6 Earned Value Week 2.

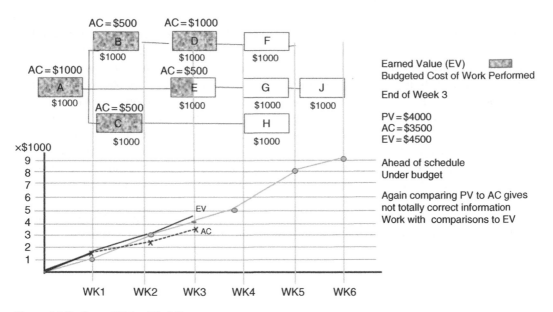

Figure 11.7 Earned Value Week 3.

change, quality management is important. It is often overlooked in project planning, but that can cause problems later in implementation. The Quality Management Plan must specify the measures and techniques for the team to use to manage and control quality.

Managing quality is the process of taking the expectations from documentation such as the quality plan and quality policies in place in the organization in which the project is undertaken, and creating the right activities to ensure that these can be met.

Controlling quality is the monitoring of the results of these activities to ensure that they are being met. Control specifications should be set for these so that the team can notice when they are heading in a poor direction, and make decisions on how to get back on track. The control specs should be somewhat tighter than the quality requirements so that the expectations are always within reach.

There is a cost in time and/or dollars to set up to meet the quality requirements within allowable limits, but of course, there may be a much higher cost in failing to meet the quality goals. The team needs to trade off these costs and select the route that will provide the best results for the least cost.

A famous example is that of an airline company designing a new aircraft. The team located a new metal which could be used in place of conventional materials, which is less expensive (super), lighter (very important for a plane), and stronger (clearly critical for a plane). After extensive testing, the team found that this metal, with constant use under flight conditions, would show wrinkles which looked like cracks, even though they were not actually cracks, and the strength remained. But how could this be explained to all technicians worldwide, who are mandated to fix all cracks before the plane can be allowed to fly. This metal with all its benefits is not good enough quality for this purpose.

In the Risk Management section, we discuss known unknowns – things that might be anticipated to derail the project. These are the risks. We build in contingency time and contingency funding to cover for these. In addition to known unknowns, we should always keep in mind that sometimes things go wrong which were never anticipated. These would be "unknown unknowns." The obvious current example is the 2020 COVID 19 global pandemic. It is very unlikely that anyone was prescient enough to have a pandemic on their risk list, and this pandemic derailed thousands of projects. Few unknown unknowns have this kind of impact, but they can certainly cause problems to project teams. Many product projects are derailed by things that no one thought to put on the risk list. Since these were not considered, there is no contingency to cover for them. In fact a strict rule for contingency is that it should be used to deal with the risks, and not for anything else. Anything else includes unknown unknowns and scope changes. For the unknown unknowns, it is necessary to negotiate with management and possibly also with customers to get the resources (time, money, or anything else needed) to deal with these. If not all resources can be found, the project must be changed or canceled, because without sufficient resources the path leads only to failure.

11.3.5 Project Resource Management

At the core of every project are the people. People make the project successful; people cause the project problems, people make it enjoyable to work on project teams – or not. The PM must be able to work through the people involved to make things happen and obtain results. The PMBOK lists six Resource Management Processes:

- Plan Resource Management
- Estimate Activity Resources
- Acquire Resources
- Develop Team
- Manage Team
- Control Resources

11.3.6 Project Communications Management

Excellent communication is possibly the most important attribute of successful project management. Where communication is good, project teams can manage more effectively – maneuvering through the many problems and issues that arise during project planning and implementation, and sometimes even finding solutions for serious issues that might otherwise completely derail the project. Through communication, people can formulate the plans for the project, they can be kept aware of what is planned, kept up to date on the status, and the overall performance of the tasks can be coordinated. Communication is also a core factor in managing and motivating the team.

Projects required constant, continued, clear communications of many types. These include:

General communications
- two-way sharing of information, speaking, writing, and listening

Project communications
- the project Charter
- the project Scope description
- the WBS
- the logic diagram showing the workflow
- the Gantt chart showing the schedule
- an RFI and/or RFP
- Supplier bids and proposals
- Requests for funding
- Requests for personnel
- Status reports
- Contracts
- Minutes of meetings
- Customer presentations
- Invitations to meetings and agendas
- Product descriptions
- Product manuals of various types
- The project budget
- Project funding requests
- Project account tracking reports

Status reporting
- Meetings
- Communication for motivation of team members and suppliers
- Electronic tools for communications

It is important to develop a communications plan to identify the source and the recipient, plus the timing of each of these types of communications, to ensure that nothing is missed.

11.3.7 Project Risk Management

As discussed earlier, risks are denoted in projects as known unknowns. These cover things that could be anticipated to impact the project, but they are not part of the plan because it is not certain that they will happen. If the probability of occurrence is high, they should be included as an activity included in the plan, with the risk in this case being a low probability that this will not occur. Some risks impact the timelines of the project, some impact the budget, and some impact both.

For every risk the team should identify the probability that this will occur, and the impact it will have, if it does occur, on the time or the cost of the project. The plan must include contingency to allow the flexibility to deal with the risks.

There are many different methodologies for dealing with risk. One preferred method for determining the amount of money to include in the budget or the amount of time to build into the schedule to cover the risks is as follows.

1) List all of the risks.
2) Quantify these. If you start by estimating the probabilities or the impacts in general terms such as Low, Medium, and High, then assign a numerical value to each gradation so that you will have a numerical probability and numerical dollar impact on the project for each risk in the list. Do this twice, once using the impact on time and once for the impact on the cost.
3) For each risk, multiply the probability times the impact to get a working dollar value for each risk to the budget, and time value for each risk to the schedule. The resulting number is a statistical number. In itself, it is not meaningful. However, the sum of all these statistical numbers does give a value that has been found to be close to the amount of contingency needed, as some of the risks will occur while others will not.
4) Add the working dollar and time values. The total number is the number of dollars and time units that you should add to the budget as contingency.

This is shown for monetary contingency as follows:

Risk name	Probability	Impact ($)	Value ($)
Risk 1	0.1	1,000	100
Risk 2	0.1	5,500	550
Risk 3	0.1	10,000	1,000
Risk 58	0.4	4,000	1,600
Risk 59	0.4	1,000	400
Risk 60	0.4	8,000	3,200
Risk 410	0.7	2,000	1,400
Risk 411	0.7	1,000	700
Risk 412	0.7	6,000	4,200
Totals		$260,000	$80,000

The methodology calculates a single number for the total amount of contingency that should be included in the project budget or schedule. The question then is how this amount should be included in the project. Again, there are multiple different methodologies in use. Generally, the total is broken into components with individual components added at strategic points in the budget or schedule, with more contingency added in the riskier areas. The PM and team should determine how best to include this from each specific project, ensuring that everyone is aware that contingency should be used only to cover for risks that do occur, with any unused contingency moved to later spots where it might be needed.

Risk management is complex, and it takes thought, effort, and time. But that cost is usually far less than the cost of the project if the risks are not managed.

11.3.8 Project Schedule Management

A major focus for almost all PMs is the schedule of the project. This must be well planned, and then monitored closely to ensure that the project can be delivered on time.

The schedule plan is built using the bottom-level elements from the Work Breakdown Structure. These should all be actions, and each should include a verb to state clearly what action must be taken. These are all elements which have nothing below them in the WBS. Anything which does show lower-level items should be a deliverable – a thing – described with noun(s) but no active verbs. The schedule is built by lining up the bottom-level elements. We'll call these activities.

The first step is to determine what activity will actually kick off the project. Every project should have one single initial activity, and one single final activity. All the other activities can then line up between these. Once the initial activity is set, it becomes a predecessor for everything else. The second step then is to determine the immediate successors to the initial activity. In the small sample network shown in Figure 11.7, activity A is the first activity. This might be the signing of the Charter, the kickoff meeting, or one of the actual activities related to building the product of the project. In Figure 11.7, activities B, C, and D are all immediate successors to activity A. We can see from the network shown that we now know the content of each activity – although we are using only letters here to simplify the explanations, and we also see the duration of each activity. Let us assume that the durations are days, as that is very common. At this point, the team needs to sit down with the list of activities and determine predecessors and successors for every activity. In programs such as Microsoft Project, the predecessors can all be entered for every activity. Some PMs enter both predecessors and successors, but many find it easier to work with only one of these. In the program these are called "dependencies," as each activity, except the first one, is dependent on its predecessors. It is important to understand the concept of dependencies here, as there are four different types, and all of these types of dependencies are quite valid for use in mapping out the project activities.

11.3.9 Dependencies

The four types of dependencies are:

Finish to Start: this one is the most common. We will write it as FS.
Start to Start: this is also fairly common. We will write it as SS.
Finish to Finish: another common one. We will write it as FF.
Start to Finish: this one is uncommon, but it still occurs. Write it as SF.

In an **FS** dependency, we are waiting for the trigger situation, the finish of the first activity, to occur, and when it does, this enables the start of the second.

Examples are quite common. In creating a program, we could say that we must finish the writing of the first feature before we start to write the program for the second one.

In an **SS** dependency, we are waiting for the trigger situation, the start of the first activity, to occur, and when it does, this enables the start of the second.

These examples are also fairly common. Perhaps for a new product development we could start the planning for our new product once the competitive analysis has started. We do not need to wait for the completion of the full competitive analysis before we start plans for our own.

In an **FF** dependency, we are waiting for the trigger situation, the finish of the first activity, to occur, and when it does, this enables the finish of the second.

An example here might relate to the establishment of the rates or price for our new service or product. We must finish the development of the initial prototype before we can finish the rate setting. But the two activities could have been occurring in parallel.

In an **SF** dependency, we are waiting for the trigger situation, the start of the first activity, to occur, and when it does, this enables the finish of the second. This sounds strange because it means that the second activity must already be in place, and in fact finished, before we can even start the first one.

Since this one is much less common it is more difficult to think of examples, but the examples are fairly understandable. In a project to build a prototype of a new electric car, we cannot start to install items such as door handles or other finishing trims until the painting of the body has completed. The handle installation start depends on the finish of the painting.

This dependency is rare, but examples do occur. A possible example of the SF dependency from the electric vehicle industry would be the activity of generating press releases about the future availability date of a new car can finish when the first car is finally delivered.

As we consider each of these dependencies a number of additional characteristics become evident. First, while activity M is dependent in one of these ways on activity L, M does not have to start immediately when the trigger state has been reached. It cannot start earlier, but it can start right away, or later, as determined by other aspects of the project.

Also, some of these dependencies are hard dependencies, which means that the first must be there before the second can start or finish. These are called "mandatory dependencies." Others are soft dependencies, which means that we prefer the trigger state to be reached, or we have rules of guidelines saying this, but if needed we can back off on a soft dependency. These are called "discretionary dependencies."

Finally, any of the dependencies could have a built-in lag or lead. If there is a lag, this means that we must wait for the trigger state to occur, and then wait an additional time, before we can start or finish the second activity. One example everyone seems to quickly comprehend is the time needed for concrete to cure when building a structure. The builders must give the concrete three days to cure before building the structure above it, otherwise at some point the structure will collapse. The additional wait time, the lag, is part of the actual dependency, not just something that happens because the materials did not arrive, or people were not available, etc. A lead is similar, with the extra time in this case falling prior to the start or finish of the second activity. For example, maybe a company has a rule that testing of a program cannot begin until 95% of program has been written. In this case, the dependency is that the second activity, testing, is dependent on the first, programming, with a lead of 5% of the programming time. As shown in these examples lags and leads can be specified in durations, or in percentages. Programs can handle either.

In Figure 11.8 all the dependencies are Finish to Start. Since FS is the most common dependency it is common for programs to assume that specified dependencies are FS unless otherwise stated. In our figure, only one is labeled FS, and this is because that one has a lag, which must also be specified.

11.3.10 Critical Path

Looking again at Figure 11.8, there is a question posed – What is the longest path through the network? It is important to find this path, which is called the "Critical Path," because every activity in the network must be completed in order to complete the project. The PM must determine how long the project will take, in order to give the customer the date at which the product will be available. The longest path through the network determines this. In our figure, there are four paths – ABF, ACF, ADEF, and ADEGF. From the numbers in Figure 11.9, we can see that ABF is the critical path.

When the network is simple, as the ones shown here, it is quite easy to find the critical path by listing all the paths and adding the durations of the activities on each. However, with a project which has 100, 200, or over 1000 activities, there will be many paths and this methodology would

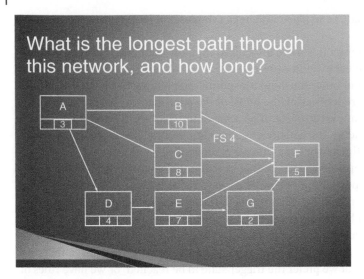

Figure 11.8   Project network.

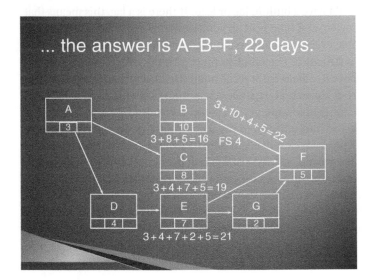

Figure 11.9   Project Critical Path.

be too cumbersome to determine the length of the project. Fortunately, there is a methodology which can be used and has been effectively programmed in project management software. Figures 11.10 and 11.11 illustrate this progress which can be done manually, or perhaps drawn out in excel, or automatically by using project management software.

Start by doing a Forward Pass, which is a walk through the network noting the start and finish of each activity. Note that the network shown in the following figures differs from that in the previous figures, so pay close attention to the information provided in order to follow the discussion.

Activity A will start at the beginning of the first day, so its bottom left box shows day 1. It's a three-day duration, so it will finish at the end of day 3. This will then allow B, C, and D to start on

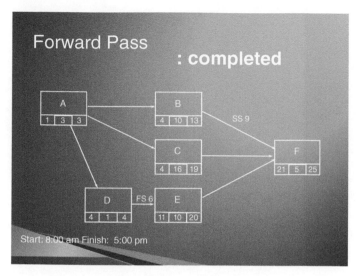

Figure 11.10 Forward Pass time calculations.

day 4. The forward pass walks through the network like this, activity by activity. Thus we can see that this project can complete at the end of day 25. The methodology just described is called "CPM." It is critical that the PM ensures that contingency time (calculated the same way as contingency costs, using time units such as days instead of cost units such as dollars). CPM does not specify where the contingency time should be included. It can be all at the end, evenly distributed across all activities, or more usefully placed in places where the most risk might occur. As with contingency cost, this is expected to be fully used, so it is up to the PM to ensure that the team uses this time only to deal with risks. If it is not needed in its initial placement, it should be moved forward to be available when a risk does occur.

The completion of the forward pass gives an indication of the time needed for the project. It is important that the team not share that information until all of the integration has occurred. Often, once the schedule has been set up and the resources assigned to each activity, it is found that someone or something is overloaded, when this happens the schedule must be revised to level the loading on all resources. Leveling usually causes the critical path to lengthen. The project duration should not be shared till this has occurred.

The next steps involve walking backward through the project from the final time, looking to the finish of each predecessor activity and calculating, where needed, new start times. The initial times found via the forward pass are called the "Early Times." The times found using the backward pass are called the "Late Times." When there is a difference between the Earlies and the Lates for any activity, that activity has float, sometimes called "slack." If issues with the scheduling require changing the schedule, one easy technique is to use up float in order to move and activity. When that is not possible, and the project needs to be shortened, it might be necessary to drop some of the activities – with approval from management and the customer – or maybe move some to a follow-up project. These would then need to be removed completely from all aspects of the plan.

Looking at Figure 11.11 we can see that B and C have float, but A, D, E, and F do not. ADEF is the critical path. If any of these activities slip, the project will be late unless the team can find ways to make up time later in the flow.

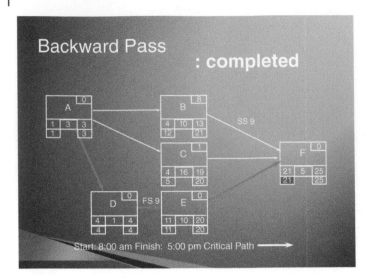

Figure 11.11 Backward Pass – float calculations.

11.3.11   Project Procurement Management

Going outside the company to hire assistance or materials from outside suppliers is called "procurement" in PMI terms. Some companies call this purchasing. Not all projects require procurement; for these, once a decision has been made to work entirely in-house, this section is not relevant.

The processes recommended in the PMBOK for procurement management include procurement planning, solicitation planning, solicitation, source selection, contract admin, and contract closeout. Instead of discussing the processes per se, we'll talk about the tools that are used, which should give a good picture of what is required of the project team.

11.3.11.1   Definition of Requirements

In order to prepare for solicitation, which is the process of obtaining information about what is available from outside to meet our needs, we need to first be clear about exactly what it is that we are looking for. Once we have decided what to go outside for, we need to prepare the specifications to clearly describe what we want.

11.3.11.2   Solicitation

There are many tools that can be used for solicitation, but probably the three most common are RFI, RFP, and RFQ. We address each of these here.

11.3.11.2.1   Request for Information   An RFI is a request for information. This is exactly what it sounds like – a request to potential vendors for information on what they might provide to meet our requirements.

This document can be issued to as many or as few potential suppliers as desired. And it is acceptable to ask for any information except for a quote on the price. If a price quote is requested, this document becomes either an RFP or an RFQ, and as such, it is subject to all the implications of those documents.

11.3.11.2.2 Request for Proposal An RFP is a Request for Proposal. This is used when the buyer is serious about his request, and skills are required to provide the desired items. This tool is used when the requirement is some professional or technical work, or when some design is required, etc. If externally sourced custom software is required for a project, an RFP may be issued, and the resulting proposals for provision of the software could represent very different products. For technology projects, the RFP could be for some type of equipment engineered for specific features, requirements, or service levels.

11.3.11.2.3 Request for Quote An RFQ is a Request for Quote and is used when purchasing off-the-shelf items or as a follow-up to a successful RFP response. It is subject to the same constraints and issues as the RFP, but the quoted capability and price will be considered binding.

Once a supplier has been selected and a contract drawn up, the project team must manage the implementation of the contract to ensure that all deliverables are met with work or material that meets the project requirements.

11.3.12 Project Stakeholder Management

Stakeholders are those who are impacted by the project. Some of these are internal to the project team while others are external. Internal stakeholders include the PM, each of the team members, who are often from different disciplines within the company, management within the project company, the client, suppliers, outside funders, or even those who might be impacted by the existence of the new technology, service, or product. This might include regulatory bodies, others operating within the same market as the product, or competitors offering products and services to solve the same problem that the project product addresses.

Some stakeholders exist for all projects, including (but not limited to):

- **Project manager:** And possibly others who assist in the management such as team leaders, and project coordinators.
- **Project Sponsor:** The sponsor is the person who pays for the project. It is important to realize that the sponsor must be someone within the company that is doing the project.
- **Management:** Senior management in the company in which the project is being done.
- **Functional managers:** Managers of team members who do not report directly to the project manager in their day-to-day position.
- **Team members**
- **Customer**
- **Suppliers if there is procurement involved.**

11.3.13 Recommended Best Use and Focused Decisions to Identify the Most Adequate Approaches and Techniques for Particular Cases

In this discussion of basic project management concepts, we have focused on CPM. Even with this method there are variations. In some cases the WBS cannot be fully broken out at the beginning so the team must work with higher-level boxes, increasing the risk of not meeting cost, time, and quality objectives. There are many other methodologies as well, such as Agile Project Management. Agile is more appropriate for software and research projects than for repetitive design, development, and implementation projects, where CPM is more appropriate. Regardless of the methodology employed, all projects have risks, and all need to define scope at some point. All require effective communication in order to be successful. Some have no procurement component, in some situations, the timelines are much less important than the cost, while the opposite is true for others. It is important that the PM work with the client and management to gain full understanding of all requirements. Then the appropriate methodologies can be selected.

For those who are required to manage projects, but have not been using any of these techniques, it is better to start implementing one or two, and practice these to get comfortable with them before trying to implement additional methodologies. In every case determining clear requirements, understanding the risks and planning mechanisms to deal with these, and building a very clear communications plan should be primary goals.

The better the teams become at following the methodologies here, the better chance they have of success.

11.4 Summary and Conclusions

The launching of a new product or service, or the development of new technology is usually best handled on a project basis. All projects can benefit from the use of effective project management. Project environments and priorities differ, but in every case, appropriate project management will be an important success factor.

Reference

1 Project Management Institute (2017). *A Guide to the Project Management Body of Knowledge PMBOK Guide*, 6e. Project Management Institute.

Further Reading

1 Desmond, C. (2004). *Project Management for Telecommunications Projects*. Kluwer Academic Publishers. ISBN: 1-4020-7728-9.
2 Desmond, C. (2010). *Managing Telecommunications Projects*. IEEE Press Wiley. ISBN: 978-0-470-28475-9.

12

Project Plan Structure: The Gantt Chart and Beyond

Andrea Molinari

Department of Industrial Engineering, University of Trento, Trento, Italy

12.1 Knowledge Area Fundamentals

At any moment of the lifecycle of a project, there is a document that helps people involved in it stay in line with planning, monitor variations, communicate with stakeholders, calculate performance, and report the progress. This document is the Project Plan (PP), also called "Master Project Plan (MPP)" or "Project Work Plan."

The PP is a document where the project's objectives are defined, the project's scope is described, where main deliverables are listed and explained, and how the project will be executed, monitored, and verified against the original plan. The PP is the main deliverable of planning, a phase of the project life cycle that precedes the execution of the project and therefore analyses many aspects of the project before it is executed.

It is fundamental inside organizations to perceive the value of a PP and understand how to standardize it to facilitate communication, exchange of information, analysis, and evaluation of results.

12.2 Managing Approaches, Techniques, and Benefits of Using Them

Several different standards and methodologies about project management define the concept and structure of a PP slightly differently. The Project Management Institute defines a PP as "a formal, approved document used to guide project execution and control." According to the PMI Body of knowledge, the primary uses of a PP are ". . . to document planning assumptions and decisions, facilitate communication among project stakeholders, and document approved scope, cost, and schedule baselines."

In the recently-released PM2 methodology, sponsored by the European Commission, the same concepts are grouped into the document called "Project Work Plan," which ". . . further elaborates the project scope and identifies and organizes the project work and deliverables needed to achieve the project goals." Within this methodology, the project work plan is helpful in various moments of the project's lifecycle. The plan (i) delimits the project scope, (ii) defines the tasks and

IEEE Technology and Engineering Management Society Body of Knowledge (TEMSBOK), First Edition.
Edited by Gustavo Giannattasio, Elif Kongar, Marina Dabić, Celia Desmond, Michael Condry, Sudeendra Koushik, and Roberto Saracco.

their schedule, (iii) provides an estimation of the needed resources, and (iv) develops the planning details in terms of work breakdown, effort, costs, and schedule. For the prosecution of the project, the PP will be updated during the planning phase. It will be the lighthouse for any tasks performed and any deliverables obtained during the execution of the project.

Another worldwide standard for Project Management, PRINCE2, defines a PP as ". . . a statement of how and when a project's objectives are to be achieved, by showing the major products, milestones, activities, and resources required on the project."

The above three definitions are taken from "predictive" project management methodologies. The PP represents the prediction about what will happen in the project, why it should happen, how it will happen, and who will be responsible for what and act on which activities.

While a PP is central in predictive project management, this document is much more controversial in adaptive project management methodologies. Indeed, predictive approaches (once called "waterfall") like those presented above have widely been criticized from different angles [1]. As a reference to these critics, we can take the Agile Manifesto, which in 2001 gave the start of the Agile Project Management movement, specifically for software development [2]. Conversely, Agile methodologies are often referred to as "project management without a plan," another common and dangerous misconception.

Agile methodologies, like Kanban, Lean, Scrum, XP, or any other flavor, focus on aspects like operating and reacting quickly, efficiently, and effectively without an overall strategic plan. Indeed, the iterations in most of the Agile approaches can be a source of waste of time and effort while moving forward through iterations without a master plan. This does not mean that Agile teams should work in the dark without any vision of the rest of the project: agile does not mean a project without a plan. A well-organized, responsible agile team has a flexible plan to manage valuable and essential changes in the original vision. Agile evangelists say that the long-term plan for an Agile organization is a vision of a possible future.

Citing the MIT Sloan School of Management [3]: "[Agility] comes in different forms, but basically, it is the ability to adapt to or even anticipate and lead change quickly. Agility in the broadest form affects strategic thinking, operations, technology innovation and the ability to innovate in products, processes and business models." As Alan Lakein summarized, failing to plan is planning to fail. The importance of planning in today's project management is out of the discussion. The relevance of an Agile mindset and product development approach is also relevant. We can hate fixed dates and rigid scopes, but it is the case for many organizations every day, so it cannot be ignored. On the other side, considering the PP as an immutable, static object is out of reality and modern times.

Even outside the predictive worlds of project planning, thus fully embracing one of the different flavors of adaptive project management and diving into product development, planning is still crucial. A document that represents the different components of a project is central because most customers will not commit a budget to a non-specified plan with non-defined dates of delivery.

Another source of confusion about a PP is the word "planning" used in specific contexts of some Agile methodologies – for example, SCRUM planning, a software development-specific part of the SCRUM framework related to sprints. Instead, an Agile plan must be a PP that adapts to more dynamic markets typical in these last two decades, where business (not software) requirements change very frequently. In this permanently changed context, more flexible planning is needed. So while a traditional, predictive PP is a step-by-step heavy planning process, Agile planning is iterative and adaptive to changes. In building a skyscraper, stakeholders want to see a detailed plan, while late requirements and specifications are dangerous, expensive, and unpopular. In settings like knowledge creation projects, changes are welcomed at any stage because they represent a clear symptom of knowledge creation. The project learns from itself so that it can be accepted at the very end. The goal, in this case, is to satisfy stakeholders, so any addition/modification to the original plan that goes in this direction is allowed. According to this, an Agile PP can be a (very detailed) plan only valid for shorter timeboxes, immediately updated when context and requirements change.

By aggregating the different approaches and definitions, we can say that a PP is a set of documents, formalized by an organization's authority, that defines the characteristics of a project, how it will execute, and the execution and control stages. While addressing scope, cost, and schedule baselines, the PP includes other elements related to other knowledge areas, like risk management, resource management, communications, quality, etc. It outlines the objectives and scope of the project. It serves as an official point of reference for the project stakeholders, so the Sponsor, the Board, and the Project team can effectively monitor and control project progress.

The PP is created during the planning phase, which is the most important deliverable of this phase. For this reason, it requires formal approval at the beginning of the project and is progressively updated throughout the project.

12.3 Realistic Examples

The PP is one of the less-standardized documents in project management. Every (structured) organization could have its template, usually devoted to presenting the project to the Board of Directors rather than being vital support to project lifecycle management. It is frequently a replica of some administrative requests for funds or authorizations when standardized. Typically, these documents are destined to remain unused or underused in the operational phases of the project.

On the Internet, there are plenty of examples of PPs. Most of them are templates that are too general or specific for that sector, that company, or that view of the world. Other PPs are covered by industrial secrets, classified information, or private data that cannot be presented publicly.

To provide some examples, we will take templates that could represent different approaches to creating a PP, representing most of the general information requested in a PP.

The first group of PPs is concentrated on a very famous output of the planning phase, i.e., the Gantt Chart. To quickly recap, a Gantt Chart is a diagram that uses horizontal bars to represent the project's activities. The duration of the activity is, on a specific temporal scale, represented by the length of the respective bar (see Figure 12.1).

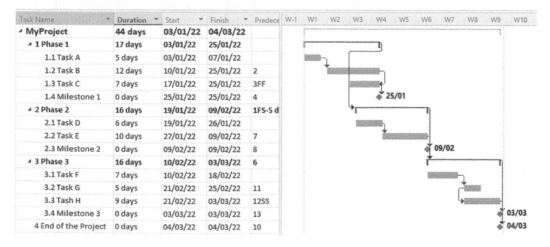

Figure 12.1 Example of Gantt Chart with durations, start/end date, and relationships.

At the end of the nineteenth century, different researchers, engineers, and entrepreneurs (like the polish Karl Adamiecki) started to use this intuitive representation of project schedules [4, 5]. However, at the beginning of the twentieth century, Henry Gantt systematically started using this schedule representation, so he is tributed as the "inventor" of this form of representation.

The popularity of Gantt charts is mainly due to the amount of information they can represent and the simplicity of comprehension. Due to these objective elements, it is pretty common to receive, as the PP, a document containing a Gantt Chart as the only documentation of the plan. We will call this group of examples "Gantt-centric project plans."

The first case is a PP based on a Gantt chart that only conveys all the project information through the graphical representation of tasks, durations, and relationships of the various tasks. The focus and message of this PP are mainly on how the project will move from the starting situation to the end and when the final deliverable will be released. It serves as a roadmap that presents various project elements, mainly focusing on time. The elements are:

- The different project phases;
- The key activities of the project, with a level of details taken from the work breakdown structure (WBS);
- The main milestones and deliverables;
- The start and end dates of each task;
- The dependencies between tasks.

These are the essential but unavoidable elements of a Gantt Chart: the Gantt Chart is no more than a sketch missing some vital information without any of these. For example, presenting a PP through a Gantt Chart that does not contain various dependencies is a bad practice that should be avoided. In this way, the project's logic, the relationships between tasks and phases, the general meaning of what we are doing and how, and the problems derived from the relationships will not be revealed. Considering the lack of other information that the Gantt Chart, by nature, is not made for (risks, quality, stakeholders management, communication mechanisms, etc.), this first Gantt-centric PP is a deplorable habit of project managers. It should be changed as soon as possible, and stakeholders should push project managers to adopt a qualitative PP.

The Gantt Chart can be enriched with more details, especially when created with appropriate and evolved project management software. The more information we add to the Gantt Chart, the more this representation can act as a PP. The first group of information we can add to our Gantt template relates to resources, thus substituting some methodologies called "responsibility matrix" (RACI), "responsibility assignment matrix" (RAM), or linear responsibility chart. In a fully-fledged PP, this information is represented in tabular format. It is commented on where and when needed. Most of all, it contains different aspects of assigning a resource to a task. This information should describe the role in completing tasks and deliverables by resources, who is responsible for what, who should be consulted in case of problems/suggestions, who should be accountable, and who should be informed about what is happening on every single task. This kind of information is vital for any project. However, it becomes mandatory when we have cross-functional or cross-departmental projects, involve resources from different companies, or have huge and distributed responsibilities on different project parts that we have to monitor.

In Figure 12.2, an example of a RASCI matrix for a hypothetical project is presented, where for the different resources of the project (on the columns), we assign five different level responsibilities:

The meaning of different levels of responsibilities are the following:

R = Responsible: the resource will do the work to complete the task.

A = Accountable: who will be answerable for completing the task and its correctness/appropriateness. Usually, this role is in charge of approving/signing off the work provided by the responsible

Tasks	Product Owner	Sponsor	Developer	Support	Analyst	Sales	Engineer
Task A	A	C	I	R		C	C
Task B	R	S	C	C	A	I	I
Task C	C	R	C		I	I	A
Task D	C	R	I	C	I	C	A
Task E	C	S	I	C	A	R	I
Task F	S	I	I	C	A	R	R
Task G	R	I	R	C	A	I	S
Task H	C	R	I	C	I	C	A

Figure 12.2 Example of a RASCI matrix.

S = Support: a resource that helps to complete the task by supporting the respective Responsible
C = Consulted: a resource whose contribution is crucial because they are experts in the subject
I = Informed: must be kept updated about the task's progress.

Gantt Chart tools cannot represent this sophisticated but crucial set of information. Assigning a resource to a task is not precisely like assigning responsibility for that task. With a Gantt, we create much simpler connections between tasks and resources, so the PP is missing many chunks of information for the receiver of the PP. It simply means that the resource will contribute to completing that task, but we do not know the effort and what we should expect from them. Responsibilities, permissions, power-to-act, and supervision are all possible actions by a resource that is precious information for a PP. We should also consider that the same resource can play different roles, especially in small organizations or small projects, so a good PP should present this information. However, the Gantt chart is not an adequate tool.

Typically, the Gantt Chart represents the pure assignment of the resource to the task. This attribute will be precious information, in any case, to determine another fundamental element of a PP that could be represented on a Gantt Chart: the final cost of the task. Aggregating all the tasks' costs will determine the project's total cost. If we enrich the Gantt Chart with this further information, the PP will start to present other charts/documents rather than the chart itself. For example, a list of resources with their costs and other attributes will be needed. Once we have assigned the resources to tasks, we can derive much other information from a Gantt Chart enriched with assignments:

- The detailed budget for each task, the total budget for summary tasks, and the total budget of the project;
- The overall allocation of each resource on the project, thus checking possible optimization of scale, re-assignments, or pathological conditions like overallocation;
- The to-do list for each resource, with indications where the resource is involved with the expected effort;
- The project's cash-flow in every time slot we desire. Usually, it is also possible to distribute costs in different ways (uniformly, at the beginning of the tasks, and at the end of the task). This distribution allows starting economic and financial analysis on the project and its development over time, thus contributing to portfolio management analysis;
- Cross-checking the team's allocation, discovering, for example, under-used personnel, project periods where we could move unused resources to other projects, etc.

Following the example of the Gantt Chart presented in Figure 12.1, we will add these essential elements for a PP and show the various additional information that could constitute robust documentation of a PP that most stakeholders could be interested in (see Figure 12.3).

Firstly, a PP should provide a list of the involved resources with their characteristics (cost, type, location, calendar, availability, and number of units available). Resources can be of different types, not only human resources. A resource for the project is any entity that, with its effort, will contribute to the project's results: so we have our employees, of course, equipment, supplies, partners, and external consultants.

As we have provided a breakdown of the project's work through the WBS, it is essential to estimate the cost of a project in a PP. For this, we need another instrument for the planning phase, i.e., the resource breakdown structure (RBS). The WBS helps us to plan the work needed to accomplish the project's objectives. The RBS helps the project manager to plan all the resources needed to perform that work. The RBS, in whatever format it will be included in the PP, will list all resources that represent a cost to the project. It provides the basis for calculating project costs and is also an input to estimating activity durations and developing the project schedule. Because of its nature, the Gantt chart represents the use of resources listed in the RBS, but no information about them is provided. Therefore, the RBS is another element that should be added to this "Gantt-centric" PP to better document the project during the planning and execution phases.

In the example presented (see Figure 12.4), we imagine having a generic team (see Section 12.4 for details about scope, time, quality, cost, and resource management) with the costs for each resource. Here follows an example with a few essential attributes:

The PP would benefit significantly from having a list of components of the project's team and seeing how these resources are used and allocated inside the project. The assignment is the

Project ID	
Project Title	
Project Manager	
Client	

		RBS ID		Resource Name	Units	Type	Rate	Calendar	Notes
R1.				Project					
	R1.1			Labor					
		R1.1.1.		Project Management					
			R1.1.1.1	Project Manager	40%	Work	120€/h		
			R1.1.1.2.	Assistant Project Manager	100%	Work	80€/h		
		R1.1.2.		Software Development					
			R1.1.2.1.	Solutions Architect		Cost	150€/h		
			R1.1.2.2.	Development Lead	100%	Work	100€/h		
			R1.1.2.3.	Developer	100%	Work	60€/h		
			R1.1.2.4.	Developer	100%	Work	70€/h		
			R1.1.2.5.	Tester	100%	Work	40€/h		
			R1.1.2.6.	Trainer		Cost			
	R1.2.			Equipment					
		R1.2.1.		Hardware					
			R1.2.1.1.	Laptop High Performance	5 units	Material	1200€		
			R1.2.1.2.	Laptop Office	12 units	Material	600€		
			R1.2.1.3.	Projector	2 units	Material	800€		
			R1.2.1.4.	Printer	1 units	Material	1600€		
		R1.2.2.		Software					
			R1.2.2.1.	Visio	300%	Material	200€		
	R1.3.			Supplies					
		R1.3.1.		Ink	200 units	Material	12€		
		R1.3.2.		Paper	100 units	Material	3€		
		R1.3.3.		Binders	5 units	Material	50€		
	R1.4			Locations					
		R1.4.1		Chicago	200%	Work	30€/h		

Figure 12.3 Example of an RBS with basic information about resources.

Resource Name ▼	Type ▼	Max. Units ▼	Std. Rate ▼	Base Calendar
R1	Work	40%	120,00 €/hr	Standard
R2	Work	100%	80,00 €/hr	Standard
R3	Work	100%	100,00 €/hr	Standard
R4	Work	100%	60,00 €/hr	Standard
Consultant	Cost			
Printer	Material		1.600,00 €	

Figure 12.4 An example of a team with different resources of different types.

component of creating the PP, where the project manager should make many different decisions. As project managers, we should not only assign resources to tasks, but we should (i) check the appropriateness of this assignment, (ii) evaluate the competencies and the consequences of this assignment on this project and other projects, (iii) discuss with other project managers the allocation in case of overlapping projects, (iv) present the results of our assignments to the Project Management Office (PMO) to solve possible overallocation and priority problems, etc.

A PP without the information related to the assignments is poor, unprecise, source of risks and conflicts within and outside the organization with partners and stakeholders in general. Gantt-centric PPs without this set of information are feeble in terms of value for the Project manager and the organization and increase enormously the risks associated with the project and resource management.

Suppose we assign the appropriate resources to the different tasks of the project. Conversely, we will provide a Gantt chart with much different qualitative information that can be presented in the same chart or in separated parts of the PP to have the space to explain and detail the budget. An example of our test project is where we assigned our fictitious resources (see Figure 12.5).

The assignments of resources to respective tasks produced many relevant results in detail in other sections of the PP. In Figure 12.5, we can see an example of such information condensed in the Gantt chart: the results are in place, but it is hard to read, a bit confusing, and many other details can not be presented on the Gantt in order not to weigh down the graph and make it even

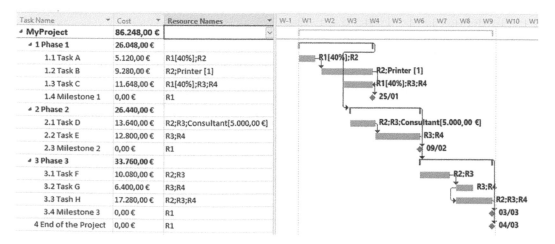

Figure 12.5 A Gantt chart attempting to represent resource allocations and respective costs.

more illegible. What can be appreciated in any case is the idea that, if we provide in the PP the RBS. The associated costs, once we assign the members of the project's team to the tasks, we can obtain (i) the cost for each task, (ii) the cost for each phase (as summarizing the various dependent tasks), and (iii) the cost of the whole project.

This fundamental result for any PP allows us to extract many other charts and data to integrate and document our PP using state-of-the-art project management software. Here are some examples.

The first example of the output is the overall allocation of the various resources on the project, with an overview of potential overallocation problems. From Figure 12.6, for example, we can see that this resource is overallocated concerning 100% of the time available from R3 (s/he will be working full time on our project) in weeks 3, 4, and 8. In particular, R3 is working for 200% of the available time, probably being assigned to two tasks simultaneously. This situation must be solved before releasing the final PP, or we should ask the resource manager to have a second resource compatible with R3 available to work with here in those weeks. Another piece of information crucial for the plan derived from this chart is that, during weeks 1 and 2 and weeks 10 and 11, R3 is free from any duty with our project so that she can allocate elsewhere.

Instead, in Figure 12.7, we have resource R1 that has been correctly allocated to the project according to its availability. However, this resource is substantially underused because, except for Weeks 1, 3, and 4, she is free from any commitment for the rest of the project's time.

Once the over-allocations are solved, these analyses can be included (or not) in the PP based on the project manager's decision, according to the level of details requested by the Sponsor and

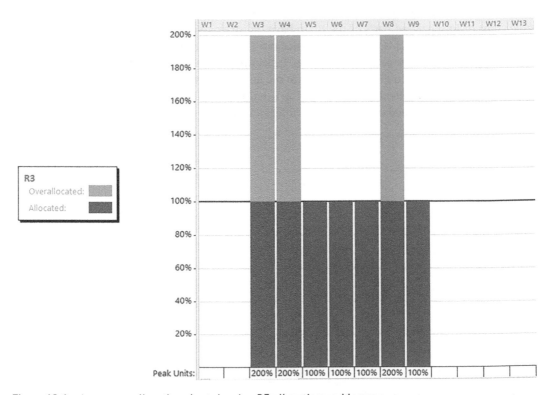

Figure 12.6 A resource allocation chart showing R3 allocation problems.

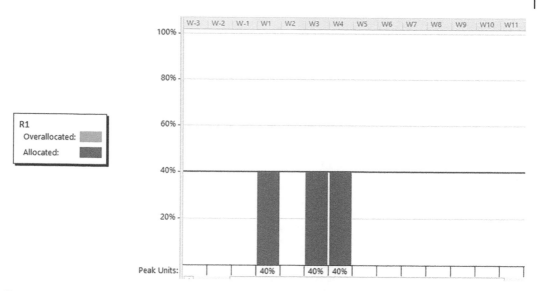

Figure 12.7   A resource allocation chart showing R1 unused slots of time.

stakeholders. In general, they are beneficial for documenting the details of the PP and controlling possible deviations from the presented plan during the execution.

The previous graphical view is very efficient in double-checking allocations. However, it is not providing crucial information about the PP, i.e. the details of what every resource is expected to do. Suppose charts can be unhandy in projects involving many resources. In this case, an overview of the allocation details (with respective costs) can provide some insights while giving appropriate information about cost distribution. In Figure 12.8, resource usage on our test project is provided.

Data in Figure 12.8 are examples of the expressivity we can extract from allocations for our PP. The chart is organized by resource: for each resource, we have:

- the list of tasks where this resource is assigned;
- for each task, the cost of the resources;
- for each task, the details of allocation for the period (in our case, the week);
- the indication of the precise tasks and period where overallocation have been.

Once we have this information, it is straightforward to obtain other crucial information about the project, vital information for any PP, i.e., a cash-flow that shows how and when the project costs materialize, for each resource, and for each task they are assigned (see Figure 12.9).

If detailed effort distribution over time could be too much detailed information for the project plan's readers, the same data converted into money could have a radically different appeal, primarily because of the various totals we can find. This view is an excellent representation of the project's data and valid for the project manager during both planning and executing the project. It is similarly helpful for stakeholders who want to check project details and progress during its lifecycle at different levels.

Another contribution to the PP comes from a simple pivoting operation of the previous data. Instead of looking at each resource and the assigned tasks, we can be interested in the orthogonal view, i.e., details about which resources for each task. A different view based on the same data can

Resource Name	Work	Cost	Details	W-1	W1	W2	W3	W4	W5	W6	W7	W8	W9	W10	W11
◢ R1	38,4 hrs	4.608,00 €	Work		16h		16h	6,4h		0h			0h		
Task A	16 hrs	1.920,00 €	Work		16h										
Task C	22,4 hrs	2.688,00 €	Work				16h	6,4h							
Milestone 1	0 hrs	0,00 €	Work					0h							
Milestone 2	0 hrs	0,00 €	Work							0h					
Milestone 3	0 hrs	0,00 €	Work										0h		
End of the Pr	0 hrs	0,00 €	Work										0h		
◢ R2	312 hrs	24.960,00 €	Work		40h	40h	64h	40h		16h	40h	40h	32h		
Task A	40 hrs	3.200,00 €	Work		40h										
Task B	96 hrs	7.680,00 €	Work			40h	40h	16h							
Task D	48 hrs	3.840,00 €	Work				24h	24h							
Task F	56 hrs	4.480,00 €	Work							16h	40h				
Task H	72 hrs	5.760,00 €	Work									40h	32h		
◢ R3	352 hrs	35.200,00 €	Work				64h	56h	40h	40h	40h	80h	32h		
Task C	56 hrs	5.600,00 €	Work				40h	16h							
Task D	48 hrs	4.800,00 €	Work				24h	24h							
Task E	80 hrs	8.000,00 €	Work					16h	40h	24h					
Task F	56 hrs	5.600,00 €	Work							16h	40h				
Task G	40 hrs	4.000,00 €	Work									40h			
Task H	72 hrs	7.200,00 €	Work									40h	32h		
◢ R4	248 hrs	14.880,00 €	Work				40h	32h	40h	24h		80h	32h		
Task C	56 hrs	3.360,00 €	Work				40h	16h							
Task E	80 hrs	4.800,00 €	Work					16h	40h	24h					
Task G	40 hrs	2.400,00 €	Work									40h			
Task H	72 hrs	4.320,00 €	Work									40h	32h		

Figure 12.8 Resource usage, with details on costs and hours.

Resource Name	Work	Cost	Details	W-1	W1	W2	W3	W4	W5	W6	W7	W8	W9	W10	W11
◢ R1	38,4 hrs	4.608,00 €	Cost		1.920,00 €		1.920,00 €	768,00 €							
Task A	16 hrs	1.920,00 €	Cost		1.920,00 €										
Task C	22,4 hrs	2.688,00 €	Cost				1.920,00 €	768,00 €							
Milestone 1	0 hrs	0,00 €	Cost												
Milestone 2	0 hrs	0,00 €	Cost												
Milestone 3	0 hrs	0,00 €	Cost												
End of the Pr	0 hrs	0,00 €	Cost												
◢ R2	312 hrs	24.960,00 €	Cost		3.200,00 €	3.200,00 €	5.120,00 €	3.200,00 €		1.280,00 €	3.200,00 €	3.200,00 €	2.560,00 €		
Task A	40 hrs	3.200,00 €	Cost		3.200,00 €										
Task B	96 hrs	7.680,00 €	Cost			3.200,00 €	3.200,00 €	1.280,00 €							
Task D	48 hrs	3.840,00 €	Cost				1.920,00 €	1.920,00 €							
Task F	56 hrs	4.480,00 €	Cost							1.280,00 €	3.200,00 €				
Task H	72 hrs	5.760,00 €	Cost									3.200,00 €	2.560,00 €		
◢ R3	352 hrs	35.200,00 €	Cost				6.400,00 €	5.600,00 €	4.000,00 €	4.000,00 €	4.000,00 €	8.000,00 €	3.200,00 €		
Task C	56 hrs	5.600,00 €	Cost				4.000,00 €	1.600,00 €							
Task D	48 hrs	4.800,00 €	Cost				2.400,00 €	2.400,00 €							
Task E	80 hrs	8.000,00 €	Cost					1.600,00 €	4.000,00 €	2.400,00 €					
Task F	56 hrs	5.600,00 €	Cost							1.600,00 €	4.000,00 €				
Task G	40 hrs	4.000,00 €	Cost									4.000,00 €			
Task H	72 hrs	7.200,00 €	Cost									4.000,00 €	3.200,00 €		
◢ R4	248 hrs	14.880,00 €	Cost				2.400,00 €	1.920,00 €	2.400,00 €	1.440,00 €		4.800,00 €	1.920,00 €		
Task C	56 hrs	3.360,00 €	Cost				2.400,00 €	960,00 €							
Task E	80 hrs	4.800,00 €	Cost					960,00 €	2.400,00 €	1.440,00 €					
Task G	40 hrs	2.400,00 €	Cost									2.400,00 €			
Task H	72 hrs	4.320,00 €	Cost									2.400,00 €	1.920,00 €		

Figure 12.9 Resource usage, with detailed cash-flow.

explain each task's detailed planning (see Figure 12.10). Here, we can present to the project plan's readers which resource will perform every task.

To conclude the examples on how to increase the information of a Gantt-centric PP, we can take advantage of the features of modern software supporting project managers. Indeed, we can see an expanded Gantt chart, where the Gantt is presented not from a task perspective but from a resource perspective (see Figure 12.11).

Task Name		Cost	Start	Finish	Details	W-1	W1	W2	W3	W4	W5	W6	W7	W8	W9	W10	W11
⊿ MyProject	44 days	86.248,00 €	03/01/22	04/03/22	Work	56h	40h	184h	134,4h	80h	80h	80h	200h	96h			
⊿ Phase 1	17 days	26.048,00 €	03/01/22	25/02/22	Work	56h	40h	136h	54,4h								
⊿ Task A	5 days	5.120,00 €	03/01/22	07/01/22	Work	56h											
R1		1.920,00 €	03/01/22	07/01/22	Work	16h											
R2		3.200,00 €	03/01/22	07/01/22	Work	40h											
⊿ Task B	12 days	9.280,00 €	10/01/22	25/01/22	Work		40h	40h	16h								
R2		7.680,00 €	10/01/22	25/01/22	Work		40h	40h	16h								
Printer		1.600,00 €	10/01/22	25/01/22	Work		0,42	0,42	0,17								
⊿ Task C	7 days	11.648,00 €	17/01/22	25/01/22	Work			96h	38,4h								
R1		2.688,00 €	17/01/22	25/01/22	Work			16h	6,4h								
R3		5.600,00 €	17/01/22	25/01/22	Work			40h	16h								
R4		3.360,00 €	17/01/22	25/01/22	Work			40h	16h								
⊿ Milestone 1	0 days	0,00 €	25/01/22	25/01/22	Work				0h								
R1		0,00 €	25/01/22	25/01/22	Work				0h								
⊿ Phase 2	16 days	26.440,00 €	19/01/22	09/02/22	Work				48h	80h	80h	48h					
⊿ Task D	6 days	13.640,00 €	19/01/22	26/01/22	Work				48h	48h							
R2		3.840,00 €	19/01/22	26/01/22	Work				24h	24h							
R3		4.800,00 €	19/01/22	26/01/22	Work				24h	24h							
Consultant		5.000,00 €	19/01/22	26/01/22	Work												
⊿ Task E	10 days	12.800,00 €	27/01/22	09/02/22	Work					32h	80h	48h					
R3		8.000,00 €	27/01/22	09/02/22	Work					16h	40h	24h					
R4		4.800,00 €	27/01/22	09/02/22	Work					16h	40h	24h					
⊿ Milestone 2	0 days	0,00 €	09/02/22	09/02/22	Work							0h					
R1		0,00 €	09/02/22	09/02/22	Work							0h					

Figure 12.10   Task usage and details for each task on resources.

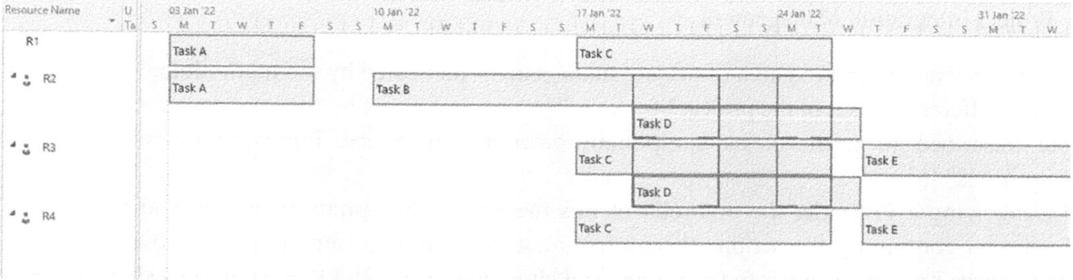

Figure 12.11   A Gantt Chart from a resource perspective.

Again, this view is an excellent completion of the PP. It allows a view of tasks assigned to resources. It identifies and fixes overall, un, and under-allocated tasks, presenting a graphical workload view.

In concluding this digression on Gantt-centric PPs, a Gantt Chart is essential for any PP. It is an excellent communication means for the project manager because it is well-known and straightforward to understand. However, it should be considered as a sort of minimal stage. The more proactive stakeholders of the project, like the Sponsor or the customers, will probably be unsatisfied with this content. For this target, the Gantt chart can be too long or detailed for their interests in the project while lacking other details they are interested in, missing some vital information from their perspective. A PP where the Gantt Chart is an influential but first-level information aggregator and the starting point for many other reports/documents is an excellent way to conceive a Gantt-centric PP. We prefer to imagine other information concerning the typical constraints of a project (Scope, Time, Cost) and other areas where the project will have to be planned.

The second type of PP structure can be derived from the organization's project management methodology. Any predictive or adaptive project management methodology or method is aware of the relevance of a PP document, whatever the name assigned to it. Consequently, the PP can follow the dictates of that methodology, with the common objective of clarifying the project to stakeholders. With internal people educated to apply the methodology extensively, producing a PP where stakeholders can find all the answers to the knowledge areas that the methodology requires should be straightforward. For example, in a company that used to apply the PMI PMBOK™ approach, it is natural to conceive a PP template that contains the answers to the 10 knowledge areas of the methodology. PMI has divided the field of project management into 10 knowledge areas, which can be interpreted and used in many different ways, but indeed are areas that interest any project management process. Consequently, any project will have something to deal with these areas, and a PP should contain the answers from the planning phase to these areas. The 10 areas are the following:

- Project Scope Management
- Project Schedule Management
- Project Cost Management
- Project Quality Management
- Project Resource Management
- Project Communications Management
- Project Risk Management
- Project Procurement Management
- Project Stakeholder Management
- Project Integration Management

A Knowledge Area is an area of project management identified by the methodology as relevant for the different phases of the project life cycle. The area is defined by "its knowledge requirements and described in terms of its component processes, practices, inputs, outputs, tools, and techniques" [6].

Conceiving a PP under the umbrella of this methodology is pretty straightforward: a PP is a document containing the actions the project manager wants to implement to provide answers about these knowledge areas to the project stakeholders. In such a PP, stakeholders should therefore find answers about:

- what the scope of the project is, what will be faced during the project, and what will not be part of the project's activities;
- which is the schedule of the project to manage the timely completion of it;
- the cost of the project, with an estimation of costs for any task of the project;
- the quality used to meet the stakeholders' expectations about the results;
- how resources of the project will be acquired, managed, and paid to reach the project's objectives;
- how the different communication processes of project's details and results will be managed, from collecting and distributing project information to storing and retrieving it, from controlling and monitoring communication inside the project to giving ultimate disposition about the project closure;
- the risks associated with the project, who is involved, the costs associated with each risk when it becomes real and the project suffers its effects, and what is the response plan for each risk;
- how the project manager and the team intend to procure the various resources for the project, like purchasing/acquiring products and services needed from outside the project team;
- which actions will be put in place to identify the people or the organizations that could impact or be impacted by the project, to manage their needs and to face their potential attacks, and what kind of communication style and media will be used to interact with them, and how we will meet their expectations regarding the project and their engagement;

- finally, how all these nine areas will be coordinated together to be a unique homogeneous and coordinated corpus of knowledge used by the project manager to lead the project to achieve the final results.

If stakeholders find adequate answers to these questions for their needs, we can consider that PP a well-formed one.

In conclusion, there is no precise structure for a PP. However, if we refer to a specific Project Management methodology or enrich the typical Gantt Chart (or equivalent temporal representation), we can create a comprehensive plan representation. This representation contains information about how we want to achieve the final results and all the details a stakeholder needs to approve or sponsor our project.

12.4 Recommend the Best Use and Focused Decisions to Identify the Adequate Approaches and Techniques for Particular Cases

The PP is essential to the project manager; therefore, it should leave no room for last-minute improvisations or inventions. Its structure must be clear, and contents must be complete, readable, and organized according to rules shared with the other project managers and the PMO. We suggest developing our PP structure referring to internal standards or external methodologies. Following a methodology of Project Management for creating and completing a PP helps project managers under different perspectives, from clarity to comparability to project marketing. From this point of view, the PP is the answer to the requests of the methodology.

If we do not want to follow a specific template, we must list what we consider a minimum set of information that should be provided inside a PP. Therefore, the project manager should perform the actions.

The first step is to identify all stakeholders, whatever methodology, method, best practice, or golden rule we want or must follow. The project will always have multiple stakeholders. Not all of them will be involved in day-by-day activities. Not all of them will push in the same direction the project manager wants to go, as the position of stakeholders concerning the project can be positive or negative. In any case and for both reasons, they must be identified and classified, and adequate reactions should be studied and codified in response to stakeholders' activities. Stakeholders have different interests in the project, from the customer to the company and its leaders. Ignoring them means accumulating serious troubles (or losing good opportunities) for the project. Describing stakeholders' knowledge in a PP is an optimal start. For each stakeholder, define the roles and responsibilities of stakeholders. Once identified, stakeholders must have a specific role in the project. A specific role implies defining their involvement in the project, optimizing their role, using their power for the project's objectives, building a profitable relationship with them, and establishing optimal communication channels depending on their role.

A second step in creating the PP is to define what deliverables the project will produce. Project deliverables derive from the project goals and are a building pillar of any PP. A standard definition [6] of a deliverable is "any unique and verifiable product, result, or capability to perform a service that is produced to complete a process, phase, or project." A PP without clear identification, presentation, and location in time of deliverables is missing vital information for all the project stakeholders, the project manager included.

A third step, once we have clarified the previous points, is to set priorities on goals. We should explain why phases are executed before/after others and why these priorities are needed to meet those objectives. Task prioritization allows explaining individual goals and tasks according to importance and dependencies. Setting priorities gives us another burden: putting a system in place to ensure corrective actions when goals are not met on time. Priorities and goals are also fundamental when we have to adjust the project during the execution for any reason, and impacting decisions must be taken.

Having performed these activities to create the PP, we have enough elements to create the project schedule. We have to detail the project timeline to estimate tasks' duration, connections, and organization. We can adopt different approaches if we have a predictive approach to managing the project or are more on the side of adaptive project management techniques. In a PP, we must indicate the project schedule. Otherwise, stakeholders (specifically customers and sponsors) will not probably be willing to sign a blank check. Scheduling is heavily conditioned by the resources available in the project, so we will indicate the resources required to complete each task and any other information that clarifies how we will use the project team.

All the processes identified so far are iterative and not sequential, so we need to repeat our analysis to refine the results, understand the context and its characteristics better, and make adjustments in roles, scope, budget, etc. The processes can even be conducted in parallel because information found in one step can be helpful and used in other steps. Information about stakeholders and their role, for example, can be a good source of information for schedule creation. Conversely, during project schedule definition, it could be helpful to know stakeholders to involve them in helping to define specific parts of the project and their attributes (duration, cost, resources).

While creating the PP is proceeding, another step attracted the attention of project managers over the year, i.e., doing a risk assessment. Problems that may or may not arise throughout the project are called "risks." We must plan reactions to these potential problems from a project manager's perspective. According to this evaluation, we should study the risk, define its reification's probability and impact on the project, and prepare a reaction with appropriate tasks and budget. It is much better to identify mitigation techniques for our risks during the planning phase rather than be caught off guard during the project's execution when it is probably too late. Risks can be everywhere in the project, from resource allocation to tasks failing. It is impossible to control all potential risks, but thinking ahead of time can save our project from failure.

We have a final recommendation in this general list of suggestions for creating a PP. A project is a phenomenal communication media from different perspectives and for different stakeholders with various and contrasting interests in our project. Accordingly, we must prepare a solid and well-structured communication plan that should be integral to the PP. Especially technical project managers do not like this part too much because communication is seen as far from the core, high-tech activities of the project. In the end, we have to produce deliverables, and communication activities are mostly perceived as a loss of time.

The recommendation is to ensure that the project manager has communicated the project contents to the team and all other stakeholders at the end of the project planning phase. The communication plan will also be used during the project's execution to clarify how to communicate the project's progress and changes, to whom this communication should be addressed, and what to do with the various reactions from stakeholders.

12.5 Summary and Conclusions

The PP is the main result of the project planning process, where the project manager defines the project scope, objectives, and steps needed to get the work done. It represents one of the most critical processes in project management. The PP maps the steps and resources needed to complete a project on time and within budget by outlining the project's processes to meet stated goals. The PP is a vital communication instrument where the essential elements of the project are reported. The Gantt chart is the most commonly used timeline representation and can be enriched with many other project elements. Too often, the Gantt chart is used as the primary (if not the only) representation tool. However, it is insufficient to document all the project's aspects and support the project manager during the other phases of its life cycle.

There is no pre-determined structure for a PP, and organizations tend to have their format. Project management methodologies can provide valuable suggestions on how to structure this document because the plan should support the implementation of methodological suggestions about managing a project. The PMO is another significant source of PP templates to standardize its structure and content. Whatever structure the organization chooses, the PP should support the interaction with stakeholders whose needs are understanding how the project will be executed and how the project's progress will be monitored to ensure goals' achievement.

References

1 Söderlund, J. (2004). Building theories of project management: past research, questions for the future. *International Journal of Project Management* 22 (3): 183–191. ISSN: 0263-7863. https://doi.org/10.1016/S0263-7863(03)00070-X.

2 The Agile Manifesto (2023). Agile Alliance. https://www.agilealliance.org/agile101/the-agile-manifesto/ (accessed 21 March 2023).

3 Cusumano, M.A. (2011). How to innovate when platforms won't stop moving. https://sloanreview.mit.edu/article/how-to-innovate-when-platforms-wont-stop-moving/ (accessed 21 March 2023).

4 Morris, P.W.G. (1994). *The Management of Projects*, 18. Thomas Telford. ISBN: 0-7277-2593-9, Google Print.

5 Marsh, E.R. (1975). The Harmonogram of Karol Adamiecki. *The Academy of Management Journal* 18 (2): 358–364. https://doi.org/10.2307/255537.

6 Project Management Institute (2017). *A Guide to the Project Management Body of Knowledge (PMBOK Guide)*. Newtown Square: Project Management Institute. ISBN 978-1-62825-184-5.

Section 7

Digital Disruption

13

The Evolution of Smart Sustainable: Exploring the Standardization Nexus

Cristina Bueti and Mythili Menon

Telecommunication Standardization Bureau, International Telecommunication Union, Geneva, Switzerland

Abbreviations

FG-SSC   Focus Group on Smart Sustainable Cities
ICTs   Information and Communication Technologies
IoT   Internet of Things
ITU   International Telecommunication Union
KPIs   Key Performance Indicators
QoL   Quality of Life
SDGs   Sustainable Development Goals
SSC   Smart Sustainable City
U4SSC   United for Smart Sustainable Cities

13.1 Digital Transformation and Smart Sustainable Cities

13.1.1 Driving Digital Transformation in the Urban Domain

Rural-to-urban migration continues to contribute to population growth in cities. By the year 2050, it is expected that nearly 70% of the world's population will be living in urban areas. This exponential demographic growth has been the driving force behind global urbanization, characterized by social, economic, and environmental opportunities as well as challenges. Cities are expected to account for approximately 80% of the world's GDP but also are responsible for 90% of the pollution problems [1]. To improve the quality of life (QoL) of their inhabitants, cities have been increasingly investing in digital infrastructure to become "smarter," to deliver urban services using digital technologies, while ensuring access to basic amenities for all [2].

5G, blockchain, Internet of things (IoT), cloud computing, robotics, virtual and augmented reality machine learning, artificial intelligence (AI), and quantum computing are among the digital technologies that are increasingly being utilized for reshaping the way services are offered in the urban domain. These digital technologies are transforming urban management with the promise

IEEE Technology and Engineering Management Society Body of Knowledge (TEMSBOK), First Edition.
Edited by Gustavo Giannattasio, Elif Kongar, Marina Dabić, Celia Desmond, Michael Condry, Sudeendra Koushik, and Roberto Saracco.

of automating urban operations, leveraging on big data to improve the QoL, and tackle challenges associated with climate change, health, education, sanitation, and energy conservation.

The ongoing COVID-19 pandemic also exemplifies the need for technological convergence and innovation in order to enhance cities' responsiveness and agility to global emergencies and other emerging challenges. The implementation of these digital technologies requires the adoption of strategies for creating new business opportunities, and securing for future applications across verticals, while supporting sustainability [3].

Digital technologies can play an important role in the pursuit of urban sustainability, which is a critical component for the attainment of various important global commitments including:

- Sustainable Development Goal 11 recognizes that while cities are the center of economic growth, they are also responsible for 60–70% of global carbon emissions. It further identifies that current urbanization trends will overburden the existing infrastructure, generate housing problems, and contribute to environmental pollution.
- Sustainable Development Goal 9, aiming to *build resilient infrastructure, promote sustainable industrialization and foster innovation*, can facilitate the introduction and utilization of new technologies, and promote new businesses by leveraging these technologies along with judicious use of resources [4].
- The New Urban Agenda[1] encompasses an overarching approach to sustainable development in the urban context [5]. The New Urban Agenda also references the concept of "smart cities" which are premised on digital technologies and big data to respond to urban challenges.

13.1.2 Smart Sustainable Cities: An Introduction to the Concept

The International Telecommunication Union (ITU) through its Focus Group on Smart Sustainable Cities (FG-SSC) contributed to the development of a concrete definition along with the required indicators which can be used by stakeholders to monitor their respective smart city progress. Furthermore, to ensure that *the sustainability* aspect is not undermined, the newly coined term (based on smart cities), "Smart Sustainable Cities" was utilized for the activities of FG-SSC. The definition for Smart Sustainable Cities (SSC was formulated based on extensive research and analysis of over 100 definitions for smart cities and sustainable cities and other related terms from various resources.

> A smart sustainable city is an innovative city that uses information and communication technologies (ICTs) and other means to improve quality of life, efficiency of urban operation and services, and competitiveness, while ensuring that it meets the needs of present and future generations with respect to economic, social, environmental as well as cultural aspects.

The incorporation of digital technologies (IoT, digital twin, AI, robotics) lays the foundation for leveraging on data to drive urban operations more efficiently and on a larger scale. IoT-based sensors can equip the cities to monitor ongoing activities in real-time with the possibility of making timely interventions. The data derived from these sensors can form the basis for quick

---

1 The New Urban Agenda was embraced in October 2016 during United Nations Conference on Housing and Sustainable Urban Development. It sets out the urbanization good practices and highlights relevant tools required for.

decision-making using AI. As a part of the intervention, drones and robots can be deployed. Therefore, to achieve the required improved QoL within smart sustainable cities, performance and prediction is essential; this is pertinent to maintain a dynamic and continuously adaptative system. Digital technology can also ensure long-term efficacy of the urban ecosystem while enabling sustainability and adaptability. Additionally, the concept of "digital twin" can provide an accurate model/representation of urban processes, helping to drive simulations as a part of the planning process to improve transparency, reduce overheads, disseminate information before integrating any specific technology, and serving as a feedback channel for citizens to provide inputs on smart city planning endeavors.

13.1.3 The Role of International Standards in Smart Sustainable Cities

One important aspect to be considered is the connection and communication between devices and networks which is not feasible without international standards. As IoT is expected to connect billions of devices within this decade, a common protocol for functioning is mandatory to ensure interoperability, compatibility, and reducing costs of manufacturing and implementation. Voluntary consensus-based standards formulated by international standards organizations (SDOs) present best practices for making a city function better, providing guidelines for the implementation of digital technologies (for energy efficiency, education, health, security, manufacturing, environmental monitoring) and a pathway for attaining important global commitments such as the United Nations' Sustainable Development Goals. Several SDOs, including (but not limited to) the International Organization for Standardization (ISO), International Electrotechnical Commission (IEC), ITU, and Institute of Electrical and Electronics Engineers (IEEE), have been working on the development of standards in various sectors related to IoT and smart cities in a concerted manner.

In line with the need to explore the role of next-generation ICTs within the smart city ecosystem, complementary standardization work on smart cities is being carried out by ITU-T Study Group 20 on IoT and Smart Cities and Communities (SG20).[2] The activities of this Study Group have elaborated on how emerging technologies such as IoT, AI, blockchain, and virtual reality enable smart city ecosystem to offer relevant services pertaining to intelligent transportation, healthcare, education, provision of utilities, and maintenance of urban green spaces – with the aim of improving QoL and attaining the targets stipulated within the Sustainable Development Goals to make cities more interconnected, livable, safer, and resilient [6].

13.1.4 Smart Sustainable Cities in the Post-COVID-19 Era

Currently, as cities are on the front-line of managing the COVID-19 pandemic, it would be appropriate to also ensure that existing as well as envisioned smart cities are capable of curbing the spread of COVID-19 by leveraging digital technologies for contact tracing, hospital admissions, emergency medical facilities, and maintenance of databases for the administering of the required vaccinations [7]. While cities across the globe have developed apps and employed relevant technologies for limiting the transmission of COVID-19, even leading smart cities have not been able

---

2 The ITU-T Y. Series Recommendations are the main ITU standards on the topic of IoT and Smart Cities. The foundational work was initiated within the FG-SSC. Following the closure of FG-SSC, the baton was passed to ITU-T Study Group 20 to continue the standardization work within this domain.

to adequately ensure public health security for its citizens. The COVID-19 pandemic has inadvertently exposed the limitations of smart city deployments in their current literation [8]. While there has been certain progress pertaining to the development of COVID-19 vaccinations, the ongoing research related to their efficiency along with the relatively slow-paced delivery of these inoculations, has brought about the need for focusing on prevention methods to reduce the spread of the virus through social distancing protocols, lockdowns modalities, and travel limitations/restrictions. Studies have noted that social distancing can significantly reduce the number of COVID-19 cases, if implemented sufficiently toward the initial phases of the pandemic. This would also reduce the overreliance on medical facilities in cities, especially during the pandemic [9]. Under the current circumstances, it is essential to provide effective mechanisms to limit the spread of COVID-19 and have effective surveillance techniques which are capable of monitoring and enforcing social distancing [10]. Countries like Switzerland have launched an app known as SwissCovid [11] to support contact tracing, while Singapore has its FluGoWhere [12], which provides a thorough search list of Public Health Preparedness Clinics (PHPCs). The advent of COVID-19 has therefore indicated that there is a need to strengthen the capacity of smart cities in dealing with public health crises. While currently there may be no model city to be emulated, Figure 13.1 provides an overview of the envisaged actions to be undertaken in smart cities in the event of an epidemic or pandemic. As depicted in Figure 13.1, situational intelligence and automated response are imperative to foster and ensure public health protection against future epidemics and/or pandemics. Therefore, the envisioned smart cities of the post-COVID-19 period will need to implement IoT devices, including thermal cameras, to enforce social distancing rules as and when required along with the detection of fever-related symptoms. In addition to monitoring social distancing protocols, IoT sensors can also be used to survey quarantine violations with limited human intervention.

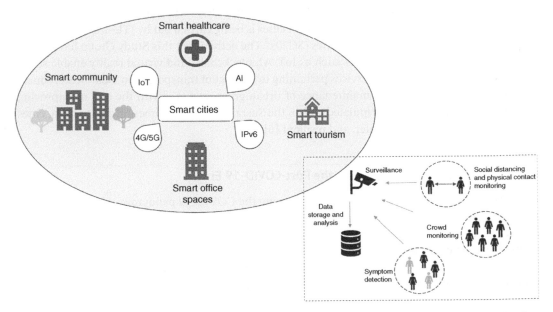

Figure 13.1 Overview of envisioned activities to be undertaken in the advent of the COVID-19 pandemic in smart cities. *Source:* Adapted from Shorfuzzaman et al. [10].

13.2   Smart Sustainable Cities Standardization Requirements: Reimagining Data-Driven Practices

13.2.1   Stepping-Stone for Cities of the Future: The Smart Sustainable City Cycle

While the success of a smart city project could be measured by the level of integration of fundamental services, predicated on technology, the COVID-19 pandemic has made the world aware that the integration of digital technologies needs to cater to the needs of all citizens. Therefore, the transition to a smart and sustainable city should be envisioned as a journey, not as a destination. Accordingly, smart city transitions need to be *adaptable*. ITU developed the Guide for City Leaders (as contained in Y.Suppl.32 to ITU-T Y.4000 series), which introduces the Smart Sustainable City cycle (see Figure 13.2 and Table 13.1) [13].

13.2.2   Key Verticals in Smart Sustainable Cities

The main subsystems within the smart and sustainable city ecosystem include (see Figure 13.3) [15]:

1) **Smart transportation (Intelligent Transport Systems):** This refers to the utilization of new and emerging technologies to make transportation systems more efficient, safe, and cost-efficient. With the application of IoT, 5G, AI, and others, transportation systems are veering toward automated operations including self-driving vehicles. Intelligent transport system also

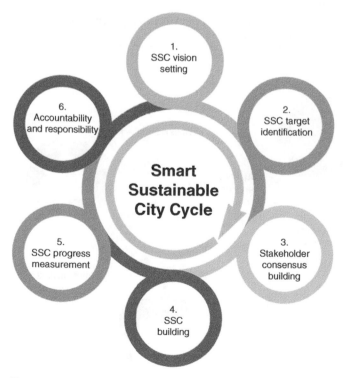

Figure 13.2   Smart and Sustainable City Lifecycle. *Source:* Adapted from Y.Suppl.32 to ITU-T Y.4000 Series [13].

Table 13.1   Smart Sustainable City Cycle – key steps.

---

1) **Set the vision for your SSC venture**

   This involves identifying a specific SSC vision and adhering to it based on the availability of resources, political commitment, location, and stakeholder nexus along with an assessment of the feasibility of becoming a Smart Sustainable City.

2) **Identify your SSC targets**

   This step entails elaborating on the key objectives, priorities, initiatives, and actions needed in the short, medium, and long term for the smart city transition. Available key performance indicators to monitor SSC transformations can also be selected at this stage.

3) **Achieve political commitment**

   This step is essential for gaining political goodwill accompanied by cooperation with local authorities in developing a strategic planning document, which will form the basis for the next stage of the process.

4) **Build your SSC**

   This step involves updating the urban infrastructure by integrating new and emerging technologies (IoT, machine learning, virtual reality, digital twin) to support the planning process, and assist with the delivery of services and data management.

5) **Measure your city's progress**

   For any smart city project, it is important to have a feedback loop to improve the existing processes. Additionally, United for Smart Sustainable Cities has developed an assembly of Key Performance Indicators (KPIs) to monitor progress and evaluate advancements in line with the Sustainable Development Goals.

6) **Ensure accountability and responsibility**

   This step is focused on reporting and learning from the SSC process and related experiences in order to further streamline the SSC planning document prepared in Step 3.

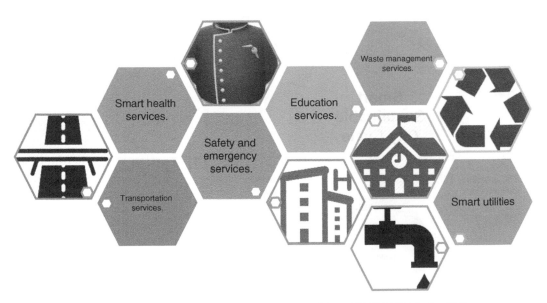

Figure 13.3   Smart city verticals (ITU-T, 2016). *Source:* Adapted from ITU-T Y.4400 Series [14].

seeks to use more renewable forms of energy as a fuel source. Therefore, it also promotes the utilization of electric cars [16]. This domain corresponds to various SDGs including SDG11 (Sustainable Cities Communities), SDG9 (Industry and Innovation), and SDG7 (Renewable Sources).

2) **Smart waste management:** This entails the implementation of technologies for the collection, storage, sorting, and recycling of waste to "close the loop" and transition to a circular economy [17]. It is particularly important for electronic waste or e-waste that contain hazardous elements, which need to be adequately disposed of. E-waste can be a double-edged sword as it can also serve as the source of valuable metals, including gold and silver, that supports the economy and could drive social transformation [18].

3) **Emergency services:** This encapsulates key services which allow for the detection of hazards and provides the required responses to ensure limited danger to human life and property. For example, in the event of a house-fire, IoT sensors can detect sudden changes in temperature and initiate sprinkler systems. Automated warnings could also be sent to the fire-safety services to commence the evacuation action. The activities correlating to this domain are closely linked to SDG11 (Sustainable Cities and Communities) [19].

4) **Smart health services:** These services focus on the use of technologies including wearable devices, IoT, and AI for the dissemination of health information along with the management and response to urgent medical needs [20]. This service structure is specifically essential during the ongoing COVID-19 pandemic, which warrants and responsive medical care system to support contact tracing and social isolation (see example elucidated in Figure 13.1). This domain corresponds to SDG1 (No Poverty), SDG2 (Zero Hunger), and SDG3 (Good Health and Well-being).

5) **Education services:** This involves the optimization of the use of technology to deliver courses and impart training for various levels of education. During the pandemic, these services have played a pivotal role in ensuring that children's education continues. The main bottleneck for the provision of these services is related to the lack of access to ICT devices including mobile phones, tablets, and laptops, particularly in rural areas [21]. This domain corresponds to SDG 4 (Quality Education).

6) **Smart utilities:** The smart utilities nexus intends to ensure the smooth supply of basic amenities including water and energy. This system may avail the use of smart meter to assess usage and prevent leakage as well as theft [22]. This domain corresponds to SDG7 (Renewable Energy) and SDG9 (Industry and Innovation).

These verticals have been elaborated on in Y Supplement 27: ITU-T Y.4400 series. Following the smart sustainable city conceptualization phase (including the identification of verticals), it is essential to explore the processes associated with the operational implementation of building a smart city from scratch or upgrading the existing infrastructure to assist an existing city's transformation to an SSC. For this, an effective planning mechanism is essential which can capitalize on the expertise of the stakeholders, while ensuring accountability, transparency, inclusiveness, and efficiency [23]. As a part of its standardization work on smart cities, ITU had undertaken the process for stakeholder identification as given in Y Supplement 34: ITU-T Y.4000 series (see Figure 13.4) [24].

Smart city endeavors can provide space for exploiting new and pioneering entrepreneurial opportunities, given the diversity of services required for the urban ecosystem and the availability

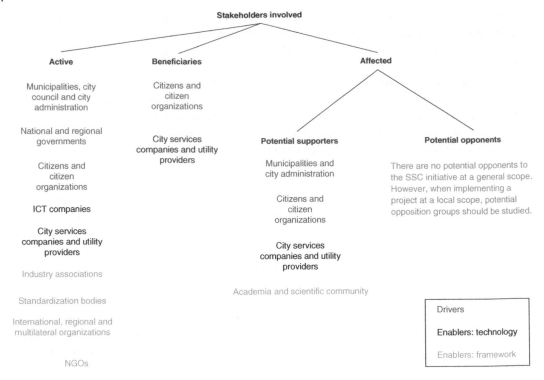

Figure 13.4 Smart city stakeholder identification. *Source:* ITU, 2020 / https://www.itu.int/rec/T-REC-Y. Sup34-202007-I/en/ITU.

of data derived from IoT sensors (see Figure 13.5) [25]. Public Private Partnerships (PPPs) have been embraced for smart city transitions to provide new business opportunities, ensure cost-effectiveness, and access to expertise. As technologies advance, new applications of digital technologies will also be determined within the smart city framework.

13.2.3 Data-Powered Smart Sustainable Cities

The integration of IoT within the smart city ecosystem, across verticals has generated a high volume and high variety of data [26]. In order to be able to effectively unlock the value of the available data streams, efficient data management methodologies involving the use of machine learning (see Figure 13.6) need to be employed to enable big data analytics [28].

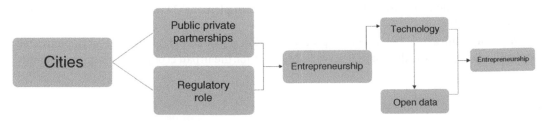

Figure 13.5 Smart city entrepreneurial ecosystem. *Source:* Kummitha et al. [25].

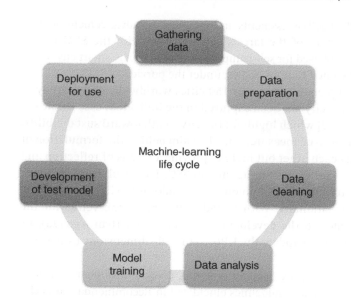

Figure 13.6 Machine-learning lifecycle for data management. *Source:* Adapted from [27].

With reference to the data privacy challenges, blockchain (see Figure 13.7) can be used to store data and ensure its security, privacy, and transparency. Several types of transactions of smart cities can be recorded in a blockchain.

13.2.4 Smart Sustainable Cities Assessment Tools: Key to Driving to Digital Transformation

Following on from the incorporation of relevant technologies in smart cities, the next important step is to be able to assess and monitor progress. In line with this understanding, the United for Smart Sustainable Cities initiative (U4SSC) [30], has developed an array of Key Performance

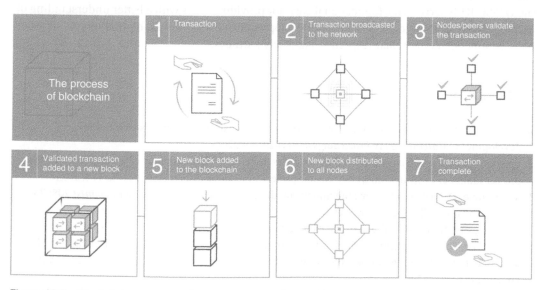

Figure 13.7 Blockchain process to data management. *Source:* Nascimento and Pólvora [29]/Publications Office of the European Union/CC BY 4.0.

Indicators (KPIs) to help cities conduct self-assessments and measure the advancement of their local goals together with the achievement of the targets elaborated on with the SDG framework [31]. These KPIs also serve as a tool used for self-comparisons based on the smart city ontology research work undertaken by 17 United Nations bodies, under the purview of U4SSC.

These KPI Indicators have been implemented in over 150 cities worldwide. The journey of the adoption of these KPIs within each city has been captured in the individual case studies as well as city factsheets and snapshots [32] which highlight the city's path toward sustainability, smartness, and SDGs based on standardized measurement. The aim behind the formulation of these KPIs is not to put cities against each other but more to provide a means of reflection and examination of existing actions taken for transforming into an SSC. The results and feedback from the cities are also leveraged to refine the KPI framework. Ecuador is the latest country on the block to implement these KPIs nationally. With Ecuador also being the venue for the Habitat III meeting (which culminated in the development of the New Urban Agenda), its participation in the KPI pilot will also help update the KPIs to include dimensions from other international instruments [33].

In keeping with its *Innovative and Competitive Ecuador program*, Ecuador also plans to bridge the digital divide, and has also adopted the key guidelines elucidated in Recommendation ITU-T Y.4904, to help identify the city's goals, and ascertain its current status and determine critical capabilities needed to progress toward becoming a smart and sustainable city.

13.3 Conclusion

Smart Sustainable Cities are complex systems, the establishment of which involves multiple stakeholders, together with the application of technologies including IoT and AI sensors for managing the heterogeneous data streams to ensure interoperability between domains and platforms. International standards play a key role in ensuring connectivity between sensors and devices within the urban ecosystem, which in turn enables communication and the adoption of emerging technologies to foster trans-domain, digital transformation.

Smart cities need to be adaptable in order to cater to the immediate needs of urban populations. The U4SSC KPIs serve as the foundational basis for providing cities with a better understanding of the progress in relation to the Sustainable Development Goals as well as internal targets set prior to the commencement of their respective smart city journeys.

References

1 Akhtara, I. and Kirmania, S. (2021). Impact of sustainable design on India's smart cities development. *Material Today Proceedings* 46: 11020–11022.

2 Bjørner, T. (2021). The advantages of and barriers to being smart in a smart city: the perceptions of project managers within a smart city cluster project in Greater Copenhagen. *Cities* 114: 103187.

3 MacKenzie, D. (2010). Interview: the millenium development goal pioneer. *New Scientist* 208: 23.

4 Rozhenkova, V., Allmang, S., Ly, S. et al. (2019). The role of comparative city policy data in assessing progress toward the urban SDG targets. *Cities* 95: 102357.

5 Giles-Corti, B., Lowe, M., and Arundel, J. (2020). Achieving the SDGs: evaluating indicators to be used to benchmark and monitor progress towards creating healthy and sustainable cities. *Health Policy* 124 (6): 581–590.

6 ITU (2021). ITU-T study group 20 on IoT and smart cities and communities. https://www.itu.int/en/ITU-T/about/groups/Pages/sg20.aspx (accessed 21 February 2021).

7 Yang, S. and Chong, Z. (2021). Smart city projects against COVID-19: quantitative evidence from China. *Sustainable Cities and Society* 70: 102897.

8 Adil, M. and Khan, M.K. (2021). Emerging IoT applications in sustainable smart cities for COVID-19: network security and data preservation challenges with future direction. *Sustainable Cities and Society* 75: 103311.

9 Kim, H.M. (2021). Chapter 16 – Smart cities beyond COVID-19. In: *Smart Cities for Technological and Social Innovation*.

10 Shorfuzzaman, M., Shamim, M.H., and Alhamid, M.F. (2021). Towards the sustainable development of smart cities through mass video surveillance: a response to the COVID-19 pandemic. *Sustainable Cities and Society* 64: 102582.

11 Federal Office of Public Health – Switzerland (2021). Coronavirus: swissCovid app and contact tracing. https://www.bag.admin.ch/bag/en/home/krankheiten/ausbrueche-epidemien-pandemien/aktuelle-ausbrueche-epidemien/novel-cov/swisscovid-app-und-contact-tracing.html#-1601404801 (accessed 9 February 2021).

12 Singapore Government (2021). Looking for PHPC near you? https://flu.gowhere.gov.sg (accessed 19 February 2021).

13 Y.Suppl.32 to ITU-T Y.4000 Series (2020). Smart sustainable cities – a guide for city leaders.

14 ITU-T Y.4400 Series (2016). Smart sustainable cities – Setting the framework for an ICT architecture.

15 Y.Sup27 (2016). ITU-T Y.4400 Series – Smart sustainable cities – Setting the framework for an ICT architecture.

16 Dimitrakopoulos, G., Uden, L., and Varlamis, I. (2020). Chapter 18 – intelligent transport systems and smart mobility. In: *The Future of Intelligent Transport Systems*. Elsevier.

17 Fatimah, Y.A., Widianto, A., and Hanafi, M. (2020). Cyber-physical system enabled in sustainable waste management 4.0: a smart waste collection system for indonesian semi-urban cities. *Procedia Manufacturing* 43: 535–542.

18 Mor, R.S., Sangwan, K.S., Singh, S. et al. (2021). E-waste management for environmental sustainability: an exploratory study. *Procedia CIRP* 98: 193–198.

19 De Nicola, A., Melchiori, M., and Luisa Villani, M. (2019). Creative design of emergency management scenarios driven by semantics: an application to smart cities. *Information Systems* 81: 21–48.

20 Ghose, B. and Rehena, Z. (2020). A mechanism for air health monitoring in smart city using context aware computing. *Procedia Computer Science* 171: 2512–2521.

21 Molnar, S. (2021). Smart cities education: an insight into existing drawbacks. *Telematics and Informatics* 57: 101509.

22 Venketraman, A., Thatte, A., and Xie, L. (2021). A smart meter data-driven distribution utility rate model for networks with prosumers. *Utilities* 70: 101212.

23 Y.Suppl.45 to ITU-T Y.4000 Series (2017). An overview of smart cities and communities and the role of information and communication technologies.

24 Y.Suppl.34 to ITU-T Y.4000 Series (2020). Smart sustainable cities – setting the stage for stakeholders' engagement.

25 Kummitha, R.K.R. (2019). Smart cities and entrepreneurship: an agenda for future research. *Technological Forecasting and Social Change* 149: 119763.

26 Ben Atitallah, S., Driss, M., Boulila, W., and Ben Ghézala, H. (2019). Leveraging deep learning and IoT big data analytics to support the smart cities development: review and future directions. *Computer Science Review* 38: 100303.

27 AI platforms in 2021: guide to ML life cycle support tools. https://research.aimultiple.com/ai-platform.

28 Rathore, M., Paul, A., Hong, W. et al. (2018). Exploiting IoT and big data analytics: defining smart digital city using real-time urban data. *Sustainable Cities and Society* 40: 600–610.

29 Nascimento, S. and Pólvora, A. (2019). *Blockchain Now and Tomorrow: Assessing Multidimensional Impacts of Distributed Ledger Technologies*, EUR 29813 EN, Publications Office of the European Union, Luxembourg.

30 ITU (2021). United for smart sustainable cities initiative. https://www.itu.int/en/ITU-T/ssc/united/Pages/default.aspx (accessed 19 February 2021).

31 U4SSC (2021). Collection methodology for key performance indicators for smart sustainable cities. https://www.itu.int/en/publications/Documents/tsb/2017-U4SSC-Collection-Methodology/files/downloads/421318-CollectionMethodologyforKPIfoSSC-2017.pdf (accessed 19 February 2021).

32 ITU (2021). ITU's implementation of the U4SSC KPIs. https://www.itu.int/en/ITU-T/ssc/united/Pages/publication-U4SSC-KPIs.aspx (accessed 19 February 2021).

33 Using Key Performance Indicators project for Smart Sustainable Cities to reach the Sustainable Development Goals (SDGs).

14

Wireless 5G (The 5G Mobile Network Standard)

Cristina Emilia Costa[1] and Fabrizio Granelli[2]

[1] *Consorzio Nazionale Interuniversitario per le Telecomunicazioni (CNIT), Genoa, Italy*
[2] *Department of Information Engineering and Computer Science (DISI), University of Trento, Trento, Italy*

14.1 Introduction

This chapter describes the main features of the 5G Mobile Network Standard. Indeed, mobile communications currently represent the most used technology to access to the Internet and related online services. The following sections provide a brief description of the fundamentals of such technology, which are the main functionalities expected by 5G Networks, some deployment scenarios and recommended best practices.

14.2 Knowledge Area Fundamentals

The 5G Mobile Network Standard (3GPP Release 15 and beyond) defines the current state-of-the-art architecture and protocols related to mobile cellular communications. As such, the basis of the knowledge area is related to the basics of mobile communications. For a comprehensive illustration of such knowledge base, the reader is suggested to check the well-known book [1]. A good reference to understand the basics of 5G is a most recent book [2].

In a cellular network (see Figure 14.1), mobile devices (i) connect to a base station (ii), which connects to a backhaul network (iii). The backhaul, typically wired, is used to connect mobile devices to the Internet (iv). A cellular network standard specifies the two sections of the mobile network: the Radio Access Network (RAN), interconnecting mobile devices to the base stations (gNodeB, in the 5G standard), and the Core Network (CN), where networking and control functionalities are implemented.

The RAN specifies the protocols and architecture for interconnecting mobile devices to the base station.

Digital cellular networks evolved starting from GSM (Second Generation) in the 1990s. A brief roadmap of the main milestones of cellular communications includes:

- 2G (GSM): first cellular standard to transmit voice using digital modulations, and to support mobility and security;

IEEE Technology and Engineering Management Society Body of Knowledge (TEMSBOK), First Edition.
Edited by Gustavo Giannattasio, Elif Kongar, Marina Dabić, Celia Desmond, Michael Condry, Sudeendra Koushik, and Roberto Saracco.

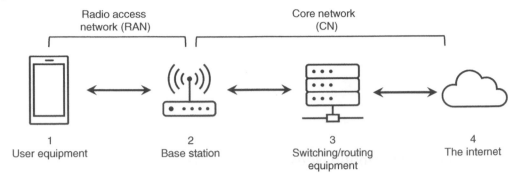

Figure 14.1 Simplified block diagram of the cellular network architecture.

- 3G (UMTS): data-oriented communications and mobile Internet connectivity, with limited capacity (in average, around 100–200 kbps per user), introduction of packet switching for data transmission;
- 4G (LTE): transition to a full IP native architecture, high throughput (10s of Mbps) – making mobile Internet video a reality, monolithic Base Stations (incorporating control and resource allocation).

Traditional cellular networks (up to 4G – LTE) had access to three portions of the wireless spectrum: 800, 1800, and 2100 MHz. Moreover, all frequencies of operation were subject to geographical licensing and exclusive usage.

5G New Radio (NR) (e.g. the specification of the wireless interface of the 5G System) is the first radio system designed to support different spectrum bands between 400 MHz and 90 GHz, in order to enable different combinations of high capacity, high data rates, ubiquitous coverage, and ultra-high reliability. Figure 14.2 provides an overview of the main frequency bands used by 5G NR standard.

Low bands below 6 GHz (mostly those already being used by traditional mobile systems) are useful for wide area coverage and data rates up to a few Gbps. Those bands are used to plan the macro-cells and micro-cells coverage, in order to provide stable service and easy handover. However, it should be noted that 5G operation can be extended to low bands formerly used by terrestrial television broadcasting and currently being released due to the shift to digital television. Indeed,

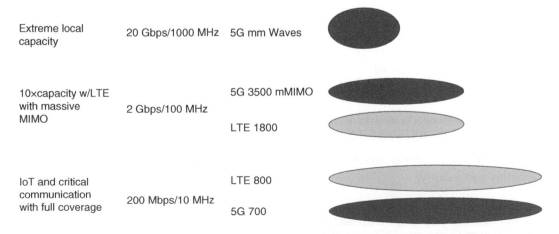

Figure 14.2 Conceptual diagram illustrating the main frequency bands of operation of 5G NR.

reliable coverage is an important factor in providing connectivity, e.g. for IoT devices and for critical communication such as remote control or automotive communication. In those cases, low frequencies are most favorable (especially in the range 700–800 MHz).

Higher frequencies (higher than 6 GHz) are used to offer extreme local capacity (up to 20 Gbps) in short range. Those frequencies are used to provide capacity in the presence of high density of users (picocells).

A relevant novelty of 5G is that it can also be deployed on shared spectrum, such as the 3.5 GHz band in the USA, and in unlicensed spectrum, like 5 GHz. This opens new possibilities for enterprises and industries to benefit from 5G technology without the need for licensed spectrum. An example of this scenario (Private 5G Networks) is presented later in the chapter.

The CN of a cellular network consists of the backhaul used for data plane switching and routing, and the network management components. In previous standards, the CN included switching nodes (Packet Gateway and Serving Packet Data Network Gateway) as well as components for connection management (such as Mobility Management Entity) and user authentication/billing (Home Subscriber Service).

The 5G system will be deployed in two ways: standalone (SA) and non-standalone (NSA) modes. In the NSA mode, the 5G RAN will use pre-existing 4G CN for signaling and management functionalities. This will represent the architecture used in the first 5G deployments. The NSA mode will instead exploit the full benefits of the evolution of the CN, and basically implement the 5G Service Based Architecture.

The resulting complete architecture of the 5G System is illustrated in the Figure 14.3.

The 5G Service Based Architecture represents a step forward into the integration of emerging technologies such as Software Defined Networking (SDN) and Network Function Virtualization (NFV) within the mobile network architecture. Indeed, the entire Service Based Architecture

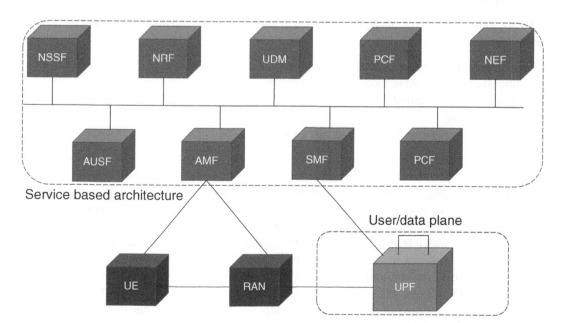

Figure 14.3 The 3GPP block scheme of the 5G Service-Based Architecture. The 5G Service-Based Architecture consists of the following network functions (NF): AUSF: Authentication Server Function; AMF: Access and Mobility Management Function; NEF: Network Exposure Function; NRF: Network Repository Function; NSSF: Network Slice Selection Function; PCF: Policy Control Function; SMF: Session Management Function; UDM: Unified Data Management; UPF: User Plane Function.

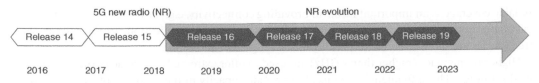

Figure 14.4 Overview of the 3GPP 5G roadmap.

consists of a set of virtual functions that interact through Rest APIs and can be installed, provisioned, and maintained using cloud technologies.

Moreover, external parties (over the top service providers or other mobile operators) will be offered the possibility to directly and programmatically interact with the 5G system by using two dedicated components: the Network Exposure Function (NEF) and Application Function (AF).

Figure 14.4 provides an outline of the roadmap defined by 3GPP for the development of 5G and further releases.

First official release of the 5G standard is labeled by 3GPP as Release 15 [3]. The following releases are focusing on improvements and enhancements to extend and perfect its functionalities [4].

In brief, 5G NR Release 16 [5] focuses on the following features:

- Integrated access and backhaul (IAB).
- NR operation in unlicensed spectrum.
- Features related to Industrial Internet of Things (IIoT) and ultra-reliable low latency communication (URLLC).
- Intelligent transportation systems (ITS) and vehicle-to-anything (V2X) communications.
- Positioning.

5G NR Release 17 will produce enhancements to eMBB, URLLC, and mMTC, and introduce new functionalities, such as:

- Supporting NR from 52.6 to 71 GHz.
- Multicast and broadcast services for eMBB.
- Support for non-terrestrial networks.
- Sidelink relaying.
- Support of reduced-capability NR Devices for mMTC.

5G will represent a technology finally bridging the gap between the mobile and Internet worlds, bringing softwarization in the field of mobile networks governed by massive hardware devices. For this reason, managing 5G technology will require knowledge of the above two main areas. In particular, in the area of mobile networks, the best references are the 3GPP and ITU-approved standards. However, as the cloud and virtualization technologies are heavily involved in the 5G SBA and in the interaction with it from the OTT perspective, certifications focusing on cloud and virtualization technologies will also be necessary. Example of useful certifications in the latter area is those offered by international cloud operators, such as Google and Amazon. For an overview on 5G technology, and in general softwarized networks, the interested reader might refer to [6].

Indeed, general approaches that enable to manage the area of knowledge of modern cellular networks include:

- **Wireless communications:** spectrum allocation and management, efficient waveform planning, OFDM.

- **Wireless networking:** cellular network planning, network dimensioning, capacity planning, cellular networks emulation, and simulation.
- **Network softwarization:** Software Defined Networking and Network Function Virtualization.
- **Service deployment in softwarized environments:** cloud computing, Open Stack, Kubernetes, containers' deployment, Rest APIs.

Moreover, 5G-specific knowledge is mostly focused on

- 5G SA/NSA deployment.
- Edge-cloud continuum.
- Network slicing.
- Zero-touch automation.

Such four approaches are detailed in the following section.

14.3 Managing Approaches, Techniques, and Benefits to Using Them

14.3.1 Transition Between 4G and 5G: 5G SA/NSA

In order to support the transition from 4G to 5G standard, various solutions are available. Indeed the adoption of the 5G standard is foreseen as a gradual transition, that includes different phases and heterogenous environments, were both 4G and 5G co-exist.

3GPP defines both a new 5G CN (5G Core or 5GC) as well as an NR access technology, the 5G NR, that are meant to substitute the EPC and 4G radio. To consent a smooth transition, 3GPP has specified two main deployment options for 5G development, named NSA and SA.

With hybrid NSA deployments, 3GPP standards allow the integration of elements from both 4G and 5G generations. The EPC CN from 4G is used in this setup, while dual connectivity combines NR and LTE radio cells. The NSA is central in the 4G–5G transition. It depends entirely on existing LTE network for all control functions and add-on services and involves heavy integration between 4G and 5G radio.

NSA relies on the LTE infrastructure for grating service continuity and continuous coverage, and, before that, CN enables the LTE network interwork with 5G NR. It is expected that 5G NR will be deployed selectively in areas with an expectedly high demand for data services at the beginning.

The so-called SA deploys a full 5G Core, relaying on 5G Service Base Architecture, with the NR cells handling both the control plane and user plane. As 5G deployment goes on, 5 G-only deployments, i.e. SA mode, are expected.

The SA option simplifies management by allowing an independent network using normal, inter-generational handover between 4G and 5G. Indeed, it is expected that heterogenous environments might emerge, such as mobile and WiFi dual connectivity (and multi-connectivity in general). The SA network will still interoperate with the existing 4G/LTE network to provide service continuity between the two network generations. Interoperation with LTE network takes place in order to cover areas not yet covered by 5G and to connect 5G users with non-5G users. Interoperation, in this case, is confined to session handover and service continuity.

With 5G SA, operators will be able to offer full 5GC-dependent use cases and perform the E2E network slicing necessary to serve new verticals.

14.3.2 Edge-Cloud Continuum

Another novelty introduced in the 5G infrastructure is the support of computational and storage resources near the RAN, inside the operator's network. Also called "Edge cloud," it has been introduced to support the network in providing advanced services such as lower latency with computational resources at the network edge, e.g. near the RAN. Even if some 4G deployments include Edge computing, in 5G, Edge Computing becomes an enabler technology supported by the 3GPP and ETSI standard in an integrated cloud-native architecture.

Edge Computing plays a relevant role in 5G. It supports latency requirements of URLLC use cases, but it also has other applications and benefits. Even if it is often pictured near the gNodeB, there are different types of deployments, depending on its position in the infrastructure. Terms as Far Edge, Near Edge, and Central Office Edge are commonly used to distinguish different options.

Complementary to the 3GPP standardization, ETSI has initiated the ETSI MEC ISG, to providing a standard-based framework for Multiple access Edge Computing and a 3GPPP synergized architecture [7].

The multi-access edge computing (MEC) architecture by ETSI is designed to offer developers and content providers standardized access to cloud computing resources and services at the edge, so that applications can take advantage of low latency and high bandwidth brought by the proximity to the user. Applications may also have access to data and information regarding radio network, such as quality of connection, user location, and handovers, that can be exploited to provide a better service to the end users.

ETSI MEC introduces various Service APIs to support the development in the Edge of network/edge-aware applications. Plus, it provides mechanisms for the migration of edge applications in case of handover.

The MEC architecture comprises a set of individual MEC hosts and a higher level managing the network connections among the hosts built on top of the virtualized connect-compute infrastructure. MEC applications are hosted in the MEC host, together with a MEC platform which provides services to the applications. The infrastructure, platform, and MEC applications are managed locally at each host, while the MEC orchestrator, the core function of the MEC system-level management, maintains a global view of the entire MEC system and coordinates the host for maintaining an end-to-end service provisioning and handling handovers. ETSI MEC [8] introduces various Service APIs to support the development in the Edge of network/edge-aware applications. Plus, it provides mechanisms for the migration of edge applications in case of handover.

A central piece of innovation at the core of the edge computing paradigm is its interaction with networking and NFV. MEC architecture relies on the flexibility of a virtualized network infrastructure for deploying the MEC platform and applications. Relying on a shared infrastructure allows to improve and dynamically adapt resource utilization and service performance.

In NFV architecture envisaged by ETSI MEC, a unified virtualization layer for connect-compute infrastructure is provided. This vision, also shared by the design of the SBA in 5G networks, implies enabling the connect-compute convergence through the adoption of joint management of both virtual network and compute resources [3].

Acknowledging the relevance of virtualization in MEC and network infrastructures, ETSI MEC ISG defined the MEC-in-NFV that clarifies the interconnections and interaction between the elements of these two architectures and shows how MEC entities can be integrated into the NFV architecture, also proposed by ETSI [9]. In this architecture, the NFV MANO (the NFV management and network orchestration framework) has a unifying role, being responsible for orchestration and management not only of regular VNFs but also of the MEC platform and applications.

14.3.3 Network Slicing

Network Slicing is an important technology that has been introduced in 5G to meet the diverse requirements of different industries or verticals, such as automotive, utilities, media, and entertainment.

It relays in network and computing resource virtualization to deploy in the same mobile network services with different and often divergent requirements of 5G innovative network services, such as enhanced Mobile Broadband (eMBB), Ultra-Reliable Low-Latency Communications (URLLC), massive Machine-Type Communications (mMTC), and Vehicle to Everything (V2X). These services have specific requirements in terms of latency, number of users, reliability, and bandwidth, and they need to be managed and orchestrated at an End-to-End (E2E) level on top of a shared physical infrastructure, i.e. the RAN, the CN, and the Transport Network (TN). Although network slicing in 4G networks was not standardized, it has been still provided in the limited form of isolating a service within a common infrastructure (e.g. Access Point Name [APN] Routing, Multi Operator Core Network [MOCN], and Dedicated Core Network [DECOR]). In 5G, network slicing allows allocating resources to create isolated virtual data networks and services for each of these services, defining for each slice specific QoS parameters, thus ensuring that data transmission requirements of even highly demanding services, such as time-sensitive, mission-critical ones, are met.

14.3.4 Zero Touch Automation

New technologies introduced in 5G came at the price of increased complexity. To maintain the maximum performance, 5G requires an optimal resource allocation. The introduction of advanced automation tools and mechanisms is thus required to guarantee the services' performance, reliability, and efficiency and often involves dynamic reallocation of shared resources. To this aim, new tools and approaches have been introduced to support both network and application deployment. In this context, artificial intelligence and machine learning (AI/ML) are increasingly gaining attention and are critical for implementing closed-loop solutions. These approaches are often referred to as zero-touch automation since they aim at no direct human intervention.

Operational data is needed to be fed AI/ML models training to perform its predictions and optimization tasks while supporting intelligent and autonomous network operations and service management. The quality of the model depends on the quality of the input data. This extensive use of AI/ML tools has implications for the operational data collection that should be done timely and efficiently and needs to be pre-processed. Network Data Analytics Functions (NWDAFs) have been introduced in the 5G SBA as an evolution of the RAN Congestion Awareness Function (RCAF) in 4G. NWDAFs collect and process information from SBA modules to provide network analysis information to other 5G Core network functions. The information provided, for example, the load level of a particular network slice, is used for making decisions about network operations and management actions.

14.4 Realistic Examples

This section provides an overview of some examples of deployment of the 5G technology, in order to provide practical examples on how to implement and materialize services to mobile users and service providers.

14.4.1   Heterogeneous Network Slicing for Effective Service Support

A distinct key feature of the 5G system architecture is network slicing. Such feature enables the creation and provisioning of virtual overlay networks that can be programmatically configured to satisfy the requirements of a specific class of applications, such as enhanced Mobile Broadband, Ultra Reliable Low Latency Communications, and massive Machine Type Communications [6] (see Figure 14.5).

5G network slicing is a much more powerful concept and includes the whole PLMN (Public Land Mobile Network). Network slicing allows for controlled composition of a PLMN from the specified network functions with their specifics and provided services that are required for a specific usage scenario.

Earlier system architectures enabled a single deployment of a PLMN to provide all features, capabilities and services required for all wanted usage scenarios, addressing the system requirements in a "one-fits-all" static network architecture.

5G network slicing enables the network operator to deploy multiple, independent PLMNs.

Thanks to the 5G architecture modularity, each sliced PLMN can be customized by instantiating only the features, capabilities, and services to satisfy the heterogeneous requirements coming from the diverse usage scenarios.

Figure 14.6 illustrates the concept of network slicing and an example of three network slices built on top of a single networking infrastructure [6].

14.4.2   Multi-Access Edge Cloud

Service providers' interest is currently focused on extending the service deployment within the network, and especially in the edge of the network. This is made possible by the capability of 5G to support the edge-cloud continuum vision.

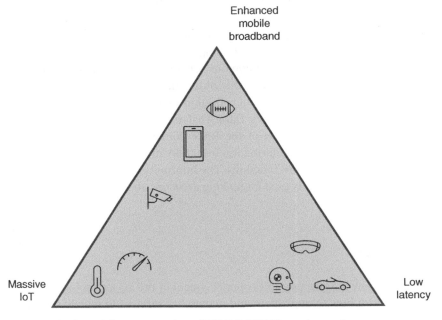

Figure 14.5   Pictorial representation of ITU-R IMT 2020 requirements.

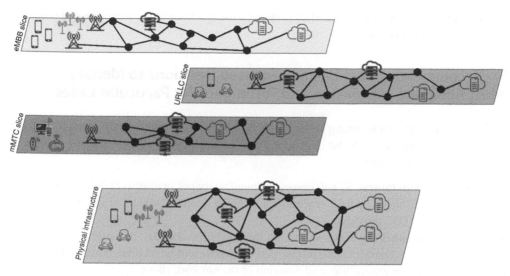

Figure 14.6 Network slicing scenario.

In particular, MEC, or mobile edge computing, is an ETSI-defined network architecture concept that enables cloud computing capabilities and an IT service environment at the edge of the cellular network and, more in general at the edge of any network. The basic idea behind MEC is that by running applications and performing related processing tasks closer to the cellular customer, network congestion is reduced, and applications perform better. MEC technology is designed to be implemented at cellular base stations or other edge nodes, and it enables flexible and rapid deployment of new applications and services for customers. Combining elements of information technology and telecommunications networking, MEC also allows cellular operators to "open" their RAN to authorized third parties, such as application developers and content providers.

Figure 14.7 provides a conceptual vision of the MEC system architecture, as it can be integrated into the 5G Service Based Architecture. The MEC platform interacts with the 5G networks by

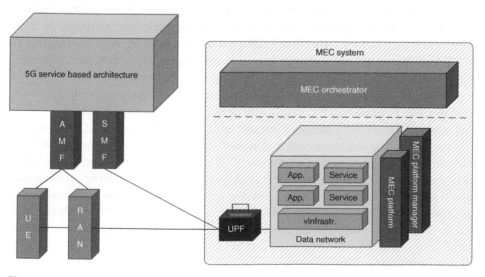

Figure 14.7 Overview of the ETSI conceptual MEC architecture and interconnection with 5G SBA.

steering the traffic from the edge to the MEC system by using a User Plane Function positioned at the edge of the 5G network (near to the RAN).

14.5 Recommended Best Use and Focused Decisions to Identify the Most Adequate Approaches and Techniques for Particular Cases

5G enables tighter cooperation among an increased number of different stakeholders with respect to the previous generations of cellular networks, and it is capable of providing added-value benefits to all of them. Those include:

- **Mobile Network Operators:** 5G will empower mobile network operators (MNOs) with higher flexibility and performance, as well as with the capability to share "slices" of the infrastructure with other MNOs or OTT providers.
- **Mobile Virtual Network Operators:** through the introduction of 5G, the interaction between mobile virtual network operators and infrastructure owners will be facilitated and enhanced by the run-time adaptation capabilities and isolation of the network slices.
- **Service Providers:** service providers will gain more degrees of freedom in deploying their services to mobile users, and maintain improved control over their virtual overlays.
- **Mobile users:** mobile users will benefits by an improved network, with superior performance, capable of connecting also personal "smart objects."
- **Industry:** companies will have easier access to ICT services and the capability to introduce mission-critical communication services within their workflows, including ultra-reliable and low-latency communications.

The following are some selected relevant scenarios where the application of 5G technology brings potentially new opportunities for business, opening novel scenarios for industry and network operators.

14.5.1 Private 5G Networks and Industry 4.0

The advent of 5G is paving the way to revitalize the concept of private mobile networks. Indeed, several innovation-oriented companies are interested in managing their own networks, in their office buildings as well as in factories. However, traditional solution of using 4G LTE (Long Term Evolution) WiFi deployments did not enable the proper levels of reliability, speed, and communication coverage needed for smart manufacturing (also known as Industry 4.0).

5G is expected to finally provide a reliable solution for private networks, due to its unprecedented performance and programmability, jointly with flexibility and spectrum-agility.

A private 5G network is pratically a company Intranet built using 5G technology. The advantage of such solution is to use dedicated bandwidth and infrastructure to reach the required Quality of Service for delivering carrier-grade services. Not only 5G is capable of transmitting speeds up to 100 times faster than 4G LTE, but it is also capable of supporting mission-critical or carrier-grade services. Indeed, private 5G networks aim to address mission-critical wireless communication requirements in public safety, industrial operations, and critical infrastructure. The newly released 3GPP Standad for 5G networks[1] supports the business-critical communication requirement of the industrial world.

---

1 https://www.3gpp.org/news-events/2200-sa2_artcle.

To reach its goal, a private 5G network needs spectrum and proper equipment [10]. Interested companies might opt to lease spectrum directly for the government, from the MNOs or third-party spectrum providers. Then, hardware equipment for 5G needs to be acquired by network infrastructure providers and deployed. This includes base stations, radio towers, and small/pico cells that create the necessary architecture to connect mobile devices (smartphones, gateways, sensors, etc.). However, due to the flexibility of the 5G architecture, different solutions might be employed, such as using unlicensed spectrum in the RAN and partially "lease" the hardware components by exploiting network slicing and virtualization.

Private 5G networks offer compelling and irresistible benefits to manufacturers because 5G-enabled technologies are the foundation of smart manufacturing and smart factories. This includes advanced technologies where private 5G networks are essential, such as collaborative mobile robots, self-driving machines, swarm intelligence, automatic guided vehicles (AGVs), augmented reality (AR) predictive maintenance, AR/VR headsets, digital twins, etc.

14.5.2 O-RAN

In 5G, virtualization has also been introduced in the RAN, with the concept of Disaggregated RAN, where the RAN is split into different components, namely the centralized unit (CU), the distributed unit (DU), and the radio unit (RU). This transformation at the RAN level calls for open and standard interfaces to allow interoperability between the RAN components and their programmability. While disaggregated RAN breaks the former monolithic architecture into modules that can be virtualized and run in separate general-purpose hardware, interoperability enables a multivendor ecosystem, where it is possible to choose and connect modules provided by different vendors. The O-RAN Alliance has introduced a unifying standardization framework for supporting this RAN transformation. In the view of the Alliance, RAN should be open, intelligent, virtualized, and fully interoperable. The O-RAN architecture reflects these choices and brings them into the 3GPPP architecture. O-RAN approach comes with various benefits. The cloud-native nature of O-RAN builds upon the ETSI NFV reference architecture and allows to take advantage of all benefits of a virtualized environment, allowing the scale in and scale out of resources. The embedded intelligence introduced by O-RAN supports network operations' automation: performance and resource usage are continuously monitored and adjusted through near real-time closed-loop control supporting efficient, optimized radio resource management, even for complex setups.

Figure 14.8 illustrates the main elements of the O-RAN architecture and the overall architecture, as well as the interfaces defined by 3GPP to the 5G System.

14.6 Beyond 5G, Toward 6G

Research and standardization are already looking beyond 5G as they define the roadmap to the future next generation of mobile networks. Beyond 5G networks are expected to capitalize on the flexibility and adaptability of 5G to support a whole new set of applications and services, including:

- Holographic-type communications
- Extended reality
- Tactile internet
- Multi-sensorial experience
- Digital twin
- Pervasive intelligence

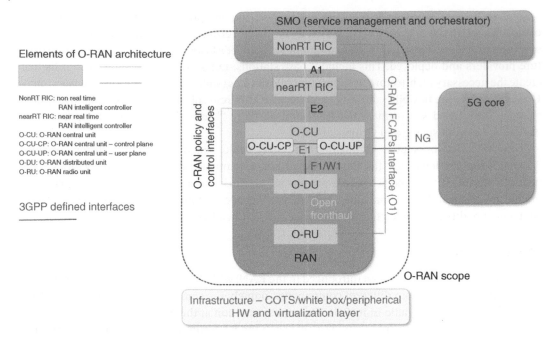

Figure 14.8 The O-RAN architecture.

- Intelligent transport and logistics
- Global ubiquitous connectivity.

Enabling technologies to support 6G include new spectrum (e.g. visible light communications, millimeter wave communications), new networking technologies (e.g. post-quantum communications and security), new air interfaces and modulations, a heterogeneous architecture (integrating satellites, high altitude platforms, unmanned aerial vehicles) and emerging novel paradigms, such as artificial intelligence, blockchain, and digital twins.

The interested reader might refer to [11] for a survey on requirements, roadmap, and technological enablers of 6G.

14.7 Summary and Conclusions

The 5G mobile network standard represents a milestone in the evolution of future mobile networks, and it will surely represent a disruptive technology. This chapter provides a brief overview of the different Releases of the 5G-related 3GPP standards (Rel. 15–17), the roadmap to the future next generation (6G), and introduced the most relevant technological components of such architecture. Related skills to successfully manage such technologies and corresponding certifications were also illustrated to help practitioners and companies understand their requirements and potential knowledge gaps.

The chapter also included some examples of the deployment of 5G technologies and solutions to deploy modern ICT services, including network slicing and MEC.

Finally, best practices and recommendations on the 5G technology as well as emerging markets and business cases were outlined.

References

1 Goldsmith, A. (2010). *Wireless Communications*. Cambridge Press.

2 Peterson, L. and Sunay, O. *5G Mobile Networks: A Systems Approach*. Morgan & Claypool Publishers. ISBN: 978-1681738888.

3 3GPP TS 23.501 (2020). *System Architecture for the 5G System*, version 15.12.0. Sophia Antipolis: 3GPP.

4 Peisa, J., Persson, P., Parkvall, S. et al. (2020). *5G Evolution: 3GPP Releases 16 & 7 Overview*. Ericsson Technology Review.

5 3GPP TS 23.501. *System Architecture for the 5G System Stage 2*. Rel.16.

6 Fitzek, F., Granelli, F., and Seeling, P. (2020). *Computing in Communication Networks – From Theory to Practice*. Springer Academic Press. ISBN: 9780128204887.

7 Taleb, T., Samdanis, K., Mada, B. et al. (2017). On multi-access edge computing: a survey of the emerging 5G network edge architecture & orchestration. *IEEE Communication Surveys and Tutorials* 19: 1657–1681.

8 ETSI (2019). *GS MEC 003 Multi-Access Edge Computing (MEC) Framework and Reference Architecture*, version 2.1.1. Sophia Antipolis: ETSI.

9 ETSI (2018). *GR MEC 017 Mobile Edge Computing (MEC) Deployment of Mobile Edge Computing in an NFV Environment*. Sophia Antipolis: ETSI.

10 McClean, T. (2021). Private 5G networks are on the rise, fueling the Industry 4.0 drive. *Forbes*. http://forbes.com.

11 Jiang, W., Han, B., Habibi, M.A., and Schotten, H.D. (2021). The road towards 6G: a comprehensive survey. *IEEE Open Journal of the Communications Society* 2: 334–366. https://doi.org/10.1109/OJCOMS.2021.3057679.

15

The Internet of Things and Its Potential for Industrial Processes

Francesco Pilati, Daniele Fontanelli, and Davide Brunelli

Department of Industrial Engineering, University of Trento, Trento, Italy

15.1 Knowledge Area Fundamentals

15.1.1 Internet of Things Definition and Its Potential for Industrial Processes

The term "Industry 4.0" is a widely used term with conflicting definitions in the literature. While it is sometimes regarded as indicating the beginning of the Fourth Industrial Revolution, the most agree upon a set of guidelines to evolve the current industry and manufacturing processes to herald the Fourth Industrial Revolution. The basis of the whole movement is centered on the heavy digitalization within factories and within society and the ever-developing interest in future-proofing technologies to optimize the control of manufacturing processes.

As described by Lasi et al. [1], due to the socio-political and socio-economic environment developed, especially in the last decade, the field progress is subject to two main development directions: an application pull and a technology push. The "application pull" is derived mainly from customer requirements that are rapidly changing, with great focus on shorter development periods, individualization on demand, flexibility, decentralization, and resource efficiency. The "technology push" comes from within the industrial field itself, with the growing need of enhancing mechanization and automation of processes through digitalization and cyber-physical systems. The realization of these development directions brought to the definition of one of the fundamental concepts of Industry 4.0, the "Smart Factory": manufacturing processes will be completely automated and self-organizing, allowing businesses to focus more resources into the fast development of new products and services.

The main technology advancement that allowed for these concepts to be born in the first place is the vast field of the Internet of Things (IoT). The term IoT, coined in 1999, refers to a network of interconnected computing devices through the Internet protocol, and communicate, collect, and exchange data autonomously, without the need for human intervention. While this is a very general definition, we will explore the significant implications and impact that such a network of devices can have in the manufacturing processes and some quality-associated fields like

IEEE Technology and Engineering Management Society Body of Knowledge (TEMSBOK), First Edition.
Edited by Gustavo Giannattasio, Elif Kongar, Marina Dabić, Celia Desmond, Michael Condry, Sudeendra Koushik, and Roberto Saracco.

metrology [2]. It is necessary to differentiate between the Consumer Level IoT (which includes uses such as smart home connectivity in household devices) and the Enterprise Level IoT (or IIoT, "Industrial Internet of Things"). The latter is a field with a sharply increasing adoption rate with new technologies such as machine learning (ML), real-time analytics, blockchain, and electronic embedded systems [3].

There are now plenty of companies offering installation services and software suites that make the transition relatively painless and seamless. The possibility of deep integration with other software suite offerings, such as data analysis software such as Power BI from Microsoft, furthers the accessibility of advanced technologies without investing the majority of the business resources in the IT department.

Bringing legacy systems online is not the only and most useful application in this field, as it is only a transitional step into full deep integration with manufacturing processes. The main reason for this is the considerable amount of resources that must go into rethinking traditional processes and all the requirements needed to adopt state-of-the-art technology. This radical change is not limited to the shopfloor or to the boosting of the IT department, but it also encompasses the management section: innovative leadership and talent management are necessary to bring the most out of the Industry 4.0 concepts [4]. The latter is especially important, as the phenomenon of IIoT (and IoT more in general) is increasing the importance of data scientists.

In the manufacturing sector, according to the aforementioned socio-economic changes, the concept of "batch size one" manufacturing lines is gaining traction [1]. A most interesting application useful to achieve that goal is the field encompassed by the definition of "Digital Twin" (DT). The DT definition is not precisely known nor fully developed, as it is still in development and has undergone numerous changes since its first introduction in 2002 by Michael Grieves. According to new developments in literature, three separate categories can actually be identified under the umbrella of the DT denomination [5]: Digital Model (DM), Digital Shadow (DS), and finally Digital Twin. The DM is a digital representation of an existing or a planned physical object. It does not possess any connection or data exchange with the physical object it represents. The DS is a digital representation of a physical object, much like the DM, but with an added one-way data flow between the physical and digital object. The DT is the further and final step, where the data flow both ways (i.e. a change in the state of the digital object reflects a change in the physical object).

The vast majority of cloud-connected sensors are embedded in machines, devices, or any object to be made into a DT, and they produce real-time data. Edge analytics [6] are then used to monitor any function the object is expected to perform. Indeed, the growing potential of the aforementioned application is strategic to improve the manufacturing processes themselves and completely revolutionize the design process.

We show that edge computing provides processing of data in places physically closer than a data center to allow for faster data transfers. It is not a replacement or competitor of the cloud but a booster.

For example, by co-designing computing and data storage into industrial equipment and machines, data can be generated in real-time. They can allow for better predictions (e.g. condition monitoring, predictive maintenance), allowing a remarkable cost reduction while maintaining better reliability and productive uptime.

Edge-Cloud continuum computing can also provide remarkable advantages when industries operate in regions where bandwidth is low or absent.

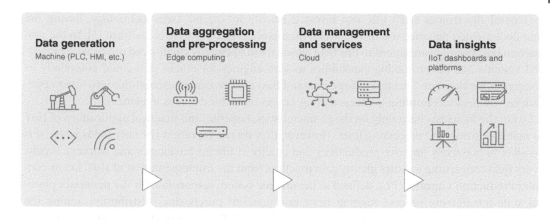

We summarize the four main benefits of edge-to-cloud computing using the IIoT architecture in the figure above:

1) Minimizing the amount of data at data generation
   The collected big-data from the industrial equipment is pre-processed at the edge. Thus, raw data from industrial devices are distilled to just the essentials.
2) Lower latency by data aggregation and processing
   One great driver for edge computing is the speed due to the short distance between data generation and the first place for massive pre-processing. After calculation and processing, the edge device sends the data to the cloud. Ensuring low bandwidth and high speed because only the relevant information is transmitted.
3) Balance processing by data management and services
   The goal of an edge device is to offload the work from the cloud and to provide backup capability in case of network failure.
4) Maintaining continuity via data insights
   IoT devices usually require a constant internet connection for continuous and reliable data logging. Edge computing can manage the massive amounts of data produced by industrial devices in local storage until the internet connection can be re-established, ensuring service continuity.

Having such a tightly knit interaction between physical and digital environments allows the prototyping phase to be much faster, more accurate, and eventually AI-based automated. Moreover, advancements in the Mixed Reality field seem to bring the industry development toward the same direction and reduce the limitations present in the interaction between the physical world and the digital one.

Practically, the main applications currently achievable lie more in the realms of DM and DS, which already allow for an extremely precise control of the manufacturing process through data analytics and the aforementioned predicting capabilities regarding resources' maintenance. There are some examples of Digital Twin application though, notably in the automotive industry by Tesla: a DT of every car sold is created, and the software updates are manufactured and catered to each car based on the sensor readouts of each individual vehicle. This means a greatly improved user experience for vehicle owners and a more efficient resource allocation in this data-driven software development aftermarket process [7].

General Electronics (GE) has also invested heavily in Digital Twin technology, having the first documented patented use of DT, which then evolved into the Predix platform [5]. In the production line, their advancement in this field allows them to employ small-sized robots with embedded systems and AI processing capabilities to scan engines for cracks that could potentially be dangerous for the user. This technology can (and has) also expanded capabilities into other applications where defect hunting is necessary (e.g. cracks on turbine blades in planes).

So far, the focus has been only on the characteristics, benefits, and practical applications of IIoT in manufacturing and processing lines. However, they do not represent the only fields where it is possible to massively improve productivity and quality of life for businesses and workers. Firstly, every field concerning logistics greatly gain practices from the implementation of IIoT. Let us consider Production Logistics (PL), defined as the organic system responsible for the material's physical state-transferring process ranging from raw material purchasing, distribution among the workshops, and the flow along with production cells [8]. IIoT enables the tracking of every asset in each step of its processing life even after it has been sold. One option is to employ recognizable patterns or codes on objects that are then kept track of when checked at every processing step as it is most common (e.g. QR codes). The other one is benefitting from visual sensing units combined with AI-trained algorithms for object recognition [9]. Effectively, deep integration of PL and other logistic systems with IIoT allows for the production scheduling activity to work cooperatively with the logistics scheduling activity. Therefore, new cyber-physical systems can enable companies and businesses to overcome inefficiencies and discrepancies in schedules related to unexpected dynamic changes [10].

In a manufacturing setting, it is of extreme importance to rightly consider the wellness and safety of the workers. There are plenty of current and emerging technologies that monitor shopfloors or offices through sensing units and IIoT integration. While many actions can be taken to reach the same result, they can be divided into two main categories: (i) preventive actions and (ii) corrective actions. "Preventive action" involves running algorithms that are able to find potentially dangerous interactions between the workers and a critical point in the production line. This can take place in the design phase as well as a renovating phase in case the production line needs to be modernized with IIoT capabilities. "Corrective actions" have a more direct approach by directly monitoring the factory plant about any kind of hazards (e.g. environmental and health). Fire detection is the most commonly found safety IIoT. While it is now a fairly mature technology that can help in detecting fires (or dangerous conditions that may lead to a fire breaking out) almost instantaneously, it is still being researched and advanced, especially with the great strides in late computer vision techniques [11]. Other solutions are also implemented and research to prevent unhealthy practices from the workers' perspective: sitting down for an extended period is proven to be detrimental to a person's health. For instance, IoT systems can be implemented to monitor the sitting time and, hence, notify to the worker to get up at regular intervals or after a certain limit has been reached [12].

The last-mentioned application for IoT brings to light one of the main current limitations of adopting this new technology, mainly in the form of privacy invasion and trust issues. For example, the use of intelligent cameras instead of radar-based sensors as discussed in [12]. While it could provide better performances, such usage requires a constant visual monitoring of the worker. More in general, many of the "Smart Factories" applications can and do represent an invasion of privacy on the shopfloor in some cases. Therefore, such adoption will have to be thoroughly thought out and developed with the support of local and international regulations regarding privacy rights. This may result in additional expenditures of resources.

Furthermore, the implementation of IIoT naturally implies that the entirety of the manufacturing line (if not the factory online) will be constantly broadcasting data. This can represent a huge security risk for a business. As cybersecurity is gaining more and more traction over the last years, the vulnerabilities of an IoT system get more and more investigated. Research is always active to find new systems and technologies to prevent the exploitation of such vulnerabilities. Firstly, potential malicious physical interaction must be considered when designing and monitoring an environment with IIoT-capable sensing units [13]. Secondly, the plethora of communication disruption techniques such as DDoS attacks, man-in-the-middle attacks, and backdoor exploitation are a definite issue in all connected systems [14]. The more we advance the spread of IoT applications, the more surface area for external attacks are exposed [15]. Many breakthrough technologies, such as the implementation of a blockchain to decentralize the processed information and sensitive data are recently proposed [16, 17].

Taking such measures to increase internet-side security, highlights another limitation that comes with (and may hinder) the adoption of IIoT. A sizable IT department with a discrete amount of experts is fundamental, as well as analyzing the effect of the energy consumption of the hardware required to sustain an IIoT system, which is set to represent 3.5% of the global energy consumption by 2027 [3].

15.1.2 The Fourth Industrial Revolution: Digital Data as the Raw Material of Twenty-First Century

The progressive increase of physical devices connected to the internet such as Radio Frequency IDentification (RFID), sensors, gateways, etc. results in vast acquisitions of data transmitted to the IoT [18]. Indeed, data is becoming an important and extremely asset in the manufacturing sector. Since data acquisitions involve vast quantities of data, from different sources and formats at a fast pace, digital data are more commonly known as big data. From early days, big data were perceived as a key factor of competition, underpinning new waves of productivity growth, innovation, and consumer surplus [19]. Though, raw data are barely useful without a proper understanding of their value and adequate algorithm to leverage such information. Subsequently, in the acquisition stage, different processing methods such as data cleaning, integration, and normalization are performed in order to improve data quality (e.g. noise removal), aggregate heterogeneous datasets, and reduce dimensionality [20]. Porter and Heppelmann [21] described this trend as "the most substantial change in the manufacturing sector since the Second Industrial Revolution." According to a survey released by McKinsey in September 2019, consistent percentages of earning before interests (EBIT) over the past three years are directly attributable to data and analytics [22]. While the role of big data and how to mine value inside them through analytics and algorithms is widely understood, several definitions of big data have been proposed over the last few years. For instance, the Gartner IT glossary [23] suggests that "Big data is high-volume, high-velocity, and high-variety information assets that demand cost-effective, innovative forms of information processing for enhanced insight and decision making." While Bekmamedova and Shanks [24], suggest that "Big data involves the data storage, management, analysis, and visualization of very large and complex datasets. It focuses on new data-management techniques that supersede traditional relational systems and are better suited to the management of large volumes of social media data."

Beyond the different understandings of big data driven by a specific scope, the concept of Volume, Variety, and Velocity, most commonly known as three Vs, which later on added more features, has emerged as a shared framework [25–27]:

- **Volume:** Sensors acquire a huge magnitude of data in the order of multiple terabytes and petabytes. Moreover, it is worth noting that two datasets of the same dimension may require different data management technologies.

- **Variety:** It suggests the physiological heterogeneity in datasets. Data acquired by sensors can be conceptually divided into structured (e.g. tabular data spreadsheets or relational databases), semi-structured (e.g. Extensible Markup Language), and unstructured (e.g. text, images, audio, and video). The real challenge is the flexibility of capturing and analyzing in order to produce valuable insights.
- **Velocity:** All the activities monitored by sensors generate data at a fast pace. This aspect is strictly related to the need to rapidly processing information and devise buffering at devices to be sure the cloud backend is able to handle the flow of data
- **Veracity:** Data collected from sensors are inherently unreliable (e.g. customer sentiment). Hence, analysis and modeling processes need to deal with unprecise and uncertain raw data.
- **Variability:** It refers to possible variations in the data flow rates and how future insights may be interpreted in various ways.
- **Value:** Despite big data are often characterized by relatively "low-value density," this feature represents the extent to which data generates, after the leveraging process, economically worth insights and or benefits.

In manufacturing, massive amount of data is generated from a set of heterogeneous sources. For instance, the manufacturing execution systems deals upon data acquired from warehousing management systems that schedule the production processes to fulfill the customer demand. Furthermore, enterprise resource planning (ERP), with respect to the inbound and outbound activities, updates the inventory stock. Considering the aforementioned IIoT concept, technology-driven data sources (e.g. RFID, real-time locating systems (RTLS)) increase the traceability of indoor and outdoor logistics flows and ensure better monitoring processes of assembly lines, job shops, etc. Moreover, a key role is played by the partial and/or the entire automation of productive processes (e.g. Flexible Manufacturing Systems) and in-plant logistics flows (e.g. Automated Guided Vehicles). Consequently, the manufacturing processes increased exponentially their efficiencies and maintained the economic concept of mass production by reducing consistently batch sizes, ideally, to one. This irreversible digital transition, characterized by an exponential data acquisition, triggered the start of the "Fourth industrial revolution" most commonly known as "Industry 4.0." This concept appeared first in an article published by the German government in November 2011 from an initiative concerning the high-tech strategy for 2020 [28]. The factory of the future as outlined by Industry 4.0 results in a pervasive integration where manufacturing entities autonomously exchange information, trigger actions and control themselves independently. Indeed, this approach toward data created the so-called "smart factories" whose processes are shaped by decentralized, automatic, and digital production networks. Besides, the described attitude to not consider manufacturing entities as a stand-alone was transposed to dynamic value-creation networks, collaborations between factory branches, and even among different industries. Therefore, data generated by every single player are shared throughout supply chains [28, 29]. For successful implementation of Industry 4.0, it is worth mentioning the most adopted enabling technologies which have progressively reduced the boundaries between the digital and physical environments [30–32]:

1) **Adaptive robotics:** Flexible robots combined with artificial intelligence techniques provide more efficient and less costly manufacturing processes. Assigned tasks are divided into simpler sub-problems but after their completion, they are bundled together in order to find the optimal solution.
2) **Cloud systems:** Cloud platforms are a viable solution to store a vast amount of data collected and, consequently, performing tools to leverage such data through fast computing times.

These systems reduce costs, eliminate infrastructure complexity, extend work area, protect data, and provide access to information at any time.

3) **Embedded systems:** They consist of supportive technologies for the entire organization to, upon interactions between the physical and digital environment, integrate, control, and coordinate in real-time processes and operations. Examples of widely established embedded systems are RTLS, RFID, connected actuators and controllers, etc.

4) **Additive Manufacturing:** It uses data computer-aided design software to produce 3D objects by depositing material such as polymers, ceramics, or metals, layer upon layer, in precise geometric shapes.

5) **Virtual reality and Augmented reality:** Virtualization technologies replicate the physical environment in order to enhance human perception, which results in better monitoring of manufacturing processes.

6) **Simulation tools:** Upon data acquired from the field, simulation tools can predict future states of manufacturing systems. For instance, testing different queueing approaches (e.g. LIFO, FIFO, etc.) to reduce the impact of bottlenecks in a given productive sequence [33].

While the outlined implications of the fourth industrial revolution have disruptively changed the manufacturing environment, according to the survey carried by Motyl et al. [34], the practitioners need to create broader and better-structured knowledge of concepts. The aim is to leverage through adequate skills such complexity increase. Furthermore, over the past two years due to the COVID-19 pandemic, the role of industry 4.0 has further gained traction. While players benefiting from digital solutions weathered better the storm, as the companies think how to restore their operations, a recent McKinsey survey found that 90% of them planned to invest in digitalization [35].

The relevant scientific literature offers different contributions embracing the adoption of such technologies. For this purpose, Faccio et al. [36] developed a unique architecture called "Human Factor Analyzer (HFA)," based on cameras to monitor the working times and motion of activities performed by human operators. Moreover, thanks to the dynamism of the proposed technology, motion capture is beneficial also to determine angles assume by upper and lower arms during the shifts [37]. Feeding the data collected into the ergonomic index (e.g. NIOSH, REBA, etc.), evaluation of potential musculoskeletal disorders follows. What is interesting is that during this digital transition different technologies can target similar scopes. For instance, Kersten and Fethke [38] leverage data acquired from the combination of the inertial measuring unit (IMU) and the RFID technology to evaluate ergonomic aspects related to the upper arms. The scalability of the RFID combined with affordable hardware costs provides a huge and diversified range of applications. Gotthardt et al. [39] digitalizing the milk run system for an assembly line collected different insights to better monitoring the process. While the working times and productivities of the different stations are tracked, replenishment orders and their shortest route from the supermarket to the out-of-stock station are autonomously issued and computed. The RFID solution finds a lot of application in optimizing in-plant logistic flows. According to Choy et al. [40], its adoption results in a more structured allocation of inbound goods and, consequently, in a consistent reduction in distance traveled and times required to retrieve stockkeeping units (SKUs). Intelligent sensors in smart factories make it possible to obtain an ever-increasing amount of data about the health of the machines. As a result, the predictive maintenance, upon degradation rates of tools, defines and, eventually, devises the maintenance schedule in order to minimize unexpected downtimes and maximize the reliability of the entire process. According to the mathematical models applied by Lu et al. [41] in a boring mill, the data-driven predictive approach not only saves cost in comparison to traditional maintenance policies, but it avoids quality losses.

15.2   Managing Approaches, Techniques, and Benefits to Using Them

15.2.1   Customized Algorithms to Value Mining from Locating Raw Data

The term "IoT Value Mining" refers to the value an IoT application may bring to the business case at hand. Indeed, despite the indisputable technological improvement and IIoT may bring to the efficiency and effectiveness of the industrial phases, it is always not predictable what are the added values, how they are reached, and to what extent to a specific business case. One possible way to answer this relevant question may be to identify the technological benefits the IIoT brings into the picture, identifying the impact of automated operations like monitoring, tracking, and teleoperations may have on the specific working area. To this end, it is useful to start with a preliminary description of the problem, identifying what is the problem to tackle (*what*), what are the parameters adopted to quantify the effectiveness (*performance metric*), what are the operations that should be automated or monitored (*which*), and what are the providers of the service and their availability (*how*). For instance, in a warehouse logistic area, we may be interested in tracking the location of goods. Hence, the *what* is the correct tracking of boxes and containers in the warehouse, the *performance metric* is the number of untracked or lost objects compared to the current situation (usually comprising human worker only). While the *which* is the automated detection, identification, and relocation with an error less than 1 m, the *how* is the set of components (network deployment, passive/active sensors attached on the boxes, detection system – e.g. an RFID tag detected by an RFID antenna and the algorithms necessary for this purpose), who is the providers and what are the technical issues to be considered (e.g. battery life, data rate, real-time communication requirements, latency, reliability, etc.) [42, 43]. To actually apply the decision framework previously described and thus build upon an effective IIoT deployment to generate revenues (computed on the *performance metric*), the first step is to build a model, possibly a digitalized model of the process to automate (what), and demonstrate in the digital world the effectiveness of the application of the desired technological solution to the problem at hand (*which*). In this phase, it is probably just sufficient to demonstrate that the foreseen solution is feasible and may improve the efficiency. This digital model can then become a Digital Twin (DT), where data goes back and forth and further improve the benefits of the application of the IIoT. Once the four steps above are clearly defined, the value for the business can be clearly identified. Of course, the immediate consequence of the application of IIoT solutions will be on the cost reductions and the revenue increase. These can be directly measured by translating the performance metric into costs. Besides, other indirect advantages are the increase in the production as well as the possibility to make additional revenues from the digitalized data coming from the field (e.g. by selling data, the DT or other insights coming from data analysis). Intangible benefits are instead related to the efficiency of the process, which may reduce the environmental footprint or the reduction of pollutant emissions.

Having clear where the value resides in IIoT applications and how to quantify it is a fundamental step toward the definition, design, and adoption of specific algorithms for managing and interpreting raw data. Indeed, there is a need to reduce the complexity of raw data to extract the meaningful information and then find an answer to the application requirements and questions, toward which one step to take is to link them with the raw data and identify the links between the data themselves. This is a crucial step that can be readily explained with a metaphor: as some researcher refers to (big) data as the new gold for the information society, some others look at them as the new oil, that is a resource of value that becomes evident and useful after being processed to obtain petrol or plastic materials. Similarly, data in an IoT application comes from different sources, e.g. sensors and actuators deployed in the environment, and with different format and uncertainties in a quantity that was never witnessed before. However, data not adequately analyzed and processed, are not of relevance for the foreseen application nor can increase the knowledge and the insights to let the business grow. The fact that this huge amount of data comes from different locations connected through secure and scalable connections, makes locating raw data a necessary precondition for a correct data analysis. Indeed, knowing when (data *timestamps*) and where (data *location*) a particular event occurs gives the organizations the ability to understand where, e.g. bottlenecks happen. Moreover, it is possible to extract trends that may help in optimizing the production processes and improve decision-making. There are several examples where data location is as important as the data per se: optimization of routes in smart cities to reduce pollution and travel time; allows to identify problem locations when a machine gets faulty or breaks in large production areas; increase the perceived customer care; verify correctness in contracts, e.g. using blockchain algorithms by understanding where a specific object is moving; track goods in, e.g. warehouses to reduce losses. Similarly, locating raw data becomes important for tracking motion of vehicles inside production and logistic areas to increase safety by, for instance, avoiding certain sequences of maneuvers that has been identified through data analysis and location tracking. Another example is given by the location of objects of particular relevance (e.g. expensive instruments that are of specific value for the production process) inside widespread production sites: finding in a timely and efficient manner the needed equipment turns out to be vital for the business revenue.

An interesting application example, where the value is related to the number of users that can be treated is the heal-care system. Imagine a hospital or a retirement house where a certain equipment is shared among the patients. Using location algorithms, the equipment can be easily located inside the area in order to increase the number of patients treated and, thus, increase both the efficiency on the use of the machinery and the quality of service perceived by the users. However, the more a certain tool is used, the higher is the probability to incur in a malfunction. In such a scenario, using IIoT allows to set-up preventive maintenance algorithms that, once combined with location information, reduce the time for the human operator to act on the system and optimize the quality of service offered to the user.

To summarize, IoT revolution in the industrial processes generates a huge amount of data that can be roughly regrouped in three categories: data related to the actual working condition of the equipment (to enable predictive maintenance algorithms); data related to the automation status (to control production processes); data related to location (to monitor and track goods, equipment, services). Locating raw data is usually carried out using a combination of positioning and localization algorithms, which may work indoor and/or outdoor depending on the specific application needs. The most common techniques and algorithms will be discussed in the next sections of this chapter.

15.2.2 Approaches and Benefits of Industrial Internet of Things Adoption

The definition of IIoT can be given in terms of a set of heterogeneous and connected devices, and machineries (through dedicated channels or via networked solution) that using different hardware/software solutions produce, share, store, and elaborate information about a certain plant or system [44]. Considering the IoT as the component that turns a city, a building, a hospital into a smart city, a smart building, a smart hospital, the IIoT solutions enable smart factories, smart production plants, smart warehouses, etc. [45]. This captivating goal is reached by the IIoT architecture, which can be roughly divided into three layers:

- **Edge layer:** This includes the field components (sensors, actuators, and edge nodes) that communicate with the field controller, which is responsible to take the correct course of actions on the field and to steer the traffic from and to the platform layer using an IoT gateway.
- **Platform layer:** This is the core of the IIoT solution and comprises the IoT platform adopted and the server or cloud services adopted to store and analyze the data, as well as the communication capabilities to communicate with the lower edge layer and the upper enterprise layer.
- **Enterprise layer:** This layer is the responsible to turn the data into meaningful information and, consequently, turn the factory into a smart factory.

In order to derive a classification of the benefits and the drawbacks that such a complex layer organization carries through the different possible approaches, we take into account that the successful application of a technology goes through the identification of the constraints it must face. These include the context in which the technology has to operate, the nature of the technology adopted and the characteristics of the organization using the specific technology [46]. To identify the benefits and the limits of IoT adoption, we move along the lines of analyzing the Big, Open, and Linked (BOLD) data characteristics of the IIoT data [47]. This analysis takes into account the fact that most often the choice of the application of IIoT is made considering the new opportunity and the positive impact it may have for companies (e.g. generation of timely, located, granular, heterogeneous, and variable data, or, briefly, big data [48]). At the same time, it disregards the effect it inevitably has on the preexistent company structure (e.g. data security [49, 50]).

For what concerns the big data, the adoption of multiple sources of information and of several communication paradigms allows to generate smart solutions such as identification and tracking [51], smart objects [52] or effective distributed control strategies [52], as well as reducing statistical data analysis errors [53]. However, major issues arise for big data once the privacy of data comes into the landscape, which may generate unbearable data leaks [54]. Similarly, security issues are becoming more and more relevant, due to the novel severity of the IoT attacks (e.g. malicious nodes, jamming or dos attacks [49], which asks for specific protocols or resort to ML algorithms [50]. For example, [55] describes one of the most popular protocol adopted for IoT communication, the Message Queue Telemetry Transport (MQTT), which builds on top of the Transport Control Protocol, and that has potential weaknesses in terms of data security that should be addressed with specific countermeasures [56]. Another, maybe less evident, the problem is related to difficulties in assessing data quality, while the time variability of data may lead to incorrect interpretations and findings [57].

The main benefits related to data openness in IoT are mainly related to the possibility to share sensor readings and the gained knowledge to the open public without human interventions. This may be relevant for companies to increase transparency and potentially improving the business revenues [58] as well as to increase social awareness about specific issues [59]. Nevertheless, this information sharing in a public domain may lead to data usage in proprietary applications that uses data in a way that was unforeseen by the data owner [60]. Moreover, the difficulties in locating the data and the need of new regulations for information sharing may impair the evolution of IoT data toward a full and useful openness.

Linking the data among IoT platforms represents another way to improve revenues and reduce labor costs. For example, by pushing self-service paradigms from users, e.g. automatic check-out from shops, may reduce the costs, be more prompt to restock, and gaining additional insights in consumer behaviors [61]. Another example of enforcing data link is to share information and connect different sources of information to avoid frauds [62]. The downside of linking data together from different IoT sources is related to data, protocol, and platforms heterogeneity that, for industrial applications, is usually a challenge [63]. This is particularly relevant for IIoT because usually these systems are integrated with different heterogeneous components just to operate properly [64].

While the BOLD aforementioned analysis offers an overview of the benefits and the pitfalls in IIoT adoption (which has also a remarkable impact on the overall company organization [65], we can better clarify the benefits in the following:

- **Cost reduction and improved revenues:** One remarkable example is related to the reduction of waste in the New Green Deal.
- **Improved working conditions for the human workforce:** The possibility of an IoT platform to monitor and control, e.g. the air quality and conditioning inside an area populated with human workers ameliorates the working condition and improves productivity.
- **Process optimization and automation:** Large production areas can be easily automated and optimized using data coming from the field; similarly, the safety of human operators can be monitored and improved; the flexibility and the reconfigurability of the production areas can be improved as well.
- **Real-time data access:** All the previous benefits are basically enabled by the possibility to gain access to the data timely and in a secure way, which may be an important plus for specific applications.

Having clearly identified the improvements that an IIoT adoption may bring in industrial applications, it is also relevant to outline where the current research results focus for an effective application of IIoT.

- **Real-time issues:** A relatively high effort is currently spent to evaluate and improve the performance of the technological solutions, especially regarding the real-time requirements that a platform may offer. For this purpose, communication latency and jitter have a primary role, since estimation and control algorithms rely on timely sensor readings and actuator actions to reach determined application goals. Latencies play a major impact when the data has to travel back and forth from the field to the cloud, especially when nodes are not geographically co-located. Typically, protocols based on MQTT may rely on about 300 ms latency, with one order of magnitude less jitter. In the same direction is the current trend of application of 5G technologies or of the Open Platform Communications Unified Architecture (OPC UA) or of the Bluetooh Low Energy (BLE) for real-time and reliable communications.
- **Security:** As previously mentioned, security issues are considered as one of the main points of potential failure for IIoT. Hence, there is a non-negligible effort in this direction to develop ML algorithms to cope with data security. Among them, Block Chain technologies are currently gaining a lot of attention.
- **Protocols:** The industrial needs toward reliable communication protocols is pushing a lot of developments in this direction, especially for monitoring applications. In this respect, Low-Power Wide-Area Networks (LPWANs) are gaining a lot of attention wherever real-time requirements are not so stringent (albeit, quite recently solution adopting RT-LoRa for real-time requirements are currently under development). Moreover, machine to machine protocols (e.g. ZeroMQ) are currently facing investigations to enable ubiquitous data access. New message scheduling algorithms to be embedded on standard protocols are currently investigated to ensure timely provision of data to multiple clients.

15.3 Realistic Examples

15.3.1 Indoor Locating Technologies: Features and Applications in Industrial Environment

The rise of IoT in recent years and the increase of interest in its application in various fields (from industrial and environmental to household applications) is certainly an effect due to the great potential that its adoption entails. However, the progressive miniaturization and cost efficiency (reducing hardware cost) of the emerging technologies have undoubtedly played a big role [1]. Other technologies that existed before the concept of IoT/IIoT have therefore gained great benefits from its widespread use. Some notable examples are Mixed Reality technologies, which since their conception in the 1970s have seen a spike in interest during the last decade in both consumer and industrial markets. Indoor locating technologies have also been on the rise [66] and can sometimes be considered parallel in their development with some Mixed Reality applications, such as Advanced Reality/Virtuality.

This section will focus on existing indoor locating technologies with a strong interest in its industrial applications to improve manufacturing processes.

Indoor localization is the process of finding and tracking the location of whichever asset in an indoor setting or environment [67]. Its place as technology inside the industry environment can be found mainly in two sectors, which are in-house logistics and warehousing, and manufacturing. In logistics and warehousing, due to the increase in automatic and unmanned warehouses that employ moving robots, the main benefit of indoor localization is tracking robots to improve their route optimization and collision avoidance. Additionally, inventory management optimization is also a significant benefit that exploits this technology. In manufacturing, indoor localization can be useful for the marked improvement in safety in factories and quality assurance and workflow optimization due to the added control of the production line.

Before describing the main technologies used for indoor localization systems, it is necessary to briefly present the most common localization techniques, which are algorithms that allow signal processing and gathering data useful to locate the tracked assets. For a more thorough explanation and a complete list of localization techniques, the reader can refer to the available literature [66, 68]. The first and most straightforward approach is the Received Signal Strength Indicator (RSSI), which measures the power strength at the receiver. If more receivers or more transmitters are used, it is possible to triangulate the position of an object.

Directly connected to the RSSI approach is Fingerprinting, which compares the measured strength of a signal with a pre-recorded measurement in a database. Therefore, this approach requires a survey (offline) phase, in which the measurements are taken and saved on said database in vector maps. During the query (online) phase that allows for the object's location, the measurement taken is compared using similarity metrics with the data from the vector map. It offers high accuracy, but due to the innate variance in tracking operation and signal strengths, especially in a manufacturing environment, this application presents considerable computational challenges that are still being researched [69].

Since triangulation and trilateration are an essential part of algorithmic procedures for localization techniques, it is important to mention and describe the most common approaches. As a concept, they take advantage of the geometrical properties of a triangle to localize a target. Angles in the case of triangulation, and distance in the case of trilateration. The main methods involve analyzing the measurement of either the distance of the object from at least three base stations or the angle measurements of the signal compared to base stations (or in some cases both).

Trilateration is a localization technique that is based on the received signal strength or time of arrival (ToA), which is the time that the signal takes to move from the base station to the tracked object, and by knowing how much time it took for the signal to reach the target, its distance from the base station is known and then compared to that of the other base stations: the intersection of the spheres (of radius equal to the distance measured by each base station) gives the location of the tracked asset. Triangulation instead uses the angle of arrival (AoA) of the signal, and the intersection of the lines drawn following the angle measured by the base stations gives the object's position.

The way signals are distributed and recognized (communication protocols) to perform the described localization techniques can be divided into two broad categories of technologies identifying the location of an asset, as they can be either active or passive [66]. Active localization requires the target to perform an action to be identified by using either a device or a tag, which will, in turn, communicate with dedicated servers with the sole purpose of location tracking. Passive localization does not require target action, but it can still take advantage of devices or tags. Therefore, it can be further divided into device-free and device-based passive localization, the former of which has seen great improvements due to the introduction of the concept Industry 4.0, Smart Factories, and IoT.

Several different broadcasting technologies are explored and used for localization, and they can effectively be differentiated by the technology used to perform the localization and by the properties exploited in order to rely on known localization techniques described in the previous paragraphs: mechanical and electrical sensors (MEMS), magnetic sensors, acoustic technologies, light (LEDs), and radio frequencies (RF) are the most commonly available options [66, 68].

MEMS-based localization systems are mostly based on Inertial Measuring Units (IMUs), which consist of a gyroscope and/or an accelerometer that can determine the position of a target from angular velocity and linear acceleration measurements, respectively. The advantage of IMUs is the high Mean Time Between Failure (MTBF) and resource efficiency. On the other hand, its invasiveness and the need for filtering methods due to drift render the technology not usable for most needs.

Localization systems that are based on magnetic technologies take advantage of the vectorial nature of magnetic fields. Indeed, base stations are scattered strategically in the indoor environment and generate a periodic magnetic field. Magnetic sensors placed on the asset to be tracked will measure the strength of the magnetic fields coming from different base stations, and they will be localized thanks to different techniques (mainly trilateration). The accuracy is relatively high, being around a few centimeters, and the sensing rate sits around 10 Hz.

Acoustic technologies for localization have different applications based on the devices used and the frequency range of the audio signal sent by the emitters. If the acoustic signal is in the audible band (<20 kHz), in general, a transmitter and a receiver are employed. The microphone of the receiver picks up the signal, and the device will then process it, which will use techniques like Time of Flight (and sometimes even accounting for the Doppler effects in signal post-processing) to perform the localization. For audio signals outside the audible frequency band (>20 kHz), the technology used takes the name of ultrasound. The main attractiveness of ultrasound is the ability to penetrate walls and the cheap cost of implementation for relatively small locations, maintaining the same high accuracy. The main limitation for acoustic technologies comes from the fact that the transmission needs to happen at low powers to not contribute to noise pollution, as human ears should not perceive it. Furthermore, the requirement for synchronization between transmitter and receiver to perform localization techniques (ToF) adds another complicated layer to the signal processing and requires more hardware (wires for electrical synchronization pulses or RF for wireless applications). The high sound pollution and environmental disturbances (i.e. humidity and temperature) in some indoor environments can also degrade accuracy or render the localization process impractical.

A recent addition to indoor localization technologies is the use of optical-based location identification. All applications of this type take advantage of infrared wavelengths since they are non-intrusive and mostly invisible to human eyes in all conditions. They can be divided into two main categories, based on the role of the tracked object: IR light detection and IR light emission. In the first case, the asset that needs to be tracked has photosensors that can tell when IR light emitted from a base station hits them while the location of the object can be found by analyzing the time it took the light to reach the sensors after a synchronization pulse (HTC Vive visor Lighthouse System [70]). In the case of IR light emission, the object itself emits light with IR LEDs in a specific pattern which detectors can recognize (e.g. Oculus Rift Constellation System [71]). In less sophisticated systems, object recognition can also happen through identification of the LEDs via unique light modulation. Optical-based localization technologies can be highly accurate and have some advantages over RF technologies (for instance light suffers from negligible reflection and it is not subjected by multipath effects). However, they are characterized by some clear limitations: line of sight between the object and tracker device is always required.

Radio Frequency (RF) based technologies are the most widely and commonly available technologies for indoor localization, and there is a great interest in researching advancement for such applications. Different communication protocols take advantage of RR. The main accepted ones are Wi-Fi, Bluetooth, Ultra-Wide Band technology, and RFID.

WLAN or Wi-Fi (IEEE 801.11 standard) is a communication protocol that operates in the frequency band between 2.4 and 5.0 GHz, and it is mainly used for networking capabilities to connect different devices to the internet. Its specialization in networking and access point capabilities make its implementation in localization technology not straightforward, despite the great worldwide coverage. The localization is mainly performed with either RSS or fingerprinting methods, and as such, it offers good accuracy (within a few meters) and a very low cost of entry, due to the availability of cheap transceivers. The only limitations come from the fact that the device must support a Wi-Fi communication system, but it does not represent too great an obstacle to overcome with the large mainstream spread of Wi-Fi.

Bluetooth (IEEE 802.15.1) is very similar in concept and application to Wi-Fi, as it operates in the unlicensed frequency band (2.4 GHz), and it can be used in a wide variety of applications without encountering issues. The large availability of Bluetooth modules also makes its implementation very cheap. Its application in localization technologies was spread by Apple's iBeacon protocol, which is specifically designed for proximity-based services, with the limitation for its application in real-time locating technologies of a slow response time imposed by Apple.

Ultra-Wideband Technology (UWB) is a short-range RF technology widely used for indoor localization purposes. While Wi-Fi and Bluetooth measure signal strength (RSS), UWB employs transit-time techniques (ToF) in conjunction with triangulation or trilateration, performed by at least three receivers. Differently from the other radio-based technologies, line of sight is necessary between the tracked devices and receivers. It sends short pulses with low-duty cycles in a wide range of frequencies (3.1–10.6 GHz), achieving relatively low power consumption and immunity to interference from other signals, given that it has a different type of signal and frequency bands compared to other available standards. However, it can generate interference in other devices. Therefore, UWB-powered tracking systems have to operate under a strict low-power consumption limit, which can make it infeasible in some cases.

RFID is a radio-based technology that uses radio waves to transmit and identify the characteristics of an object using RFID tags. It is an interesting technology for various reasons. Firstly, it is a passive system, as in it does not require the tags to be powered, as it is the locator node that provides power wirelessly to the tag, and the tag then transmits data. Secondly, the tags have a very limited size, making them easy to use and apply in most cases. The characteristic that separates RFID from other localization technologies is that it is a short-range localization system, as it has a range of less than a meter. This makes it unsuitable for most needs in industrial environments, but it is the technology of choice in many fields for the identification of goods and inventory management.

There is a particular interest in industrial applications for indoor locating technologies, as businesses that introduced them as a product into the market saw huge returns on investment (ROI) [66, 72, 73]. In the case of Infsoft [38], it is reported that their main service as a product is "indoor tracking and analysis of components in manufacturing," and the example of their work with Sterman Technische Systeme GmbH is provided on their website: by using Bluetooth Low Energy (BLE) beacons technology in conjunction with the AoA technique as well as their software suite, the client can visualize the historical and current position data with the accuracy ranging between 1 and 3 m on a dashboard at a quick glance. This gives information about production status and production times, thus documenting the entirety of the production line and monitoring performance, allowing great control and optimization of throughput times. Automation through geo-fencing (creation of virtual perimeters on a digital map) and geo-based assignment is also available as an additional service.

15.3.2 The Factories of the Future with Localized Robots, Humans, and Goods

During the last decades, the manufacturing sector has been dealing with the progressive increase of logistic flows all over the globe and the reconfiguration of production processes according to the mass personalization concept. For this purpose, import and exports volumes in developed countries, between 2005 and 2019, faced both an increase higher than 20% [74]. Moreover, while 83% of consumers expect personalized products with a shorter lead time, according to a McKinsey survey solely 15% of companies believe to be on the right track with personalization [75, 76]. To meet such challenges, several technologies appeared and, consequently, evolved their features and reliabilities along the Industry 4.0 paradigm. In this context, the progressive implementation of embedded systems with real-time functions connected through the IoT provides to modern and complex manufacturing systems high degrees of flexibility. The new technologies offer accurate digital representations of ongoing physical processes enhancing the context-awareness to assist people and machines during the execution of their respective tasks [77]. The awareness involves a solid knowledge of the employed technological equipment and the development of adequate and intelligent algorithms responsible to leverage value from raw data. Therefore, the possibility to compare historical key performing indicators (KPIs) with the ongoing state of the system dramatically increases the consistency of a decision-making process progressively shifted from human experience to data analysis. This manufacturing approach to combine reliable connected sensors with intelligent algorithms to serve specific purposes is known as smart manufacturing. While this term is widely adopted among industries, there are several other expressions for this purpose such as factories of the future, ubiquitous factories, and factories of things [78]. Despite this proliferation of terms, a comprehensive definition describes smart factories as "a manufacturing solution that provides such flexible and adaptive production processes that will solve problems arising on a production facility with dynamic and rapidly changing boundary conditions in a world of increasing complexity. This special solution could, on one hand, be related to automation, understood as a combination of software, hardware and/or mechanics, which should lead to optimization of manufacturing resulting in a reduction of unnecessary labor and waste of resources. On the other hand, it could be seen in a perspective of collaboration between different industrial and nonindustrial partners, where the smartness comes from forming a dynamic organization" [77]. Indeed, the smart factories have different design principles which serve different purposes [79]:

- **Modularity:** It expresses the capability of single components, set on a plug-and-play principle, to interact with each other easily and quickly.
- **Interoperability:** It results in the ability to share information to different entities involved in complex manufacturing processes (e.g. between customers and manufacturing enterprises).
- **Decentralization:** Data acquired from connected sensor and leveraged by tailored algorithms let each component of the system address decisions autonomously according to the overall organizational target.
- **Virtualization:** The vast acquisition of data from the field enables to simulate in a digital environment the physical processes and, eventually, compare discrepancies between these two.
- **Real-time capability:** The real-time flows of information results in dynamic systems able to weather unexpected changes in, for instance, production schedules, driven by customer requirements.
- **Human-computer interaction:** The tasks performed by workers are increasingly supported by innovative digital tools (e.g. augmented reality glasses).

In the outlined scenario, the traditional barcode systems are utterly inadequate due to their static measure and a relevant dependence on human commitment and accuracy. Consequently, RTLS due to their ability to track, with the smallest time delay, whichever industrial entity widen their applications [80]. While raw data acquired, in the spatial and temporal dimensions, by RTLS are relatively straightforward, if properly leveraged by adequate algorithms, they can serve different purposes aimed to enhance the efficiency and the degree of safety in the monitored systems. Indeed, the dynamism and scalability of RTLS constitute a solid and viable path to let manufacturing processes increase resilience toward stressful periods. For this purpose, the following presents different scientific contributions which target specific research questions. Kelepouris and McFarlane [81] propose a method to quantify the benefits of an asset location information system, based on expected time savings. In the monitored case study, the production process takes place in two interconnected buildings that use the same subset of tools. Despite shared tools have their own "parking place," it happens that they are frequently misplaced and, consequently, a significant amount of time is spent searching for them. To reduces these downtimes, each shared tool is equipped with an RTLS tag, and 15 anchors are displaced in the facilities to cover 95% of the total area. Overall, the designed and installed system provides accuracy up to 50 cm. Using the data collected it is expected to save approximately eight minutes each time a worker is looking for a specific tool. Considering that these items are used 4–5 times/day, daily downtimes are reduced by 40 minutes. According to the design principles of the smart factories, it is possible to integrate raw data acquired by RTLS with the warehouse managing system (WSM). Pilati et al. [82] integrating to the spatial database the expected time in stock of each inbound SKU develop a set of innovative and intelligent algorithms to evaluate the efficiency of in-plant flows in a real cross-docking warehousing system. While predefined ranges of the expected time in stock enable to divide the layout of the warehouse into three distinctive areas up to store specific goods, through the interactions between tags, embedded in each SKU, and the displaced anchors it is possible to determine the position in stock of inbound goods. Consequently, the consistency in placements is determined by the comparison between the real area in which an SKU is stocked and the target one driven by the mentioned clustering algorithm. At this point, the thresholds of acceptance, with respect to the different typology of misplacements, have to be set in order to display to the plant manager in real-time any underperformances of the monitored system. Moreover, once an SKU is delivered, tailored algorithms determine the distance traveled from the position in stock to the outbound dock. After having collected a relevant dataset, it is evaluated three distributions of meters traveled from a specific area to the dock. Doing so, according to the authors' contribution, whenever meters traveled from a generic SKU does not belong to the 95% of instances of its respective distribution an early warning should notify the plant manager of an unexpected increase of meters traveled. The opportunity to promptly display to the qualified personnel unexpected underperformances may result in a continuous enforcement of the internal best practices and in this case a consistent reduction in traveling and square meters occupancy costs. Besides, the distinctive characteristics of the RTLS provide a more accurate implementation of management tools that previously relied on human accuracy and commitment. The dynamic spaghetti diagram aimed to identify redundancy in the workflow and opportunities to enhance the process is an example of it [83]. Tagging forklifts results in a punctual monitoring of operations and solid evidence of interdependencies among the different tasks to be performed. The unique possibility to analyze in real-time unexpected routes makes it possible to assess, at first, their frequency and magnitude on daily routines and, consequently, define the root causes. These valuable insights combined with constantly updated KPIs such as the utilization and availability of forklifts can enhance the consistency of the decision-making process. Indeed, the management deals works in the ideal scenario to promptly address deviations from the scheduled scenarios and, eventually, to redesign partially or even entirely processes. Moreover, duties of the dedicated operators can be eased by targeted dashboards that exploit visual management principle in order to show, for instance, the nearest position of a handling system.

Regarding the production processes, orders of customers, generally divided into batches, start the production cycle with a specific due date. Though, without specific tools and approaches to monitoring the evolution of such complex ongoing processes the system overall may result inefficient and unreliable. For this purpose, the visibility of whichever production process is greatly enhanced by tagging WIP, parts, and components. Furthermore, the possibility to feed positioning data acquired by RTLS into simulation software triggers relevant comparison between the real scenario and the scheduled one [84]. Weak spots in the process may be easily identified along with the respective root causes. Upon the generated performing KPIs, the plant managers may quantitatively and analytically evaluate different productive scheduling algorithms in order to reduce downtimes such as bottlenecks and set-up times. Focusing on another viewpoint, indoor tracking technologies have a relevant impact on blue collars working conditions. In line with the requirements of the monitored case study, different approaches can increase the level of safety which has huge externalities on degrees of productivities. Tsang et al. [85] develop a unique hardware and software architecture to monitor the actual health status of workers in a cold storage facility. While equipping workers with an RTLS tag provides an accurate and reliable indoor positioning, it is possible to associate to each tag ID a set of relevant and strategic parameters updated on a regular basis. These refer to physical conditions such as gender, weight, age, and heart rate. Leveraging the mentioned dataset, the architecture aims to increase the occupational health and the safety of workers. According to the tailored algorithms, the plant managers determine the customized duration of cold exposure to provide the appropriate clothing insulation and rests and to prevent that the workers stay too long in the cold storage. While the early warnings notify each worker of any safety concerns, the real-time position of the blue collar is shared among all colleagues in order to ease and speed up potential rescue plans.

Finally, regarding the crescent and vast automation brought by the Industry 4.0 paradigm, it is worth mentioning an approach to monitor and thus manage the dynamic routes of automated guided vehicles (AGV) in materials distribution. Jiang et al. [86] propose a framework, based on RTLS, to guide dispatchments routes in a discrete manufacturing environment divided in different tasks. Since data acquired by RTLS provide complete information about the current position of items, each traveling activities is controlled and integrated by real-time coordinates. Moreover, different sensors such as infrared distance and collision sensors are responsible for detecting and avoiding any kind of objects in the selected short paths. This unique configuration which combines the benefits of manual forklifts and AGV shows high degrees of flexibility and maneuverability. Indeed, the absence of a fixed path may result in reduced traveling times and labor costs. Tackling this scientific field from another viewpoint, it is possible to smooth in complex and modern manufacturing systems the mutual collaboration between plant workers and AGV. Mathematical forecasting methods in the short term may temporarily stop potentially dangerous AGV routes with respect to future positions assumed by tagged workers. The overall safety may be further increased through geofencing practice. For this purpose, as long as a tagged entity is detected inside a critical area, defined in the digital environment, the autonomous handling system interrupts the traveled route [87].

15.4 Recommended Best Use and Focused Decisions to Identify the most Adequate Approaches and Techniques for Particular Cases

15.4.1 The Emerging Technologies of Indoor Locating Systems to Create Value for Industrial Processes

New communication technologies and communication protocols are being more and more researched in current times, as not only the businesses have an interest in indoor locating systems and IoT but also individual customers and users who have a heightened interest in smart-home applications.

With the current and great advancements seen in ML algorithms and their successful deep integration and implementation in a vast number of fields, it seems trivial that they would provide great usefulness to applications in indoor locating technologies. ML algorithms have been introduced in research to overcome some of the challenges and limitations of existing indoor locating systems, such as accuracy [88], reliability, scalability, and adaptability [89], as discussed in the previous section.

The basic advantage of ML integration is the capability of training from sets of data that are gathered from any environment and then fed into the neural network (NN). Therefore, ML can be introduced in almost any different scenario, making its implementation trivial once designed and properly trained. Fingerprinting is one of the common technologies in which ML integration could show great benefits. Since the accuracy of a system that employs fingerprinting techniques is given by an extensive and time-consuming offline data-gathering phase and recording of a cloud of points onto a digital map, this phase could be instead substituted with a data-gathering phase for a recurrent NN. This reduces the required storage space and computational power needed to perform fingerprinting [89], with the possibility also to achieve a greater accuracy.

Using different and more modern types of NNs (such as policy-based NNs like Proximal Policy Optimization NN) enables the option for dynamically scaling positioning systems previously incapable of doing so.

Another useful application of ML algorithms for indoor localization that has shown some success is minimizing the inefficacy of locating technologies in case of non-line-of-sight (NLOS) applications [90]. This is achieved by eliminating the need to manually extract and analyze necessary features of the data (mainly delays) in order to identify a NLOS case. The ability to perform sensor fusion without employing filtering algorithms that require significant computational power is a great benefit for energy efficiency in which ML algorithms excel.

The implementation of ML algorithms does not come without some disadvantages, mainly in the form of cost of resources and time. In order to make them functional and accurate up to a state-of-the-art level, there is a great need for large amounts of training data. The lack of standardization for training data makes it hard to gather useful and usable data. In addition, determining the correct amount of data to feed to a NN for it to be properly trained is no easy task.

ML models are very application-specific, and they perform at their desired functionality only in the presence of predictable and repeatable events that recurrently happen in any given application. This requirement can represent a harsh limitation in some industrial environments characterized by a certain degree of unpredictability due to necessary human interaction.

Given the great need of data for ML applications, it would make sense for Indoor Positioning Systems (IPS) to work alongside IoT, whose main function is to generate large amounts of data (Big Data). While IPS have been used before the mainstream adoption of IoT, they have not yet been fully integrated. There are different reasons for the slow integration of these technologies. The broader issue is represented by the fact that IoT is not defined by standards and is characterized by a wide heterogeneity and diversity of IoT hardware and software [3]. That represents a harsh limitation for the integration of both technologies, for the reason that, as we have seen, locating technologies rely on one technology per implementation. Though, the choice of signal processing and network systems may change on the basis of each different application [91]. Designing and developing an indoor locating system that works simultaneously with different types of wireless protocols is difficult and represents a significant challenge [67].

One of the main proposed solutions is to intervene on the software level to provide localization capabilities to existing IIoT sensing units. However, implementing IIoT systems on a production plant means more often than not that it was a product coming from a third-party business providing hardware installation and software suites, as discussed in the provided examples of actual industrial applications. Therefore, accessing the software or modifying hardware may not be possible in most cases, and as such, it does not represent a feasible solution.

Another issue that may come from the merging of IoT and IPS is interference management. Information sent for localization can disrupt and collide with other communications that may happen simultaneously in a workshop or in an industry plant, which are generally noisy environments. Solutions for this problem are more accessible and are already a well-developed field of study [92], since its interest lies not only in IIoT but has a wider field of applications, from smart homes to healthcare.

While IPS and IIoT are still away from being considered and implemented in one system, research is moving toward solutions that may bring advantages to both technologies. New communication protocols are being considered, such as SigFox, LoRa, IEEE 802.11ah, and ZigBee. The mainstream adoption of these protocols can birth network systems fully designed with IoT and localization systems in mind, thus advancing their integration.

One of the latest advancements, which is currently seeing wide adoption, is the ZigBee protocol for wireless sensor networks [93], as mentioned in the previous paragraph. It is based on the IEEE 802.15.4 standard, and it is suited for low data rate, compared to Bluetooth, whose main use is for high data rate, and Wi-Fi whose first and foremost function is to connect devices to the internet. For this reason, the power consumption of ZigBee driven devices can be much lower with regards to devices that use other protocols, leading to longer battery life or lower energy harvesting requirements. Additionally, it creates a dynamically changing network mesh with at least one device acting as the coordinator (the bridge between sensing units and the network), and others acting either as routers (helping expand the range of the coordinator) or as end devices (sensing units). As a communication technology, ZigBee has also proven to be very effective for indoor localization applications [94] using fingerprinting and RSSI, with an overall accuracy based on the accuracy of the fingerprint database.

Another current trend that is noticing a rise in relevancy in recent years and that is considered one of the main paths toward integration is that of hybrid positioning systems, that employ a collection of different algorithms instead of focusing on one technology [95–99]. The most promising hybrid approach being researched extensively in late years is Collaborative Indoor Positioning Systems (CIPS) [100]. They differ from non-collaborative IPS in the techniques and technologies used to estimate the position, as endpoints of the network are not referring to a central locating unit or node but collaborate and participate in the positioning algorithm [101]. Therefore, the main characteristic of such systems is the interaction and exchange of information between devices with the intention to derive the relative location between each other, to then be localized on a digital map that is freely consultable by users and supervisors.

CIPS take advantage of the already existing localization techniques and technologies and use them to localize other nodes with respect to their own position, and they additionally take advantage of existing technologies used for communication and data exchange in IoT and IIoT to communicate with the other nodes. This means that the CIPS network is decentralized. In normal IPS applications, as we have seen, the sensors or receivers/transmitters on tracked assets only signal back their raw data to a central unit/node or a remote server, which is then responsible for the processing and identification of the location. In collaborative systems, there is no central unit, instead, each device is responsible for the sharing of the data and the determination of its own position relative to the other devices, obtained through data fusion (Extended Kalman Filtering) and estimation algorithms mostly based on non-Bayesian methods (Least Square and Maximum Likelihood estimation approaches) [100].

The decentralization of such systems presents great benefits. The first noticeable advantage is the disappearance of the need to have complex and especially expensive infrastructure that can support IPS implementation, which can sometimes require a re-organization of an industrial plant due to line of sight requirements, as we have seen in the previous section. Another advantage is the dynamical scalability (which is one of the main limitations of traditional IPS systems) of a CIPS system. There is no current limitation on the number of devices that can join the network, given that each device is responsible for self-location. Based on the employed communication technologies, the adoption of collaborative systems may also be able not to require line of sight, thus working in NLOS cases [102], which are sometimes unavoidable. The coverage area may also be greatly expanded, as they do not rely on pre-placed detection infrastructure with a limited range. The precision and accuracy can also see a significant improvement because each device considers data coming from close-by nodes as well. From a security and robustness viewpoint, CIPSs also present the advantage of not relying on the stability of a central computing unit: if the latter goes offline for reasons that can be common in an industrial environment, the entirety of the IPS goes offline with it. Consequently, no device can be located. This does not happen in CIPS, as one device going offline means that only that asset cannot be tracked, while leaving the rest of the network functional.

CIPS also come with relevant disadvantages. Firstly, the heavy computational process requirement and high computing time are the main current trade-off of collaborative systems. This means that each node will have to have sufficient computational power in what still needs to be a relatively small device. Moreover, due to the heavy processing and high-rate data transmission, power consumption is also significantly higher than their non-collaborative counterpart. In [100], the authors identify in the available CIPS literature a trade-off related to balancing position accuracy, fast response times, and computational complexity to optimize energy efficiency.

While CIPS are still only explored and researched in simulation environments, the growing interest in such technology and the growing need for integration between IoT nodes and IPS nodes make them one of the most interesting and exciting ways forward in both industrial and consumer environments, alongside ML applications.

15.4.2 Learning and Adaptive Factories as Foundations of the Fifth Industrial Revolution

Nowadays, modern companies have to deal with different external threats in order to keep being profitable in the long term. For this purpose, the globalized markets are continuously pushing to a higher standard of processes' productivities and products' quality while reducing labor costs. The rising demand for individual products in combination with technical innovations leads to a consistent shortening of the product life cycle, known as the timeframe between the introduction of an item in the market to its degeneration. At the same time, the decline in birth rate combined with a longer life expectancy is changing the structure of most European nations' populations and, thus, is slowly leading to different needs of the customers. Finally, buyers must be able to weather limited access to raw materials and their floating prices [103]. To meet such challenges, flexibility, and changeability result in essential drivers. These abilities that should be shared among all hierarchy levels are provided by innovative learning approaches.

In 1994, a consortium led by Penn State University developed the concept of a "learning factory." This innovative model emphasizes the key role of the learning-by-doing process, which leads to greater retentions and application than traditional approaches, in real and functioning manufacturing environments equipped by the mentioned sensors connected to the IoT. While during the early stages this approach was adopted solely in academia to deepen the hard skills of future practitioners and explore new scientific research areas, it followed applications in the field to assist the daily routines of every stakeholder involved and thus, at the management level enhance the decision-making process. In the learning factories, the integration between the physical environment and its digital counterpart allows to notify, in real-time, feedback on the monitored process. To this extent, data leveraged in the digital environment along with continuous monitoring of the operations allow emitting early warning to notify unexpected deviation of performance and the respective consistency of the selected corrective actions. Consequently, benefitting from the real-time functionality and the prompt evaluation of corrective actions, factories work in the best condition to achieve high levels of productivity and adapt to volatile market demand. In addition to the heterogeneous set of practitioners that may target, the learning process can take place in the planning and the ramp-up phase but also in the improvement of existing processes with unchanged efficacy [104, 105].

In literature, it emerges different applications aimed to exploit the learning outcomes of practitioners. The target is not solely reduced to improving the performance or being able to promptly react to unexpected scenarios, but there is also great attention on workers' health. According to the previous section, the mentioned dynamism and scalability of the RTLS provide a wide range of applications even with learning outcomes. For this purpose, RTLS can localize products, resources, and blue collars during the working shift providing useful insights to better understand and thus manage such complexity and interdependencies between different processes. According to lean principles, value stream mapping (VSM) is a key tool to identify materials flow and products' lead-time along with potential wastes and inefficiencies. Though, increasing the number of families of products the VSM may result inaccurate and waste itself. RTLS allows reducing the human effort in collecting data from the field and simultaneously identify the weak spots of in-house operations. Potential inefficiencies may be quickly identified and addressed through predefined corrective measures. For instance, a buffer can be dynamically dimensioned according to the flow rate of the families of products requested by customers. Moreover, the production schedule can vary upon the group of the manufactured products in order to minimize the downtimes (e.g. set-up). The real strength of a learning factory is driven by its real-time, or close to, dynamism. The learning algorithms define targeted KPIs and related ranges of acceptance. Whenever the selected KPIs record an unexpected value, the developed system notify through early warning inefficiencies of the tagged entities. As a result, plant managers may promptly implement a corrective action to keep performances at the desired level. The possibility of quantitative comparing the two scenarios may let the stakeholders involved in the process learn about, at first, the root causes and then the positive and negative externalities of the implemented corrective actions [106]. Pilati et al. [82] provide an example of this innovative approach in a real cross-docking warehouse. After having defined three consistent distributions of distances traveled by tagged SKU from a given area to the outbound docks, it is possible to evaluate the efficiency in the outbound routes. Authors suggest recording as "inefficient handling" all distances traveled which do not belong to the 95% of instances related to the respective mean value. Consequently, an early warning has to notify the qualified personnel of this underperformance. Plant managers in addition to assess the impact on costs of inefficiencies may redesign the entire layout whether necessary. Thanks to this approach, the warehousing system is able to learn upon its inefficiencies over time.

Despite their wide range of applications, RTLS does not represent the only enabling technologies for the learning factories. Umeda et al. [107] developed in an automated factory the digital triplet which consists of a physical world, cyber word, and the intelligent activity word where engineers execute their activities through cyber-physical systems. While the physical system is an automated line of Lego cars, the cyber word is represented by a digital twin, which replicates with the highest degree of detail the physical word. Finally, the digital activity world is connected to the digital and physical environments to supports the decision-making process. Authors induce different inefficiencies such as stoppages of the line at different stages, reductions of tact time, and errors in inspections. After receiving unexpected feedback, the related learning opportunities are to identify the root causes and react as quickly as possible to minimize the impacts on the overall equipment effectiveness (OEE). Despite this approach is tested in academia, it can be applied in real factories as well. The idea would be to enhance the consistency and the response time of the decision-making of plant managers. Besides, the same approach could be shifted to tackle inefficiencies directly attributable to human accuracy and commitment. The unique possibility to track errors and the respective frequencies may generate tailored or group training programs.

Finally, the ergonomic viewpoint represents a major area of interest in the learning factory context. For this purpose, plant workers are equipped with sensors and actuators to collect real-time data about angles assumed by upper and lower limbs during the working shifts. In addition, to achieve an accurate motion capture, camera settings, and room calibration take place before the assembly process takes place. Information acquired by the hardware architecture is leveraged by AI algorithms to detect dangerous postures in the workflow. At this point, processed information feeds several ergonomic indices. While the Rapid Upper Limb Assessment (RULA) criteria is adopted to display the load on limbs, the Key Indicator Method and the Ergonomic Assessment WorkSheet (EAWS) criteria describe schematically the common incorrect postures assumed along with the corresponding risks. The proposed framework can identify which worker is operating and in a given working station. By doing so, it is possible to emit early warnings whether a specific accumulation of dangerous postures is detected within a timeframe. Moreover, the detected mal-position is also related to external circumstances such as work speed and daytime. Upon these dangerous movements, specific workshops and training are generated for small groups or, if necessary, even personalized. Plant managers would be able to assess whichever improvement in the working conditions of blue collars and thus assess the efficacy of the implemented training sessions. Along with this enhancement, the management may also achieve an increase of the productivity of the monitored system [108].

15.4.3 Upcoming Algorithms with Superior Performance for IoT Data Management in Industry

Data is the major driver of innovations of the application of IoT solutions for industries. Due to the heterogeneity of the representation of data types, which is a consequence of the diversity of the technological solutions adopted, a taxonomy into different categories is inevitable to define which management algorithm is actually more suitable for their analysis.

Probably, the most straightforward type of data is related to the sensors. In this case, besides the peculiar characteristics of the system to be sensed, sensory data need to be specified in terms of sampling frequency (e.g. event-based, time-triggered, or continuous) and number of repeated measurements (e.g. to achieve a good "quality" of the phenomenon of interest, reducing the uncertainty with built-in estimators). While in a centralized application (in which the sensors, the actuators, and the control and/or monitoring unit are co-located in the same area), high frequency and high number of measurements are usually foreseen, this is not trivial for IoT applications where a shared (wired or wireless) communication means is adopted. Indeed, more data means more traffic on the shared network and, hence, complicated message scheduling schemes and higher probability of packet losses. In such a case, the quality of service offered by the communication system may become so low that all the performance increase due to IoT may be simply lost.

RFID data are generated by active or passive tags, with different levels of miniaturization, that can be attached to many different objects. Once the tag detects a wireless connection with the reader antenna, it sends back the data stored in it to the reader (i.e. the ID). While an active RFID can work autonomously (but needs a battery), passive RFID are only activated by the reader antenna. The use of RFID is nowadays pervasive in many applications, ranging from healthcare to logistics or supply chain monitoring.

As standards for industrial communication systems, such as FieldBuses and their ability to enable interchangeability and interoperability [109], data may be just descriptive of the specific node, specifying the component characteristics (e.g. for a sensor it usually report the feasible range, temperature of operations, resolution, etc.) and self-monitoring data (e.g. current battery level, possible failures, etc.). Besides the single components, this data (and metadata) may be related to the entire system or on a specific process.

Since the IoT components may be static or moving inside an environment, location data is another type of IoT information. Usually, GPS data are adopted, which offers a lot of information about the global location of a component. Similarly, there are a lot of efforts in using cellular base station, Wi-Fi access points and TV towers to solve the problem of global positioning when the GPS signal is not available (e.g. like in natural or urban canyons). Due to the peculiarity of the IIoT, local positioning systems are usually adopted, which are based on various RF techniques and messages from environmental sensors, such as RFID, RSSI, ToA, TDoA, etc.

In the previous taxonomy, the main difference is dictated by the characteristics of the data that is shared among the components. Nonetheless, since the data has also a time feature, after a while (that can be milliseconds or days depending on the application) the data may become obsolete for control or monitoring purposes. However, it is still of worth because it constitutes the knowledge base from which new insights can be gained (e.g. for predictive maintenance or to identify possible points of failure in advance). This idea is very much the same as for SCADA systems [110], where historical data are retained for future analysis (indeed, SCADA systems have some major commonalities with IIoT applications [111]). Borrowing the same terminology, we can refer to this type as historical data. Notice, however, that despite the intrinsic value of storing data, the storage is usually limited, and queries become unpractical for huge amounts. Therefore, one key question is when, how, and which part of the data should be deleted periodically.

The management system for IoT applications works as an interconnecting layer between the field devices (sensor and actuators) and the application [112]. Data management is usually carried out in standard computer science applications using relational databases. However, this common approach hardly works for IoT applications, because of the heterogeneity of the sources and the nature of the data, as aforementioned. Moreover, databases are very efficient in managing data and updating records with a relatively small number of queries, which is not the case for IoT applications, where data arrives continuously together with a lot of queries that should have real-time requirements and shows quite different purposes. Finally, relational databases use data normalization and strict storing schemes, while for IoT it is better to renounce to strict storing paradigms in favor of flexibility to adapt to different types of data and queries.

As such, data management systems comprise different components (data storage, data indexing, and data middleware), which needs to be defined. Data storage solutions conceived for IoT renounce to centralized storage units, but uses distributed components to reduce the query time and increase the updating frequency [113]. Moreover, distribution of storage units comes handy to manage heterogeneity: depending on the data type, the storage is performed using different approaches [114].

The data indexing is extremely relevant for simple and fast data access, which has to be necessarily distributed according to the nature of an IoT system [115]. Distributed indexing can be implemented using, for instance, Geographic Hash Tables [116], which associates to a certain event value (e.g. power surcharge) a location data in order to implement a semantic in the hash table (that is, events that may happen geographically far apart are semantically connected). Dynamic indexing is instead implemented using approaches similar to cache memory management in standard computers, enforcing locality or time-based paradigms [117], or by using distributed gateways organizing IoT resources in clusters [118].

Finally, data management systems benefit by the application of middleware components, whose main purpose is to handle data heterogeneity [119]. Indeed, the middleware masks the difference between the set of communication protocols adopted and simplifies the deployment of the applications on different platforms. Of course, there is always a trade-off between the capacity to deal with different platforms and the specific applications that should be executed on it [120]. The main purpose of the middleware for data management, which sits between the field devices and the data storage, is to facilitate the access to the stored data as well as to empower the analysis of heterogeneous data and their aggregation. For example, [121] uses of a middleware built on the field devices data and uses of federated databases to guarantee independence between the different IoT systems.

Due to the complexity of data management systems for IoT applications, research in the field is very active and novel solutions has been developed recently. One of the topics of active research is to integrate blockchain architectures and their inherent distributed features to efficiently manage data in a secure way [122, 123], sometimes in conjunction with specific data exchangers [124]; with the same aim, AI-oriented solutions seek to use, e.g. ML approaches, like federated learning, in combination with deep reinforcement learning for field devices selections to increase data management security and efficiency in IIoT [125]. Another relevant aspect is related to the real-time features of certain IoT industrial applications. For instance, [126] offers a resource allocation mechanism oriented to the data management for Multi-Access Edge Computing, where data management and analysis are carried out closer to the field and in a distributed fashion. Other kind of recent results focuses on specific application fields, such as smart cities, where privacy and data security can be enforced with the use of AI tools [127] or conceived for IIoT in a manufacturing context [128], while recent solutions build upon fog computing paradigm [129].

15.5 Summary and Conclusions to Takeaway

This chapter provides an extensive focus on the Industry 4.0 paradigm in the manufacturing environment. While the first sections deal with IIoT and its challenges and applications, the followings provide a punctual explanation about indoor technologies features and potential use cases. For this purpose, the aim of this chapter is to outline from different viewpoints both the advantages and disadvantages in digitalizing whichever factory having as a reference the indoor locating technologies.

The first section presents the fundamentals of such a complex and vast area. To this extent, it is introduced the general concept of IoT along with the potentialities of acquiring such vast quantities of data. From data acquired by different kind of sensors, the DT technology represents a powerful tool to simulate in virtual environments the expected scenario. Relevant comparisons between the real and the simulated scenario provide a valid path to achieve desired in-plant performances. Subsequently, it is introduced the concept of big data and the strategic role they play in manufacturing. From the competitive advantage they may offer and from which management tools are constantly generated. This section ends with a set of enabling technologies of the fourth industrial revolution along with different case studies.

The second section describes the relevant approaches and techniques to leverage value from this vast amount of unstructured data. At first, an effective framework guides practitioners to successfully implement an indoor locating system. Acknowledged the problem and the different parameters to be monitored, the final target is to create value for the entire factory. According to the aforementioned features of the digital twin technology, its capabilities play a strategic role in this process. Furthermore, it is provided a technical description of the IIoT's three different layers and their respective functions. Having as a reference the BOLD data characteristic and the MQTT protocol, benefits and weaknesses are extensively described. To pragmatically conclude the section, a set of features describes where practitioners should focus to achieve successful and effective IIoT applications.

The third section provides realistic examples of signal processing for indoor localization and different industrial case studies. To this extent, common localization techniques such as RSSI, trilateration, and triangulation are properly presented. In addition, since the signal distribution affects the positioning techniques, both the active and passive localization are addressed. The last aspect analyzed is the different localization technologies along with their most relevant features. Subsequently, it introduced the concept of the factory of the future or smart factory underlining the most important design principles. The section ends with heterogeneous applications of indoor locating technologies in manufacturing. For instance, they mention the possibility to optimize in-plant logistic flows, manage the dynamic routes of AGV, and enhance the visibility of bottlenecks in production. Finally, in the last section, several recommended best-use cases are described. ML algorithms whether properly trained may improve fingerprinting algorithms and minimize inefficiencies in NLOS communications. Moreover, it is outlined the challenges and novel protocols considered as a feasible solution to integrate IPS with IIOT. Regarding the hardware part, another trend is represented by the CIPS, which uses different algorithms instead of one technology to localize dynamic entities. From a manufacturing perspective, the cutting-edge concept of the learning factory is introduced. Benefitting from data collected by the sensor, the management can focus on specific drivers of performance and thus emit early warnings to notify to whichever stakeholder of any unexpected performance. By doing so, it is possible to constantly keep efficiencies at target values. This work ends with the description of different approaches to optimize the data management.

References

1 Lasi, H., Fettke, P., Kemper, H.-G. et al. (2014). *Industrie 4.0*. Wirt-Schaftsinformatik.

2 Eichstädt, S. (2020). From dynamic measurement uncertainty to the Internet of Things and Industry 4.0. *2020 IEEE International Workshop on Metrology for Industry 4.0 & IoT*. Roma, Italy: IEEE, pp. 632–635. http://doi.org/10.1109/MetroInd4.0IoT48571.2020.9138250.

3 IEEE (2020). *Heterogeneous Integration Roadmap*. IEEE. Chapter 3. https://eps.ieee.org/technology/heterogeneous-integration-roadmap/2020-edition.html.

4 Shamim, S., Cang, S., Yu, H. et al. (2019). How firms in emerging economies can learn industry 4.0 by extracting knowledge from their foreign partners? A view point from strategic management perspective. *2019 International Conference on Advanced Mechatronic Systems (ICAMechS)*, pp. 390–395.

5 Fuller, A., Fan, Z., Day, C., and Barlow, C. (2020). Digital twin: enabling technologies, challenges and open research. *IEEE Access* 8: 108952–108971.

6 SAS (2014). Analytics at the edge. https://www.sas.com/en_us/insights/articles/big-data/internet-of-things-examples.html.

7 SAS *Modern Manufacturing's Triple Play: Digital Twins, Analytics and IoT*. SAS.

8 Qu, T., Lei, S.P., Chen, Y.D. et al. (2014). Internet-of-things-enabled smart production logistics execution system based on cloud manufacturing. In: *International Manufacturing Science and Engineering Conference*, vol. 45806. Detroit, Michigan, USA: American Society of Mechanical Engineers.

9 Qu, T., Lei, S.P., Wang, Z.Z. et al. (2016). IoT-based real-time production logistics synchronization system under smart cloud manufacturing. *International Journal of Advanced Manufacturing Technology* 84: 147–164.

10 Zhang, Y., Guo, Z., Lv, J., and Liu, Y. (2018). A framework for smart production-logistics systems based on CPS and industrial IoT. *IEEE Transactions on Industrial Informatics* 14 (9): 4019–4032.

11 Dua, M., Kumar, M., Singh Charan, G., and Sagar Ravi, P. (2020). An improved approach for fire detection using deep learning models. *2020 International Conference on Industry 4.0 Technology (I4Tech)*. Pune, India: IEEE, pp. 171–175. http://doi.org/10.1109/I4Tech48345.2020.9102697.

12 Cardillo, E. and Caddemi, A. (2019). Feasibility study to preserve the health of an industry 4.0 worker: a radar system for monitoring the sitting-time. *2019 II Workshop on Metrology for Industry*

4.0 and IoT (MetroInd4.0&IoT). Naples, Italy: IEEE, pp. 254–258. http://doi.org/10.1109/METROI4.2019.8792905.

13 Ding, W and Hu, H. (2018). On the safety of IoT device physical interaction control. *Proceedings of the 2018 ACM SIGSAC Conference on Computer and Communications Security (CCS '18)*. New York, NY, USA: Association for Computing Machinery, pp. 832–846.

14 Zhang, Z-K, Cho, M.C.Y., Wang, C-W. et al. (2014). IoT security: ongoing challenges and research opportunities. *2014 IEEE 7th International Conference on Service-Oriented Computing and Applications*. Matsue, Japan: IEEE, pp. 230–234. http://doi.org/10.1109/SOCA.2014.58.

15 Krishna, B.V.S. and Gnanasekaran, T. (2017). A systematic study of security issues in Internet-of-Things (IoT). *2017 International Conference on I-SMAC (IoT in Social, Mobile, Analytics and Cloud) (I-SMAC)*. Palladam, India: IEEE, pp. 107–111. http://doi.org/10.1109/I-SMAC.2017.8058318.

16 Khan, M.A. and Salah, K. (2018). IoT security: review, blockchain solutions, and open challenges. *Future Generation Computer Systems* 82: 395–411.

17 Reyna, A., Martín, C., Chen, J. et al. (2018). On blockchain and its integration with IoT. *Challenges and Opportunities, Future Generation Computer Systems* 88: 173–190.

18 Qi, Q. and Tao, F. (2018). Digital twin and big data toward smart manufacturing and industry 4.0 : 360° comparison. *IEEE Access* 6: 3585–3593.

19 McKinsey Global Institute (2011). *Big Data: The Next Frontier for Innovation, Competition, and Productivity*. McKinsey Global Institute.

20 Wang, L. (2017). Heterogeneous data and big data analytics. *Automatic Control and Information Sciences* 3 (1): 8–15.

21 Porter, M.E. and Heppelmann, J.E. (2015). How smart, connected products are transforming companies. *Harvard Business Review* 93 (10): 96–114.

22 McKinsey & Company (2019). *Catch them if you Can: How Leaders in Data and Analytics Have Pulled Ahead*. McKinsey & Company.

23 Gartner IT Glossary (2022). Information technology gartner glossary. http://www.gartner.com/it-glossary/big-data

24 Bekmamedova, N. and Shanks, G. (2014). Social media analytics and business value: a theoretical framework and case study. *2014 47th Hawaii International Conference on System Sciences*, pp. 3728–3737. IEEE.

25 Gandomi, A. and Haider, M. (2015). Beyond the hype: big data concepts, methods, and analytics. *International Journal of Information Management* 35 (2): 137–144.

26 Ameur, Y.A., Bellatreche, L., and Papadopoulos, G.A. (eds.) (2014). Model and data engineering. *4th International Conference, MEDI 2014, Larnaca, Cyprus. Proceedings* Vol. 8748 (24–26 September 2014). MEDI, Larnaca, Cyprus: Springer.

27 Mikalef, P., Pappas, I.O., Krogstie, J., and Giannakos, M. (2018). Big data analytics capabilities: a systematic literature review and research agenda. *Information Systems and e-Business Management* 16 (3): 547–578.

28 Zhou, K., Liu, T., and Zhou, L. (2015). Industry 4.0: Toward future industrial opportunities and challenges. *2015 12th International Conference on Fuzzy Systems and Knowledge Discovery (FSKD)*. Zhangjiajie, China: IEEE, pp. 2147–2152. http://doi.org/10.1109/FSKD.2015.7382284.

29 Rüßmann, M., Lorenz, M., Gerbert, P. et al. (2015). Industry 4.0: the future of productivity and growth in manufacturing industries. *Boston Consulting Group* 9 (1): 54–89.

30 Pereira, A.C. and Romero, F. (2017). A review of the meanings and the implications of the Industry 4.0 concept. *Procedia Manufacturing* 13: 1206–1214.

31 Salkin, C., Oner, M., Ustundag, A., and Cevikcan, E. (2018). A conceptual framework for Industry 4.0. In: *Industry 4.0: Managing the Digital Transformation*, Springer Series in Advanced Manufacturing, 3–23. Cham: Springer. https://doi.org/10.1007/978-3-319-57870-5_1.

32 Oztemel, E. and Gursev, S. (2020). Literature review of industry 4.0 and related technologies. *Journal of Intelligent Manufacturing* 31 (1): 127–182.

33 Chongwatpol, J. and Sharda, R. (2013). RFID-enabled track and traceability in job-shop scheduling environment. *European Journal of Operational Research* 227 (3): 453–463.

34 Motyl, B., Baronio, G., Uberti, S. et al. (2017). How will change the future engineers' skills in the Industry 4.0 framework? A questionnaire survey. *Procedia Manufacturing* 11: 1501–1509.

35 McKinsey & Company (2020). *Industry 4.0: Reimagining Manufacturing Operations After COVID-19*. McKinsey & Company.

36 Faccio, M., Ferrari, E., Gamberi, M., and Pilati, F. (2019). Human Factor Analyser for work measurement of manual manufacturing and assembly processes. *The International Journal of Advanced Manufacturing Technology* 103 (1–4): 861–877.

37 Bortolini, M., Faccio, M., Gamberi, M., and Pilati, F. (2020). Motion Analysis System (MAS) for production and ergonomics assessment in the manufacturing processes. *Computers & Industrial Engineering* 139: 105485.

38 Kersten, J.T. and Fethke, N.B. (2019). Radio frequency identification to measure the duration of machine-paced assembly tasks: agreement with self-reported task duration and application in variance components analyses of upper arm postures and movements recorded over multiple days. *Applied Ergonomics* 75: 74–82.

39 Gotthardt, S., Hulla, M., Eder, M. et al. (2019). Digitalized milk-run system for a learning factory assembly line. *Procedia Manufacturing* 31: 175–179.

40 Choy, K.L., Ho, G.T., and Lee, C.K.H. (2017). A RFID-based storage assignment system for enhancing the efficiency of order picking. *Journal of Intelligent Manufacturing* 28 (1): 111–129.

41 Lu, B., Chen, Z., and Zhao, X. (2021). Data-driven dynamic predictive maintenance for a manufacturing system with quality deterioration and online sensors. *Reliability Engineering & System Safety* 212: 107628.

42 Jia, X., Feng, Q., Fan, T., and Lei, Q. (2012). RFID technology and its applications in Internet of Things (IoT). *2012 2nd International Conference on Consumer Electronics, Communications and Networks (CECNet)*. IEEE.

43 Sun, C. (2012). Application of RFID technology for logistics on internet of things. *AASRI Procedia* 1: 106–111.

44 Lin, S.W., Miller, B., Durand, J. et al. (2015). Industrial internet reference architecture. Industrial Internet Consortium (IIC), Tech. Rep.

45 Mumtaz, S., Alsohaily, A., Pang, Z. et al. (2017). Massive Internet of Things for industrial applications: addressing wireless IIoT connectivity challenges and ecosystem fragmentation. *IEEE Industrial Electronics Magazine* 11 (1): 28–33.

46 Boyne, G.A., Gould-Williams, J.S., Law, J., and Walker, R.M. (2005). Explaining the adoption of innovation: an empirical analysis of public management reform. *Environment and Planning C Government & Policy* 23 (3): 419–435.

47 Dwivedi, Y.K., Janssen, M., Slade, E.L. et al. (2017). Driving innovation through big open linked data (BOLD): exploring antecedents using interpretive structural modelling. *Information Systems Frontiers* 19 (2): 197–212.

48 Kaisler, S, Armour, F Espinosa, J.A., and Money, W. (2013). Big data: issues and challenges moving forward. *2013 46th Hawaii International Conference on System Sciences*, pp. 995–1004. IEEE.

49 Mohanta, B.K., Jena, D., Satapathy, U., and Patnaik, S. (2020). Survey on IoT security: challenges and solution using machine learning, artificial intelligence and blockchain technology. *Internet of Things* 1 (11): 100227.

50 Hassija, V., Chamola, V., Saxena, V. et al. (2019). A survey on IoT security: application areas, security threats, and solution architectures. *IEEE Access* 20 (7): 82721–82743.

51 Rathore, M.M., Ahmad, A., Paul, A., and Thikshaja, U.K. (2016). Exploiting real-time big data to empower smart transportation using big graphs. *2016 IEEE Region 10 Symposium (TENSYMP)*, pp. 135–139. IEEE.

52 Atzori, L., Iera, A., and Morabito, G. (2010). The internet of things: a survey. *Computer Networks* 54 (15): 2787–2805.

53 Barde, M.P. and Barde, P.J. (2012). What to use to express the variability of data: standard deviation or standard error of mean? *Perspectives in Clinical Research* 3 (3): 113.

54 Skarmeta, A.F., Hernandez-Ramos, J.L., and Moreno, M.V. (2014). A decentralized approach for security and privacy challenges in the internet of things. *2014 IEEE World Forum on Internet of Things (WF-IoT)*, pp. 67–72. IEEE.

55 NASTASE, Lavinia (2017). Security in the internet of things: a survey on application layer protocols. *2017 21st International Conference on Control Systems and Computer Science (CSCS)*, pp. 659–666. IEEE.

56 Dinculeană, D. and Cheng, X. (2019). Vulnerabilities and limitations of MQTT protocol used between IoT devices. *Applied Sciences* 9 (5): 848.

57 Wahyudi, A., Pekkola, S., and Janssen, M. (2018). Representational quality challenges of big data: Insights from comparative case studies. *Conference on e-Business, e-Services and e-Society*, pp. 520–538. Cham: Springer.

58 Brous, P., Janssen, M., and Herder, P. (2019). Internet of Things adoption for reconfiguring decision-making processes in asset management. *Business Process Management Journal* 25: 495–511.

59 Gubbi, J., Buyya, R., Marusic, S., and Palaniswami, M. (2013). Internet of things (IoT): a vision, architectural elements, and future directions. *Future Generation Computer Systems* 29 (7): 1645–1660.

60 Blackstock, M. and Lea, R. (2012). IoT mashups with the WoTKit. *2012 3rd IEEE International Conference on the Internet of Things*, pp. 159–166. IEEE.

61 Bi, Z., Da Xu, L., and Wang, C. (2014). Internet of things for enterprise systems of modern manufacturing. *IEEE Transactions on Industrial Informatics* 10 (2): 1537–1546.

62 Fleisch, E. (2010). What is the internet of things? An economic perspective. *Economics, Management, and Financial Markets* 5 (2): 125–157.

63 Villagrán, N.V., Estevez, E., Pesado, P., and Marquez, J.D. (2019). Standardization: a key factor of industry 4.0. *2019 Sixth International Conference on eDemocracy & eGovernment (ICEDEG)*, pp. 350–354. IEEE.

64 Zeng, D., Guo, S., and Cheng, Z. (2011). The web of things: a survey. *The Journal of Communication* 6 (6): 424–438.

65 Brous, P., Janssen, M., and Herder, P. (2020). The dual effects of the Internet of Things (IoT): a systematic review of the benefits and risks of IoT adoption by organizations. *International Journal of Information Management* 1 (51): 101952.

66 Oguntala, G., Abd-Alhameed, R., Jones, S. et al. (2018). Indoor location identification technologies for real-time IoT-based applications: an inclusive survey. *Computer Science Review* 30: 55–79.

67 Zafari, F., Gkelias, A., and Leung, K.K. (2019). A survey of indoor localization systems and technologies. *IEEE Communication Surveys and Tutorials* 21 (3): 2568–2599,. third quarter.

68 Xiao, J., Zhou, Z., Yi, Y., and Ni, L.M. (2016). A survey on wireless indoor localization from the device perspective. *ACM Computing Surveys (CSUR)* 49 (2): 25.

69 Xing-Yu, L., Ke, H., and Min, Y. (2014). Research on improvement to WiFi fingerprint location algorithm. *IET Conference Proceedings, IET Digital Library*, pp. 648–652. http://doi.org/10.1049/ic. 2014.0173.

70 Vive. https://www.vive.com (accessed 1 July 2021).

71 Oculus. https://www.oculus.com (accessed 1 July 2021).

72 Xandem. https://xandem.com (accessed 1 July 2021).

73 Insoft. https://infsoft.com (accessed 1 July 2021).

74 United Nations (2020). *Conference on Trade and Development, Key Statistics and Trends in International rade.*

75 Forbes (2020). New research shows consumers already expect mass personalization. Time to get ready. *Forbes.*

76 McKinsey & Company (2019). *The Future of Personalization – And how to Get Ready for it.* McKinsey & Company.

77 Radziwon, A., Bilberg, A., Bogers, M., and Madsen, E.S. (2014). The smart factory: exploring adaptive and flexible manufacturing solutions. *Procedia Engineering* 69: 1184–1190.

78 Hozdić, E. (2015). Smart factory for Industry 4.0 : a review. *International Journal of Modern Manufacturing Technologies* 7 (1): 28–35.

79 Mabkhot, M.M., Al-Ahmari, A.M., Salah, B., and Alkhalefah, H. (2018). Requirements of the smart factory system: a survey and perspective. *Machines* 6 (2): 23.

80 Krishnan, S. and Santos, R.X.M. (2021). Real-time asset tracking for smart manufacturing. In: *Implementing industry 4.0. Intelligent Systems Reference Library*, vol. 202 (ed. C. Toro, W. Wang, and H. Akhtar), 25. Cham: Springer. https://doi.org/10.1007/978-3-030-67270-6_2.

81 Kelepouris, T. and McFarlane, D. (2010). Determining the value of asset location information systems in a manufacturing environment. *International Journal of Production Economics* 126 (2): 324–334.

82 Pilati, F., Regattieri, A., and Cohen, Y. (2021). Real time locating system for a learning cross-docking warehouse.

83 Gladysz, B., Santarek, K., and Lysiak, C. (2017). Dynamic spaghetti diagrams. A case study of pilot RTLS implementation. *International Conference on Intelligent Systems in Production Engineering and Maintenance*, pp. 238–248. Cham: Springer.

84 Slovák, J., Vašek, P., Šimovec, M. et al. (2019). RTLS tracking of material flow in order to reveal weak spots in production process. *2019 22nd International Conference on Process Control (PC19)*, pp. 234–238. IEEE.

85 Tsang, Y.P., Choy, K.L., Poon, T.C., et al. (2016). An IoT-based occupational safety management system in cold storage facilities. *6th International Workshop of Advanced Manufacturing and Automation*, pp. 7–13. Atlantis Press.

86 Jiang, J., Guo, Y., Liao, W., et al. (2014). Research on RTLS-based coordinate guided vehicle (CGV) for material distribution in discrete manufacturing workshop. *2014 IEEE International Conference on Internet of Things (iThings), and IEEE Green Computing and Communications (GreenCom) and IEEE Cyber, Physical and Social Computing (CPSCom)*, pp. 1–8. IEEE.

87 Löcklin, A., Ruppert, T., Jakab, L., et al. (2020). Trajectory prediction of humans in factories and warehouses with real-time locating systems. *2020 25th IEEE International Conference on Emerging Technologies and Factory Automation (ETFA)*, Vol. 1, pp. 1317–1320. IEEE.

88 Che, F., Ahmed, A., Ahmed, Q.Z., Zaidi, S.A.R., and Shakir, M.Z. (2020). Machine learning based approach for indoor localization using ultra-wide bandwidth (UWB) system for industrial internet of things (IIoT). *2020 International Conference on UK-China Emerging Technologies (UCET)*. Glasgow, UK: IEEE, pp. 1–4. http://doi.org/10.1109/UCET51115.2020.9205352.

89 Nessa, A., Adhikari, B., Hussain, F., and Fernando, X.N. (2020). A survey of machine learning for indoor positioning. *IEEE Access* 8: 214945–214965.

90 Jiang, C., Shen, J., Chen, S. et al. (2020). UWB NLOS/LOS classification using deep learning method. *IEEE Communications Letters* 24 (10): 2226–2230.

91 Koyuncu, H. and Yang, S.H. (2010). A survey of indoor positioning and object locating systems. *IJCSNS International Journal of Computer Science and Network Security* 10 (5): 121–128.

92 Na, W., Jang, S., Lee, Y. et al. (2019). Frequency resource allocation and interference management in mobile edge computing for an Internet of Things System. *IEEE Internet of Things Journal* 6 (3): 4910–4920.

93 Connectivity Standards Alliance (CSA). https://zigbeealliance.org (accessed 3 July 2021)

94 Uradzinski, M., Guo, H., Liu, X. et al. (2017). Advanced indoor positioning using Zigbee wireless technology. *Wireless Personal Communications* 97: 6509–6518.

95 Farid, Z., Nordin, R., and Ismail, M. (2013). Recent advances in wireless indoor localization techniques and system. *Journal of Computer Networks and Communications* 2013: 185138,. 12 pages. https://doi.org/10.1155/2013/185138.

96 Mehmood, H. and Tripathi, N.K. (2011). *Hybrid Positioning Systems: A Review*. LAP LAMBERT Academic Publishing.

97 Sakpere, W., Adeyeye-Oshin, M., and Mlitwa, N.B.W. (2017). A state-of-the-art survey of indoor positioning and navigation systems and technologies. *South African Computer Journal* 29 (3): 145–197.

98 Chiputa, M. and Xiangyang, L. (2018). Real time wi-fi indoor positioning system based on RSSI measurements: a distributed load approach with the fusion of three positioning algorithms. *Wireless Personal Communications* 99: 67–83.

99 Chen,W-C., Kao, K-F., Chang, Y.-T., and Chang, C-H. (2018). An RSSI-based distributed real-time indoor positioning framework. *2018 IEEE International Conference on Applied System Invention (ICASI)*. Chiba, Japan: IEEE, pp. 1288–1291. http://doi.org/10.1109/ICASI.2018.8394528.

100 Pascacio, P., Casteleyn, S., Torres-Sospedra, J. et al. (2021). Collaborative indoor positioning systems: a systematic review. *Sensors* 21: 1002.

101 Wymeersch, H., Lien, J., and Win, M.Z. (2009). Cooperative localization in wireless networks. *Proceedings of the IEEE* 97: 427–450.

102 Vaghefi, R.M. and Buehrer, R.M. (2017). Cooperative source node tracking in non-line-of-sight environments. *IEEE Transactions on Mobile Computing* 16: 1287–1299.

103 Adolph, S., Tisch, M., and Metternich, J. (2014). Challenges and approaches to competency development for future production. *Journal of International Scientific Publications – Educational Alternatives* 12 (1): 1001–1010.

104 Abele, E., Metternich, J., Tisch, M. et al. (2015). Learning factories for research, education, and training. *Procedia CiRP* 32: 1–6.

105 Wagner, U., AlGeddawy, T., ElMaraghy, H., and MŸller, E. (2012). The state-of-the-art and prospects of learning factories. *Procedia CiRP* 3: 109–114.

106 Mütze, A., Hingst, L., Rochow, N., et al. (2021). Use cases of real-time locating systems for factory planning and production monitoring. Available at SSRN 3857878.

107 Umeda, Y., Goto, J., and Hongo, Y., (2021). Developing a digital twin learning factory of automated assembly based on 'digital triplet'concept. Available at SSRN 3859019.

108 Brenner, B., Estler, D., and Hummel, V. (2021). Motion capturing in connection with human model simulation and artificial intelligence AI for employee training in the area of joint-gentle assembly workflows in production environment. Available at SSRN 3869762.

109 Mahalik, N.P. (ed.) (2013). *Fieldbus Technology: Industrial Network Standards for Real-Time Distributed Control*. Springer Science & Business Media.

110 Liu, Y. and Schulz, N.N. (2002). Knowledge-based system for distribution system outage locating using comprehensive information. *IEEE Transactions on Power Systems* 17 (2): 451–456.

111 Baker, T., Asim, M., MacDermott, Á. et al. (2020). A secure fog-based platform for SCADA-based IoT critical infrastructure. *Software: Practice and Experience* 50 (5): 503–518.

112 Pujolle, G. (2006). An autonomic-oriented architecture for the internet of things. *2006 JVA06 IEEE John Vincent Atanasoff 2006 Int. Symp. On, Mod. Comput.*, pp. 163–168. IEEE.

113 Liu, Z.C., Lin, D.S., and Ning, Y.Z. (2013). Embedded real-time database system concurrency control protocol AC-based OCC-FV. *Applied Mechanics and Materials* 1402–1406. Trans Tech Publications Ltdhttps://doi.org/10.4028/www.scientific.net/AMM.263-266.

114 Jiang, L., Li Da, X., Cai, H. et al. (2014). An IoT-oriented data storage framework in cloud computing platform. *IEEE Transactions on Industrial Informatics* 10: 1443–1451. https://doi.org/10.1109/TII.2014.2306384.

115 Gani, A., Siddiqa, A., Shamshirband, S., and Hanum, F. (2016). A survey on indexing techniques for big data: taxonomy and performance evaluation. *Knowledge and Information Systems* 46 (2): 241–284.

116 Ratnasamy, S., Karp, B., Yin, L. (2002). GHT: a geographic hash table for data-centric storage. *Proc. 1st ACM Int. Workshop Wirel. Sens. Netw. Appl.* ACM. pp. 78–87.

117 Diao, Y., Ganesan, D., Mathur, G., and Shenoy, P.J. (2007). *Rethinking Data Management for Storage-Centric Sensor Networks*, 22–31. CIDR.

118 Fathy, Y., Barnaghi, P., Enshaeifar, S., and Tafazolli, R. (2016). A distributed in-network indexing mechanism for the internet of things. *IEEE 3rd World Forum Internet Things WF-IoT*, pp. 585–590. Reston, VA, USA: IEEE.

119 da Cruz, M.A., Rodrigues, J.J., Al-Muhtadi, J. et al. (2018). A reference model for internet of things middleware. *IEEE Internet of Things Journal* 5 (2): 871–883.

120 Farahzadi, A., Shams, P., Rezazadeh, J., and Farahbakhsh, R. (2018). Middleware technologies for cloud of things: a survey. *Digital Communications and Networks* 4 (3): 176–188.

121 Abu-Elkheir, M., Hayajneh, M., and Ali, N.A. (2013). Data management for the internet of things: design primitives and solution. *Sensors* 13 (11): 15582–15612.

122 Jiang, Y., Wang, C., Wang, Y., and Gao, L. (2019). A cross-chain solution to integrating multiple blockchains for IoT data management. *Sensors* 19 (9): 2042.

123 Xiong, Z., Zhang, Y., Luong, N.C. et al. (2020). The best of both worlds: a general architecture for data management in blockchain-enabled Internet-of-Things. *IEEE Network* 34 (1): 166–173.

124 Zheng, X, Lu, J, Sun, S, and Kiritsis, D. (2020). Decentralized industrial IoT data management based on blockchain and IPFS. *IFIP International Conference on Advances in Production Management Systems*, pp. 222–229. Cham: Springer.

125 Zhang, P., Wang, C., Jiang, C., and Han, Z. (2021). Deep reinforcement learning assisted federated learning algorithm for data management of IIoT. *IEEE Transactions on Industrial Informatics* 17 (12): 8475–8484.

126 Bolettieri, S., Bruno, R., and Mingozzi, E. (2021). Application-aware resource allocation and data management for MEC-assisted IoT service providers. *Journal of Network and Computer Applications* 1 (181): 103020.

127 Chen, J., Ramanathan, L., and Alazab, M. (2021). Holistic big data integrated artificial intelligent modeling to improve privacy and security in data management of smart cities. *Microprocessors and Microsystems* 1: 81.

128 Saqlain, M., Piao, M., Shim, Y., and Lee, J.Y. (2019). Framework of an IoT-based industrial data management for smart manufacturing. *Journal of Sensor and Actuator Networks* 8 (2): 25.

129 Abbasi, M.A., Memon, Z.A., Memon, J. et al. (2017). Addressing the future data management challenges in iot: a proposed framework. *International Journal of Advanced Computer Science and Applications* 8 (5): 197–207.

16

Trends in Robotics Management and Business Automation

Gastón Lefranc

Escuela de Ingeniería Eléctrica Pontificia Universidad Católica de Valparaíso, Valparaíso, Chile

16.1 Introduction

IFR, the International Federation of Robotics, is a non-profit professional organization established in 1987, with the aim of strengthening and protecting the robotics industry worldwide and of promoting international research, development, robot adoption, and cooperation regarding industrial robots and service robots. IFR represents, through membership, 16 nations.

According to IFR, the adoption of industrial robots in factories has tripled in the last 10 years, with more than 1,381,000 units reached, thus stimulating an emerging market. Trends indicate that this increase does not only concern industrial robots but also service robots, mobile robots, and drones. Currently, there is intensive research concerning new standards, especially those intended for robots applied in medicine [1].

The use of robotic technology enables companies to increase efficiency and serve consumers better. In the fields of health, agriculture, food preparation, textiles, wood and plastic production, manufacturing, and in the military, an intense interest in robotics is found. Robots can be differentiated by where and how they are to be applied; among the many categories, there are robotic manipulators, mobile robots, service robots, and robotic drones. Robotic manipulators and mobile robots are used in manufacturing companies and, specifically, those that rely on manipulation or translation of objects, pieces, or parts. Robotic manipulators may be stationary or mobile (Figure 16.1). Drones are used in delivery, surveillance, and other such aerial activities. Mobile robots (Figure 16.2) have wheels (WMR), legs (walk), skids (glide), or tracks, which allow movement on flat or smooth surfaces. Drones are mobile robots with air as working space and are used in delivery, surveillance, and aerial activities (Figure 16.3).

There are mobile robots that can fly (drones), dive or be hybrid. These robots are UAS (Unmanned Aerial Systems) or UAV (Unmanned Aerial Vehicles).

Robots can act autonomously, react with intelligence, and make decisions based on environmental perception to act with mechanical, electronic, and computer technologies. New trends prioritize artificial intelligence, robotic vision, autonomous driving, communication networks,

IEEE Technology and Engineering Management Society Body of Knowledge (TEMSBOK), First Edition.
Edited by Gustavo Giannattasio, Elif Kongar, Marina Dabić, Celia Desmond, Michael Condry, Sudeendra Koushik, and Roberto Saracco.

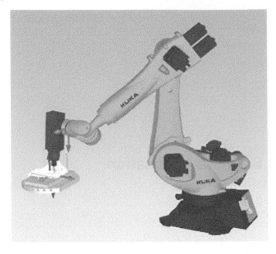

Figure 16.1 Robotic manipulator. *Source:* Ahmad et al. [2]/MDPI/CC BY.4.0.

cooperative working, nanorobotics, safe robot-human interactions; and, among many others, the expression and perception of emotions. The most important fields of application are in the fields of health, sports, product distribution, and in the field of service robots [3].

Drones are remotely programmed and controlled to take pictures, record video, transport cargo, inspect bridges and industrial chimneys, aid agricultural activity, etc.; they are becoming more accessible and profitable and allow obtaining data from industrial facilities, allowing quick business decision-making. In mining operations, drones are used in stockpile management, tailings dams, inspections,

Figure 16.2 Mobile robot.

Figure 16.3 Robot drone. *Source:* Dean Andy. quadcopter-drone-in-the-air-near-corporate-industrial-building. https://stock.adobe.com/ar/images/unmanned-aircraft-system-quadcopter-drone-in-the-air-near-corporate-industrial-building/312384583.

and more. There is already high demand for drones mostly due to the advances in technology. This makes them more accessible and profitable [4]. Some countries in Europe, the United States, and Canada, already have laws and regulations for the use of drones and remote controls (RCs).

There are many differences between industrial service robots and personal service robots. The first kind is aimed at industrial automation, often performing heavy, repetitive, or tedious tasks. The personal service robot and collaborative robot (the cobot) perform non-professional and non-remunerative tasks such as dirty, boring, distant, dangerous, or repetitive jobs, including housework. These robots can work safely alongside humans, are more customized, and have greater compatibility with the cloud.

Forecasts indicate that the revenue of the manufacturing industry would reach a figure of $24 billion by 2030, and that spending by consumer-packaged goods (CPG) on data and analytics services will reach $4 billion by that year, compared to $500 million in 2021 [5].

The drone market is growing at an annual rate of 57.5% this year (2021) and is forecasted to reach figures of $501.4 billion by 2028. Some industries that heavily rely on drones are the film, agriculture, media, oil, and gas industries. Some of the disadvantages of using drones are that simple rotor drones are more difficult to fly than multi-rotor drones; they are more expensive and can be dangerous due to the heavy rotating blade.

The drones are autonomously operated by a control system, with the option of manual override. Examples of personal service robots are lawnmowers; vacuum cleaners; entertainment robots, such as kits and toys; and training and education robots. Some of these are the Aibo robot dog, the Roomba vacuum cleaner, and A.I. equipped robots (Figure 16.4).

A special case is that of robots used in clinics and hospitals, with an emphasis on the elderly. Some are used in high-precision surgeries, and others to administer vaccines and medications. The COVID-19 pandemic has boosted their use in the areas of disinfection, decontamination, delivery of medicines and food, transport of goods, monitoring of patients, companions, and preventive support [6].

In the case of patients paralyzed due to strokes or spinal cord injuries, robotic exoskeletons are now used in rehabilitation. Other types of robots give therapeutic help to patients with mental health problems [7, 8].

A robot can perform disinfection measurements, food and medicine distribution, and vital sign measurements [9, 10]. The automation benefits for health care during the pandemic are analyzed in [11], and the conclusions are that infection risk and transmission of the virus decrease. The experience obtained in the Smart Field Hospital of China, replicated in other countries, shows that some robots work well in rooms (hotels and airports) [12, 13]. The Lio robot presents skills in handling products in the hospital context [14].

Figure 16.4 Drone.

A growing trend is that of human–robot interaction (HRI). This is observed in Industry 4.0, where robots interact with humans or other robots. Collaborative HRI is prominent in Europe, Australia, Japan, and the United States [15, 16] (Figure 16.5).

The perception of the robots must be robust, to detect environmental changes and plan adaptive movement and action [17].

Navigation and robot movements (mobiles and drones) are important tasks, already solved traditionally with route planning. Recent research highlights a trend of

Figure 16.5 Cobots. *Source:* Yaskawa America, Inc.

reinforcement learning and unsupervised learning methods to solve difficult problems in navigation and movement, which supersedes traditional methods [18].

In [19], cooperation among robots is considered, must have architecture, protocols to integrate collaborative robotics and machine learning, based on the RAMI 4.0 model (Reference Architectural Model Industrie 4.0), performed by humans and robots.

This article displays the trends in robotic management and business automation that have gained relevance in recent years and that have been affected by the COVID-19 pandemic. Suggestions and recommendations are provided to engineers and entrepreneurs on how to integrate this technology to improve efficiency, flexibility, productivity, and safety to increase the added value in their processes.

This technology is used in automation with a variety of sensors and actuators which, combined with artificial intelligence, are easy to program and install, following industry 4.0 criteria. Robots are efficient for providing better quality in products and services, and for decreasing production rejection and failure, and, if the use of renewable energy is considered, the result is ecological and more profitable. It is possible to innovate business models, where producers can easily diversify in different sectors of the economy such as the fields of manufacturing, food and beverage, textiles, wood and plastics, healthcare, and in many other sectors that entrepreneurs could venture in.

The work is organized as follows: Industries that use robotics; trends in robotic manipulators; trends in mobile robots; trends in robotic drones and specifications. All these sections include suggestions and recommendations to business engineers and entrepreneurs on how to integrate and choose robots for better production, and on how to make acquisition and management decisions to obtain greater benefits.

16.2 Industries and Enterprises Utilizing Robotics Management

Forecasts for the use of robotics are in the fields of health, agriculture, food preparation, textiles, wood and plastic products factories, manufacturing, military, disasters, and other emerging areas that entrepreneurs could venture into. Every business requires management and could profit from automation [20].

1) Health care
   An increasing use of robotics is observed in medical care, surgery, rehabilitation, therapy, in aiding patients, and in other daily activities. The robots used are intended to facilitate the work of healthcare professionals.
   For example, in the da Vinci surgical system, teleoperation through the surgeon's hand movements is performed to control instruments in surgery. This makes such surgeries less invasive, being used in cardiac, colon rectal, gynecological, head and neck, thoracic, and urological operations [15]. In the case of patients paralyzed due to strokes or spinal cord injuries, robotic exoskeletons are now used in rehabilitation. Other robots can give therapeutic help to patients with mental health problems [16].
2) Agriculture
   The agricultural industry is beginning to use robotics to increase productivity and reduce costs. GPS-guided tractors and harvesters are already in use; automatic systems in pruning, thinning, cutting, spraying, and removing the grass are also common. The use of sensors in robots makes it possible to control crop pests and diseases [21].

3) Food Preparation

Robots can prepare and cook meals for you at home. Typically, the user chooses a recipe and sets pre-packaged containers of cut and prepared ingredients. The robot then can cook certain foods quickly and efficiently. The company Moley Robotics has developed an interactive robot that includes the use of an integrated smart dishwasher and refrigerator [22].

4) Manufacturing

Robotics has been used in manufacturing, assembly, quality control, parts, and product storage for years, resulting in increased productivity and efficiency, as well as in reducing production costs. Robotic manipulators already work collaboratively with workers to perform repetitive, monotonous, or complex tasks under the guidance and control of the worker. These machines have high precision, speed, and the ability to be reprogrammed for specific tasks of different extents and complexity, and they are designed for human safety; they use cameras, sensors, and automatic shutdown capabilities to detect and stay away from humans in the workplace [23].

5) Military

Robotic technology is also applied in many areas of the military and of public safety organizations. Unmanned drones are used for surveillance and support on the battlefield, flying over conflict zones and hostage situations.

6) Disasters

Robots are being used in dealing with natural or man-made disasters, helping by accessing dangerous areas with great speed and precision without endangering first responders. These drones can assess risk and provide important information to responders in real-time. Robotic devices already have a presence in several commercial industries, growing more affordable and versatile over time.

16.2.1 Trends on the Impact Areas of Robotics

This new connectivity is caused by the digital transformation that allows producers to use new business models and to diversify easily. This new business environment requires robotics and automation management.

Robotics is applied in manufacturing, assembly, quality control supply chains, and in production of all manufacturing sectors, of the automotive industry, and of other sectors. In smart factories, different products are assembled with the same equipment. An estimated 80% of businesses depend on a scarce and unreliable workforce, which the Covid-19 pandemic has further weakened. This creates a great opportunity for growth in the world of robotization and process automation. Demand through electronic commerce grew before the Covid-19 pandemic by 12% annually. Now a large increase in demand is expected until 2023. E-commerce is difficult to predict as the products ordered daily and simultaneously are not well recorded. To respond to this demand, one requires more intelligence, digitization, and optimization [24].

There are countless technological solutions to make stores smarter and more efficient. The challenge is finding the right solutions for a specific warehouse. Today's automation systems must be more flexible, adaptable, and efficient. Traditional technologies like WMS (Warehouse Management System) add autonomous mobile robots (AMR). The dilemma for executives is choosing which one of the two alternatives best suits the needs of a company, WMS or WES (Warehouse Execution Systems).

Robotics is now applied to supply chains, which are globally disrupted due to the pandemic and most severely in countries with high wages. Something similar occurs with the automation of container handling, and production in the automotive industry and with all manufacturing sectors, where robotic manipulators are used, which are easy to install and to program (with industry 4.0 criteria). These manipulators are handled with intelligent software, which feeds on the information provided by various sensors and computerized vision. Using newer robots can improve flexibility, productivity, and safety, as they are more efficient and of better quality. The result is fewer rejected products, more profitable production, reduced energy use, and reduced carbon footprint [3].

Some of the types of robots used in manufacturing are robotic manipulators (that manipulate parts and pieces), mobile robots (that move from one place to another), drone robots (that transport by air), and mixed types. These robots are now being used in new markets such as those of food and beverages, textiles, wood products, and plastics.

According to ABI Research [5], manufacturers of CPG analyze the market to anticipate customer demand and thus optimize their processes. These manufacturers use digital information to perform production planning; to work with retailers, distributors, and supply chain partners; and to share sales information throughout the chain, right down to the checkout line. For this reason, these manufacturers monitor with sensors and perform IoT analysis.

Due to the Covid-19 pandemic (2020–2022), most economic sectors have been negatively affected (e.g. manufacturing, distribution, transportation, retail, restaurants, and other types of personal and government services). The main areas of opportunity for robotics, in this situation, are medical assistance, therapeutical assistance, and disease prevention.

According to the World Economic Forum, 50% of employers plan to step up automation in their companies. It is observed that several industries are now buying small mobile robots, of which the food and consumer goods sectors represent the fastest growing market. This means that millions of people will remain inactive due to COVID-19. This new push for automation goes hand in hand with the increased use of robots, which tends to increase inequality by displacing low-skilled workers from tasks that do not require skill and from routine and repetitive tasks. It is unclear if this is solely due to COVID-19 [24]. The demand for more automated equipment results in more companies of greater production in the United States and the European Union. Factories in Asian countries like China and South Korea make more use of robots than in the United States, giving them a competitive edge [25, 26].

The main trend is to use collaborative robots (cobots) and the intensification of the use of artificial and computational intelligence. This implies the use of robotic vision, neural networks, machine learning, etc., which involve different areas of mechanical, electronic, and computer technology. New trends incorporate the methods and applications of autonomous driving, network communication, human–robot cooperation, nanorobotics, friendly human–robot interfaces, safe HRI, and the expression and perception of emotions.

There are four trends in collaborative robots [27, 28]:

1) Cobots perform repetitive and dangerous tasks, and they are safe for humans.
2) Cobots collaborate, without competing with humans. Odense Robotics cluster example, in Denmark.
3) Cobots multitask while robotic manipulators ride on mobile robots and perform tasks in different parts of the plant.
4) Cobots are found in emerging sectors alongside humans. For example, in hotels (Food by Robots company) and painted doors and windows (Incerco collaborative)

The average cost of cobot robot is about US$35,000 (Universal Robot, ABB, Fanuc, Kuka . . .) varies from US$10,000 to US$50,000, depending on the application.

16.3 Some Trends on Robotics Manipulators

The trends in robotics are embodied in manufacturing applications such as industrial robots; in applications of services to society, such as service robots; and in the new role of robots in healthcare, considering the current pandemic. To achieve this, intense research, innovation, and development in robotics are ongoing. Networks of robots (Figure 16.6) and AMR are now used in the automotive industry; new navigation equipment, with artificial intelligence, is also added to these AMRs [29].

On production lines, mobile robots are more flexible. The driverless bodies are individually assembled, creating a new production line, and reprogramming the robots and AMRs; those lines integrate seamlessly with human–robot collaboration workstations.

Different varieties of robots are used in the automation of an industry depending on the application. The most common one is the robotic manipulator with articulated structure, attached to a rotating base. These robots usually have four to six axes. On production lines, mobile robots are more flexible. The driverless bodies are individually assembled, creating a new production line, reprogramming the robots and AMRs. And those lines seamlessly integrate with human-robot collaboration workstations.

A robot manipulator is depicted in Figure 16.7. In the wrist end-effector, a tool such as gripper, welding torch, or powered screwdriver among others, can be used. Figure 16.8 shows different types of robot manipulators. The most common applications are welding and material handling [30]. The robotic manipulator effector allows to manipulate, weld, and move parts. Manipulators perform welding processes with skill and precision. Welding robots can be classified as arc welding or resistance welding robots. An example of arc welding is the FANUC Arcmate 120ic and Motoman MA1400 manipulators. Resistance welding includes spot welding, with the FANUC R-2000ic/210F being ideal for this type of welding. Both examples can be used for laser and electron beam welding (Figure 16.9).

The product manufacturing process implies the following: the manufacture of the individual parts of a product, the assembly of those parts (Figure 16.10), the loading of the machine, the injection molding, the transfer of parts and pieces, the picking and placing, the quality control processes, packaging, palletizing, dispensing, and the storage procedure.

These processes involve the use of robots to perform the aforementioned tasks and others such as stock removal, drilling, deburring, sanding, grinding, and polishing. In laser, plasma, oxyfuel, waterjet, and ultrasonic cutting applications, these include material removal. The type of effector to be used depends on the cutting and material removal processes [31].

Figure 16.6 Robot networks. *Source:* Photocreo Bednarek/ Adobe stock.

Articulate Robots   **Figure 16.7**   Parts of the robot manipulator.

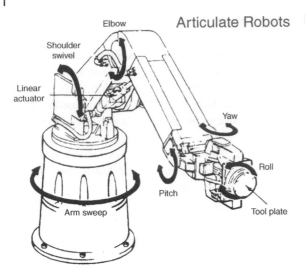

Figure 16.8   Types of the robot manipulator.

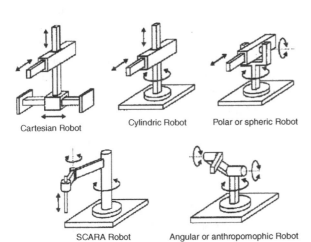

These manipulators can also be used in coating and gluing applications, including painting, thermal spraying, sealing, and bonding, which involve the use of harsh chemicals; automation keeps workers from being exposed to these chemicals. The accuracy and precision of the robots ensure uniform applications for higher-quality parts.

There are also newer applications such as inspection, 3D printing, and worker training. Integrating the FANUC LR Mate 200ic, which has a robotic vision system, allows one to perform quality control by inspecting the parts to detect defects. In worker training, articulated robots play an important role, especially when tasks cannot be fully automated [32]. Robotic manipulators mounted on rails or on a mobile robot carry out the transfer and assembly of parts (Figures 16.11 and 16.12).

Figure 16.9 Welding robot. *Source:* bobo1980/ Adobe Stock.

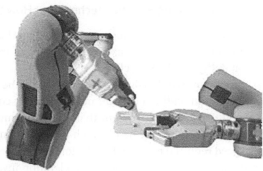

Figure 16.10 Robot for assembly. *Source:* Robotic Biz Portal. Robotic sensors: Fundamentals of robotics technology. 13 March 2020. https://roboticsbiz.com/ fundamentals-of-robotics-technology-robotic-sensors/.

Figure 16.11 Robot manipulator on mobile robots. *Source:* RealAws/ Wikimedia Commons.

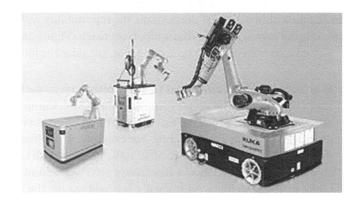

Figure 16.12 Robot manipulator on rail. *Source:* WINKEL GmbH.

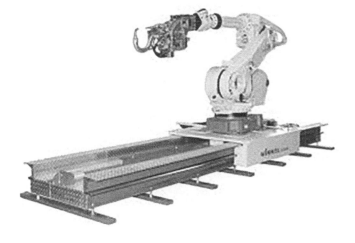

16.3.1 Managing Robotics Technology: How to Choose a Robotics Manipulator

Once the purchase of robotic manipulators has been decided, the following characteristics should be considered: number of axes (DOF degree of freedom), payload, speed, acceleration, precision, repeatability, workspace, etc. [33].

Three degrees of freedom are required to define the position of the effector in three-dimensional space, and an additional 3° are needed to fully control the orientation of the effector: yaw, pitch, and roll. Payload is the weight that the robot can lift, which will vary depending on the distance from the effector to the base, which is the maximum horizontal reach of the arm. Robot speed is the speed required to move the effector from a start point to a desired endpoint. This speed depends on the motors used in the joints. Acceleration involves how fast a joint can be accelerated.

Accuracy is the measure of the difference between the commanded position and the absolute position of the robot. Repeatability is when the robot returns to a certain position consistently. Accuracy and repeatability may appear similar, but they are different measures. Repeatability is usually the most important thing for a robot and is like the concept of precision. For example: if the robot is commanded to move to a certain position, and the arm only reaches 2 mm from that position, its accuracy is 2 mm. Calibration can improve this precision. If it is sent to the position and came within 0.2 mm of the position, the repeatability will be of around 0.2 mm. The workspace (area) is the region of space that a robot can reach (Figure 16.13).

To make an informed purchase of a robot, consider the following recommendations [34]:

- **It is necessary to know what to automate:** For companies that have many processes to automate, to start with the simplest applications is best. Complex applications involve more personalized integration, which reduces the return on investment. The most common applications are in material handling and welding. For example, if a company wants to automate a straight-line welding process, a FANUC Arc Mate 100ic 6L robot with an integrated FANUC R30ia controller could be used with a Lincoln Powerwave i400 welder and could be installed between two fixed tables to do the welding. Additionally, it would require the purchase of Lincoln welding supplies, possibly a robot lift, an operator station, and tables.

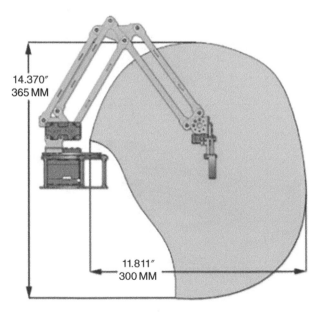

Figure 16.13 Workspace of a robotic manipulator.

- **Determine the condition the robot must be in:** One could buy it new, used and tested and working, or reconditioned. New robots are generally purchased directly from the manufacturer at higher costs, thus obtaining the highest quality product and taking the longest time to recover the investment. Buying a used robot "as is" will be more profitable, but it presents considerable risks; the seller generally does not provide guarantees on the robot. Buying a tested and functional robot can reduce the initial investment and give a better idea of the condition, but it could present problems in the future if proper preventive maintenance is not performed.
- **Delivery time:** The delivery time of a robot depends on the state in which it is purchased (used or new) and from where and to whom it is purchased. Some new robots may have a delivery time of 8–10 weeks after payment, not including any other extras that are required (like a welding effector, etc). Robots can be shipped on the same day, but effectors and add-ons usually take longer. Time to start up and test the robot must be considered in planning.
- **Price comparison:** Check if additional equipment is required for the chosen robot to perform the task and compare different brands that are compatible with the base robot [32–35].

Industrial robots have an average base selling price ranging from $50,000 to $80,000. When including peripherals for specific applications, the cost can increase to between $100,000 and $150,000, which may be expensive for small-scale industries and low-volume production [26].

16.4 Trends on Mobile Robots

Mobile robots serve as versatile platforms, including self-guided vehicles or automated guided vehicles (AGV), AMR, or intelligent guided vehicles (IGV). A mobile robot is designed to perform various activities autonomously.

Mobile robots exhibit various modes of land movement, including those with wheels (WMR), walking or legged robots, gliding or skidding robots (on tracks), robot hybrids, air-based mobile robots, and water-based mobile robots. Depending on the form of movement, there are different types of mobile robots: with wheels, which are used on smooth surfaces; with tracks, used on uneven surfaces and with legs, used on very uneven surfaces (see Figure 16.14).

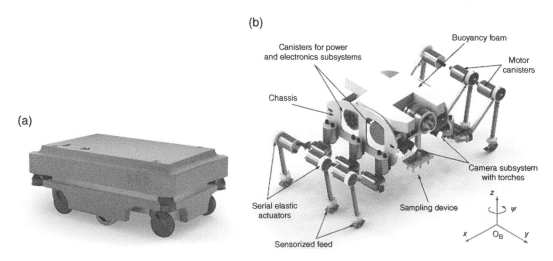

Figure 16.14 Mobile robots. (a) In manufacturing. (b) Legged robots.

AGV and AIV robots are specifically designed for Industry 4.0 and multifunctional applications, seamlessly integrated into systems. These devices have a robust frame and a command-and-control system. The AGVs' movement is guided by a conductive cable installed underground, which constitutes a simple method, not very flexible and with limited travel capability, depending upon the installed cable; this allows the load to be transported autonomously. AIVs are next-generation robots, capable of learning and moving naturally, without a driver.

The AMR robot autonomously navigates along a programmed route, using sensors to create trajectories, avoiding fixed objects (buildings, racks, etc.) and moving objects (people, forklifts, and debris). They are like AGVs but have more flexibility and can work collaboratively with humans in collection and sorting operations. An AMR can perform inventory tasks in a facility; assist in the selection process; be a flexible sorting solution; and perform the harvesting operation. The move operation is expensive and consumes about 75% of the total time of a task [36]. The different models are equipped, for example, with handling technologies such as tilting tray conveyor rollers and transverse belt systems [37].

16.4.1 Managing Robotics Technology: Mobile Robot Applications in Enterprises

Mobile industrial robots are currently used in diverse applications, including those of healthcare, home and industrial security, space, and ocean exploration, and the food service industry, alongside distribution applications.

In manufacturing, robots help increase productivity and efficiency, thus reducing production costs. Many robots in manufacturing work collaboratively with humans on repetitive, monotonous, or complex tasks under the guidance and control of the worker.

Warehouse and distribution center robotics can be automated to increase productivity and efficiency, using AGV, AS/RS, and AMR. The trend is to take advantage of robotic picking and thus reduce labor costs and increase efficiency. AMRs offer the flexibility to create routes between locations within a warehouse and execute order fulfillment operations. AMR transport, collect and deliver the product so that workers can perform other tasks such as picking up, checking, or packing an order (Figure 16.15).

Mobile robots use their properties of locomotion, perception, cognition, and navigation to perform a task. Locomotion enables movement using the proper mechanics, kinematics, dynamics,

Figure 16.15 Mobile robot for warehouse. *Source:* chesky/Adobe Stock.

and control theory. Perception allows it to acquire signals perceived by sensors to know its work environment and to perform signal analysis. With this, it is possible for it to know its destination and current location of the robot. In some cases, the sensor is a vision camera. The control system analyses the data received from the sensors, and decides, based on the data analysis, the actions to send to the actuators, that is, to achieve the objectives commanded to the mobile robot. Robots have planning algorithms, which enable navigation from a starting position to a final one [38].

The most essential of sensors in mobile robots are those of touch, torque, encoders, infrared sensing, ultrasonic sensing, sonar sensing, active beacons, accelerometers, gyroscopes (to decide navigation), and of computer vision (recognition and tracking of objects) [39]. For further precision, the robot utilizes non-linear control algorithms. The most popular control strategies are calculated torque control methods; robust control; slider mode control; adaptive; neural networks (approximate functions and model uncertainty); fuzzy logic (used to deal with imprecision and uncertainty); zero point of the moment; control based on approximate linearization: parametric uncertainty of the model [40]; and Control based on Lyapunov's theory [41].

There are robots that use maps of the environment to plan trajectories and, together with other algorithms, to plan how they should interact with the environment. The planned movement must be carried out without collisions or other problems. The navigation of a mobile robot allows it to move to a destination point using sensor measurements with which decisions are made and signals are thereafter sent to actuators. Motion planning techniques are used to route. If there is accuracy the location problem is solved. Localization implies knowing the absolute and relative position with respect to a target. The sensors are essential in the location task. This location helps the robot to map its surroundings and itself within them; the techniques to do that are presented in reference [36, 42, 43].

Robot position techniques are classified as: (i) Probabilistic map-based location, (ii) Markov location, (iii) Kalman filter location, (iv) Monte Carlo location, (v) Reference point-based navigation, (vi) Globally unique location, (vii) Positioning beacon systems, (viii) Route-based location, (ix) Autonomous map construction, (x) The stochastic mapping technique, (xi) Cyclical environments, and (xii) Dynamic environments [36, 44–53]. The navigation of the mobile robot must generate a model as a map and determine a route without obstacles from the starting to the destination position [54].

16.4.2 Map Representation

The mapping techniques in route planning are of the following varieties: map-based probabilistic, Markov, Kalman filter, Monte Carlo, waypoint-based, unique location around the world, positioning beacon systems, route-based location, construction of autonomous maps, stochastic map technique, cyclical environments, and dynamic environments [55, 56].

Motion planning techniques are used for the planning of trajectories, which is shared. In route planning, the best feasible route for the robot to navigate on without collisions is found, which ranges from an initial to a final location [57]; for drones [58] and underwater robots [59–61]. There are algorithms and techniques for classic, probabilistic, heuristic, sensor-based, and evolutionary-algorithm – based planning methods. These first algorithms consider the time required to complete the trajectory, and the current ones optimize parameters of applied torque, energy consumed, over-acceleration, etc. through genetic algorithms (GA) [62, 63], particle swarm optimization (PSO) [64–66]; optimization of ant colonies (application to mobile robotics [67–70], simulated annealing (SA), numerical technique [71], and mobile robotics [72, 73].

Track planning: In many applications, mobile robots must follow a predefined trajectory; this tracking is considered part of motion planning, which, in turn, consists of planning and following a trajectory in the presence of noise and uncertainty. For route tracking, a control law must be established so the difference between the planned route and the actual route followed by the mobile robot is negligible [74].

Obstacle avoidance: A mobile robot navigates by following a path to the target without hitting any obstacle. This process requires a map and a motion planner to detect collisions between the robot and obstacles and command path changes or stops before collisions occur. The most popular methods for avoiding obstacles are those that depend on geometric models or topological maps of the environment. Other map-making methods depend on sensor measurements, some others use the error algorithm, which traces the contour of the obstacles encountered in the robot's path and circumnavigates them [75]. The proximity diagram (ND) method uses a "divide and conquer" approach [76]. The flex speed method considers the dynamic limitations of the vehicle, allowing it to move quickly in a dense environment. The vector field method obtains distances from the sensors that it uses in the route planner to detect obstacles and avoid collisions. Range-based sensors (ultrasonic and laser rangefinders) are well-suited for these tasks [77]. In yet another method, the mobile robot is considered a particle immersed in a potential field created by the target and obstacles in the work environment [57].

16.4.3 Managing Robotics Technology: How to Choose Mobile Robots

Mobile robots are suitable for various applications. AMRs constitute an efficient, simple, and cost-effective solution to automate material handling, internal transportation, and heavy routine tasks with good hygiene standards; this helps reduce the workload of the employees while avoiding contamination [33].

Consider the following when selecting a mobile robot for load transportation:

1) Consider the loads to be moved. Lower payload capacity is typically less expensive and more agile than heavier payload models.
2) Maximize investment when implementing the robot. Choose a flexible mobile robot that moves without predetermined routes and can reprogram its own trajectory in real-time. At work, the robot must be enabled for collaboration with humans, other robots, and other machines and must be safe for humans. To maximize investment, a fleet management system must be in place to coordinate AMRs, share tasks, and minimize the number of robots required.
3) Battery life. When selecting a mobile robot, choose a powerful battery with up to 11 hours of activity time and 9000 charge cycles. This guarantees 10 years of 24/7 operation with a charging time of just 36 minutes.
4) Add-ons for mobile robots. Consider the ability of robots and machines to interact smoothly; this allows modification of production runs quickly and easily. AMRs can recognize and learn from their environment and make independent decisions. An example of the use of mobile robots in food and beverage manufacturing is the handling and recycling of garbage containers.

The average AMRs cost is in the range of $30,000–$50,000, plus peripherals for a given application, it rises to between $70,000 and $90,000 [78].

16.5 Trends on Drone Robots

Drones are UAV or UAS, which have an RC and can be programmed. The former, the UAV, was originally designed for recreational purposes but is now used professionally in photography and video, to transport cargo, to inspect bridges and chimneys, to observe wildlife, etc.

The drone industry is growing due to advances in this technology that have made it more accessible and profitable. As a result, there are now thriving UAV communities [79, 80]. In the United States, Canada, and other countries, there already are laws and regulations on the use of remote-controlled drones, which need to be improved.

Drones have batteries that last as long as smartphone batteries. Drones have brushless motors, a camera, a headless mode, built-in GPS, follow-me mode, a gimbal, and obstacle avoidance. The best way to learn about UAVs is to fly a drone. A ready-to-fly (RTF) pre-built quadcopter is worth less than $180, which has a battery life of 30 minutes, no stabilization system, a lower quality camera and is light-weight (which implies difficulties with the wind). Examples of UAVs are the Scoot Mini Drone, UFO 4000 LED Mini Drone, Holy Stone HS270 2.7K Drone, DJI Tello, and the U49WF FPV Camera Drone. Among high-quality camera drones are the following: DJI Mavic 2 Pro, DJI Mavic 2 Zoom, DJI Mavic Mini, and the U49WF FPV Camera Drone.

Attributes of drones: Quadcopters usually have a square X or H frame, and they are stable and reliable. The four propellers of most quadcopters generate enough lift for 1 kg, maneuvering smoothly against winds of up to 25 km/h. They have a gyroscope to measure the pitch, roll, and yaw of the quadcopter. With this information, the quadcopter can determine its position in space and can adjust each motor, thus being enabled to float in place. The pilot has a transmitter control to operate the quadcopter; additionally, quadcopters can rotate 360°.

Another trend is the use of semi-autonomous drones to execute flights, which reduces dependency on operators. Drones have a simple structure with aerodynamic efficiency and the ability to withstand long flights at high speeds (Figure 16.16).

The NE-SAC (Northeast Space Applications Centre, USA) has incorporated UAVs to facilitate large-scale real-time mapping, especially in areas affected by landslides, and in assessing damage to infested crops. Mapping will keep a significant growth rate until 2027, primarily due to the expansion of the agricultural sector (these will perhaps be used for monitoring agriculture). These can provide, in real-time, information on crop damage, with which farmers can make more effective decisions. This will drive the expansion of the commercial drone market in various areas [80].

European companies are developing innovative drone technologies, boosting the commercial drone market, which is estimated to represent more than 10% of the global market share by 2027. An example corresponds to FlyNex GmbH in collaboration with KSI Data Sciences in March 2021 to enable real-time data transfer from drones by integrating the FlyNex Enterprise Suite with KSI's Mission Keeper platform. Companies operating commercial drones such as Drone Deploy, DJI, Delair, Cyberhawk, BAE Systems, and Airobotics Ltd., are focused on developing new solutions for the global market. For example, BAE Systems received an IDIQ (Indefinite Delivery/Indefinite Quantity) contract in October 2020 to develop a digital design for UAVs that will be capable of autonomous functions under the Skyborg program.

Since 2016, the US drone industry's growth has been aided by the decision of the Federal Aviation Administration (FAA) to grant exemptions in insurance, construction, and agriculture and a regulatory framework in the operation of drones in the United States. This means that many companies are now capitalizing on these exemptions and turning drones into a thriving part of the industry [80] (Figure 16.17).

Figure 16.16 Drone carrying a box. *Source:* Denys Rudyi/Adobe stock. https://stock.adobe.com/images/drone-delivering-a-package/89418474.

Figure 16.17 Drone in agriculture.
Source: kinwun/Adobe Stock.

The drone services market is estimated to be valued at $63.6 billion by 2025, and Insider Intelligence predicts that drone shipments to consumers will amount to 29 million this year and that global sales will reach 2.4 million in 2023, with a compound annual growth rate (CAGR) of 66.8%. According to Statista, sales of US drones to distributors were, in 2020, valued at $1.25 billion. According to Goldman Sachs, it is forecasted that the total size of the drone market will be of $100 billion, driven by demand from the commercial and government sectors. Drone growth will occur in five industry segments: agriculture, construction and mining, insurance, disasters, law enforcement, media, and telecommunications [81, 82].

The UN projects that the world population will reach 9.7 billion by 2050, causing an increase of 69% in agricultural consumption between the years 2010 and 2050. Since many of these farmers and agribusinesses manage thousands of acres of land, it is estimated that they will use drones for crops and livestock, irrigation management, and for fertilization. Drone Fly company estimates that drones can spray fertilizer 40–60 times faster than humans. The use of drones in construction and mining could constitute a global market of $28.3 billion. Drones are used to enforce laws and regulations on worker safety; in construction, regular inspection to ensure standards is carried out. These inspections typically take 10 hours to a few days, but, using drones, these can be done in about 15 minutes [83].

The global average annual cost of insurance claims for natural disasters is estimated to have increased eightfold since 1970. Insurance companies are likely to use drones to perform property appraisals more quickly and are interested in assessing disaster damage with images and videos and in making real-time assessments. The global average annual cost of insurance claims for natural disasters is estimated to have increased eightfold since 1970 [79].

Police forces are using drones to enforce the law, patrol areas, negotiate hostage situations, find armed suspects, and investigate bomb threats. Drones are less expensive than helicopters, which are not always available. Most importantly, these allow the police to navigate dangerous situations, ensuring the safety of their officers.

Drone Industry Outlook. Parrot is the fastest growing company in the drone market, offering cheaper drones for recreational use. There is currently great competition, in the field of high-end drone models that take photos and stream live video.

The Parrot company launched Anafi Thermal in 2019 in response to DJI's Mavic 2 Enterprise Dual 2018, but $700 cheaper. Both are drones with miniature thermal imaging units. The Mavic 2 Enterprise Dual is a better product overall, offering a longer flight time and twice the range. On the other hand, the Parrot product is lighter, and the camera has zoom capabilities (Figure 16.18).

Figure 16.18 Drones applications.
Source: Jale Ibrak/Adobe stock.

16.5.1 Managing Robotics Technology: How to Choose Drone Robots

In choosing a drone, the first thing to consider is its intended use, such as aerial photography and videography; transportation freight; industrial bridge and chimney inspection; wildlife tracking; and others [84].

When deciding on a drone, the following specifications should be considered:

1) **Type of drone (quadcopter or others) and payload capabilities (to lift the load):** It is advisable to choose a flexible drone that can reprogram its own trajectory in real-time. The robot must be capable of collaboration with humans and with other drones and must be safe for humans. To maximize investment, a fleet management system could coordinate drones, share tasks, and minimize the number of drones required.
2) **Battery life:** When selecting a drone, choose a powerful battery with the necessary activity time per day.
3) **Brushless motors:** Consider quantity, type (brushless motor), and power.
4) The stabilization system, the follow-me mode, the gimbal, and the drone's weight (risk with the wind).
5) The gyroscope or accelerometer, which helps determine the drone's position in space, and the drone's ability to automatically adjust each motor.
6) Integrated GPS.
7) Camera quality.
8) Transmitter controller for personal and automatic operation.
9) **Add-ons for mobile robots:** Obstacle avoidance and included software. One should also consider if the drone is easy to fly, if it can move freely without predetermined paths, and if it can reschedule a task by calculating its trajectory in real-time. The quadcopter units most used by aerial videographers are the DJI Mavic 2 Pro, DJI Mavic 2 Zoom, DJI Mavic Mini, U49WF FPV Camera Drone, DJI Inspire 2, DJI Phantom 4 Pro V2.0, and the Skydio 2. This list is not comprehensive, but, nonetheless, these are the most popular on the market.

The average drone cost depends on the amount of weight a drone can carry. It can be 5–200 kg. The range is $3000–$20,000, plus peripherals for a given application, it rises to between $3000 and $30,000 [78].

16.6 Trends in Robots Applied to Health

There are two main applications for robots in healthcare: therapeutic assistance; contagion reduction; and, also, improving quality of life in healthcare facilities [85–87]. The objective for these robots is to help health professionals in the areas of telemedicine, decontamination and disinfection, delivery of medicines and food, transportation of goods, patient monitoring, patient companionship, and preventive support.

In relation to patients, robotics can help hospitals transport elderly or immobile patients and comfort patients with mental health problems. Robotic exoskeletons can help and guide patients who have suffered strokes or spinal cord injuries or who are otherwise paralyzed. Airports are already using mobile robots to spray disinfectant chemicals on their premises [14].

Regarding contagious diseases, robotics can help manufacturing assembly lines maintain social distancing. In various industries, robots can cooperate and collaborate with humans, allowing workers to keep distance from sick people. Manufacturers of detergents, diapers, toilet paper, and other household items have already incorporated robots and cobots into their production [84].

Robotics and artificial intelligence technologies have been present in healthcare for a long time in surgery with teleoperation; in the management of harmful substances, disinfection, and prevention; and in assisting patients, doctors, and health personnel with miscellaneous tasks. During the Covid 19 pandemic, these technologies have made it possible to find new solutions through research with new approaches and adaptation of existing ones; these technologies have supported studies on viruses and vaccine experimentation [7–13]. The International Federation of Robotics indicates that operational service robots and cobots are increasing by the year, particularly in health, to help doctors and professionals, to reduce contact between people, to prevent humans from performing dangerous operations, and to help disinfect patient rooms [3].

16.7 Robotics Technology Management: How to Choose Robots for Health

Management in the selection of robots (cobots) for health [84, 88, 89].

1) Identify risky areas to be disinfected; consider if they may control temperature, humidity, and other variables of interest in healthcare. Consider food and medicine delivery, and infection risks for staff.
2) Consider autonomous disinfection of hospital rooms. An example of a robot with such purpose is the Danish UVD robot for autonomous disinfection without the intervention of medical personnel. Mobile robots like Fetch Robotics can perform autonomous disinfection. The SmartGuardUV robot can disinfect up to 99.9% of viruses and bacteria.
3) Consider if the robot can deliver medications and food, run blood tests, and help with general hospital logistics.
4) Consider if the robot can monitor the vital signs of patients, can monitor rooms, and it can communicate these to health personnel.
5) Consider if the robot can accelerate health tests, automating them.

The average cost of cobot varies from US$10,000 to US$50,000.

16.8 Conclusions

In this chapter, trends in robotic management and business automation have been presented, as well as the uses and trends corresponding to each type of robot, including robotic manipulators, mobile robots, and robotic drones; specifications have been presented. Besides, some suggestions and recommendations for engineers and entrepreneurs on purchase decisions, on possible applications, and on the integration of different types of robots have been presented.

This technology relies on sensors and actuators, devices that, combined with artificial intelligence, are easy to program and install and comply with the criteria of industry 4.0. These robots are effective for providing better quality in products and services with less product rejection and failures, and, if renewable energies are considered, the result is a more cost-effective and ecological production. It is possible, then, to implement new business models with which producers can diversify more easily than ever, and which are applicable to the different sectors of the economy.

References

1 Heer, C. (2022). World Robotics Report: "All-Time High" with Half a Million Robots Installed in one Year – International Federation of Robotics (ifr.org). IFR Press. https://ifr.org/.

2 Naveed, A., Sikandar, K., Muhsan, E. et al. (2022). Estimating the total volume of running water bodies using geographic information system (GIS): a case study of Peshawar Basin (Pakistan). *Sustainability* 14: 3754.

3 MH&L Staff. Top 5 robot trends in 2021. Material handling and logistics (mhlnews.com). https://www.mhlnews.com/technology-automation/article/21155725/top-5-robot-trends-in-2021.

4 Rubio, F., Valero, F., and Llopis-Albert, C. (2019). A review of mobile robots: concepts, methods, theoretical framework, and applications. *International Journal of Advanced Robotic Systems* 1–22. https://doi.org/10.1177/17298814198395.

5 ABI Research. The enduring risk of robots in the public sphere. 1Q 2022|IN-6492. https://www.abiresearch.com/market-research/insight/7780521-the-enduring-risk-of-robots-in-the-public-/.

6 Cohen, Y., Shoval, S., Faccio, M., and Minto, R. (2021). Deploying cobots in collaborative systems: major considerations and productivity analysis. *International Journal of Production Research* https://doi.org/10.1080/00207543.2020.1870758.

7 Martiñon, S. and Hernández-Miramontes, R. (2021). *Use of Exoskeletons in the Treatment and Rehabilitation of Paraplegia Patients*. https://doi.org/10.5772/intechopen.94920.

8 Yang, G.-Z., Nelson, B.J., Murphy, R.R. et al. (2020). Combating COVID-19 – the role of robotics in managing public health and infectious diseases. *Science robotics* 5 (40): eabb5589. https://doi.org/10.1126/scirobotics.abb5589.

9 Khan, Z.H., Siddique, A., and Lee, C.W. (2020). Robotics utilization for healthcare digitization in global COVID-19 management. *International Journal of Environmental Research and Public Health* 17 (11): 3819.

10 Jecker, N.S. (2020). You have got a friend in me: sociable robots for older adults in an age of global pandemics. *Ethics and Information Technology* 1–9. https://doi.org/10.1007/s10676-020-09546-y.

11 Tavakoli, M., Carriere, J., and Torabi, A. (2020). Robotics, smart wearable technologies, and autonomous intelligent systems for healthcare during the covid-19 pandemic: an analysis of the state of the art and future vision. *Advanced Intelligent Systems* 2: 2000071.

12 Chen, B., Marvin, S., and While, A. (2020). Containing COVID-19 in China: AI and the robotic restructuring of future cities. *Dialogues in Human Geography* 10 (2): 238–241.

13 Wei, S. (2020). From virus-slaying air purifiers to delivery robots, how university inventions are fighting COVID-19. *World Economic Forum*. https://www.weforum.org/agenda/2020/03/how-can-universities-fight-coronavirus (accessed 25 February 2021).

14 Mišeikis, J., Caroni, P., Duchamp, P. et al. (2020). Lio-a personal robot assistant for human–robot interaction and care applications. *IEEE Robotics and Automation Letters* 5 (4): 5339–5346.

15 Prindiville, M. (2015). Surgical robots for minimally invasive procedures. https://www.automate.org/tech-papers/surgical-robots-for-minimally-invasive-procedures.

16 Anandan, T.M. (2015). Robots and healthcare saving lives together. https://www.automate.org/industry-insights/robots-and-healthcare-saving-lives-together.

17 Ortenzi, V., Cosgun, A., Pardi, T. et al. (2021). Object handovers: a review for robotics. *IEEE Transactions on Robotics* https://doi.org/10.1109/TRO.2021.3075365.

18 Dugas, D., Nieto, J., Siegwart, R. and Chung, J.J. (2021). NavRep: unsupervised representations for reinforcement learning of robot navigation in dynamic human environments. *2021 IEEE International Conference on Robotics and Automation (ICRA)*, pp. 7829–7835. https://doi.org 10.1109/ICRA48506.2021.9560951.

19 Lins, R.G. and Givigi, S.N. (2021). Cooperative robotics and machine learning for smart manufacturing: platform design and trends within the context of industrial internet of things. *IEEE Access* 9: 95444–95455. https://doi.org/10.1109/ACCESS.2021.3094374.

20 Staff of Master of Science In Electrical Engineering Ohio University (2021). 5 industries utilizing robotics. https://onlinemasters.ohio.edu/blog/5-industries-utilizing-robotics.

21 Ball, D., Ross, P., English, A. et al. (2017). Farm workers of the future: vision-based robotics for broad-acre agriculture. *IEEE Robotics & Automation Magazine* 24 (3): 97–107. https://doi.org/10.1109/MRA.2016.2616541.

22 Huen, E. (2016). The world's first home robotic chef can cook over 100 meals. *Forbes*.

23 Velfl, Z., Humphreys, P., Trifone, A. et al. (2020). Robots: the next generation. www.smc.eu. ROB-WP-A-UK. https://static.smc.eu/binaries/content/assets/smc_be/solutions/robotics_whitepaper_en.pdf.

24 Zhuo, Z., Yang, M., Adeel, M. et al. (2021). Applications of robotics, artificial intelligence, and digital technologies during COVID-19: a review. *Disaster Medicine and Public Health Preparedness* https://doi.org/10.1017/dmp.2021.9.

25 Francesco, C., Ekkehard, E., and Enzo, W. (2018). Robots worldwide: the impact of automation on employment and trade. International Labour Office ILO Working Paper No. 36. http://www.ilo.org/wcmsp5/groups/public/---dgreports/---inst/documents/publication/wcms:648063.pdf.

26 Markets and Markets (2019). Collaborative Robot Market, Market Research Report: Nov 2019. *Report Code: SE 4480*. https://www.marketsandmarkets.com/Market-Reports/collaborative-robot-market-194541294.html (accessed November 2020).

27 Østergaard, E.H. (2017). The role of cobots in industry 4.0. https://info.universal-robots.com hubfs/Enablers/Whitepapers/Theroleofcobotsinindustry.pdf (accessed 25 May 2020).

28 Billard, A. and Kragic, D. (2019). Trends and challenges in robot manipulation. *Science* 364: eaat8414.

29 Ed, R. (2021). *Types and Applications of Autonomous Mobile Robots*, vol. 11. AMRs. https://www.conveyco.com/types-and-applications-of-amrs.

30 Bernier, C. (2021) Articulated Robot Applications. https://howtorobot.com/expert-insight/articulated-robots.

31 Editorial Staff (2020). Robotic sensors – fundamentals of robotics technology. https://roboticsbiz.com/fundamentals-of-robotics-technology-robotic-sensors.

32 Cherry, J. (2020). Fanuc vs. Yaskawa in the battle of the robots. https://seekingalpha.com/article/4353757-fanuc-vs-yaskawa-in-battle-of-robots.

33 Adam, B. (2020). *Future of Manufacturing: How to Select Mobile Robots for Heavy Loads*. Omron Adept Technologies. https://www.eetimes.eu/. future-of-manufacturing-how-to-select-mobile-robots-for-heavy-loads.

34 Bélanger-Barrette, M. (2021). How to choose the right industrial robot? https://blog.robotiq.com/how-to-choose-the-right-industrial-robot.

35 Elazary, L. Warehouse Automation (2021). The paradox of too many choices. https://www.mhlnews.com/warehousing/article/21167323/. warehouse-automation-the-paradox-of-too-many-choices.

36 Castellanos, J.A. and Tardos, J.D. (2012). *Mobile Robot Localization and Map Building: A Multisensor Fusion Approach*, 1–6. Berlin and Piscataway, NJ: Springer Science & Business Media and IEEE Operations Center.

37 Romaine, E. (2020). Types and applications of autonomous mobile robots (AMRs). https://www.conveyco.com/blog/types-and-applications-of-amrs/.

38 Mozaffari, M., Saad, W., Bennis, M. et al. (2019). A tutorial on UAVs for wireless networks: applications, challenges, and open problems. *IEEE Communication Surveys and Tutorials* 21 (3): 2334–2360. https://doi.org/10.1109/COMST.2019.2902862.

39 Peng, L. and Xiangpeng, L. (2019). Common sensors in industrial robots: a review. *Journal of Physics Conference Series* 1267: 012036. https://doi.org/10.1088/1742-6596/1267/1/012036.

40 Behal, A., Dixon, W., Dawson, D.M. et al. (2009). *Lyapunov-Based Control of Robotic Systems*, Automation and Control Engineering Series. CRC Press, Taylor and Francis Group. ISBN 10: 0849370256, ISBN 13: 9780849370250.

41 Rigatos, G.G. (2017). *State-Space Approaches for Modelling and Control in Financial Engineering*. Berlin: Springer.

42 Ye, C. and Borenstein, J. (2002). Characterization of a 2D laser scanner for mobile robot obstacle negotiation. *Proceedings 2002 IEEE International Conference on Robotics and Automation* (Cat. No. 02CH37292), Washington DC, USA (10–17 May 2002), Vol. 3, pp. 2512–2518.

43 Sayed, L.A.A. and Alboul, L. (2014). Vision system for robot's speed and position control. *2014 13th International Conference on Control Automation Robotics & Vision (ICARCV)*, pp. 1010–1014. IEEE Catalog Number: ISBN: CFP14532-POD 978-1-4799-5200-7 2014-ICARCV.pdf (ppns.ac.id). https://doi.org 10.1109/ICARCV.2014.7064444.

44 Ye, C. and Borenstein, J. (2003). A new terrain mapping method for mobile robot's obstacle negotiation. *Proceedings of the Unmanned Ground Vehicle Technology*, Orlando, FL, USA.

45 Gutmann, J.S., Burgard, W., and Fox, D. (1998). An experimental comparison of localization methods. *Proceedings IEEE/RSJ International Conference on Intelligent Robots and Systems*, Vol. 2, pp. 736–743. Victoria, BC, Canada (17 October 1998). IEEE. https://doi.org 10.1109/IROS.1998.727280.

46 Betke, M. and Gurvits, L. (1997). Mobile robot localization using landmarks. *IEEE Transactions on Robotics and Automation* 13 (2): 251–263.

47 Kuffner J, Nishiwaki K, Kagami S, et al. (2005). Motion planning for humanoid robots. In: *Robotics Research. The Eleventh International Symposium. Springer Tracts in Advanced Robotics* (eds. P. Dario and R. Chatila), Vol. 15. Berlin, Heidelberg: Springer. https://doi.org 10.1007/11008941_39.

48 Osorio-Comparan, R., De Vasquez, E.J., Peña, M. et al. (2018). Object Detection Algorithms and Implementation in a Robot of Service. *IEEE ICA ACCA*, Chile.

49 Osorio-Comparán, R., Lopez-Juarez, I., Peña, M. et al. (2016). Mobile robot navigation using potential fields and LMA. *IEEE ICA ACCA*.

50 Osorio, R., Peña, M., López-Juárez, I. et al. (2013). Using background and segmentation algorithms applied in mobil robots. *IFAC MCPL2013 International Conference on Management and Control of Production and Logistics*. Brazil 2013.

51 Daniel, R., Ginno, M., Fernando, P. et al. (2014). Algorithms for maps construction and localization in a mobile robot. *Studies in Informatics and Control* 23 (2): 189–196. ISSN 1220-1766, Publicación ISI WEB of Science.

52 Fredes, D., Cubillos, C., and Lefranc, G. (2012). Mobile robot with multi agent architecture. *IEEE CESA, Conference International on Engineering and Systems Applications*, Santiago de Chile.

53 Latorre, H., Harispe, K., Salinas, R., and Lefranc, G. (2012). Multi-agent robots' model for collaborative and cooperative work. *IEEE CESA, Conference International on Engineering and Systems Applications*, Santiago de Chile.

54 Hughes, I., Millán, G., Cubillos, C., and Lefranc, G. (2014). Colony of robots for exploration based on multi-agent system. *International Journal of Computers, Communications & Control* 9 (6): 103. Publicación ISI WEB of Science.

55 Moravec, H. and Elfes, A.E. (1985). High resolution maps from wide angle sonar. *Proceedings of the 1985 IEEE International Conference on Robotics and Automation*, pp. 116–121. St. Louis, MO, USA (25–28 March 1985). IEEE. https://doi.org 10.1109/ROBOT.1985.1087316.

56 Simhon, S. and Dudek, G. A global topological map formed by local metric maps. *Proceedings of the IEEE/RSJ International Conference on Intelligent Robots and Systems (IROS'98)*, Victoria, BC., Canada (13–17 October 1998).

57 Lee, U.J. and Bien, Z. (2002). Path planning for a quadruped robot: an artificial field approach. *Advanced Robotics* 16 (7): 609–627: https://doi.org/10.1163/15685530260390746.

58 Soto, M., Nava, P., and Alvarado, L.E. (2007). Drone formation control system real-time path planning. *AIAA Infotech@Aerospace 2007 Conference*, Rohnert Park, California (7–10 May 2007). https://doi.org 10.2514/6.2007-2770.

59 Kosari, A., Maghsoudi, H., Sabatian, S.M. et al. Trajectory planning and trajectory tracking of a submarine robot based on minimization of detection capability. *22nd Iranian Conference on Electrical Engineering (ICEE)*, Iran (May 2014). https://doi.org 10.1109/IranianCEE.2014.6999729.

60 Petres, C., Pailhas, Y., Patron, P. et al. (2007). Path planning for autonomous underwater vehicles. *IEEE Transactions on Robotics* 23: 2.

61 Yigit, K. (2011). Path planning methods for autonomous underwater vehicles. Thesis. MIT, Cambridge.

62 Yang, Z.Q., Liu, L.B., and Yang, W.D. (2008). Flexible inspection path planning based on adaptive genetic algorithm. *Proceedings of Control and Decision Conference* 1558–63: 104–111.

63 Abu-Dakka, F., Valero, F., Mata, V. et al. (2007). Path planning optimization of industrial robots using genetic algorithm. *Proceeding's 16th International Workshop on Robotics*, pp. 424–429. Ancona, Italy.

64 Qin, Y.Q., Sun, D.B., Li, N. et al. (2004). Path planning for mobile robot using the particle swarm optimization with mutation operator. *Proceedings of 2004 International Conference on Machine Learning and Cybernetics* 4: 2473–2478.

65 Niehaus, C., Rofer, T., and Laue, T. (2007). Gait optimization on a humanoid robot using particle swarm optimization. *Proceedings of the Second Workshop on Humanoid Soccer Robots*, pp. 1–7. Pittsburgh.

66 Sun, B., Chen, W.D., and Xi, Y.G. (2005). Particle swarm optimization based global path planning for mobile robots. *Control and Decision* 20 (9): 1052.

67 Garcia, M.P., Montiel, O., Castillo, O. et al. (2009). Path planning for autonomous mobile robot navigation with ant colony optimization and fuzzy cost function evaluation. *Applied Soft Computing* 9 (3): 1102–1110.

68 Brand, M., Masuda, M., Wehner, N. et al. (2010). Ant colony optimization algorithm for robot path planning. *International Conference on Computer Design and Applications* ICCDA 2010, Vol. 3, pp. V3–436. Qinhuangdao, China (25–27 June 2010). IEEE.

69 Cong, Y.Z. and Ponnambalam, S.G. (2009). Mobile robot path planning using ant colony optimization. *IEEE/ASME International Conference on Advanced Intelligent Mechatronics*. AIM 2009, pp. 851–856. Singapore (14–17 July 2009). Piscataway, NJ: IEEE.

70 Liu, C.A., Yan, X.H., Liu, C.Y. et al. (2011). Dynamic path planning for mobile robot based on improved ant colony optimization algorithm. *Acta Electronica Sinica* 5: 042.

71 Zhu, Q., Yan, Y., and Xing, Z. (2006). Robot path planning based on artificial potential field approach with simulated annealing. *Sixth International Conference on Intelligent Systems Design and Applications*. ISDA'06 (eds Y. Chen and A. Abraham), Vol. 2, pp. 622–627. Jinan, China (16–18 October 2006). Korea: Chung-Ang University.

72 Miao, H. and Tian, Y.C. (2013). Dynamic robot path planning using an enhanced simulated annealing approach. *Applied Mathematics and Computation* 222: 420–437.

73 Masehian, E. and Sedighizadeh, D. (2007). Classic and heuristic approaches in robot motion planning-a chronological review. *World Academy of Science, Engineering and Technology* 23: 101–106.

74 Jimenez, P., Thomas, F., and Torras, C. 3D collision detection: a survey. Computers and graphics 2001; 25 Ed Romaine. Guide To Autonomous Mobile Robots (AMRs) License.

75 Wei, S. and Zefran, M. (2005). Smooth path planning and control for mobile robots. *Proceedings. 2005 IEEE Networking, Sensing and Control*, Tucson, AZ, USA, pp. 894–899. http://doi.org/10.1109/ICNSC.2005.1461311.

76 Arkin, R. (1999). *Behavior-Based Robotics*. Cambridge: MIT Press.

77 Kunchev, V., Jain, L., Ivancevic, V. et al. (2006). Path planning and obstacle avoidance for autonomous mobile robots: a review. In: *KnowledgeBased Intelligent Information and Engineering Systems*, Lecture Notes in Computer Science, vol. 4252 (ed. B. Gabrys, R.J. Howlett, and L.C. Jain), 537–544. Berlin: Springer.

78 Sugla, S., Gupta, A., and Garg, A. (2021). Markets and markets (2021–2027). Collaborative Robot Market Size & Share, Industry Report. http://www.marketsandmarkets.com/Market-Reports/collaborative-robot-market-194541294.html.

79 Huang, H. and Savkin, A.V. (2020). A method of optimized deployment of charging stations for drone delivery. *IEEE Transactions on Transportation Electrification* 6 (2): 510–518. https://doi.org/10.1109/TTE.2020.2988149.

80 Constantine, D. (2020). The future of the drone economy. A Report of Levitate Capital. Levitate-Capital-White-Paper.pdf. http://levitatecap.com.

81 Hegde, A. (2021). Commercial drone market 2021–2027, top 3 trends enhancing the industry expansion; global market insights Inc. https://www.prnewswire.com/news-releases/commercial-drone-market-2021-2027-top-3-trends-enhancing-the-industry-expansion-global-market-insights-inc-301341067.html.

82 Clark, D. (2023). Drone market outlook in 2021: industry growth trends, market stats and forecast. *Insider Intelligence*. https://www.businessinsider.com/drone-industry-analysis-market-trends-growth-forecasts.

83 Illiushkin, N., Ismagilov, L., Korol, M. (2021). Skygauge robotics. The drone. Reinvented for industrial inspections. https://www.skygauge.co/the-skygauge.

84 Alan, P., Lana, A., Judy, H., and Mike, M. (2021). How to buy a drone. https://uavcoach.com/buy-a-drone.

85 Hariri-Ardebili, M.A. (2020). Living in a multi-risk chaotic condition: pandemic, natural hazards, and complex emergencies. *International Journal of Environmental Research and Public Health* 17: 5635.

86 Hussein, H.A., Hassan, R.Y.A., Chino, M., and Febbraio, F. (2020). Point-of-care diagnostics of COVID-19: from current work to future perspectives. *Sensors* 20: 4289.

87 Kaplan, E.H., Wang, D., Wang, M. et al. (2020). Aligning SARS-CoV-2 indicators via an epidemic model: application to hospital admissions and RNA detection in sewage sludge. *Health Care Management Science*. https://doi.org/10.1007/s10729-020-09525-1.

88 Anandan, T.M. (2020). *Reducing COVID Risks with Robots*. Association for Advancing Automation. https://www.automate.org/industry-insights/reducing-covid-risks-with-robots.

89 Tisaire, J.E. Rodijk-Rozeboom, E., Bisset, D. et al. (2020). 10 ways robots fight against the COVID-19 pandemic euRobotics. The European Commission promotes robotics use in the fight against COVID19. euRobotics Aisbl. https://old.eu-robotics.net/eurobotics/newsroom/press/the-european-commission-promotes-robotics-use-in-the-fight-against-covid19.html?changelang=1.

17

Healthcare Through Data Science – A Transdisciplinary Perspective from Latin America

Parag Chatterjee and Ricardo Armentano

Department of Biological Engineering, CENUR Litoral Norte, University of the Republic (Universidad de la República), Paysandu, Uruguay

17.1 Introduction – Data Science in Healthcare

The field of data science has seen a sharp push in recent years, especially through its multifarious applications in different domains. Spanning from smart transport to smart education, smart healthcare to smart services, data science has found its use extensively. Healthcare is one of the key areas where data science holds immense potential. The inclusion of paradigms like artificial intelligence and data analytics has initiated a new era of "smart healthcare," where medical decisions do not rely only on the expertise of the healthcare personnel but also on the insights obtained from the health-data. Intelligence from health-data is harvested using different types of algorithms and analytical tools, providing a deeper understanding of patterns, correlations, and other aspects of the health-data.

Predictive and personalized healthcare has an innovative approach to the use of data science in healthcare since parameters like family history, individual differences in genes, lifestyle, and environment along with various biomarkers can be used toward the prediction of diseases. One of the key objectives is to treat patients in a personalized manner but also to transfer the spectrum of healthcare to healthy people as well. This is a fundamental paradigm shift in the way healthcare services are being provided. Predictive healthcare includes the aspect of offering continuous monitoring to people who are not even sick, to identify risks with timely anticipation and subsequent care if needed. Instead of considering healthcare as only an event-driven system, data science has helped elaborating the outreach of healthcare beyond the boundaries of healthcare facilities along with the timeframe of these healthcare services, which, thanks to the smart monitoring devices, can be offered in a continuous basis.

This work highlights the importance of data science along with other new paradigms like Internet of Things (IoT) in healthcare toward providing more efficient, comprehensive, and personalized healthcare services. In this respect, a case study of the National Liver Transplantation Program in Uruguay is presented, illustrating the application of predictive analysis and machine learning in early detection and prevention of cardiometabolic diseases.

IEEE Technology and Engineering Management Society Body of Knowledge (TEMSBOK), First Edition.
Edited by Gustavo Giannattasio, Elif Kongar, Marina Dabić, Celia Desmond, Michael Condry, Sudeendra Koushik, and Roberto Saracco.
© 2024 The Institute of Electrical and Electronics Engineers, Inc. Published 2024 by John Wiley & Sons, Inc.

17.2   e-Health – A Transdisciplinary Approach to Management

The domain of eHealth is an emerging field in the intersection of medical informatics, public health, and business, referring to health services and information delivered or enhanced through the Internet and related technologies. WHO defines eHealth as a cost-effective and secure use of information and communications technologies in support of health and health-related fields, including healthcare services, health surveillance, health literature, and health education, knowledge, and research. However, the significance of eHealth has surpassed the primary technical areas and is considered more of a collective concept through networking, aimed at a comprehensive improvement of healthcare powered by digital tools [1, 2].

Considering the extensive reach of eHealth, it has two fundamental pieces – the technological aspect, and the aspect of human resources. Just as to build up a comprehensive eHealth system, a wide variety of technological tools need to be joined together, a diverse human resource also plays a key role in its successful implementation. With respect to human resources, eHealth has opened its frontiers far beyond the usual reach of the medical domain. Recent projects related to eHealth include not only medical professionals, but also people from different disciplines like biology, statistics, computer science, mathematics, and other allied fields. Data science or more specifically artificial intelligence plays a crucial role in this respect. Applying artificial intelligence in healthcare to generate insights from health-data and interpreting that needs a coordinated transdisciplinary approach, that includes professionals from medicine as well as other allied fields.

In the technological wing of eHealth, the sharp rise in emerging trends like the IoT has facilitated the conglomeration of technologies. The IoT is basically the network of physical objects that are embedded with sensors, software, and other technologies that are used for the purpose of connecting and exchanging data with other devices and systems over the Internet. Especially in a healthcare scenario, IoT enables even small devices to be converted into monitoring devices, that send data to the cloud, which could be further processed and analyzed for detailed insights (Figure 17.1).

Data science has a fundamental role in analyzing the enormous amount of health-data generated from smart eHealth devices. One of the key objectives in present-day healthcare is the prediction of diseases and risks. As a part of the new healthcare paradigm which advocates preventive care through continuous monitoring, healthcare is no more centralized in healthcare centers like hospitals; rather it is extended virtually to people beyond the closed institutional framework. This approach, on one hand, reduces the primary and auxiliary costs involved with hospital visits, and on the other hand, fortifies preventive measures considering the predictive analysis on the continuously monitored health-data.

From a Latin American perspective, the importance of digitization of healthcare is high. Especially in a scenario where about 30% of the population does not have access to healthcare and not many countries in the region meet international indicators on doctor-to-patient ratio, digital tools have a highly promising role [5]. As the regional expenditure in healthcare counts only 6.6% percent of its GDP [6], technology in healthcare needs to be ubiquitous, and permeated through already existing infrastructure, like smartphones. The region has an average of two doctors per 1000 population, and most countries stand well below the Organization for Economic Co-operation and Development's average of 3.5, with only Cuba, Argentina, and Uruguay having more. The scarcity of healthcare resources also leads to utilizing eHealth options, which, due to the recent COVID-19 pandemic, has seen a sharp rise, especially in the aspect of virtual healthcare [7].

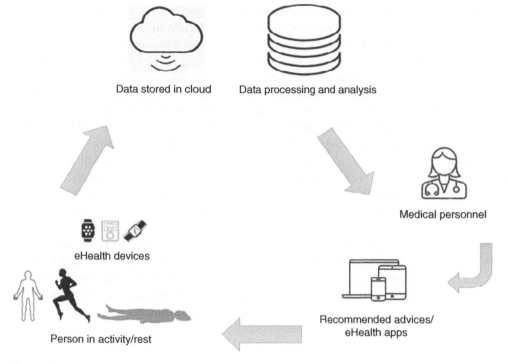

Figure 17.1 Holistic view of IoT-based healthcare system. *Source:* [3, 4].

Considering the transdisciplinary approach for *smart healthcare*, we present a case study with respect to a transdisciplinary group primarily from Argentina and Uruguay. The implementation of eHealth consists of various elements, including medical professionals and experts from other fields as well. Digital healthcare involves different stakeholders; as in the case of IoT-based eHealth systems, engineers and data scientists constitute an important component of the workforce. Similarly, the implementation of eHealth involves a continuous and seamless collaboration between the different stakeholders, which includes students as well. This transdisciplinary research group is constituted of medical doctors, engineers from biomedical, electronics, and biological fields, researchers in computer science, along with students. Through institutional partnership that includes universities, hospitals, and research centers, students along with professionals play a strong role in the construction of the human resource for seamless functioning of the eHealth systems (Figure 17.2).

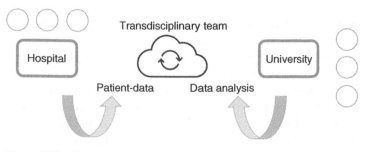

Figure 17.2 Transdisciplinary management between hospital and university resources.

The transdisciplinary approach holds a strong importance in providing a fused platform, especially in the domain of smart healthcare. From a technical point of view, management of the shared resources between the entities like hospital and university are crucial to develop a smart healthcare system. The hospital is usually the source of crude data, containing detailed information about the patients. In a shared transdisciplinary environment, the hospital shares its data resources. On the other hand, the university through its human resource provides services in efficient data analysis and applying computational intelligence on the health-data. This involves a feedback loop, where the interpretations and insights of the results obtained through data analysis are driven back to the hospital. Considering a bigger environment, this holds the base of a smart database, that receives crude data from the hospital end, passes through data processing and analysis using machine learning algorithms, and sends back the final results and insights to the hospital-end, to be applied in treatment. The development of a connected network between all the stakeholders, along with a shared transdisciplinary space is fundamental for the development of large-scale functional systems of smart healthcare.

Also, from a populational perspective, this facilitates a comprehensive and holistic view of the cohorts in terms of their health risks. Instead of an event-based approach, predictive healthcare is redefined as a continuous process, that involves a robust management of all the stakeholders along with the periodic update of the treatment strategies based on the insights obtained from the clinical data. Harnessing the power of artificial intelligence, the transdisciplinary model facilitates the early prediction of risks along with stratification of the entire population in terms of their risks and vulnerabilities.

17.3 Machine Learning and Predictive Analytics – Case Study of Cardiometabolic Diseases in Liver Transplantation

In Latin America, cardiovascular diseases produce almost a million deaths a year, becoming the principal cause of death in recent years, with projections that the number of deaths in the region attributable to cardiovascular diseases will increase in the near future [8]. Several factors like obesity, smoking, and other lifestyle aspects increase substantially the risks of chronic diseases in this region. Preventive healthcare addresses this aspect specifically, offering a constant monitoring to the people, followed by continuous analysis of the health-data. Artificial intelligence plays a significant role in data analysis, highlighting risks and anticipating the same, which when properly communicated to the people, offers the possibility of preventive measures. This case study is being presented in the context of the National Liver Transplantation Program in Uruguay, where machine learning is applied to intelligent prediction of cardiometabolic risks in patients.

For patients with end-stage liver disease, liver transplantation stands as the last therapeutic option. It involves a complex procedure of treatment and consequent healthcare, requiring an interdisciplinary team of experts, and a thorough monitoring of the patients during the timeline of the treatment, as well as in the post-transplantation period. This entire procedure generates a large volume of multifarious data and information. In this respect, a robust clinical management of transplant patients sheds important impact on their vital prognosis, and clinical decisions taken on the basis of multiple health parameters [9]. The eHealth approach consists of the key technological part, as well as the human resource responsible for carrying out the entire process.

In the first case, the patients undergo through a thorough analysis of their health indicators at the entry point of the transplantation program. The health-data obtained from these patients are analyzed using machine learning algorithms. Primarily, supervised, and unsupervised models are used. In case of the prediction of transplant event, supervised models showed a better accuracy. With the objective of predicting the transplant event and assessing the impact of cardiometabolic features in the transplantation, through predictive model using logistic regression (Figure 17.3), a precision of 0.70 was reached, along with test accuracy 0.76, balanced accuracy 0.73, AUC 0.74 and AIC 81.25. The logistic regression being represented as $p = 1/(1 + e^{(-y)})$ where $y = (0.33 + 0.07*\text{sex values} + 0.07*\text{pre-transplant diabetes} + 0.05*\text{glycemia} + 0.04*\text{hepato-cellular carcinoma} + 0.04*\text{vascular age} + 0.04*\text{diastolic pressure} + 0.03*\Delta\text{age} + 0.03*\text{MELD score} + 0.03*\text{age} + 0.02*\text{statins} + 0.02*\text{lymphocytes} + 0.02*\text{basophils} + 0.02*\text{BMI} + 0.01*\text{total bilirubin} - 0.01*\text{creatinine} - 0.01*\text{INR} - 0.02*\text{neutrophils} - 0.03*\text{albumin})$, ignoring the features with coefficient rounded off to zero, points out to the significance of cardiometabolic features in the transplant event.

Similarly, unsupervised algorithms like k-means clustering were used on the cohort, to identify the similarities between the cohort members and to harness the power of unsupervised algorithms since the internal relationships between the patients were unknown in the beginning. The cohort was separated with respect to their distinguishing properties into two optimal clusters based on silhouette analysis (Figure 17.4) [9].

As clustering was performed on all patients considering all the health indicators, the smaller cluster showed higher cardiovascular risk (8) compared to the bigger cluster (6). Similarly, the cluster having the smaller cohort demonstrated substantially higher percentage of death (32%) and higher percentage of transplanted patients (68%) than the other cluster with the bigger cohort (death: 14%, transplantation: 58%). Additionally, though both the clusters had almost similar mean age, body mass index, and blood pressure, with respect to HDL, the smaller cluster showed significantly lower values, whereas demonstrated higher values for LDL, total cholesterol/HDL, platelets, triglycerides, neutrophils, lymphocytes, monocytes, basophils, eosinophils, and glycemia, implying the cluster with smaller cohort at a higher risk than its bigger counterpart.

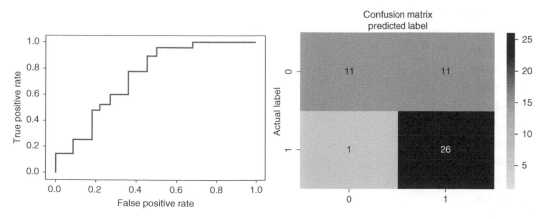

Figure 17.3 Logistic regression to predict the transplant event in the cohort.

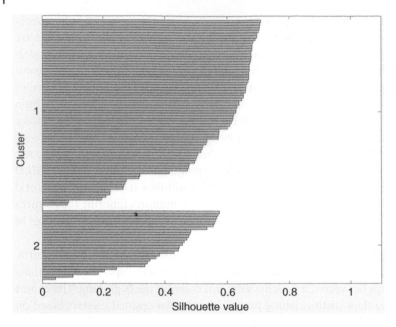

Figure 17.4 K-means clustering on the cohort – Silhouette analysis.

Parameters	Cluster 1	Cluster 2
Size of the patient cohort	61 Patients	36 Patients
Age at the moment of evaluation (years)	46 ± 16	48 ± 14
Sex (%)	Male: 49% Female: 51%	Male: 55% Female: 45%
BMI (kg/m²)	27 ± 5	27 ± 5
Systolic blood pressure (mm Hg)	117 ± 12	116 ± 12
Diastolic blood pressure (mm Hg)	67 ± 8	68 ± 8
Total cholesterol (mmol/l)	148 ± 58	175 ± 136
Triglycerides	91 ± 40	134 ± 102
HDL (mmol/l)	42 ± 24	31 ± 24
LDL (mmol/l)	88 ± 45	109 ± 79
Total cholesterol/HDL	9 ± 18	20 ± 40
Platelets (×1000)	102 ± 59	165 ± 86
Lymphocytes	1100 ± 679	1667 ± 961
Neutrophils	2616 ± 1237	5808 ± 1735
Monocytes	475 ± 274	825 ± 383
Eosinophils	187 ± 202	233 ± 323
Basophils	11 ± 32	36 ± 59
Glycemia	100 ± 40	97 ± 42

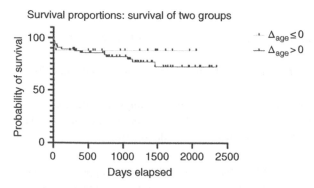

Survival proportions: survival of two groups

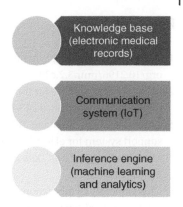

Figure 17.5 Kaplan–Meier post-transplant survival curves ($\Delta_{age} \leq 0$ and $\Delta_{age} > 0$).

Figure 17.6 Clinical decision support system.

This is a clear illustration of the capability of machine learning algorithms to stratify the risks of the cohort and could be used for detailed analysis and interpretation to take preventive actions. However, the factors of the cohort including the complex nature of the data need to be considered for further scaling of the study to other cohorts or an extended population.

Similarly, on a different approach, where vascular age was considered as the key feature, the cohort was separated into two groups, where vascular age was higher than the chronological age, and where the chronological age was more than the vascular age. The separated cohorts showed distinguishing characteristics of risks and survival for a period of five years after transplantation (Figure 17.5).

The predictions performed by the machine learning algorithms build the ground for a decision support system (Figure 17.6), to support the medical personnel in taking clinical decisions or offering more personalized healthcare to the patients. On one hand, it relies on digital tools for connected healthcare, and on the other hand, it uses the predictive power of artificial intelligence in obtaining insights from the health-data, interpreting, and using the same for providing better healthcare services.

17.4 Conclusions

Healthcare is one of the key sectors that has been reengineered in recent years thanks to technological areas like the IoT, along with the power of data science. In a Latin American perspective, where eHealth could be the answer to several existing issues like low access to quality healthcare or the high costs related to it, this work highlights the dual element of the same – the technological fusion, and the transdisciplinary human resource. This work highlights the important entities for a comprehensive system of smart healthcare, that includes not only technical aspects but human resources as well. This involves a robust management of all the parameters, especially in the transdisciplinary space. Harnessing the power of data science, clinical decision support systems would play an important role in stimulating a paradigm shift in healthcare by assisting the medical personnel to provide predictive and preventive care, along with more personalized and efficient treatments to the patients based on the monitoring and continuous analysis of their respective risks.

References

1 Eysenbach, G. (2001). What is e-health? *Journal of Medical Internet Research* 3 (2): e20. https://doi. org/10.2196/jmir.3.2.e20.

2 WHO Regional Office for the Eastern Mediterranean. eHealth. http://www.emro.who.int/health-topics/ehealth (accessed 24 March 2023).

3 Chatterjee, P., Armentano, R.L., and Cymberknop, L.J. (2017). Internet of things and decision support system for eHealth – applied to cardiometabolic diseases. *2017 International Conference on Machine Learning and Data Science (MLDS)*, Noida, pp. 75–79. https://doi.org/10.1109/ MLDS.2017.22.

4 Chatterjee, P., Cymberknop, L.J., and Armentano, R.L. (2017). IoT-based decision support system for intelligent healthcare – applied to cardiovascular diseases. *2017 7th International Conference on Communication Systems and Network Technologies (CSNT)*, Nagpur, pp. 362–366. https://doi.org/ 10.1109/CSNT.2017.8418567.

5 Elsas, C. How technology can make healthcare widely accessible in Latin America. *Industry Stories.* https://www.infosys.com/insights/industry-stories/how-technology-can-make-healthcare.html (accessed 23 August 2021).

6 The World Bank (2020). Latin America & the Caribbean countries need to spend more and better on health to be better able to face a major health emergency like COVID-19 effectively. https://www. worldbank.org/en/news/press-release/2020/06/16/latin-america-caribbean-health-emergency-covid-19 (accessed 23 August 2021).

7 Chatterjee, P., Tesis, A., Cymberknop, L.J., and Armentano, R.L. (2020). Internet of things and artificial intelligence in healthcare during COVID-19 pandemic – a South American perspective. *Frontiers in Public Health* 8: 600213. https://doi.org/10.3389/fpubh.2020.600213.

8 Fernando, L., Pamela, S., and Alejandra, L. (2014). Cardiovascular disease in Latin America: the growing epidemic. *Progress in Cardiovascular Diseases* 57 (3): 262–267. https://doi.org/10.1016/j. pcad.2014.07.007.

9 Chatterjee, P., Noceti, O., and Menéndez, J. (2019). Predictive risk analysis for liver transplant patients – eHealth model under national liver transplant program, Uruguay. *2019 IEEE 9th International Conference on Advanced Computing (IACC)*, Tiruchirappalli, India, pp. 75–80. https://doi.org/10.1109/IACC48062.2019.8971514.

Section 8

Digital Transformation

18

Digital Transformation Enabled by Big Data

Mariel Feder Szafir[1,2]

[1] *Engineering Department, Satellogic, Montevideo, Uruguay*
[2] *Software Engineering Department, ORT University, Montevideo, Uruguay*

18.1 Introduction

In the next sections, you can find information regarding Big Data Digital Transformation organized following these concepts:

First, I present an extensive list of basic definitions that will help you understand what Digital Transformation is and what it is not, what is big data and how it is an important player as an enabler of digital transformation processes, and a description of different available technologies involved in big data solutions, on a high level that will allow you to have a general understanding on where to focus and what the challenges are when facing the implementation of these solutions. This includes data analysis approaches, data collection and data storage alternatives, processing models, infrastructure, hardware, tools available in the market and architectural components of the solutions.

Following technological challenges, we next face management challenges. These new types of projects differ from traditional software development or implementation projects and present challenges of their own, so new alternatives are created or evolve from previous ones. I then submerge into possible strategies to adopt big data solutions and share concerns regarding about big data projects and data governance. Since ethics is also an important aspect of this new world of interconnected data, a section follows with a brief description of the main concerns regarding this topic.

Nothing is complete with only theoretical concepts, therefore prior to closing the chapter you will find some real examples of Big Data solutions that generated major Digital Transformations milestones in the world, followed by some conclusions of my own giving the reader the prerogative to agree or hopefully question and disagree with me.

IEEE Technology and Engineering Management Society Body of Knowledge (TEMSBOK), First Edition.
Edited by Gustavo Giannattasio, Elif Kongar, Marina Dabić, Celia Desmond, Michael Condry, Sudeendra Koushik, and Roberto Saracco.

18.2 Basic Fundamentals

18.2.1 Digital Transformation

Digital Transformation refers to the changes associated with the application of digital technology to radically improve the performance of an organization. Digital Transformation is synonymous with innovation and creativity in a given domain, rather than improving or supporting traditional methods of working or doing business. Digital transformation is generating a democratization of trade, new price models for similar products, new businesses in traditional companies, modifications in business ecosystems, the retirement from market of traditional actors, new consumer habits, new work cultures, and even new roles for traditional workers. The challenges to be solved have to do with defining with clarity business objectives, understanding the competition, the coexistence of new and legacy technology, breaking up the "silo syndrome," and managing change in an agile way.

A framework for digital transformation might group objectives into these categories:

User experience:
- **Customer understanding:** Segmentation based on analytics; knowledge based on social behavior
- **High-level growth:** Digital sales, predictive marketing, customer-centric processes
- **Customer entry points:** Customer services, omni-channels, self service

Operational processes:
- Process digitalization
- Worker productivity
- Performance Management

Business model:
- Digitally modified business
- New digital businesses
- Digital globalization

All these objectives are based on the digital management of information, which occurs over increasing volumes of data.

18.2.2 Data Analysis

Data analysis is the process by which available data is analyzed to try to understand reality in a quantitative way. Data is examined in search of patterns, trends, or relationships that allow to interpret the surrounding environment and improve the decision-making process through the added value of objective information. An example could be to find the relationship between investment in advertising and increased sales in order to find the breakeven point of how profitable the investment is. Or going one step further, what are the media or the combination of media where the investment in advertising impacts more on the sales of the company.

We can group these analyses into four large groups, in the historical order in which they began to be used which matches the order by complexity.

18.2.2.1 Descriptive Analysis

This analysis attempts to answer questions about what has already happened. How much was last quarter's sales? How many sales were made associated with publicity published in medium X? Which product was the best seller? Which product was the most profitable? The vast majority of data analyses are of this type and correspond to the simplest and most direct processes. The most common questions are "what," "how much," "when," or "where."

18.2.2.2 Diagnostic Analysis

This type of analysis attempts to answer cause-effect questions. It seeks to explain the origin of a certain phenomenon, situation, or result. Tries to solve "why" questions. Why is product A better selling than product B in a certain region? Why do certain promotions have more effect than others? This analysis is at a higher level than descriptive analysis and is a bit more complex.

18.2.2.3 Predictive Analytics

The next level is predictive analytics. It tries to predict what will happen or the magnitude of certain values based on how data or events have been in the past. It is based on the previous relationships between known data and the data to be predicted. Tries to help solve situations such as: what is the most convenient advertising for this particular target audience? If a customer purchased product A, how likely is it that he will buy product B? How likely is that this patient will respond well to a certain treatment?

This analysis is based on the relationships between historical data. If the data or its relationships change, the predictive models must be updated.

18.2.2.4 Prescriptive Analysis

Prescriptive analysis is based on predictive analysis and goes a step further by suggesting (or performing) the most convenient actions. It is used to try to get better results or mitigate risks. This is the analysis that adds the most value of all, but it is also the most complex and the one that requires the most specialized tools and software. Normally many alternatives are evaluated and compared, until finding and suggesting the one that provides the best result. The approach goes from an analysis into a recommendation, and often includes the simulation of different concrete scenarios generating multiple recommendations, such as: What is the best treatment for a patient? What kind of advertising should I invest in?

18.2.3 Business Intelligence

Business Intelligence is software (and hardware) solution whose objective is to integrate data from multiple sources of an organization, organize them in a centralized and structured way, and generate analysis on that data with different tools for visualization and distribution of information. In practice, most Business Intelligence solutions are based on a structured central repository where data from different sources is unified, known as a Datawarehouse. On these Datawarehouses, different analytical queries are executed and present the information in different formats (tables, reports, graphs, indicators, and dashboards). Some Data Mining algorithms allow to search for relationships between the data stored in the Datawarehouse (see Figure 18.1).

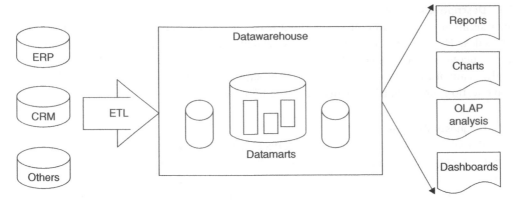

Figure 18.1 Datawarehouse.

18.2.4 Artificial Intelligence

Artificial Intelligence is a term used to describe the simulation of human intelligence in machines trained to perform as human brains. The most relevant feature in this regard is the ability to generate software that learns and evolves from the data it collects from the real world, using this learning to solve new problems, instead of executing pre-programmed statements.

Artificial intelligence is a set of technologies that includes Machine Learning, a set of strategies by which algorithms learn from relationships between existing data and generate mechanisms to infer possible outcomes when faced with new information. They seem to emulate the human way of learning in the sense that they learn by prior association, although they are still a long way from the human capacity for learning in the sense that models must be trained with a large amount of data before they are able to "learn" from a situation and are not reliable when they need to apply this learned knowledge on exceptional situations (at least not yet). However, for the cases for which they were trained they can make decisions much faster than a human being, and even more accurately because they do not have the subjective or emotional element that is part of the human decision-making process.

The concept of Big Data that is explained later is closely related to the implementation of Machine Learning models or Artificial Intelligence solutions, because normally large amounts of data are necessary to train a model so that it yields reliable results.

18.2.5 Cloud Computing

In simple words, cloud computing is the strategy of using computing capacity, servers, software, and complete solutions in third-party infrastructure, accessing via the Internet. This allows companies to access large resources that they would not be able to own, taking advantage of scale economics. It reduces the need to have your own infrastructure, minimizing the needs of physical storage, investment in fixed assets with a high level of obsolescence, the need to control elements such as the physical security of the facilities and the highly trained personnel required for their administration, control, and repair. Large cloud computing providers ensure up to 99.99% availability of their solutions, with replicated and distributed facilities around the world, fault tolerance, and disaster recovery capabilities.

Typically, the use of cloud solutions implies less initial investment, faster start-up, and the need for fewer infrastructure experts, as well as a greater capacity for infrastructure innovation and renewal than individual companies could have.

As a counterpart, the dependence on the internet connection for the use of the solutions is increased (which is more or less reliable and/or performant in different parts of the world), and the data is hosted on third-party servers, so it is very important to choose reliable cloud providers known for their reputation. The cost is per use of the service and may be as variable as the volume of data, processing, queries, and/or users changes.

Today, some of the big cloud computing providers are Amazon Web Services, Microsoft Azure, Google Cloud Computing, and Alibaba (the first in China), among others.

18.2.6 Edge Computing

Edge Computing is a new strategy related to the concept of the Internet of Things (IoT), whose application today implies that billions of devices constantly generate data that are transferred via the Internet to be processed. Edge Computing is a new paradigm that moves the processing of such data from the cloud or central servers into their place of origin, reducing latency, power consumption, and the information that circulates, also allowing to have less complex centralized entities that handle a smaller volume of information (e.g. the data of a sensor is processed within the device itself and only those measurements that deviate from their expected values are sent to the central system to be processed). This implies the need to provide some processing capacity to the devices where the data is generated, so that the data is processed as close to its origin as possible, either on the device itself, or at the edge of the network (cloud Edge).

This technology allows to:

- Diversify and scale of IoT solutions without collapsing core systems.
- Improve cybersecurity, as less data is transferred over the network.
- Increases the speed of response, eliminating the time of communications to a central entity for the data to be processed, and decisions can be made locally (e.g. perform a maneuver in an autonomous vehicle).

18.2.7 Big Data

Big data is a term created by the industry to define a massive volume of structured and unstructured data. The size, quantity, and speed of data generation mean that it cannot be processed by traditional technologies (e.g. BI in its traditional DW-based model), and usually exceeds the processing capacity of companies (this is where the cloud computing concepts mentioned above come in).

18.2.8 Relevance and Characteristics

The size and amount of information generated in the digital world have evolved dramatically. When we talk about storage related to the world of Big Data, we must think of petabytes (1 followed by 15 zeros bytes of data) or more.

This information comes from traditional sources such as companies' transactional systems, which are increasingly digitalizing their production, commercial, and internal processes. But also, new internet solutions generate a huge number of transactions that come from the global market through the transactions that are available over the Internet.

Additional information is generated in the different social networks, which can be text, images, photos, and videos, for which the size grows exponentially.

Internet of Things is another important player in generating an enormous amount of information in short periods of time from a huge number of devices scattered around the globe (and sometimes in space too).

We are not only interested in storing this information, but also to be able to use it in an integral way, combining these multiple data sources to obtain richer information that allows us to evolve the analysis to the levels of predictive and prescriptive.

The main characteristics of data when we talk about Big Data are commonly known as the five Vs: Volume, Velocity, and Variety in terms of its generation, to which we add Veracity and Value as relevant additional characteristics (see Figure 18.2).

18.2.8.1 Volume

The amount of data to be processed is enormous and usually tends to continue to grow exponentially. This implies that both infrastructure and solutions must be prepared to handle these huge volumes without collapsing and in reasonable response times, and even commensurate presentation formats that can be consumed by a human reader.

Example: Google receives approximately 4.5 million searches per minute, all of which are stored.

18.2.8.2 Velocity

Data in the world of Big Data can be generated at a vertiginous pace. This data must be processed at a rate that prevents the collapse of the processing mechanism and data loss. Depending on the type of business we must also consider peak moments (bursts of data at specific times) and events that trigger activity (for example: Black Friday in E-Commerce sales).

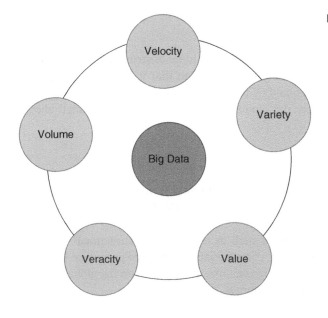

Figure 18.2 Five V volume.

18.2.8.3 Variety

The above two concepts could be considered as "doing more of the same, only faster and on a larger volume" which in itself is a very important technological challenge. To this the concept of variety is added, which involves developing new ways of processing information to include diverse and unstructured data sources, such as plain text files, images, videos, audio, sensor data, metadata, and others, which must be processed, interpreted, stored, and combined with traditional data.

18.2.8.4 Veracity

The reliability of a piece of data is given by its veracity. A huge set of incorrect, incomplete, or inaccurate data will not allow us to meet the goal of making better decisions. Veracity requires that the data must be verified in terms of its fidelity and quality before allowing its access to the Big Data environment, which implies additional processing needs (a process that serves as a gatekeeper of each datum before allowing it to enter the general solution), to maintain the integrity of the data repository. Regarding veracity, a piece of data can be part of a signal or noise. Noise is data that cannot be converted into truthful information, while a signal is data that can become information when considered by itself or as part of a set of received data. The relationship between signal and noise depends very much on the source of the data. For example, data obtained from transactional systems that apply a series of controls prior to the data being confirmed will have a good signal-to-noise ratio, compared to data obtained, for example, from the natural language analysis of the entries in a set of blogs.

18.2.8.5 Value

The value is defined as the potential usefulness of a piece of data to an organization. Obviously, the value of a piece of data is closely related to its veracity. Less reliable data generate less value, and it is necessary to draw the limit of what level of veracity a datum must have not to be dismissed, or what level of certainty the definitions that are taken from this less reliable data have. In some contexts, less truthful data may be supported. For example, to carry out targeted marketing, if the advertisement that is sent to most of the target audience is correct, the campaign could be considered successful even if some people receive information that is not of interest to them. This same criterion would not be acceptable to define what is the medicine and the dose that a patient should receive derived from the analysis of his symptoms and vital signs.

The value is also affected by other elements such as processing time. For example, a stock exchange listing information that requires 30 minutes of processing will no longer have value when it becomes available. Or the processing of a patient's vital signs in UCI that warns that it requires attention for problems in the heart rate or breathing level does not add value if it takes a few minutes to be processed. In some contexts, the value is inversely proportional to the processing time.

The relevance of the data itself also speaks of its value. Sometimes there is data that does not seem relevant and that one might think about not storing. For example, the brand and model of a cellular device from which a buyer makes their purchases on an online clothing store might be considered of no interest to the company, who is more interested in knowing what their customers buys, at what times, from which country, and as much information about "the customer" as they can get. However, another interpretation could be that information related to the socio-economic level of the customer can be inferred if the brand and model data of the source cell phone are analyzed in comparison with data from other customers in the same region, and maybe that information becomes a good behavioral predictor.

18.2.9 Data Sources

The data that is processed in Big Data solutions can be generated by Humans (interacting with devices) or by Devices (machines).

Human-generated data typically corresponds to digital transactions that a person can make using transactional systems, social networks, or other digital services available on the network. Individuals typically generate a limited amount of data per minute.

Device-generated data is typically generated by software or hardware devices that respond autonomously to real-world events. For example, logs of operation systems and servers, sensors that monitor temperature and humidity in a factory, or even the censors of position, rotation, and location within people's cell phones, which can continuously report their location and signal level, among others.

The data can be represented in one of these formats.

18.2.9.1 Structured Data

Structured data responds to a schema where all the elements that the data contains are clearly defined, including its format and constraints, and the data always meet these specifications. A very common source is transactional systems (e.g. CRM, ERP, Online Banking), where the databases hosting the information ensure that the data complies with the restrictions imposed. In these cases, there is little processing that must be done when receiving a piece of data in relation to its structure.

18.2.9.2 Unstructured Data

Unstructured data is data that does not follow the pattern of a default schema, but each datum has its own internal structure. In today's world, they have a much higher rate of growth than structured data. It can be formatted as text or binary files. Textual content files can contain texts of various origins, in narrative format and free text. Binary files, usually individual, include materials such as images, audio, or video, among others.

18.2.9.3 Semi-Structured Data

They are in between the two categories above. They usually have a structure that organizes information, but not all similar data have the same structure, and not all its contents are alphanumeric. This category includes Json or XML files, among others.

18.2.9.4 Metadata

Unstructured or semi-structured data needs additional information to be interpreted. This information that describes the content and structure of the data is called metadata. Big Data solutions when working with unstructured or semi-structured data rely on this metadata to interpret and operate with it.

18.2.10 Technological Fundamentals

Certain technological evolutions have made Big Data solutions possible, allowing available technologies and/or technological strategies to cope with the characteristics of the information that must be managed in terms of volume, variety, and speed, including the ability to manipulate structured, semi-structured and unstructured data to meet the objectives of analysis (particularly predictive and prescriptive) and feed artificial intelligence models for training or execution. Some of the strategies and technologies are described in the following section.

18.2.10.1 Clusters

A cluster is a set of interconnected computers or servers (referred as nodes) that work together to achieve a goal. They are often perceived by the solutions that use them as a single server, and there are mechanisms by which nodes are organized to divide the work and complete tasks transparently to the client.

Clusters are widely used in Big Data solutions, since they allow both the processing and storage of large volumes of data to be distributed among many servers. Clusters might vary from less than a dozen nodes, up to clusters of hundreds or thousands of interconnected servers depending on the context. In this way, it is possible to distribute the task of receiving, processing, storing, and querying the data in a distributed way, taking advantage of the computational and storage capacities of each of the computer that makes up the cluster.

The basic idea is that a large task is divided into several smaller tasks that are performed by several of the nodes in the cluster in parallel, and then integrate the results into the final response.

There are different strategies to manage a cluster, identify the nodes that compose them, distribute the tasks, and know where to find the data when it is needed.

18.2.10.2 Distributed File Systems

A File System is the way in which data is stored on a physical device, such as hard drives, pen drives, DVDs, etc. For example, operating systems manage data by storing it in files that group content and metadata (e.g. name, creation date, permissions of files), and sorts them in a tree-like structure.

A distributed file system is a mechanism for performing the same data management but on multiple nodes in a cluster. This allows to greatly increase the storage capacity given by the sum of the capacities of the devices that make up the cluster. For the end customer, the location of the file is transparent, and when read or write access to a given file is requested, it is the cluster the responsible for knowing where the file is (on which node) and transmitting the corresponding read and write instructions to that node (see Figure 18.3).

Some examples of Distributed File Systems commonly used in Big Data solutions are Apache HDFS (Hadoop File System), or GFS (Google File System).

18.2.10.3 Replication and Sharding

Sharding is the process by which a large set of data is spread across several nodes of a cluster to obtain smaller and more manageable partitions. Each of these concepts is a "shard" that contains only a portion of the whole data. Each shard is stored on a node that is only responsible for the data stored on it. Sharding can be transparent to the customer through mechanisms in the solution that know in which shard should record or query for a specific data, or the customer can be the one who defines the sharding criteria (e.g. one shard per country).

Another benefit of sharding is that it provides partial fault tolerance, in the sense that if a node fails, only the information recorded in it will not be available, but the rest of the data will.

Querying and integrating data from multiple shards results in performance penalties (Figure 18.4).

Replication is the process by which the same data is stored more than once on different nodes in a cluster. This increases the availability of the solution at the query level, because although the data contained in a node becomes unavailable, redundancy ensures access to it through one of the nodes where the information is replicated.

The disadvantage of replication models is the possible lack of consistency. When a data needs to be generated or updated, it is not done on all nodes simultaneously, therefore some query can

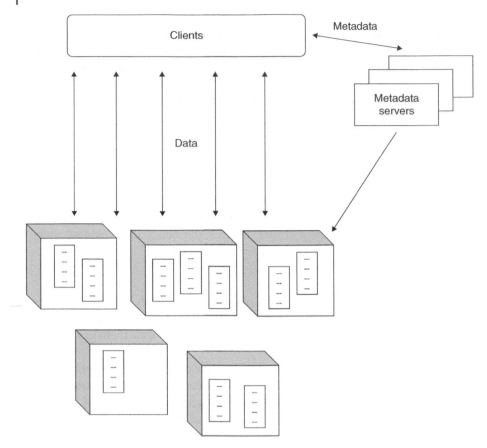

Figure 18.3 Distributed file system.

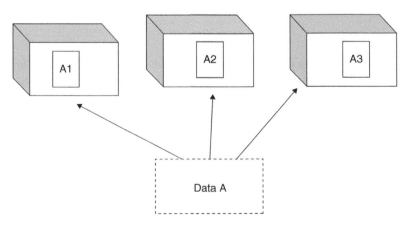

Figure 18.4 Sharding.

access the data when there is already a newer version on another node. This inconsistency usually occurs for a short period of time, and in a short time the data is synchronized. This is called "eventual convergence." Different implementations try to minimize this inconsistency with different strategies on how they allow data to be updated and queried in their different replicas (Figure 18.5).

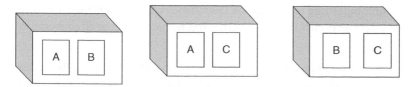

Figure 18.5 Replication.

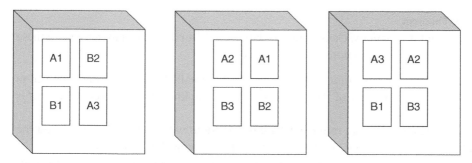

Figure 18.6 Sharding and replication.

Sharding and Replication can be combined in such a way that the data can be distributed in smaller groups (shards on different nodes), and at the same time these shards can be replicated. Some databases I will mention later include Sharding and Replication as a native part of their implementation (Figure 18.6).

18.2.11 CAP, ACID, and BASE Theorems

18.2.11.1 CAP Theorem
An acronym for Consistency, Availability and Partition, the CAP theorem also known as Brewer's theorem, states that in distributed databases it is only possible to ensure two of these three aspects:

- **Consistency:** The reading of a piece of data is always the same no matter which node it is read from.
- **Availability:** A read/write request must always be answered.
- **Partition tolerance:** In case there are communication problems between the nodes, the database should always provide a result (Figure 18.7).

1) If consistency (C) and availability (A) are required, all nodes must always be able to communicate to verify that the data is the same in all of them, therefore it is not possible to operate with the partitioned model.
2) If consistency (C) and partitioning (P) are required, the nodes will not be available until they eventually become consistent.

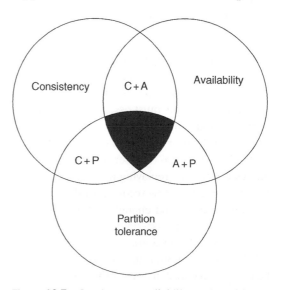

Figure 18.7 Consistency, availability, and partition.

3) If availability (A) and partitioning (P) are required, then consistency cannot be ensured, because the database must always respond, even if there is data that has not yet been able to converge to its new value on some node.

In distributed databases, scalability and fault tolerance can be improved by increasing the nodes in a cluster, although this could penalize consistency. Therefore, it should be kept in mind that incorporating replication improves some of these parameters and penalizes others. This restricts the type of solutions in which it makes sense to apply active replication.

18.2.11.2 ACID Theorem

ACID is an acronym for Atomicity, Consistency, Isolation, and Durability. Databases ensure these characteristics by handling transactions and locking records.

Atomicity: Ensures that a transaction is totally completed or totally fails (e.g. It is not allowed to save a purchase order with half of the products it includes, either it is saved entirely, or it is rejected.)

Consistency: Ensures that all data in the database is consistent with its type definition and the data it relates to (e.g. there are no purchase orders related to items that do not exist).

Isolation: Ensures that a half-done transaction is not visible to others until it is confirmed or rejected.

Durability: Ensures that the operations are permanent, that once they are confirmed they cannot be undone and become available to other users.

Traditional relational databases are built to respect the ACID model and are therefore the most used in transactional solutions (e.g. banking, billing, etc.).

18.2.11.3 BASE Theorem

According to the CAP theorem, databases with replication cannot meet the ACID model. Instead they provide a solution of type BASE:

- Basically available
- Soft state
- Eventual consistency.

A BASE database prioritizes the A + P version of the CAP theorem, favoring availability over consistency, accepting eventual consistency as part of its characteristics. Being "basically available," it will always respond even if it is not with the last exact data ("soft state"), but one quite close that in a short time will tend to consistency ("eventual consistency").

In each case you must assess which is more important in a database solution. For example, Netflix periodically records the exact location of the movie or serial episode a user is watching. In this way, if the user interrupts their activity (in an orderly or unexpected way), there is always data quite close to where the user was before it stopped. When the user resumes, the BASE database allows you to start replaying in a very similar and always prior place in the movie, but maybe not the exact one for a few seconds. In this case, a BASE solution might be good enough. However, it will not seem a good enough solution if I check the balance of my bank account and on some device the latest transactions appear and on another not (an outdated or approximate balance is not an acceptable option).

When building Big Data solutions, it is important to select the most relevant attributes that will be the architectural drivers of the solution.

18.2.12   Data Lake

A Data Lake is a centralized storage repository that contains big data from multiple sources in a granular, raw format. It can save structured, semi-structured, or unstructured data, which means that the data can be retained in a more flexible format for future use.

Each item in a data lake is assigned a unique identifier and tagged with a set of extended metadata tags. When a business issue arises that needs to be resolved, we may ask the data lake for data that is related to that issue. Once obtained we can analyze the returned dataset to help the production of a meaningful answer.

It is useful in scenarios where a large volume of data that you do not know in which context it will need to be used later is generated, so it is stored in its original format.

The main benefit of a data lake is the centralization of several content sources. Once gathered (in their "information silos"), these sources can be combined and processed by big data searches and analyses that would otherwise have been impossible.

The main difference between a traditional Datawarehouse and a Data Lake, is that in the DW the data is housed in static and predefined structures that determine the type of analysis that can be executed on the data. Therefore, its analytical capacity (reports, tables, boards, or charts), is limited by the very structure of the sources. This approach has begun to be insufficient and here is where the Data Lake concept comes in. Storing the "raw" data without schema or predefined structure, allows unlimited access to them with different mechanisms depending on the type of analysis that is required.

18.2.13   NoSQL Databases

NOT ONLY SQL databases are NON-relational databases that are used to manipulate large volumes of information.

Unlike traditional relational databases, which are designed to ensure the integrity of the information, and to allow all updates and combinations of data access, NoSQL databases are less generalist and are designed targeting the specific solution they will support. This makes these databases less efficient for general non-predefined purposes, but much more efficient at accessing data in the predefined way for which they were created, thus allowing the handling of larger volumes of information more efficiently. In addition, some of them natively support the handling of unstructured information, access to data stored outside its internal structure (for example in a DFS), and others natively implement replication and sharding strategies.

The goal of NoSQL databases is to be scalable, fault-tolerant, and natively support unstructured data. Many of them provide APIs (access interfaces) that must be invoked programmatically, others have their own languages for querying or accessing data, while one group also supports SQL (Standard Query Language) for the portion of data that is stored in a simile-structured way.

NoSQL databases focus on high availability, information distribution (the cluster concept applied to database data), flexible schemas, and lax consistency.

The main families of NoSQL databases are: see Figure 18.8.

18.2.13.1   Key Value-Oriented Databases

The information is stored in simple tables in key-value format, where the data is identified by its key, and the rest of the information is stored as value. They are simple to scale out, typically providing in-memory processing. In general, they are useful to access information through their key. Some well-known examples are Google Big Table, Redis, Riak and Amazon Dynamo. They are the equivalent of a distributed and persistent Hash dictionary.

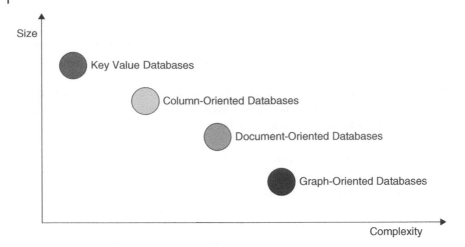

Figure 18.8   Families of NoSQL databases.

The advantages for this type of database are that their API is simple to use, it is easy to replicate and scale on data volumes making efficient use of sharding and other strategies to ensure scalability. As a disadvantage the content scheme in value must be interpreted by the client, they have limited support for transactional handling (typically limited to a key), and it is more complex to ensure content consistency.

Some examples of possible uses are Facebook storing users' settings in real time against each request, Twitter to record on a timeline every time a tweet is created, or Amazon on the implementation of its shopping cart.

18.2.13.2   Column-Oriented Databases

Column-oriented database technologies inherit Google's Big Table concepts, based on columns for which a name, value, and timestamp are defined. In these columns, information is stored by rows that are sorted and uniquely identified. The key determines the location of the data on the corresponding node in the cluster. Each row can have a variable number of possible columns.

The best-known implementations are Hbase and Cassandra.

The main advantages are the focus on performance and scalability, forcing the denormalization of data models. They are very efficient for replication and sharding. Consistency levels can be managed. They provide high availability because there is no single point of failure. The disadvantages are that they occupy larger sizes (self-describing formats lead to an increase in storage volume), and that they have a very limited capacity for queries (only by the row keys, without joins), and that the elements to store are dependent on the columns design.

Some applications where they are recommended can be for example content management systems for blogging or large-scale counts (e.g. visits on web pages) (see Figure 18.9).

18.2.13.3   Document-Oriented Databases

In this model, objects are saved as documents. They do not force a default schema. Each document writes its contents, and collections can have different formats. Documents have a unique identifier provided by the developer.

They are similar to key-value model, but there is structure in the value with a queryable schema.

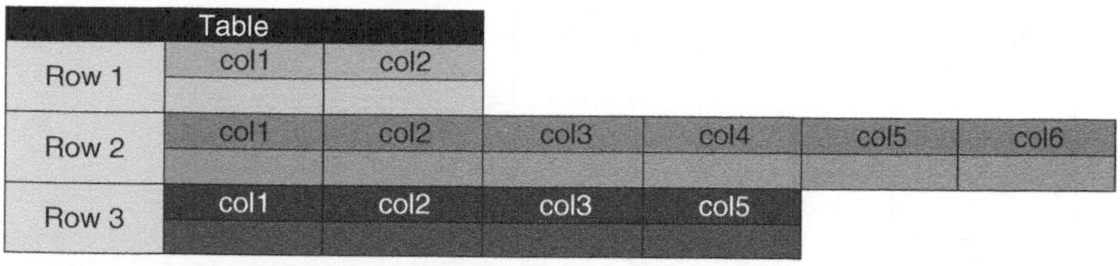

Figure 18.9 Column-oriented database.

Documents must be self-contained and self-describing, they can be complex, and each one has its own outline. These databases are flexible and, can be queried by different indexed attributes, but the entire document is also identified by a unique key. Information is grouped in a natural and logical way. The consistency and atomicity of operations are per document.

Examples include MongoDb, Couchbase, or RavenDB.

The main advantage is flexibility, since you can incorporate documents of various formats, it is easy to incorporate new types of data and documents can have different structures. When querying, documents can be indexed and consulted by any field. The main disadvantages are the size since self-describing fields lead to duplication of information (binary formats are used to mitigate this), and if a larger number of indexes are defined to enable searching, overhead can occur when recording.

Examples of use: storing event information (event logging), content management systems, Web and/or Analytics in real-time or E-Commerce Applications.

18.2.13.4 Graph-Based Databases

A graph-based data structure consists of a finite set of ordered pairs called arcs and certain entities called nodes or vertices (see Figure 18.10).

Each node represents a feature and each arc a relationship between nodes. Each node and arc are identified by a unique identifier. In the node all the information of the entity is saved, and the arcs represent links to the other nodes to which it is related.

Queries to the database are based on the paths that relate the nodes to each other. Queries to the database can traverse the entire structure. The requirements for queries can be changed without affecting the database model. The relationships between the nodes are not computed at query runtime but are persisted as a relationship. Nodes can have different types of relationships between them, there are no limits to expressing relationships.

Relationships are first-order elements in this model. The value of the database will be given by the quality of the relationships defined between its nodes

Examples of use: Important data connection requirements (managing friends on social networks), location, routing and logistics services, and recommendation engines.

Known implementations; OrientDB or Neo4J.

18.2.13.5 Polyglot Persistence

There is no one database superior to the others in all contexts, each one is more efficient in each situation.

In this sense, a possible strategy is to persist the information in more than one database, depending on how the information will be used later. This is known as polyglot persistence (PP).

PP is the use of different storage models to solve different data management requirements. PP can be applied in a context of specific applications or across the entire portfolio of an organization.

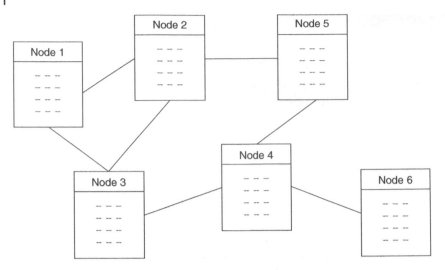

Figure 18.10 Graph-based data structure.

New models add complexity to the development efforts and this replication requires greater governance efforts, so the benefits of using PP should be carefully analyzed.

18.2.14 Processing Models

18.2.14.1 Transactional Processing

Transactional processing is normally used in transactional systems, where a certain action is performed in a productive system. Examples such as confirming an online purchase or registering the payment to a supplier in an accounting system. These transactions must be completed and processed at the time they are entered by the user and are usually referred to as online transactions. They generally involve the entry of a small volume of data.

18.2.14.2 Batch Processing

There are processes where the execution can be deferred. Most cases where data is prepared for subsequent queries (for example, importing data from transactional systems into a central repository), can be done asynchronously, at certain times of the day. This deferred processing is known as batch processing.

It typically involves processing large volumes of information, and deferred execution for the most convenient time. In general, processes that prepare data for analysis (for example, in traditional BI) perform this process in batch form, leaving the information available for later consultation. Batch processing, where processing takes place at certain times of the day, can last from minutes to hours.

18.2.14.3 Near Real-Time Processing

In Big Data, this usually refers to the processing of information received by streaming. Streaming is a continuous stream of data (constant or bursts) that is sent to a receiver who receives it and processes or stores it for later use. In the case of Near Real Time Streaming Processing, the data is received, and a batch component (micro-batch) runs constantly processing quickly the data as it is received. The ability to process the data is typically expected to be at least the same as the rate at which the data is received. An example of this type of streaming and processing generation can be the information generated by IoT devices.

18.2.14.4 Real-Time Processing

Each data is expected to be processed immediately upon reception, and acceptable response times are on the order of milli or microseconds. For example, an air traffic control system should have this processing capability.

18.2.14.5 Distributed Processing

When you have more than one node that can process information, the distributed processing strategy allows you to assign tasks to the different processors available at the same time, which work separately to produce portions of the result, that are then re-integrated into a final whole solution if required. This makes it possible to make the most of the available computing power and minimize the total working time to obtain a certain result.

18.2.15 Map-Reduce

Map-Reduce is a highly popular processing framework in the world of Big Data. It leverages the benefits of data distribution through sharding and the multi-node solution available in cloud solutions. It is a highly scalable and reliable algorithm, based on the "divide to conquer" principle of distributed processing. Its initial development was carried out by Yahoo, but now it is maintained by the Apache project in open-source mode, so it has been integrated into different implementations of many commercial Big Data solutions.

The algorithm is based on three steps:

- **Map:** Each node where part of the data resides applies a map function that looks out for the relevant data and writes its result to temporary storage.
- **Shuffle:** The data generated on each node is redistributed so that all the data in the same key are grouped, and each group is assigned to a node.
- **Reduce:** The nodes work on each of these groups of data performing the processing on their part, which is then unified into a single final result.

Not all problems in Big Data can be solved using Map-Reduce strategies, but it is very common to use it to search, organize, count, group, and sum data in large, distributed volumes. Therefore, when working with distributed information systems where it is expected that this type of solutions will be needed, it is important to select an architectural solution with support for Map Reduce algorithms. In fact, many predefined solutions (e.g. visualizers, report generators, etc.), are based on the map-reduce services of the underlying infrastructure from which they obtain the data to be presented.

18.2.16 Architectural Concepts

An architecture represents the high-level components of a system, their functionality, and how they are connected to each other. When designing a solution for a Big Data project and its architecture, at least the following components must be considered:

Data sources: What is the source, shape, content, and speed with which the data that will feed the solution is produced.

Data acquisition: Which components are implemented to receive and process the data that arrives as it is received (data ingestion).

Storage layer: Where the data will be saved for later use, either temporarily or persistently.

Processing layer: How the received data is processed to fulfill solution objectives. This is where data becomes valuable information for the business. The type of processing (batch, real-time, etc.), must be defined according to the identified needs.

Analysis layer: The analysis layer provides an efficient interface to analyze the data contained in the storage layer and consists of two engines: Query and Analysis. Depending on the type of analysis being performed, this layer may consist of a query engine, as in the case of descriptive and diagnostic analysis. However, when performing deep analytics, as in the case of predictive and prescriptive analytics, an analytics engine is mandatory. This is the layer that turns large amounts of data into information on which business-critical decisions are made.

Display layer: The goal of the visualization layer is to give analysis activities a way to display the results that facilitates the interpretation of the result in the most efficient way. How analysis results are displayed has a direct impact on decision-making.

Utilization layer: The visualization layer provides an opportunity to develop an understanding of the results of the analysis in a graphical way. However, to get the maximum benefit from these results, it is necessary to incorporate them into the critical processes of the company. The utilization layer incorporates the results of the analysis and visualization into the critical applications/processes of the company so that proactive actions can be taken and/or mitigate or prevent risks. At this point we are no longer talking about a component of the technological solution, but part of the organization processes, which must be considered when incorporating some of these digital transformation projects. This is the step where the technological output of the journey becomes the business Digital Transformation.

In a transversal way, all layers must consider how they solve the requirements of security, management, and governance of the data that passes through them (Figure 18.11).

What elements should be considered when choosing an architecture to support a Big Data solution?

Storage needs: Capacity required for the storage of large volumes of data from various sources, both structured and unstructured. Current volume and future volume (scalability). Ease of growth without compromising performance.

Analytical capacity needs: Ability to perform analytical calculations on the data stored on the platform. Required data latency. How much story do you want to have online? Is it possible to transfer a certain history to a cold area with lower storage costs (redirecting data from the table of a DBMS to the HDFS)? How many users would need to access concurrently? How often ...?

Open source vs. owner: What options are available? Which one best suits your needs? What support and operations are required from your team in each option? What are your capabilities to provide it?

Adaptation to new requirements: Ease of development of new functionalities, effort and level of expertise required, learning curve, level of reuse. Possibility to implement new functionalities not initially foreseen. Implementation time. How long will it take to have an operational solution? How gradually can the solution be deployed?

Infrastructure price: Cost of the infrastructure on which the platform would be installed. Cost of the machines, licenses, cost of implementation, and subsequent cost of use including support and maintenance. When evaluating on-premise vs. cloud allocation, consider total cost of ownership.

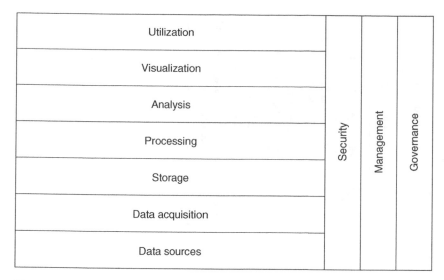

Figure 18.11 High-level components of a system.

18.2.17 Hadoop

Apache Hadoop is a working environment for software, under free license, for distributed applications that handle Big Data. It allows applications to work with thousands of networked nodes and petabytes of data. Hadoop was inspired by Google's documents on MapReduce and GFS.

Hadoop is a project of the Apache organization that is being built and used by a global community of contributors. Yahoo has been the largest contributor to the project and uses Hadoop extensively in its business.

The Hadoop infrastructure has established itself as the "*de facto*" platform in the industry. Most of the available Big Data solutions (free and commercial) are based on Hadoop, either using it as a base for their components, or making available their own products with similar characteristics.

The Hadoop ecosystem has a set of components based on the previous concepts described at the beginning of this chapter.

Depending on the solution you are looking for, the combination of components you must include will be different. That is, a solution is not the application of the complete platform, but a subset of components that must be chosen according to business requirements and can be gradually added to the solution (see Figure 18.12).

HDFS, Hadoop File System: It is the distributed file system basis of the Hadoop Big Data solution.
YARN, Yet Another Resource Negotiator: It is a fundamental piece in the Hadoop ecosystem. It is the framework that allows Hadoop to support multiple runtimes. YARN can be thought of as the brain of the ecosystem and the traffic director, which processes activities by allocating resources to them and scheduling tasks.
Apache Kafka: Is a distributed publish-subscribe messaging system that receives data from diverse sources and makes it available in near real-time. Kafka is often associated with near real-time event stream processing for big data. Kafka is used to stream data into data lakes, applications, and real-time stream analytics systems.
Apache Storm: Apache Storm is a distributed, fault-tolerant software used to process streams of data in real-time. Storm solutions can also provide guaranteed processing of data, with the ability to replay data that was not successfully processed the first time. You use Apache Kafka as

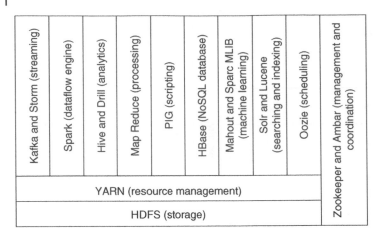

Figure 18.12 Hadoop.

a distributed and robust queue that can handle high-volume data and enables you to pass messages from one end-point to another. Storm is not a queue. It is a system that has distributed real-time processing abilities, meaning you can execute all kinds of manipulations on real-time data in parallel.

Map-Reduce: It is the core component of processing in a Hadoop Ecosystem as it provides the logic of processing, based on the Map-Reduce algorithm.

Apache Spark: Spark is mainly a real-time data processing engine developed to provide faster and more easy-to-use analytics than Map-Reduce. Spark is extremely fast and is currently under active development. It is a very powerful technology as it uses the in-memory processing of data. It was originally developed at the University of California, Berkeley.

HBase: Is a NoSQL database which supports all kinds of data. It can expose data stored in your cluster which might be transformed in some way by Spark or Map-Reduce. It provides a very fast way of exposing those results to other systems. HBase was designed to run on top of HDFS and provides BigTable like capabilities. It provides a fault-tolerant way of storing sparse data, which is common in most Big Data use cases.

Apache HIVE: Facebook created HIVE for people who are fluent with SQL. Thus, HIVE makes them feel at home while working in a Hadoop ecosystem. Basically, HIVE is a data warehousing component that performs reading, writing, and managing large data sets in a distributed environment using SQL-like interface. The query language of Hive is called Hive Query Language (HQL).

Apache Drill: As the name suggests, Apache Drill is used to drill into any kind of data. It is an open-source application that works in a distributed environment to analyze large data sets. It supports different kinds of NoSQL databases and file systems, which is a powerful feature of Drill. For example, Azure Blob Storage, Google Cloud Storage, HBase, MongoDB, MapR-DB HDFS, MapR-FS, Amazon S3, Swift, NAS and local files. Basically, the main aim behind Apache Drill is to provide scalability to process petabytes and exabytes of data efficiently (minutes).

Pig: Pig Latin is a language which has SQL-like command structure. The compiler internally converts Pig Latin to Map-Reduce jobs. It produces a sequential set of Map-Reduce jobs as an abstraction (works like a black box). PIG was initially developed by Yahoo.

Apache Mahout and Spark MLIB: Apache Mahout and Spark MLIB are two options that provide an environment for creating machine learning applications which are scalable. These libraries implement most popular Machine Learning algorithms accessing distributed data and computing in parallel taking advantage of the processing power of the cloud.

Apache Solr and Apache Lucene: Are the two components which are used for searching and indexing in Hadoop Ecosystem. Solr is a complete application built around Lucene. It uses the Lucene Java search library as a core for searching and full indexing. Lucene is a library to be included in code when programming, while Solr is a service to be invoked that uses Lucene underneath. If Apache Lucene is the engine, Apache Solr is the car built around it.

Apache Oozie: Oozie is a way of scheduling jobs on the cluster. If a task needs to be performed on a Hadoop cluster involving different steps and maybe different systems, Oozie is the way for scheduling all these things together into jobs that can be run following a certain order. Consider Apache Oozie as a clock and alarm service inside Hadoop Ecosystem. It schedules Hadoop jobs and binds them together as one logical work.

Zookeeper: Before Zookeeper, it was very difficult and time-consuming to coordinate between different services in Hadoop Ecosystem. Initially, the services had many problems when interacting and sharing common configuration while synchronizing data. Even if the services are configured, changes in the configurations of the services make it complex and difficult to handle. Grouping and naming was also a time-consuming factor. Due to the above problems, Zookeeper was introduced. It saves a lot of time by performing synchronization, configuration maintenance, grouping, and naming. Although it is a simple service, it can be used to build powerful solutions. It is used for keeping track of the nodes that are up and the ones that are down. It is a very reliable way of keeping track of shared states across your cluster that different applications can use.

Apache Ambari: Ambari is an Apache Software Foundation Project which aims to making Hadoop ecosystem more manageable. It includes software for provisioning, managing, and monitoring Apache Hadoop clusters. It is an alternative to Zookeeper.

Additional components:

Apache Flume: Ingesting data is an important part of Hadoop Ecosystem. Flume is a service which helps ingesting unstructured and semi-structured data into HDFS. It gives a solution which is reliable and distributed and helps collecting, aggregating, and moving large amount of data sets. It helps to ingest online streaming data from various sources like network traffic, social media, email messages, log files, etc. into HDFS.

Apache Sqoop: The major difference between Flume and Sqoop is that Flume only ingests unstructured or semi-structured data into HDFS, while Sqoop can also import and export structured data from RDBMS or Enterprise data warehouses to HDFS or vice versa.

Flink: Is from a similar academic background as Spark. Spark came from UC Berkley, Flink came from Berlin TU University. While Spark is essentially a batch with Spark streaming as micro-batching and a special case of Spark Batch, Flink is essentially a true streaming engine treating batch as a special case of streaming with bounded data. Though APIs in both frameworks are similar, they do not have any similarity in their internal implementations.

Hadoop Ecosystem owes its success to the whole developer community. Many big companies like Facebook, Google, Yahoo, University of California (Berkeley), etc. have contributed their part to increase Hadoop's capabilities.

Inside a Hadoop Ecosystem, knowledge about one or two tools (Hadoop components) would not help you in building a solution. You need to learn and set up a set of Hadoop components, which work together to build a solution. Based on the use case, you can choose a set of services from Hadoop Ecosystem and create a tailored solution for your organization.

18.2.18 Main Distributions of Hadoop-Style Solutions

Apache Software Foundation: Apache is the main creator of Hadoop open-source components that can be installed on on-premises servers or in the cloud. You have the original components that must be installed using your own resources, a process that can become an extremely complex project.

To simplify matters, distributions were created that are composed of commercially packaged and supported editions of open-source Apache Hadoop-related projects. Distributions provide access to applications, query/reporting tools, machine learning, and data management infrastructure components. First introduced as collections of components for any use case, distributions are now often delivered by vendors as part of a specific solution for data lakes, machine learning or other uses. Most of these vendors also offer Managed Services that include the cloud infrastructure, Hadoop components and additional proprietary components to interact with the platform in a Software as a Service (SAAS) model.

Most popular distributions (not a complete list) are:

Amazon Web Services Hadoop Distribution: AWS renders a data analytics platform built on top of the HDFS architecture. With major focus on map-reduce queries, it exploits Hadoop tools to a great extent by providing a highly scalable and secure infrastructure platform to its users.

Hortonworks Hadoop Distribution: Hortonworks is a pure Hadoop company that drives open-source Hadoop distributions in the IT market.

Cloudera Hadoop Distribution: Cloudera, founded by a group of engineers from Yahoo, Google, and Facebook, is focused on providing enterprise-ready solutions of Hadoop with additional customer support and training. Cloudera has a history of incorporating new components to Hadoop like Apache Spark, Apache HBase, and Apache Parquet, that later on were eventually embraced by the community at large.

MapR Hadoop Distribution: MapR has been recognized extensively for its advanced distributions in Hadoop. MapR has made considerable investments to get over the obstacles to the worldwide adoption of Hadoop, which include enterprise-grade reliability, data protection, integrating Hadoop into existing environments with ease and infrastructure to render support for real-time operations.

IBM Infosphere BigInsights Hadoop Distribution: IBM Infosphere BigInsights is an industry-standard distribution that combines Hadoop with enterprise-grade characteristics that can be used on a service base.

Microsoft Azure's HDInsight Cloud-based Hadoop Distribution: Hadoop as a service is offered by Microsoft as its big data solution through a solution particularly developed to run on Azure.

Hadoop Ecosystem in Google Cloud Platform (GCP): Google Cloud Platform (GCP) offered by Google, is a suite of cloud computing services that run on the same infrastructure that Google uses internally for its end-user products. Alongside a set of management tools, it provides a series of modular cloud services including computing, data storage, data analytics, and machine learning.

Big Data is a new permanently and rapidly evolving technology. As we speak, it is very probably that new components and solutions are becoming available that are not mentioned in this chapter.

18.3 Managing Approaches, Techniques, and Benefits of Big Data

Most traditional structured sources restrict queries to the structure with which the data is stored. In these solutions, volume management also has a high cost and little ability to scale out. In addition, they have not been designed to efficiently handle the volume, variety, and speed of data in the world of Big Data.

Native Big Data solutions were created considering structured and unstructured data, and large data volumes handling, taking advantage of the horizontal growth capabilities of cloud installations, and have new analytical capabilities to meet demands on large volume at lower costs.

When is it convenient to have a traditional BI solution, and when should you opt for a Big Data solution?

If you have ad-hoc queries, the queries and analytical needs are known in advance and the data is less than 600 TB, the recommendation would be a Datawarehouse.

If your company's strategy is data-driven, then a Hadoop style option is recommended, to save as much data as possible.

If the volume of data to be managed is greater than 600 TB, the recommendation would be to go toward a hadoop-based model but considering the complexity of the queries required.

If you have your own business needs and need to respond as soon as possible, the recommendation might also be to move toward a Hadoop-based model.

If the data volumes are moderate to large, no analytics is required, and high speed is required the recommendation could be NoSQL databases.

18.3.1 Strategy for Adopting a Big Data Solution

The logical steps for the process of deploying a Big Data solution are:

- Identify the **need** for a Big Data solution, especially considering the strategic objectives of the organization in the short, medium, and long term, so that the decision is based on real needs and does not become a technology implementation project because "it is fashionable."
- **Valuate** the business case, analyzing how the solution will support the business case.
- **Understand** the benefits and risks of incorporating Big Data technology into your organization.
- Identify the **necessary capabilities** for a successful project, understanding that it is not only about incorporating technology, but that it is the technology that often involves a paradigm shift (from on-premises to cloud, from centralized to distributed, from Acid to CAP), and that it requires trained personnel and subsequent support.
- Identify critical stakeholders, adding to the usual ones:
  - Data scientists
  - Users of information
  - Technical
- **Evaluate technologies** based on well-defined **drivers**. As seen before, Big Data solutions are based on the Hadoop model, but there are several solutions from different providers from which you can choose, and the ecosystem has an important set of components and you must evaluate which elements are necessary and most appropriate for the criteria and drivers established. Some are proprietary and some come from the open-source world, but most of them are based on, or resemble the Hadoop solution already described. In both cases it is important to execute a well-formalized evaluation and design based on the business requirements and the quality

attributes of the solutions to be implemented. Evaluating, selecting, and designing using scenario-based methods is a good practice. Do not rule out hiring external technical experts for this activity.

- **Deploy** the solutions incrementally. The very concept of cloud computing allows the infrastructure to be gradually increased. The same recommendation should be considered to implement the functionality gradually, iterating and increasing the solution based on the lessons learned from the previous stages, especially in organizations where a Big Data project is implemented for the first time.

18.3.2 Big Data Governance

Big data governance is part of a broader data governance program that formulates policy relating to the optimization privacy and monetization of big data by aligning the objectives of multiple functions.

Sunil Soares

The goal is to ensure that information remains accurate, consistent, accessible, secure, and available to meet business objectives.

In the case of big data, the governance of this data presents a significant challenge since the data must be categorized, modeled, and mapped as it is captured and stored, with the disadvantage that it is data whose nature, for the most part, is unstructured, from diverse sources, and with new storage and consumption strategies.

Why do we need to Govern Data? Information is a vital enterprise resource and is critical to business success. Consistent and accurate information is fundamental for financial reporting, business intelligence, and measuring the execution of corporate strategies. Regulatory compliance has come to the forefront of information initiatives. Historically, data has not been governed with the same rigor given to other vital assets, as evidenced by the lack of quality, timeliness, transparency, and reuse. The introduction of unstructured data ("big data") has increased risk and exposure if not managed properly.

Particularities of the world of Big Data generate differences over the traditional data governance approach:

- **Asset variety:** The first big difference is the number of different types of data assets, and the fact that they continue to grow constantly and rapidly in size, shape, and origin.
- **Lack of physical separation between data classes:** The second difference is more subtle, but it is a byproduct of the way we have used our information management topologies to segregate data. We often rely on the physical separation of some data to identify it as something sensitive in some way, and to manage the controls around that data. In the world of big data, while data may be distributed, that physical separation often does not exist, and other means must be used to identify which data is sensitive and who is responsible for it. Governance processes need to maintain this information.
- **Create value by combining data that has not been related before:** The goal of a data lake is to create an environment where all the data can be easily used. This means that the different parts of the organization that own the data must agree to provide and use it in a controlled manner. In addition, data can now be shared with many parts of the organization, often without much effort on their part. This means that data exchange requirements must be explicitly negotiated, so that all users of the data understand what they should and should not do with the data.

In addition, the scope of semantic mismatches increases, as different parts of the organization will use the same terms with different meanings.

- **More varied and flexible processes:** In the traditional world, the way data is used is usually predefined. In the world of Big Data, data is available to be used in new ways as you explore from it. This in turn means that the automation system for that governance needs to be highly flexible and collaborative, in addition to having a clear operating model.
- **Increases in variety make automation a requirement:** The increasing scale in quantity and variety demands automation. Manual processes sometimes cannot keep up with the number of data changes and new data that are introduced constantly. Software tools may be required to assist in this task.
- **This data is an operational necessity and is in constant use:** Ultimately, this data is the soul of the organization. The infrastructure and platform on which they are hosted and processed must be able to keep up with all changes, as well as the volume of requests to use and process the data. Without that level of reliability and security, the organization will not be able to use its data and will not be able to acquire new data sources and information quickly enough to be competitive.

An effective governance strategy brings many benefits to an organization, such as:

- **A shared understanding of data:** Data governance provides a uniform vision and common terminology for data, without business units having to give up the flexibility they require.
- **An improvement in data quality:** Data governance creates a plan that ensures data accuracy, completeness, and consistency.
- **A data map:** Data governance offers the advanced ability to understand the location of all data relating to key entities, which is necessary for data integration. Just as GPS can represent a physical landscape and help people locate themselves in unfamiliar environments, data governance makes assets useful by making it easier to associate them with business outcomes.
- **A 360° view of business entities:** Data governance creates a framework for an organization to reach a consensus on "a single version of the truth" in the case of critical business entities and achieve an appropriate level of consistency between entities and business activities.
- **Uniform regulatory compliance:** Data governance provides a platform to comply with the requirements of legislation such as Personal Data Protection Laws and others (depending on the country or region).

18.3.2.1 Big Data Governance Main Elements

The defined framework must contemplate people, processes, and technologies that interact with the data, always keeping in mind business objectives as a guiding element (the value of the data). It should at least cover the following disciplines:

Organization: Big Data must be incorporated into the data governance framework that is being used, including the structure, roles, and responsibilities, extending these considerations to the particularities of Big Data.

Metadata: Big Data metadata must be integrated into the metadata repositories of the organization, considering the particular terms, sharing a unified vision between the different actors and considering the metadata of the technological implementation (e.g. Hadoop).

Privacy: Sensitive information must be identified, and policies established for its proper use. The principle of security adds to this concept, ensuring that only authorized persons access the data for authorized purposes. This means protecting data from external attacks or improper access, but also means ensuring that internally in the organization, people only access the information to which they are authorized.

Data quality: Data quality management is a discipline that includes the measurement, improvement, and certification of the quality and integrity of an organization's data. Given the first three Vs (Velocity, Volume, and Variety) of Big Data, quality must be managed differently. Considering that the value of the data is directly related to its quality, it is important to be able to identify the level of reliability of the recorded data. An important concept is: "Good Enough," since data must not necessarily be perfect to allow performing some analytical studies that the organization needs.

Business process integration: You need to identify those key processes in the organization that require Big Data. The data may be required for decision-making or by state regulations. Policies regarding privacy, informed consent, and other aspects related to privacy and the authorized use of data should be considered.

Master data integration: Policies must be defined to integrate Big Data into the organization's master data management environment. Big Data is fed from many sources and often from different sources data associated with the same entity is obtained. Finding a way to identify this data, associating it with a uniquely identified master entity is a major challenge in the world of Big Data.

Information lifecycle management: Both the organization and regulatory needs in terms of lifecycle management, when and how much to retain datas, and when to eliminate it must be understood. Withholding information that no longer has value is a huge waste of capital.

18.3.2.2 Cloud Data Governance

As companies and organizations realize the benefits of moving some or all their data and processes to cloud integration strategies, the need for effective data governance multiplies dramatically.

Making the leap to the cloud is based on delegating certain tasks to external third parties, such as infrastructure management, application development, security, etc. The cloud is also based on the virtualization of technical resources, which can pose data sovereignty challenges, as is the case with regulations that stipulate that data must be located in a certain place or country. In addition, cloud prioritization strategies often encourage decentralization and allow business areas and workgroups to deploy their own system independently, which could lead to uncontrolled data expansion.

Here is where data governance makes its full sense. Moving data processes to the cloud adds a layer of complexity in terms of security and accessiblity. While a fully on-premises data solution still needs a robust data governance strategy, when that data is moved through the cloud, data governance becomes crucial.

Data governance is not optional. Today organizations have incredible amounts of data about their customers, providers, patients, employees, products, sales, transactions, etc. When this information is used correctly to better understand the market and the target audience, organizations are more successful. Data governance will ensure that such data is trusted, well documented, easy to find and access from within your organization, and that it is kept safe, confidential, and in compliance with the law.

18.3.3 BIG DATA Project Management

Big Data projects usually accompany digital transformation processes, with a strong technology implementation component and alignment with business objectives.

Some features impose certain management style requirements for these projects:

- The high level of flexibility required. The use of big data solutions gradually generates new objectives as the first analysis of the information gives results, which in turn generates new hypotheses on which to continue advancing, as a result of learning from previous achievements (evolution)

- At the architecture and infrastructure level, components and nodes are gradually added as the previous components stabilize (evolutionary growth)
- The need to consider emerging new roles in the discipline:

Business analyst: Their responsibility is to make the technological solution compatible with business needs.

Statistician: Collects, organizes, presents, analyses, and interprets data to reach valid conclusions that lead to correct decisions. They are key players in the world of Big Data.

Data scientist: One of the most in-demand specializations today. Establishes the relationship between business needs and data structures that will deliver the expected results.

Data engineer/data architect: They design, implement, and maintain the infrastructure necessary to store and exploit Big Data.

Machine learning engineer: They are the people who design, implement, and maintain Machine Learning algorithms such as clustering, anomaly detection, or predictions to meet the needs of business intelligence.

Data steward: Responsible for ensuring the application of governance procedures (data custodian, data quality).

A suggested lifecycle for such deployments should consider the following steps:

1) Understand and formalize the user's need
2) Understand user data and feeds (data sources)
3) Identify architectural drivers
4) Select a Reference Architecture
5) Do trade-off analysis and identify constraints
6) Map architecture to candidate technologies
7) Prototype and re-evaluate the architecture
8) Estimate implementation efforts
9) Use Devops from the start of the project
10) Advance in small iterations
11) Implement Data Governance practices
12) Prepare for change as Big Data is continually evolving

As for the management of the project of implementation of the Big Data solution, there are different approaches that can be applied:

- **Traditional:** Ad hoc, Cascade, CRISP-DM, and KDD
- **Agile:** Scrum, Kanban, Discipline Agile
- **Native emerging approaches to data science:** Team Data Science Process (TDSP), Domino Lifecycle, and Data-Driven Scrum

Ad-hoc: Is a nice way to say no management process. It is not bound to a rigid structure, and administrative burden is minimal; however, it will end in re-work as key considerations are missed, bad coordination can lead to chaos or non-productive behaviors, lack of documentation and poor scalability.

Traditional cascade: An easy to understand and straightforward to implement option, it requires a comprehensive upfront planning of an evolving and ever-changing objective (a moving target), will not allow the required flexibility and might end up being unrealistic.

CRISP-DM/KDD: *KDD:* Knowledge Discovery in Database (KDD) is the general process of discovering knowledge in data through data mining, or the extraction of patterns and information from large datasets using machine learning, statistics, and database systems. KDD approaches tend to focus more on the steps to execute data mining as opposed to describing a comprehensive project management approach. *CRISP-DM* was created in 1996 to standardize a data mining process across industries, Cross-Industry Standard Process for Data Mining (CRISP-DM). It is the most well-known KDD approach. CRISP-DM describes six iterative phases. Each phase has its own defined tasks and set of deliverables such as documentation and reports. It makes sense to Data scientists since it naturally follows a CRISP-DM-like process, and explicitly starts with business understanding (an often-overlooked step). On the other hand, the CRISP-DM approach requires a lot of time-consuming documentation (although most teams seem to skip much of it) and is not really a project management approach since it does not address how to coordinate a project or the team.

Agile solutions: Agile solutions are focused on rapidly delivering value to the end users while allowing changes as a natural aspect of the project. They are adaptive and iterative and do not assume that the solution is known upfront but allows the data scientists and product owners to discover the solution over the course of several sprints. They emphasize the role of proven evidence to improve processes and product delivery – something data scientists naturally respect. On the other hand, some elements like Time-boxing are not the most suitable way. Many steps in the data science workflow require an unknown amount of time and do not line up into predefined time periods. Trying to force a time-box might short-circuit necessary experimentation. The project might be making meaningful progress without being able to deliver potentially releasable work at the end of a sprint, which contradict the working product as the only way of measuring advance.

Team data science process (TDSP): Elaborated by Microsoft in 2016, it is presented as an agile and iterative methodology to deliver solutions efficiently. It is based on five phases: business understanding, data understanding and acquisition, modeling, deployment, and customer acceptance. It is more than just a process, and it also includes reusable templates on GitHub, role definitions, and more. TDSP can, optionally, be used in conjunction with Scrum (where a sprint goes through all the phases). TDSP is frequently updated, Microsoft seems to update its guide and repository every few months. On the other hand, some inconsistencies were found probably due to frequent updates, it presents a steep learning curve, and there are parts of the framework that are related to Microsoft products, especially the aspects focusing on infrastructure.

Domino data science lifecycle: Combines CRISP-DM elements with agile methodologies. It is conceptually similar to TDSP. It presents more details about processes flow, but fewer deployment resources. Based on "what works," Domino builds this process based on experiences with over 20 data science teams. It is flexible, intended as an "a la carte" guide whose practices can be mixed and matched with other approaches, with no fixed cadence. Compared to its cousin TDSP, Domino's process lacks reproducible templates and detailed definitions. Domino was created in a one-off guide and has not been updated since. Team Coordination is not defined. While the process suggests doing many iterations (through their phases), it is not clear how the team should decide what is "in" an iteration, and how to structure the dialog with the business stakeholders.

Data-driven scrum (DDS): It is similar to Scrum but incorporates variable length iterations, focused on capacity that can overlap. Milestones are not a subset of items from the project backlog, but achievements based on data science concepts (e.g. experiments, questions to be answered). The selection of items from the product backlog is done continuously. As ceremonies, there are daily meetings (daily standup), iteration reviews and retrospective meetings (at regular intervals independent of project advancement), and product item selection meetings (sprint planning) when there is free capability to accept tasks.

Capacity-based iterations is a benefit, and it fits easily within Scrum organizations. The similarity with Scrum extends beyond just the process name, which makes DDS attractive to

Scrum-friendly organizations, it focuses on management and leaves freedom to follow any specific technical life cycle. Divorcing reviews and retrospectives from the completion of an iteration enables short, frequent iterations while still maintaining a regular schedule for these ceremonies.

On the other hand, the lack of lifecycle can be a detractor to teams who are looking for defined steps like those found in CRISP-DM, TDSP, or the Domino Data Science Lifecycle. Decoupling iterations from reviews and retrospectives could make reviews and retrospectives seem stale when they occur.

In summary, many approaches can be applied to manage Big Data Projects, some generic and some topic-specific, each one presenting advantages and disadvantages.

What is clear is data Big Data Projects need to have a project management methodology selected, tailored, and implemented, to ensure project success, and since they are usually part of a Digital Transformation process, business stakeholders are very important.

18.3.4 Ethics

The management of Big Data presents new ethical challenges, especially when companies begin to use data for other uses than what it was collected for, and to cross-reference it with information of external and/or public origin. The scale and ease with which a broad spectrum of people's information can be handled completely changes the scenario. In this context there are some principles that must be respected:

- People's privacy must be respected. The data of individuals must be used in the context for which they were authorized, they must not be shared or accessed for other purposes, and in case of being used for analytical purposes their anonymization must be ensured.
- Companies that receive people's data should not share this data with other companies that use it for other purposes (practice known as cross-selling).
- People should know how their data is used at all levels.
- The unfair impact of the usage of Big Data or Artificial Intelligence algorithms can have on the decisions of an organization with respect to people must be prevented. For example, algorithms that predict people's behaviors could impact the way they are treated by an organization. It must be taken good care not to introduce biases into algorithms, that might lean toward racism, sexism, or other isms. In these cases, human review should always be added when there are decisions involving people based on these algorithms (e.g. predicting whether a person might commit a crime and preemptively detaining them).
- Take all measures regarding information security, to protect the data from any inappropriate access or usage, internal or external.

18.4 Realistic Examples

18.4.1 Google

Google is the world's most "data-oriented" company. It is a part of the largest implementers of Big Data technologies. Maps all the Internet data. Identifies what is most used, more clicked, more interacted. What is most beneficial? These are the main data tasks of Google. Based on the SEARCH service, its first product, over time, Google created a lot of other data products. Google Apps, Google Docs, Google Maps, YouTube, Translator, and so on. The business idea is to provide a free data search service, and in the turning, collects clicks based on the interests of billions of users. Google applies Big Data technologies to identify what is most relevant, displaying the best-sponsored announcement, which is paid by advertisers. How can Google handle large volumes of data based on Tera and PetaBytes? The answer is implementing computing tools and technologies equal to Hadoop and BigQuery (Google's NoSQL technology). Following Google, Twitter, Facebook, LinkedIn, and many other data-driven companies emerged, trying to repeat Google's success in the Big Data era.

18.4.2 Netflix

Intense use of Big Data and Data Science technologies in Netflix business explains its success. The numbers are enormous. Netflix dominates the traffic streaming on the Internet. Users consume huge bands of traffic monthly. The foremost application of Big Data at Netflix systems is the "recommendation system." It is the machine learning algorithms acting on large information databases, movies, and users, making recommendations such as "if you like this movie, I recommend you this other ones ...." It collects vast amounts of data and analyses subscribers' personal tastes. Based on this, it generates personalized recommendations and new movies offer.

18.4.3 Amazon

Amazon became famous for selling books online before starting to sell anything via e-commerce, a true online retailer. But how did Amazon get so good at what it does? Thanks to Big Data and the implementation of the ML recommendation systems. Besides e-commerce, Amazon has become the largest provider in the world of cloud computing. Initially selling the excess computing resources that it did not need, it turned into one of the biggest players in the cloud business known as Amazon AWS. To ensure business growth scalability, Amazon invested in Big Data and Data Science and implemented machine learning algorithms orientated to online sales and deployed huge computer farms.

18.4.4 Tesla

Tesla vehicles – whether or not they are Autopilot enabled – send data directly to the cloud. A problem with the engine operation meaning that components were occasionally overheating was diagnosed in 2014 by monitoring this data and every vehicle was automatically "repaired" by software patch thanks to this. Tesla effectively crowdsources its data from all its vehicles as well as their drivers, with internal as well as external sensors, which can pick up information about a driver's hand placement on the instruments and how they are operating the car. As well as helping Tesla to refine its systems, this data holds tremendous value on its own right.

18.5 Summary and Conclusions

Big Data and Artificial Intelligence are two key disruptive technologies that enable the digital transformation of organizations, allowing them to achieve objectives and modify their processes in a way that was not possible before. Technologies are available and within reach, facilitated by the new capabilities of cloud services, the advent of the Internet of Things and platforms such as Hadoop. But technologies are simply an enabling tool. The challenge is to identify the right projects to transform the way we do things or the things that we do in an organization, and to implement the necessary technological solutions through successful projects that achieve business objectives and generate benefits for the organization.

The key word is "transformation," not "technology."

Further Reading

1 Erl, T., Khattak, W., and Buhler, P. (2015). *Big Data Fundamentals: Concepts, Drivers & Techniques*, The Pearson Service Technology Series from Thomas Erl. Prentice Hall.

2 Sawant, N. and Shah, H. (2013). *Big Data Application Architecture Q&A: A Problem – Solution Approach*, Exper's Voice in Big Data. Apress.

3 Soares, S. (2012). *Big Data Governance: An Emerging Imperative*. McPress.

4 White, T. (2015). *Hadoop: The Definitive Guide: Storage and Analysis at Internet Scale*. O'Reilly Media.

5 Ohlhorst, F.J. (2012). *Big Data Analytics: Turning Big Data into Big Money*. Wiley.

6 Kavis, M.J. (2014). *Architecting the Cloud: Design Decisions for Cloud Computing Service Models, SaaS, PaaS, and IaaS*. Wiley.

7 Trovati, M., Hill, R., Anjum, A. et al. (2016). *Big-Data Analytics and Cloud Computing: Theory, Algorithms and Applications*. Springer.

8 Erl, T., Mahmood, Z., and Puttini, R. (2013). *Cloud Computing: Concepts, Technology & Architecture*, The Pearson Service Technology Series from Thomas Erl. Part of The Pearson Service Technology Series from Thomas Erl (8 libros). Pearson.

19

Digital Reality Technology, Challenges, and Human Factors

Nicholas Napp[1] and Louis Nisiotis[2]

[1] *Xmark Labs, LLC, Barrington, RI, USA*
[2] *School of Sciences, University of Central Lancashire Cyprus, Pyla, Cyprus*

19.1 Introduction

This chapter explores the concept of Digital Reality by reviewing its fundamentals including its history, technological components, and current trends. It discusses the broader components of a Digital Reality project and goes beyond the technological components by including business factors and related considerations. The chapter also identifies and discusses common challenges, examples of practical solutions and presents a generalized framework for how to approach Digital Reality projects. This includes how to manage the project, a summary of good practices, and key recommendations. Multiple examples are presented with use cases drawn from different disciplines. The chapter concludes with a summary of IEEE's Digital Reality Exchange and the role it can play in Digital Reality projects.

19.2 Knowledge Area Fundamentals

19.2.1 The Basics of Digital Reality

19.2.1.1 Background

Many enterprises are embracing "digital transformation." Digital transformation is the implementation of new technologies, talent, and processes to improve business operations and satisfy customers [1]. However, digital transformation is just one aspect of a far larger shift that is already well underway: the Fourth Industrial Revolution (4IR). As noted by the World Economic Forum, *"We stand on the brink of a technological revolution that will fundamentally alter the way we live, work, and relate to one another. In its scale, scope, and complexity, the transformation will be unlike anything humankind has experienced before"* [2].

The First Industrial Revolution (1IR) used water and steam power to mechanize the means of production. From the water wheel to the steam engine, a variety of new tools reduced the need for manual labor and significantly increased productivity. The 1IR began in 1760. For the previous

IEEE Technology and Engineering Management Society Body of Knowledge (TEMSBOK), First Edition.
Edited by Gustavo Giannattasio, Elif Kongar, Marina Dabić, Celia Desmond, Michael Condry, Sudeendra Koushik, and Roberto Saracco.

2000 years, methods of production had barely changed. It began in the United Kingdom and was the driving force behind the UK's rapid economic growth: "*Purchasing power in Great Britain doubled and the total national income increased by a factor of ten in the years between 1800 and 1900*" [3]. It also propelled the UK to become, arguably, the world's leading power. The 1IR was not without its challenges. "*Workplaces were often poorly ventilated, over-crowded, and replete with safety hazards. Men, women, and children alike were employed at survival wages in unhealthy and dangerous environments.*" It also drove radical changes in the UK's political and social structure.

The Second Industrial Revolution (2IR) began approximately 110 years later, in 1870. It was driven by the use of electricity, which enabled mass production. As with the 1IR, it re-wrote societal rules and laid the foundation for a new world power: the United States [4].

The Third Industrial Revolution (3IR) is widely considered to be the mass deployment of electronics and the use of information technology to enhance and automate production. It began 80–90 years after the 2IR, in the 1950s. Countries that took advantage of computer technologies again experienced significant economic growth. One example of such growth is Japan. In the 1950s, Japan's rapidly growing electronics industry made a variety of inexpensive, and often low-quality products. In many cases, "*Japanese companies pirated their designs almost directly from foreign products*" [5]. However, the Japanese were far quicker than their American competitors to embrace 3IR and adopt the transistor. Unlike manufacturers in Japan, companies in the United States had invested heavily in vacuum tube technology. They were too busy protecting this investment to see the advantages that the transistor would bring. By the 1980s, just 30 years later, Japanese electronics were arguably the best in the world. US electronics manufacturers were in steep decline. Another interesting example of 3IR transformation is South Korea. In the 1980s, the electronics industry in South Korea [6] was growing rapidly. Following a similar path to Japan before them, many of their early products were low-cost and often of low quality. Within 20 years, South Korea became a world leader in electronics. The economic impact of this transformation is hard to overstate. "*GDP per capita in South Korea in the early 1960s was below $100. Lower than Haiti, Ethiopia or Yemen, making South Korea one of the poorest countries in the world.*" In less than 30 years, South Korea built an economy to rival Japan's [7].

The 4IR began less than 60 years after the 3IR. It is driven by data, complex computing, and the blurring of the lines between digital bits and physical atoms. It is enabled directly by the connection of billions of people via mobile devices, unprecedented cheap sensing and processing power, massive data storage capacity, and broad access to knowledge via the internet. 4IR technologies include artificial intelligence (AI), robotics, the Internet of Things (IoT), autonomous vehicles, Virtual, Augmented and Mixed Reality, 3-D printing, nanotechnology, biotechnology, materials science, energy storage, and quantum computing. What makes 4IR different to everything that has come before it is the speed with which it is arriving and the breadth of its impact. It is unclear who the global winners will be. What we can be certain of however, is that the economical, and societal impact of 4IR will be faster and greater than any of the preceding Industrial Revolutions.

How much change will 4IR bring? Consider this quote about the impact of the Second and 3IRs on the United States: "*In the late 19th century, roughly half of Americans worked in agriculture. By 2000, that fraction had fallen to under 2 percent. During the last century alone, we have seen those involved in the production of goods (from mining to manufacturing to construction) fall from about a third of the population to just under one in five. Over the same period, the proportion of Americans involved in services more than doubled, from 31 percent in 1900 to almost 80 percent by the turn of the last century. Since 1900, the number of farms in the United States has fallen 63 percent, and the average farm size has grown by two-thirds*" [8]. These are enormous changes for any society or economy to absorb. As the previously cited World Economic Forum article notes: "The speed of current breakthroughs has no historical precedent. When compared with previous industrial revolutions, the Fourth is evolving at an exponential rather than a linear pace. Moreover, it is disrupting almost every industry in every country. And the breadth and depth of these changes herald the transformation of entire systems of production, management, and governance" [5].

19.2.1.2 Defining Digital Reality

Now that the scene has been set, the reader should clearly see that we are on the cusp of massive and disruptive change; the 4IR is unfolding rapidly. In 2019, IEEE launched the Digital Reality Initiative [9]. The term Digital Reality was created as an umbrella term for three disparate technology areas critical to 4IR: AI and machine learning (ML); Internet-of-Things sensors and actuators (IoT); and visualization technologies such as Augmented, Virtual and Mixed Reality (AR/VR/MR), and Digital Twins (DT).

IEEE's proposition is quite simple: if you are working in Digital Transformation, or participating in the 4IR, you will inevitably be integrating these previously disparate technologies. A simplified model of how these technologies fit together is shown in Figure 19.1.

19.2.1.3 Introducing the Digital Twin

DTs represent a typical application of Digital Reality tools. One of the first proponents of the term Digital Twin was NASA. In 2010, they published a draft roadmap of technologies relevant to the US space program [10]. It included the following definition for a digital twin for space vehicles:

A digital twin is an integrated multi-physics, multi-scale, probabilistic simulation of a vehicle or system that uses the best available physical models, sensor updates, fleet history, etc., to mirror the life of its flying twin. The digital twin is ultra-realistic and may consider one or more important and interdependent vehicle systems, including propulsion/energy storage, avionics, life support, vehicle structure, thermal management/TPS, etc. In addition to the backbone of high-fidelity physical models, the digital twin integrates sensor data from the vehicle's on-board integrated vehicle health management (IVHM) system, maintenance history, and all available historical/fleet data obtained using data mining and text mining. The systems on board the digital twin are also capable of mitigating damage or degradation by recommending changes in mission profile to increase both the life span and the probability of mission success.

There are variations between definitions, but all share common traits. At its simplest, "A digital twin is a virtual representation of an object or system that spans its lifecycle, is updated from

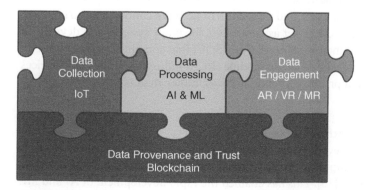

Figure 19.1 A simplified model showing how IoT, AL/ML, and AR/VR/DT fit together.

real-time data, and uses simulation, machine learning and reasoning to help decision-making" [11]. In other words, data is collected from real-world sensors (IoT), processed with AI or ML algorithms, and displayed via a virtual model that is a digital representation, or a twin, of a physical system. The sensor data is mapped directly to the digital copy, shadowing its operation so anyone viewing the digital twin can see critical, near-real-time information about the performance of the physical system. It is important to note that a Digital Twin can often display more useful data than direct observation of the physical system it represents. Critical information, such as system temperature or power consumption, is invisible to the human eye, but easy to display in a Digital Twin. Thus, a Digital Twin can reveal operational problems long before they become visible in the real world through visualization, simulation, and prediction. It is equally important to emphasize that a Digital Twin is not just the equivalent of a fancy new chart of data. The power of a Digital Twin comes from integrating the data with a virtual representation of the object or system. DT are fundamentally about making data actionable. A series of charts showing temperature data on a screen would require considerable expertise to interpret correctly. A virtual 3D model showing emerging hotspots in near-real-time is far more intuitive and actionable.

The various types of Digital Twin are discussed in other chapters of this publication and in online publications such as the blog posts and eBooks authored by Roberto Saracco for IEEE (see https://cmte.ieee.org/futuredirections/author/robertosaracco).

19.2.2 Broader Components of Digital Reality

As noted previously, Digital Reality is an umbrella term for three disparate technology areas: Internet-of-Things sensors and actuators (IoT); AI and ML; and visualization technologies such as augmented reality (AR), mixed reality (MR), and virtual reality (VR). These correspond to three broad actions: data collection (and sometimes control), data processing, and data engagement. We examine each of these areas below.

19.2.2.1 Internet-of-Things Sensors and Actuators

The steady progression of Moore's Law [12] and the mass adoption of the smartphone [13] have dramatically increased the capabilities of processors, sensors, actuators, and storage, while radically reducing their size and cost. This, in turn, has fuelled the development of innovations such as edge computing, low-cost drones, and the Internet of Things.

Fundamentally, the Internet-of-Things is defined as "*objects that 'talk' to each other. By combining these connected devices with automated systems, it is possible to gather information, analyse it and create an action to help someone with a particular task, or learn from a process*" [14]. A typical IoT device will include a sensor, a microprocessor, and some form of network connectivity. IoT devices can also include actuators allowing them to act as controllers for other systems and devices. In many cases, an IoT device will use a single System-on-Chip (SOC) that combines the microprocessor and a radio frequency transceiver into a single module. In an industrial setting, IoT devices typically only transmit data within a private corporate network rather than publicly transmitting data to the broader public internet.

19.2.2.2 Artificial Intelligence and Machine Learning

Another effect of Moore's Law is that every year, it becomes easier and cheaper to collect and process enormous amounts of data. Computing problems that previously required a supercomputer can now be solved with a desktop PC [15]. This has enabled the rapid development of AI and ML systems for an astonishingly wide range of applications. Furthermore, with the ever-increasing compute power of IoT devices, it is now possible to run some AI and ML applications directly on those devices. Libraries such as TinyML [16] support a variety of ML functions, including deep learning. Deep learning is the foundation of much of today's AI applications.

What is AI, and how does it differ from ML? AI is a broad field, loosely defined as *"the capability of a machine to imitate intelligent human behavior."* with particular emphasis on the idea of performing complex tasks *"in a way that is similar to how humans solve problems"* [17]. It allows machines to demonstrate human-like behavior and cognitive functions [18] and is arguably one of the most significant computing breakthroughs of our time. ML is a subset of AI. It was defined in the 1950s by AI pioneer Arthur Samuel as *"the field of study that gives computers the ability to learn without explicitly being programmed."* In other words, ML systems aim to intuit results from data. Such systems effectively learn to program themselves through experience and their own generated data [19]. As noted in an MIT Research Brief from 2020, *"With machine learning, human programmers don't need to write detailed instructions for solving every different kind of problem. Instead, they can write very general programs with instructions that enable machines to learn from experience, often by analyzing large amounts of data"* [20].

An ML process starts with a data set. The system then tries to derive rules, also called a *"model,"* to explain the data or predict future data. ML functionality usually falls into one of three categories:

1) Descriptive analysis that tries to explain what has happened.
2) Predictive analysis that tries to explain what will happen in the future.
3) Prescriptive analysis that tries to make suggestions about what actions to take.

In most cases, the more data the system ingests, the more refined the model will become. However, the quality of the learned model is highly dependent on the quality of the data used to train it. If the input data is biased, the output of the model will reflect that bias. This is a significant issue in the field of AI and ML. Issues of bias have led to the misdiagnosis of cancer [21], drivers being killed by autonomous driving systems that failed to recognize dangerous road hazards [22], Clinical Decision Support Systems that prescribed incorrect drugs to patients [23], strong gender bias in machine language translation [24], and a strong negative bias against black defendants in an AI system used in the US criminal justice system [25].

Within ML, there are three common subtypes: (i) supervised learning, where the system is trained using manually labeled data; (ii) unsupervised learning, where the system uses unlabeled data; and (iii) reinforcement learning, where an incentive system is created to "reward" correct outcomes e.g. assigning points based on the efficiency of a particular outcome. Creating large enough datasets of manually labeled data can be challenging. However, it is easier to address issues of bias since the data and data labels can be examined. In contrast, it can be very challenging to determine how an unsupervised system has actually processed the data and determined its output [26]. Reinforcement learning can produce amazing results, but reward systems often lead to unexpected outcomes. For example, a system may learn that it can generate bigger rewards by failing at a task rather than completing it successfully [27].

As AI and ML tools become more sophisticated, they are taking on increasingly critical roles in an ever-widening array of application areas, from business to industry to healthcare and beyond. For example, AI is being used to solve problems in the area of complex supply chains. *"The recent pandemic shined a light on the power of predictive analytics paired with AI. Data collection is crucial in the supply chain, but it is useless if it does not lead to action. We are gathering more data than ever – but we need AI to transform it into predictive and actionable insights"* [28]. AI and ML become particularly powerful when combined with other emerging technologies such as robotics and VR/AR. This provides opportunities to create intelligent and complex computing systems capable of interacting with real and digital worlds interchangeably, while supporting a plethora of application domains [29]. These opportunities are precisely why AI and ML are key elements of Digital Reality.

19.2.2.3 Visualization Technologies

The final component of Digital Reality consists of tools for visualization. These tools are designed to promote engagement and insight into the often vast amounts of data being collected and analyzed. Humans are adept at processing visual information, recognizing outliers, and seeing patterns, if data is well presented.

The commonly used techniques are AR, VR, and MR. In AR, digital content (often based on some form of data feed) is overlaid on a view of the real world in real-time. At its simplest, AR can be a heads-up display, although attaching data to a fixed real-world point of interest is generally more common. In contrast, VR replaces physical reality with a virtual one. There is no view of the real world– the user is placed in an entirely virtual environment. MR is a more recent development. It refers to AR systems that have a deeper understanding of, and greater interaction with, the real world. In an MR application, the system might have some understanding of what an object is. For example, it might recognize a particular machine and provide relevant contextual information and controls for that specific device. Alternatively, an MR system might understand that walls are solid and be able to identify surfaces within the real-world scene, allowing data and visualizations to interact more believably with their real-world surroundings. All three terms fall under the umbrella term of eXtended Reality (XR). XR refers to all real and virtual settings, hardware technologies, and human–computer interactions generated through them [30]. The different levels of reality in XR can be visualized and understood through the well-known Reality-Virtuality (RV) continuum proposed by Milgram and Kishino [31]. The RV continuum, shown below, shows that AR mostly involves the real environment (RE) and focuses on projecting and overlaying digital information on that RE, whereas augmented virtuality (AV) involves virtual elements more than the RE, and VR is a complete AV experience (Figure 19.2).

The ideas underpinning AR, VR, and MR have been around longer than you might think. The term "Augmented Reality" is attributed to Thomas P. Caudell, a former Boeing researcher who began using the term in 1990 [32]. However, Ivan Sutherland pioneered a number of AR and VR technologies in the 1960s that used early forms of computer graphics [33]. Prior to Sutherland, Morton Heilig developed the Sensorama, an immersive system that used various mechanical means to simulate an immersive motorcycle ride through New York City [34]. But the core idea goes back much further. In his 1901 short story, *"The Master Key: An Electrical Fairy Tale."*, L. Frank Baum (author of the Wizard of Oz) discusses the idea of a pair of spectacles that sound suspiciously like AR glasses:

I give you the Character Marker. It consists of this pair of spectacles. While you wear them every one you meet will be marked upon the forehead with a letter indicating his or her character. The good will bear the letter "G," the evil the letter "E." The wise will be marked with a "W" and the foolish with an "F." The kind will show a "K" upon their foreheads and the cruel a letter "C." Thus you may determine by a single look the true natures of all those you encounter.

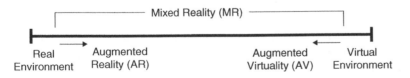

Reality-Virtuality (RV) Continuum

Figure 19.2 Milgram and Kishino's reality-virtuality continuum.

The XR Reality Continuum

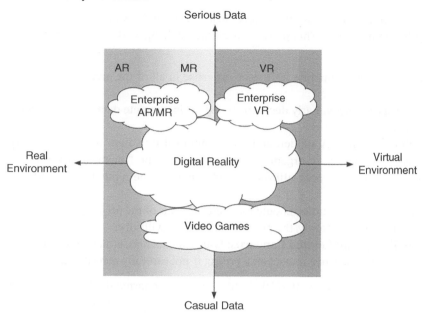

Figure 19.3 A conceptual map of the XR continuum highlighting some application areas.

It is important to understand that AR, VR, and MR also represent points on a continuum of interrelated tools and techniques, not just technologies. This is clearly implied by the RV continuum already presented. It is also part of what the term XR aims to encapsulate. However, what may be less clear to the practitioner is that the boundaries between AR, VR, and MR are often unclear and somewhat artificial. Again, this is where the term XR is often more helpful as a descriptor. It is often the case that a specific implementation will build upon ideas and techniques from many parts of the continuum. For some applications, it can be useful to delineate between "serious data," such as data generated by business systems and industrial processes, and "casual data," such as the data used to empower a video game. If we add this notion as a second dimension to the RV continuum, the result is Figure 19.3.

For example, an enterprise AR or MR application may entirely rely on serious data and will exist at the RE end of the continuum. In contrast, video games span a much broader segment of the RV continuum. Traditionally, video games have almost exclusively used casual data. However, newer games like Pokemon Go are blurring this line too. Pokemon Go combines traditional "imaginary" game elements with real-world locations, and gameplay is influenced by real-world (i.e. serious) data. As elements such as gamification are added to enterprise applications, it seems likely that enterprise AR and VR will eventually incorporate casual data too. An interesting early example of this is the Interwoven Spaces project discussed later in this chapter.

It is becoming increasingly common to find applications that do not fit cleanly into any of these less broad definitions, both in terms of their position on the RV continuum and the type of data they use. Examples include a system that combines VR technologies with traditional flat-screen monitors [35], an AR projection system combined with a deformable and tactile real-world surface [36], and a platform that combines traditional table-top gaming with stereoscopic glasses [37]. And considering the significant ongoing advancements in technology, software, and hardware, the future of XR has the potential to disrupt the way we perceive real and digital worlds and XR continuums.

19.2.3 Common Challenges in Digital Reality

From the material presented thus far, it should be very clear that Digital Reality is a deeply interdisciplinary and highly complex field. This presents multiple challenges, the most common of which are:

1) **Lack of Expertise:** Very few companies have deep expertise in all of the areas included in a Digital Reality project.
2) **Pre-judged Technology Choices:** Early decisions based on incomplete data can cause enormous problems in a project.
3) **Thoughtless Implementation:** A sudden and often unguided transition to Digital Reality solutions and implementations so an organization will not miss the "digital transformation train," typically leading to misuse, poor utilization, and damaging the reputation of both the company and Digital Reality.
4) **Poor Project Definition:** The definition should focus on the goals of the project, use cases, the expected benefits, and most importantly, the problem the project will solve.
5) **Bias, Ethics, Accessibility, and Related Concerns:** Issues of data bias, ethics, and accessibility can have a significant and negative impact on a project if not carefully considered.

This combination, together with the fact that DR is still a largely uncharted field, can often lead to project owners simply not knowing where to begin and even where to end! These challenges are discussed further in the sections that follow.

19.3 Managing Approaches, Techniques, and Benefits to Using Them

19.3.1 Suggested Approach

As previously noted, Digital Reality is a deeply interdisciplinary and highly complex field. Depending on your individual biases, you may look at a Digital Reality project and see it as mostly a software project or perhaps mostly a hardware project. In either case, you will be wrong.

19.3.1.1 Company Profile

Assessing an organization's internal skills and capabilities is a critical first step. Most organizations engaged in Digital Reality projects will fall into one of the typical profiles listed below.

A Hardware Company: This company has a background in IoT devices, M2M (machine to machine) products, or other electronic devices and really knows how to design consumer or industrial hardware products. They also understand the manufacturing and supply chain process. They've developed plenty of embedded software but have little experience developing or using web services, applying ML, and creating end-user applications or applications for data visualization.

A Software Company: This company has developed all kinds of software and has an excellent understanding of how to build software products for their target market. They understand the nuances of testing, deployment, system scalability and DevOps. They may have some experience with embedded software, and possibly even ML, but they have no experience with data visualization, or designing, building, or integrating IoT products.

A Traditional Product Company: This company has created many products for consumers and/or industrial applications, and really understands how to build products within its area of expertise. They also understand manufacturing, logistics, and branding but have little or no experience with connected devices, ML, and creating end-user applications or applications for data visualization.

A Brand: This company understands brand building, logistics, and customer management, but they have little experience in creating technical products.

A Kickstarter or Other Startup: This company or team has an idea and has raised some funding from traditional investors or a successful crowdfunding campaign. Startup teams typically fall into one of the categories already listed, but they have fewer resources and less experience working together as a team when compared to a more established company.

No matter which group your company falls into, your organization will be missing some key skills that are essential to creating a successful Digital Reality product. It is important to assess these strategic weaknesses honestly.

19.3.1.2 Key Skills for Digital Reality Success

It's time to start exploring the complexities of creating a Digital Reality project. Your first area of consideration is product related. The primary components are hardware, software, and User Experience (UX). As can be seen in Figure 19.4, each of these areas contains entire fields of expertise that need to be considered.

The second group of skills is business process related. The primary components in this category are Supply Chain Management, Business Strategy, and Project and Program Management. As with the product considerations, each of these areas contains entire fields of expertise that need to be considered. Your company profile and prior experience will determine which areas are strengths and which are weaknesses (Figure 19.5).

19.3.1.3 Key Steps and Questions to Answer
Phase 1

1) **Clearly define the project goals:** What will this project achieve for your company?
2) **Clearly define the problem:** What problem will your Digital Reality project solve?

Figure 19.4 A high-level overview of product considerations.

Figure 19.5 A high-level overview of business considerations.

3) **Identify challenges:** What do you see as your single biggest challenge in making this product? How will you overcome it?

4) **Identify your customer and market:** What is the primary market? What industry or demographic group will buy this? What does your customer look like? Is the target customer someone you already sell to? How big is the market?

5) **Identify your business model:** Will this be a one-time purchase, a subscription model, or something else?

6) **Explore pricing:** What is your target price point? What is driving your target price point? What is the total cost of ownership for your typical customer?

7) **Consider your end goal:** What does success look like? If the product is a success, what does that mean? Number of units shipped? Topline revenue? Bottom line profit? Increased customer engagement?

8) **Competition:** Are there competing products in-market? If so, how will you stand out? Are you relying on a change of behavior? Have you done a full competitive analysis and vetted your assumptions with customers?

9) **Responsibilities and funding:** Who is going to be responsible for what parts of the project? How will the project be funded?

10) **Skill gaps:** What key hires or contractors will you need to make? What skills are you missing? Do they need to be in-house, or can they be external?

11) **Intellectual property:** What intellectual property needs to be owned by your company? Can you use open-source components?

12) **Ethics, accessibility, and bias:** This is a huge area deserving of its own separate chapter. At a minimum, ensure that you create a project team that has diversity in terms of backgrounds, experience, and skills.

Many of the questions in Phase 1 will be hard to answer precisely. However, even a rough estimate or range will help identify key disconnects and potential problems. It is also important to note that **none** of the questions in Phase 1 involve any discussion of technologies or specific features. That starts to happen with the questions in Phase 2.

Phase 2

13) **Use cases and user stories** [38]: User stories are short, specific, and goal-oriented. They are a one-sentence statement with the following structure: "As a {type of user}, I want {the action they want to achieve} so that {I can attain a goal of some value}." There may be multiple types of users and hundreds of user stories for your project.

14) **Customer journey** [39]: How will the customer discover, purchase, interact with and use your product? How will that change over time?

15) **Consolidation, ranking, and insights:** Can you group, consolidate and rank the data you have collected? Which items are most important? Which are least important? Have you validated your assumptions with potential customers? What insights can be drawn from the data you have?

16) **Begin to define your User Experience (UX):** What does your user most want to achieve? How will you help them achieve it?

17) **Begin to define broad feature areas for your product:** What is essential to meet the needs of your users? Do not get too specific yet – this should not be a detailed and specific list of features. It should simply highlight the broad areas of functionality you need.

18) **Validate** your user experience and feature areas with potential customers.

As with Phase 1, Phase 2 avoids detailed discussion of specific technologies or product features. This is vital. Any choice of technology or specific features will have a downstream impact on the project. These impacts could include the ability to actually meet customer needs, cost, device size, functionality, power consumption, and many, many other factors. Literally, every decision you make will narrow the path you can go down. If you narrow the path too quickly, it may well not lead to a place that meets the needs of your customer. To avoid this, your core project assumptions must be clearly determined and validated with customers **before** you begin planning detailed features and which technologies to implement.

Phase 3

In Phase 3 you are finally able to start defining clear features and mapping those features to your user stories and the needs of your customers. Again, keep in mind that every decision you make will narrow the path you can go down. Based on experience with many complex, interdisciplinary projects, the authors strongly recommend mapping out detailed feature requirements **before** choosing any specific technologies to implement.

In Phase 1, we asked that you identify your skill gaps. It is quite likely that multiple parts of your project will require expertise you do not have. Do not make final decisions on functionality or specific technologies that are outside your areas of expertise without first meeting with multiple outside experts (contractors or organizations with the required expertise). The well-worn adage, "if all you have is hammer, everything looks like a nail" is particularly relevant in Digital Reality projects. Existing expertise can easily become a primary driver for product functionality, regardless of how well that functionality fits with a customer's needs.

19.3.1.4 Traps to Avoid

We have highlighted the complexity of a Digital Reality project, and outlined some key steps to avoid common pitfalls. However, it is worth reiterating specific traps that the authors have seen many projects fall into.

19.3.1.4.1 Lack of Expertise A Digital Reality project may require sensors and actuators with embedded software, RF technologies such as Bluetooth and Wi-Fi, cloud services, data processing, advanced ML algorithms, custom software applications, and the use of advanced visualization hardware such as AR or VR headsets. Very few companies have deep expertise in all of these areas. Failure to identify and acknowledge weaknesses and lack of knowledge can quickly lead to enormous challenges. Consider RF technologies for data communication between sensors and a central database of some kind. This may seem like a small consideration, but failing to understand the pros and cons of different wireless technologies can completely derail a project. If cellular connectivity has been specified for sensors, but there is insufficient power or not enough physical space for the required antenna, the project will encounter significant cost overruns and may even fail. Equally, if Bluetooth has been specified but the range is too short, or the RF noise in the deployed environment is too high, the project will also fail.

Key Takeaway: Companies seeking to undertake Digital Reality projects must be honest with themselves about their areas of expertise and areas of weakness.

19.3.1.4.2 Pre-judged Technology Choices Another common issue is defining a technology component before the requirements are fully understood. Using the previously cited example of RF connectivity, imagine a project where it was determined at the very beginning that the sensor device must use 4G cellular connectivity. This decision was based in part on a very early set of potential use cases for our imaginary project, and those use cases were not validated with customers. The choice of 4G will significantly impact the power consumption of the sensor. If the device is battery-powered, this will lead to an increase in cost to accommodate a larger battery. Furthermore, the minimum size of an effective 4G antenna is quite large relative to Bluetooth or Wi-Fi, which may require a bigger sensor housing. A 4G system will also require provisioning for each device, and working with a carrier partner (such as a major telco). Such a sensor may also require complex FCC and carrier approval, adding more time and cost to the project. All of these project impacts are non-trivial, and they are all driven by a choice that was based on a set of very early, and most likely incomplete, assumptions. After customer validation is complete, it is entirely possible that the primary reason for choosing 4G will turn out to be an uncommon use case and not part of the core functionality of the system. This hypothetical example precisely demonstrates the dangers of pre-judging the technology to be used in any given project.

Key Takeaway: Validate use cases and functionality with actual customers *before* choosing which technologies to implement.

19.3.1.4.3 Thoughtless Implementation Some organizations will seek a Digital Reality solution simply because it is the latest and newest technology being hyped by analysts and media. This kind of "shiny object syndrome" [40] almost always leads to disastrous outcomes. Organizations will attempt a sudden, and typically unguided, transition to a Digital Reality solution. They will fail to identify internal weaknesses, and the project definition will be weak or missing. The entire goal of an organization with Shiny Thing Syndrome is simply to chase the perceived latest and greatest thing. This inevitably leads to misuse, misunderstanding, poor utilization of technologies, and a very poor project outcome. It can also potentially damage the reputation of the company. Unfortunately, this reputational damage can also extend to the field of Digital Reality itself. The authors are aware of numerous thoughtless implementations of VR that have fallen into this trap, leading a number of companies to assume that VR is not a practical technology in a business environment.

Key Takeaway: Focus on the problem you are solving, the project definition, and your customer. Do not pursue Digital Reality just because it's a shiny object.

19.3.1.4.4 Poor Project Definition A clear project definition is essential. The initial definition should not include specific technologies to be used. Instead, the definition should focus on the goals of the project, use cases, the expected benefits, and most importantly, the problem the project will solve. A clear problem statement is vital. Consider Amazon's impressive warehouse automation system. While neither author was involved in the development of Amazon's system, it is clear that they must have had a very clear problem statement such as, "reduce the time taken for order packing." In an Amazon warehouse, millions of objects are stored across miles of shelving. Somehow, an order must be received, the items in the order must be retrieved, and they must be packaged for shipping. If Amazon had defined the problem as "we need to replace workers with robots to improve efficiency," they could have spent years trying to solve very hard problems such as object recognition and robotic object handling. Instead, they ended up implementing a system of robots capable of bringing entire shelves to the human worker. The human can very easily identify and retrieve the appropriate item from the shelf, and the robot returns the shelf to its original location (see https://robots.ieee.org/robots/kiva). The result is a highly efficient system that best utilizes the skills and abilities of all of the various system components.

Key Takeaway: Project success depends on a good project definition that is focused on the goals of the project and the problems it will solve.

19.3.1.4.5 Ethics, Accessibility, and Bias The history of engineering and technology is littered with products that failed to fully meet the needs of customers. In some cases, products have failed, or even harmed, entire groups. Examples include automated systems that fail for non-light-skinned users [41], chatbots that spewed hate speech [42], and a fitness app with default settings that put women's safety at risk by disclosing their home location [43] It is easy to dismiss these examples as unlikely to apply to your particular project. *Do not fall into this trap.* Not one of the companies involved in the examples above set out to create unethical, biased, or dangerous products. All were well-funded, and presumably believed they were being thoughtful in their product development. However, in every case, there was a failure of diversity in the design phase. A lack of diverse experiences and backgrounds in a project team creates blind spots. Furthermore, Digital Reality projects present many new opportunities for unintentional bias, from issues with bias in training data for ML systems, to a lack of support for accessibility options. One common example is voice recognition systems. They have been shown to be 13% more accurate for men than for women, and experience considerable difficulty with accents – individuals with Scottish accents can see results that are 20–30% less accurate than their American or Indian counterparts [44]. It is also well documented that a significant number of applications, "inhibit use by women and/or outright neglect factors that disproportionately affect female users" [45]. Approximately 50% of the world population identifies as female. The fact that their needs are not being considered in the design process is, and should be, shocking. Ageism in technology is also a problem. It is well documented that the world population is aging [46], and yet a great many software applications continue to use small type and low contrast colors, which can present challenges for those 40 or older [47]. Ethics, accessibility, and bias are huge topics. There are many other considerations that deserve further discussion. You ignore them at your peril.

Key Takeaway: Diversity, in every sense of the word, is essential to a successful project.

19.4 Realistic Examples

Digital Reality projects vary enormously in size and scope, from relatively simple apps that overlay a machine's operational data on a real-time view of the machine, to DT of entire cities. The examples below have been chosen to demonstrate both the broad variation in scope and the commonalities between projects.

19.4.1 A Cognitive Digital Twin

Unlike the other examples discussed below, this is a *personal* Digital Twin, designed to model and be used by a specific individual. IEEE's Digital Reality Initiative recently launched a Digital Twin project called "KaaS (Knowledge-as-a-Service)." This project aims to represent the skills and knowledge of an individual in such a way that it can act as a smart filter for the massive amounts of data available in cyberspace. Imagine a uniquely tailored combination of Google Search and Google Alerts that really understood an individual's interests. KaaS aims to understand a person's areas of knowledge and level of expertise. It then uses that data to explore information gathered from the internet, filtering for material that is relevant, understandable by the user, and presented in a language the user understands.

KaaS is an exciting development. If successful, it will help a wide range of users to discover and reinforce new, highly relevant knowledge, allowing them to be better informed and more productive.

19.4.2 The Hack Rod Project

Hack Rod was an ambitious collaboration between Autodesk and a very small Los Angeles based startup. It utilized an incredibly broad array of Digital Reality tools and techniques. The team instrumented an existing hot-rod vehicle with a broad array of IoT sensors. A trained stunt driver drove the vehicle to failure in California's Mojave Desert while recording sensor data. A scan of the vehicle chassis and all of the recorded data was fed into Autodesk's generative design tool. The generative design tool used AI to create a new chassis design with mechanical properties that met the mechanical needs implied by the IoT data from the real-world vehicle. Not only was the resulting chassis stronger, the weight was reduced from 470 to 320 lbs. The new chassis was 3D printed, and the rest of the car was assembled using off-the-shelf parts. However, the Digital Reality story does not end there. The team used VR to test the assembly and fit and finish of key components, such as the engine and the transmission. They also used cloud-based simulation tools to conduct aerodynamics studies. They even conducted initial road testing virtually, using a 3D model of the vehicle and an accurate physics simulation imported into a game engine. The "game" was played by a large group of gamers and the resulting data was used to identify and correct issues before the car was actually built. Hack Rod estimated that virtual road testing replaced 70% of real-world road testing.

Using this array of Digital Reality processes and techniques, the tiny team at Hack Rod was able to design and build a road-worthy sports car in a matter of months and at a fraction of the cost of traditional car design, a process that typically takes years and costs millions of dollars. For more details, see Hack Rod| In the Innovation Zone at AU with Mouse McCoy and Felix Holst (Figure 19.6).

19.4.3 Interwoven Spaces

Interwoven Spaces is an ongoing research project being conducted by researchers from the Centre for Automation and Robotics Research at Sheffield Hallam University and the University of Central Lancashire Cyprus Campus. The project uses an array of Digital Reality technologies to seamlessly blend real and virtual spaces using a novel combination of XR, robotics, AI, and social networks. It connects multiple users, robots, real-world spaces, and virtual environments into a single social cyberspace that can be explored dynamically. This is an example of a new type of Cyber-Physical-Social Eco-System (CPSeS) that is converging intelligent and multimodal interactive technologies to create new realities. The goal of the project is to support digital transformation in multiple application domains.

Their latest prototype is a Virtual Museum of Robotics. It is a multi-user VR environment featuring robotic exhibits, information about the history of robotics, educational games, and social spaces.

Figure 19.6 Photograph showing the chassis of the Hack Rod car. *Source:* Courtesy of Autodesk Inc.

Users can interact with exhibits and see a demo of how they operate in the real world. AI agents in the virtual world guide the user through the environment and share knowledge about the exhibits. It also features a real-world robot physically placed in one of the exhibition spaces at Sheffield Hallam University, which is represented by its Digital Twin in the virtual environment. Users can connect to the system from remote locations in VR, experience the virtual environment and choose to see the real world through the eyes of the robot in VR. As the robot moves and changes state, the Digital Twin is automatically updated. The robot can navigate autonomously, using advanced sensing to scan the environment and navigate safely using collision detection. It also supports remote teleoperation. Instructions can be issued to the Digital Twin, leading to actions in the real world. The robot can collaborate with humans for training, identifying movement and avoiding obstacles, and can pick up objects and identify them, moving them to different areas based on a combination of machine vision and instructions, e.g. pick and place operations. Multiple users can interact with the system simultaneously, and communicate via voice-over IP and chat. This digital reality project provides an example of how intelligent complex computing systems can be developed with the convergence of robotics, visualization technologies, AI, and social spaces to blend the real with digital worlds interchangeably and blur the difference between real and digital worlds (Figure 19.7).

For more details, see https://www.youtube.com/watch?v=0vO4yxQv6Go.

Figure 19.7 Still Image from "Interwoven spaces: creating new intelligent realities". *Source:* [29].

19.4.4 Unilever's Digital Transformation

Unilever PLC is a global consumer-goods company, with more than 400 brands sold in 190 countries, serving more than 2.5 billion customers [48]. The company has invested heavily in the creation of DT for its factories, using data streamed from IoT sensors and digital models of their facilities. The company has announced its intention to create DT for at least 170 of its 300 factories by 2021 [49]. Unilever's DT are used to track physical conditions in near-real-time, and test operational changes before they are implemented. The primary goal is to improve the efficiency and flexibility of production. As noted by Dave Penrith, Unilever chief engineer, "We are digitally rewiring our supply chain, focusing on generating real-time, democratized information, artificial intelligence planning, capitalizing on robotics and building digitally connected factories. All this will allow us to readily predict and respond to whatever the future throws at us."

Unilever's DT gather a broad range of data – everything from temperatures to production cycle times – and contains a virtual representation of every machine and process in each connected factory. The Digital Twin is used to predict outcomes based on historical data, in some cases even directly controlling parts of a process. The company can make real-time changes to optimize output, use materials more precisely, and help reduce waste.

Unilever has seen significant improvements in efficiency, product consistency, and overall production capacity for products as diverse as mayonnaise, soap, shampoos and conditioners, and laundry detergents. The initial pilot program at a factory in Brazil saved Unilever about $2.8 million, by reducing energy use and increasing productivity by 1–3% [50].

19.4.5 Virtual Singapore

Since 2014, Singapore has been building a Digital Twin of the entire country [51]. The project began with a 3D model but has continued to add additional data inputs and features over time. The 3D model includes detailed information such as building materials, terrain attributes, and transportation infrastructure. Integrated data sources include data from government agencies, 3D models, information from the Internet, and real-time dynamic data from a city-wide array of IoT sensors. Singapore's Digital Twin has enabled a variety of applications. Urban planners have used it to visualize design options for pedestrian bridges, and to explore a variety of sustainable and green features including LED lighting, a pneumatic waste conveyance system, enhanced pedestrian networks and an extended cycling network. Virtual Singapore has also been used to identify accessible routes for the disabled and elderly. Integrated environmental data has been used to visualize the effects of new construction, including reviews of the impact on sunlight, heat, and noise. Last, but not least, the dimensional data for buildings can be combined with sunlight data to identify optimal sites for solar panels, and estimate the likely energy output from each location.

Singapore's Digital Twin has demonstrated an array of practical applications, especially in urban planning. In densely populated, heavily urbanized locations such as Singapore, a Digital Twin can significantly reduce the likelihood of costly mistakes, and improve a wide variety of project outcomes for both the government and the country's citizens.

19.5 Summary and Conclusions

Digital Reality is a complex, interdisciplinary, and rapidly evolving field. The application of Digital Reality techniques and technologies has already delivered significant improvements in operational efficiency, cost savings, and increased productivity for many organizations. However, such significant potential brings many challenges. It should not be surprising that the barriers to creating a successful project can be quite high.

In this chapter, we have identified a basic implementation process to follow. The process strongly emphasizes the importance of putting the project definition and customer experience first. We have also highlighted the challenges and dangers of pre-selecting technologies, determining features too quickly, ignoring organizational weaknesses, and failing to introduce diversity in a project's team. We have also presented a variety of interesting Digital Reality case studies.

One of the most commonly encountered challenges is a seemingly simple one: the question, "where do I start?" Like many seemingly simple questions, the answer is often neither simple nor obvious. In order to help project owners answer this question, IEEE's Digital Reality Initiative is launching the Digital Reality Exchange (DRX). DRX is an educational marketplace that aims to connect project owners with partners, vendors, and other stakeholders while providing actionable information that will help ensure better project outcomes. DRX aims to help answer the question of where to begin and support project owners throughout their project journey. At the same time, educating project owners will help simplify and shorten the sales cycle for vendors and other stakeholders, thereby accelerating the growth and adoption of Digital Reality.

References

1 Boulton, C. (2021). What is digital transformation? A necessary disruption. *CIO*.

2 Schwab, K. Founder and Executive Chairman. The fourth industrial revolution: what it means and how to respond. *World Economic Forum*.

3 Newton, D.E. (2021). Effects of the industrial revolution – growth, workers, technological, and changes. http://science.jrank.org/pages/3574/Industrial-Revolution-Effects-Industrial-Revolution.html (accessed 13 September 2021).

4 Niiler, E. (2019). How the second industrial revolution changed americans' lives. *History.com, A&E Television Networks*.

5 Hays, J. Japanese electronics industry: history, chip development, decline and competition from apple, China and South Korea. *Facts and Details*, http://factsanddetails.com/japan/cat24/sub157/item922.html (accessed 13 September 2021).

6 Suarez-Villa, L. and Han, P.-H. (1990). The rise of korea's electronics industry: technological change, growth, and territorial distribution. *Economic Geography* 66 (3): 273. https://doi.org/10.2307/143401.

7 Georgia, Forbes Woman (2019). South Korea: from the poorest to the most developed in 30 years. *Forbes Woman*.

8 Rothkopf, D. (2012). The third industrial revolution. *Foreign Policy*.

9 https://digitalreality.ieee.org/about (accessed 13 September 2021).

10 Mike Shafto, C., Conroy, M., Doyle, R. et al. (2010). Draft: modeling, simulation, information technology & processing roadmap technology area 11. *NASA*.

11 Maggie Mae Armstrong (2020). Cheat sheet: what is digital twin? *IBM*.

12 Moore, G.E. (1965). Cramming more components onto integrated circuits. *Electronics Magazine*.

13 Silver, L. (2020). Smartphone ownership is growing rapidly around the world, but not always equally. *Pew Research Center's Global Attitudes Project, Pew Research Center*.

14 Burgess, M. (2018). What is the Internet of Things? *Wired Magazine*.

15 Leprince-Ringuet, D. (2021). Who needs a supercomputer? Your desktop PC and a GPU might be enough to solve some of the largest problems. *ZDNet*.

16 Zhuo, T.X. and Collins, H. (2020). Why TinyML is a giant opportunity. *VentureBeat*.

17 Sara Brown. Machine learning, explained. *MIT Sloan*.

18 Valin, J. (2018). Humans still needed: an analysis of skills and tools in public relations. *Chartered Institute of Public Relations*, Vol 23.

19 de Saint Laurent, C. (2018). In Defence of machine learning: debunking the myths of artificial intelligence. *Europe's Journal of Psychology* 14 (4): 734–747.

20 Malone, T.W., Rus, D., and Laubacher, R. (2020). Artificial intelligence and the future of work. *MIT*.

21 Patel, N.V. (2017). Why doctors aren't afraid of better, more efficient ai diagnosing cancer. *The Daily Beast, The Daily Beast Company*.

22 (2016). Tesla driver dies in first fatal crash while using autopilot mode. *The Guardian, Guardian News and Media*.

23 Abbrecht, P.H. (1982). Evaluation of a computer-assisted method for individualized anticoagulation: retrospective and prospective studies with a pharmacodynamic model. *Clinical Pharmacology and Therapeutics* 32 (1): 129–136. https://doi.org/10.1038/clpt.1982.136.

24 Female historians and male nurses do not exist, google translate tells its european users. *AlgorithmWatch*. http://algorithmwatch.org/en/google-translate-gender-bias.

25 Angwin, J. and Larson, J. (2016). Machine bias. *ProPublica*.

26 Fogarty, T. (2019). An introduction to unsupervised learning: clustering. *Towards Data Science*.

27 Clark, J. and Amodei, D. (2016). Faulty reward functions in the wild. *OpenAI*.

28 Beasley, K. (2021). Unlocking the power of predictive analytics with AI. *Forbes*.

29 Nisiotis, L. and Alboul, L. (2021). Initial evaluation of an intelligent virtual museum prototype powered by AI, XR and robots. *8th International Conference on Augmented Reality, Virtual Reality and Computer Graphics,* Virtual Event (7–10 September 2021).

30 Goode, L. (2019). Get ready to hear a lot more about 'XR'. *Wired*.

31 Milgram, P. and Kishino, F. (1994). A taxonomy of mixed reality visual displays. *IEICE Transactions on Information Systems* E77-D (12): 1321–1329.

32 Lee, K. (2012). Augmented reality in education and training. *TechTrends*.

33 Ivan, E. Sutherland. (1968). A head-mounted three dimensional display. *Proceedings of the December 9–11, 1968, Fall Joint Computer Conference,* 9–11 December 1968.

34 Heilig, M. (1962). US Patent #3050870.

35 Bell, M. (2017). Head coupled perspective with vive tracker. *YouTube*.

36 Kreylos, O. (2013). Augmented reality sandbox. UC Davis.

37 (2021). Introducing Tilt5. https://www.tiltfive.com (accessed 13 September 2021).

38 Muriel Garreta Domingo (2021). User stories: as a [UX Designer] I want to [embrace Agile] so that [I can make my projects user-centered]. *Interaction Design Foundation*.

39 Babich, N. (2019). A beginner's guide to user journey mapping. *UXPlanet*.

40 (2017). Do you have 'shiny object' syndrome? http://Entrepreneur.com (accessed 13 September 2021).

41 Taylor Synclair Goethe (2019). Bigotry encoded: racial bias in technology. *Rochester Institute of Technology*.

42 Schwartz, O. (2019). In 2016, microsoft's racist chatbot revealed the dangers of online conversation. *IEEE Spectrum*.

43 Spinks, R. (2017). Using a fitness app taught me the scary truth about why privacy settings are a feminist issue. *Quartz.com*.

44 Joan Palmiter Bajorek (2019). Voice recognition still has significant race and gender biases. *Harvard Business Review*.

45 Friedman, E. (2018). Wearable and immersive tech and the female workforce. *BrainXchange*.

46 (2019). World population ageing 2019. *United Nations Department of Economic and Social Affairs, Population Division* (accessed 13 September 2021).

47 Kane, L. (2019). Usability for seniors: challenges and changes. *Nielsen Norman Group*.

48 (2021). Our brands. *Unilever PLC, Unilever global company website* (accessed 13 September 2021).

49 Sokolowsky, J. (2019). Now it's personal: unilever's digital journey leads to real results for consumers and employees. *Microsoft News*.

50 Smith, J. (2019). Unilever uses virtual factories to tune up its supply chain. *Wall Street Journal*.

51 (2021). Virtual Singapore. *The National Research Foundation*, Singapore (accessed 13 September 2021).

20

Digital Reality – Digital Twins

Roberto Saracco

Consultant, Torino, Italy

20.1   Introduction

This part focuses on Digital Reality and on Digital Twin. Digital Reality has a lot in common with some interpretation of Metaverse. Digital Reality is the new ambient that is created by the overlapping of physical space with cyberspace. Notice that the physical space is more than just a geographical area including all entities (objects, people, etc.) that populate it. It comprises processes, that is the way entities interacts with one another and the organizational aspects (the rules of interaction).

Digital Twins are part of cyberspace and were born as mirror of entities but have more recently expanded to encompass the mirroring of cyberspace entities, like Digital Twin of Digital Twins. As being part of the cyberspace, they do not require any specific location (this is an implementation/ architectural decision): they may be co-located with their physical entity (like a digital twin of a machine being embedded in the machine software), may be located in an owned device, like a smartphone, in a server in the Cloud or may be distributed in several places (like part in the entity, part in the Cloud, etc.). Again, let us be clear that these are implementation/architectural decisions, the Digital Twin concept in independent of the specific implementation.

Digital Reality is first and foremost a perception space: each of us is becoming more and more familiar to act both in the physical space as well as in the cyberspace (machines are doing so too, but in this case we are not – at least yet – raising the point of perception. For a machine we simply talk about interaction, be it with another object or with a software, independently if this is or not mediated through the cyberspace).

In our growing familiarity of interacting with both the physical and the cyberspace we are reaching a point, also because of technology, where we no longer perceive a difference in the interaction. Our smartphones are both physical devices and gateways to the cyberspace, they are so seamless that information and services are seen as "embedded" in the smartphone, hence perceived as being on our hand. It does not matter (when connectivity is good) if a photo is stored in the phone or in the cloud. It is there at our fingertips whenever needed. Likewise for services, be it weather forecast, restaurant information, operation manual for the washing machine. The whole world is in our hand. The boundary between the physical and the cyber space is fading away. Let us be ready

IEEE Technology and Engineering Management Society Body of Knowledge (TEMSBOK), First Edition.
Edited by Gustavo Giannattasio, Elif Kongar, Marina Dabić, Celia Desmond, Michael Condry, Sudeendra Koushik, and Roberto Saracco.

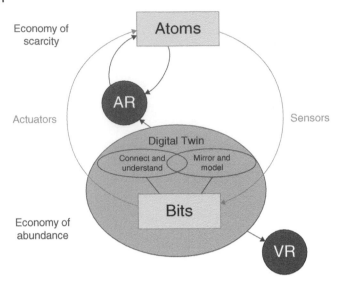

Figure 20.1 The continuum across the physical and cyberspace. By using sensors, we can synchronize the digital image of an entity in cyberspace and maintain its mirror image through the Digital Twin. This DT can be used for simulation and for intelligence gathering. When needed, using actuators, the Digital Twin can influence the physical entity. Using Virtual Reality (VR) we can observe the DT, and by using Augmented Reality (AR) we can perceive simultaneously the cyber and physical into a single Digital Reality.

for a complete disappearance once we will have seamless augmented reality gateways (like AR glasses or smart contact lenses).

This unification, from a perceptual standpoint, of the physical and cyberspace is called "Digital Reality": this is the new REAL space we are, more and more, living in.

Do not fool yourself: perception is the business space for marketing. Companies have to, and already are – to a certain extent, live in this new space and create products and services living in this space, since this is the new market space.

Once you start to consider the offer of a product in this new market space you need to make sure, and leverage, on both the features of the physical and of the cyberspace both to market, deliver and operate your product. The new Industry 4.0 paradigm is fully rooted in Digital Reality, taking most (as many as possible) processes and features to the cyberspace and delivering them, seamlessly into the Digital Reality space.

Digital Reality is the new business space and companies need to shift their mindset to this space. Digital Twins are the gateway to make the boundaries between the physical and the cyber space transparent. This is represented in the graphic of Figure 20.1.

This is what we are discussing in this chapter.

20.2 Knowledge Area Fundamentals

20.2.1 Digital Reality

We all live in a physical space and our perception and business, economic transactions, have been based on this physical space. However, the human race has early in its evolution conceptualized the physical space. Mathematics is a conceptualization of reality, and it is a useful tool to represent reality and to make predictions. Economics has started as barter but has evolved into a conceptualization of value, money and later shares, equities, etc.

The advent of computers has forced to use "digits" and because of that it has required ways to transform physical reality into digits and thereafter transform the results of computation in some ways to affect the physical reality. Sensors and actuators are the "tools" used for these transformations.

The digital reality, that is the use of digits to derive meaning, take decisions, affect other digits, and eventually impact the physical reality, has significant advantages on the physical reality because:

- it is easier to manipulate digits than atoms;
- digits can be duplicated at (basically) zero cost and the copy is exactly the same as the original;
- digits can be transported in the blink of an eye at (basically) zero cost;
- transaction cost in digit swapping is very low (it requires very little resources and effort).

All of the above is stimulating new companies to attack incumbents by leveraging on the cyberspace (as an example, e-commerce is challenging brick and mortar) and, both as a response and as a way to pursue more cost-effective business, incumbents are working to shift as much as possible of their processes and products to the cyberspace. This latter is called "Digital Transformation."

In turns, Digital Transformation is reshaping the whole business and, most importantly, it is reshaping the perception of the marketplace (to stick with the example, consumers are learning to use, and appreciate, the e-commerce).

It should be noted, as it will be expanded in a later section, that companies have started the adoption of the digital way of doing business long time ago. As an example, brick and mortar shops have been using electronic inventory, supply purchase through computer transactions, sales, and accounting using computer support. Likewise, industry has started long time ago to adopt computer for designing products. The use of CAD, computer-aided design, has become the normal way to design a product, the digital specs have been used to interact with suppliers and the digital model of the product has been used to drive CAM – computer-aided manufacturing.

Using computers, hence using data, per sé does not mean that a company has accomplished a Digital Transformation. For this such a company should have had to rewrite its processes, shift its business models and change its culture to the new digital space.

Very few companies, and businesses can exists in the cyberspace only, part of them will still remain in the physical space. One is flanking the other, and, if the digital transformation has been successful, the company will operate in this new space, the one of the digital reality. More than that. Such a company will leverage this digital reality space extending it to the market to its supply/ delivery chains and to its customers and users.

20.2.2 Digital Twins

Digital Twins, in retrospect, have been an inevitable evolution of the digitalization of products during the design phase. By using CAD companies found they had a digital copy of their products and started to use these copies for simulation purposes. In a way, the Digital Twin for these companies was born well before its physical twin was manufactured. This early stage of digital image of a (future) product represents the first stage of the evolution of the Digital Twin concept.

As shown in the figure, there has been an evolution of the concept, and implementation of Digital Twin that is still ongoing. The five stages outlined in Figure 20.2 are the ones that are implemented today (most of today's DTs fall in stage 3). We can expect further evolution in the

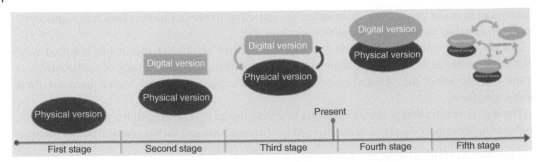

Figure 20.2 The five evolution stages of Digital Twins. On the first one the Digital Twin might pre-exist the appearance of its physical twin. In any case it is disconnected from it and being used independently of its physical twin. In the second stage, it remains disconnected, but it is used as a representation of the physical twin to perform simulation as well as to provide an understanding of its physical twin. From stage 3 onward, there is a connection with the physical twin, so that the DT remains an updated mirror image of the state of the physical twin and it might influence the physical twin state. On stage 4, functionalities are split between the DT and its physical twin so that they need to be seen as a single entity implementing and delivering the product functionality. At stage 5 the DT can take on an autonomous existence to operate in the cyberspace and interact with other DTs that only at a later stage impact the physical twin.

coming years, particularly as DTs expand their application to mirror people (personal digital twins and cognitive digital twins).

From the point of view of industry impact, we can outline the following:

- **First stage:** The digital twin is part of the design OR documentation process. In the design phase a digital copy of the (future) product is produced (most of the time automatically using CAD). There are several standards in different industry sectors. An example is the BIM[1] – Building Information Modeling – used by constructors all over the world. It has become a standard in the Digital Transformation of building construction, used in architecture, engineering, and construction (AEC) throughout the whole life cycle. The use of a digital copy of the building has rapidly expanded in use and application to cover operation and maintenance as well, thus rapidly moving into stage 3. ARUP[2] is a leader in this area.
- **Second stage:** The digital twin co-exists with its physical twin and is being used for simulation, as a starting point for upgraded design, as a training tool. Modern airplanes have a digital twin that is embedded in a simulator and this simulator is being used, extensively, for training. Notice that both in stage one and two a single digital twin, in principle, may refer to several physical entities that are "the same" from an abstract point of view. As an example, Boeing creates a digital twin of an aircraft model, say a 777, and this digital twin is used to train pilots for any 777.
- **Third stage:** The digital twin becomes a mirror image of the actual physical entity since it is connected and keeps being updated on the physical entity state (updating all parameters that are part of the modeling). It can also interact with the physical entity sending requests (to get further data AND to change its status. In this latter case, the digital twin has to await for a confirmation of execution of the request before reflecting it in its entity model). At this stage, we introduce the concept of "instance," that is the generic digital twin model of the generic instance (like the digital twin of the Tesla Model 3) is now instantiated and one instance is created for each actual entity (in the example and instance for each Model 3 car). These instances will have in common

---

1 https://www.autodesk.com/solutions/bim
2 https://www.arup.com/expertise/services/digital/digital-twin

the general model but will differ in their present status AND in their history (digital thread). This stage is the one most used today because it allows the monitoring AND operation –via interaction – through the cyberspace. Notice that at stage 3 we can have – from an implementation architecture standpoint – that the digital twin instance embeds the digital twin (that is the model), or it can be connected with the digital twin.

- **Fourth stage:** The digital twin instance (the DT has all the characteristics of the stage 3 DT) is becoming a component of the overall entity in terms of its characteristics/functionality since part of the functionality is provided through the digital twin/digital twin instance. In the absence of communication between the physical entity and the digital twin instance, the physical entity will not be able to deliver the full set of functionality. There might also be a need (this depends on the implementation/execution architecture) to maintain a connection between the digital twin (seen as the generic model of the entity) and the digital twin instance to ensure the availability of all set of functionality. The fourth stage is of interest to companies since the execution of functionality in the cyberspace, via the digital twin, may prove to be more cost-effective.

- **Fifth stage:** The digital twin can operate autonomously, interacting with other digital twins (likewise, a digital twin instance can operate autonomously interacting with other digital twin instances) to perform activities in the cyberspace. Upon the execution of an activity in the cyberspace the physical entity will either be requested to implement some action to align with the new state or the ensemble digital twin entity – physical entity will operate in a different state (because the digital twin entity has updated its state). At this level, we obtain the greater flexibility in the operation of the entity through the cyberspace. It is also worth noticing the evolution toward the creation and existence of digital twins that are clusters of digital twins. This clustering digital twin may not have a corresponding physical entity. An example is the digital twin of Singapore that mirrors the "city" of Singapore but in turns the city of Singapore is a conceptual representation (an abstraction) of an ensemble of physical entities and the infrastructures/processes connecting them. The digital twin of Singapore exists at stage 5 interacting, potentially, with digital twins of other cities, as an example in the monitoring of a pandemic.

20.3 Managing Approaches, Techniques, and Benefits to Using them

Companies approaching digital transformation need to consider that they will have to manage both the cyber and the physical space. It is indeed most unlikely that a digital transformation would result in a complete shift to the digital (cyber) space. Most likely, part of their business/process/resources will remain in the physical space and part will have to be found in (or operated from) the digital space.

However, the effects of the digital transformation are more complex than what could be assumed by simply splitting part of the business/processes/resources between the physical and the cyberspace (consider Figure 20.3).

One of the key point to consider is that the company is shifting into the Digital Reality space, and this is the NEW operation space for the company. All processes need to be reconsidered to operate in this new space. It would be a mistake to assume that those processes that have not been moved to the cyberspace will remain unaffected. To really take advantage of Digital Reality, all processes, independently from where they are residing (and part of them might end up residing partly in the physical and partly in the cyberspace, or even BOTH in the physical and in the cyberspace), need to be reconsidered in the new Digital Reality space.

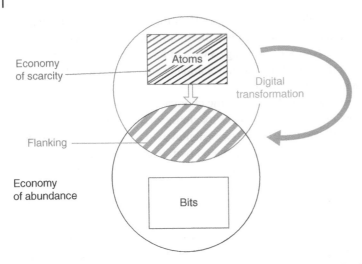

Figure 20.3 A digital transformation is usually resulting, as shown in this drawing, in part of the business being tied up in the digital space (bits) and part in the physical space. Hence, these two spaces – together – form the new business and operation space of the company. They flank one another. Notice that the part located in the physical space remains subject to the rules of the "classic" economy, the so-called economy of scarcity since physical resources are scarce. On the other hand, the part operating in the digital/cyberspace is operating under a frameworks where abundance rules, because of the very low transaction cost associated to bit manipulation (duplication, processing, transmission). Hence the company will need to re-engineer its processes both to confront the new "Digital Reality" and to leverage the "Digital Reality."

However, we have to keep in mind the different characteristics of these two spaces and the need to connect one with the other as seamlessly as possible.

This is where Digital Twins step in. Digital Twins can be seen as a tool to connect the digital and the physical space.

When planning for the digital transformation, managers and engineers should work through a checklist including the following points (there are many more as described in the IEEE/EIT Digital Course on Digital Transformation):

1) what are the existing data sets and how (to what extent) do they describe the company's
   - resources (includes both materials, tools/equipment and people, internal and external);
   - products (includes both products in the market and those in the design pipeline);
   - processes (includes internal as well as external processes);
2) what are the processes generating data and how can these data be captured? As an example design data produced by CAD systems, employees records used by HR, order processing, sales checks, trouble tickets. It is most likely that many data are being generated/managed within the company in different "silos." Leveraging, making use of, these data requires more than just providing access to them, it requires the definition of a company-wide ontology specifying the "meaning" and characteristics (static, dynamic, latency, ownership, etc.) of each data (data-set).
3) what additional data could be generated by using sensors, automated tools, monitoring processes, interfacing with suppliers/dealers?
4) what is the data ownership and how should data be shared (this includes aspects of authentication, access rights, processing and usage rights, monitoring of usage, watermarking including non-fungible tokens – NFT)?

5) how could data sets be encapsulated in a digital twin to represent a physical entity? What tools/platforms can be used to harvest data feeding the digital twins and how can these digital twins be supported (e.g. adoption of Fiware, of Mindsphere, etc.)? What standards shall be adopted (consider that in most situations the selection of a platform impose the adoption of a specific standard, like selecting Mindsphere for manufacturing implies the selection of the ISO/FDIS 23247-1 standard). Connected to d) how should a digital twin manage data to partition visibility and access? The "partitioning" aspect is crucial both in the design and in the operation of digital twins, since it provides the flexibility to use them in different situation. A digital twin of an entity is unique, however it may present different data and API to different interacting elements (other digital twins or other physical entities).

6) what is the new digital reality of the company following the execution of the digital transformation? How processes have been changed? How could they be changed (re-engineered) to benefit from the digital transformation?

7) what are the resources that are no longer needed and among these what are the ones that can be exploited in new ways? What are the new needs in terms of resources, processes?

8) what are the new business opportunities deriving from the exploitation of the digital space and the new digital reality space?

9) how is the new (reduced) physical space affecting the business (what is the loss of value deriving from the shift to the cyberspace)?

10) what should be the timeframe for executing the digital transformation (this involves cost estimate, resource availability, and constraints AND the value chain/market/competitor situation)?

Please be aware that, as noticed, this is a condensed checklist, and it is presented here to provide a sense of the macro-tasks that need to be considered and planned before initiating the digital transformation shifting the company in the new digital reality space.

Also, this should be the starting point to draw a roadmap for the transformation. You cannot execute the digital transformation in a day, nor in a month or a year. It is a long process that needs to be taken in steps. Some of these steps need an instantaneous changeover (when you move from one supporting system to a different one – although you might want to segment the execution in sectors, starting with the easier ones and then moving up to the next, fortified by the experience.

It should also be considered that lifetime in the digital space is potentially shorter than the one in the physical space; hence different processes and timelines may be required. However, some companies plan for synchronous step-by-step evolution of both physical and digital (see how Apple is releasing physical products on a yearly bases and updates the OS on a yearly bases too).

The digital twins shall be used as tools to both operate in the digital space and as a bridge to connect the physical and the digital space. Notice that this applies from stage 2 onwards. Indeed, even in stage 2, the digital twin (although not necessarily mirroring in real time the status of the physical entity) can be used to explore the physical entity using its digital model. Augmented and virtual reality can leverage on the digital twin to deliver their representations.

Further consideration shall be given to the use of cognitive digital twins and of personal digital twins. The former are digital twins mirroring/modeling knowledge (of a person, of a company, of an organization) the latter are digital twins of people, in case of companies they can be both digital twins of employees as well as digital twins of customers/users. Both cases rise issues of privacy and require, in most legislation, transparent information and opt-in/opt-out possibility. A personal digital twin may embed cognitive aspects, the knowledge/expertise of the person it is mirroring but this is not always the case. As an example a personal digital twin in a healthcare application does not need to capture the knowledge of that person but just the physiological/physical characteristics plus the record of all exams, prescriptions, ailments, DNA sequencing, etc.

The area of personal digital twin is quite new, although the use of profiling has deep roots in the past. A personal digital twin might be seen in terms of data management and up to a point in terms of technology as an extension of profiling, in the same way as a digital twin can be seen as an extensions of the model of a product deriving from the design phase.

Amazon, Netflix, and most companies having to manage a huge offer and present it as a manageable (effective) way to their customers have developed very hi-def profile of their customers. This is seen both as a service to the customer (simplifying the search and customization of products) and as a marketing lever (know your customer!).

From a company point of view the concept of cognitive digital twin is more relevant from a biz standpoint since it captures knowledge as a resource. As shown in Figure 20.4, a company's knowledge is about its products, processes, projects, and product lines. The cognitive digital twin of a product can be instantiated to reflect the actual physical product (the way it is being used). This of course raises issues on the ownership of data (the usage data are generated by the user and may not be shared with the company unless the users agree) and to a certain extent may involve privacy issues. In the software area, it has become "normal" for the software/service provider to monitor the usage, but here again there is a thin line between using the data to deliver better services and benefitting from the data to the company advantage. In any case, when engaging in the digital transformation the possibility of creating cognitive digital twins shall be considered.

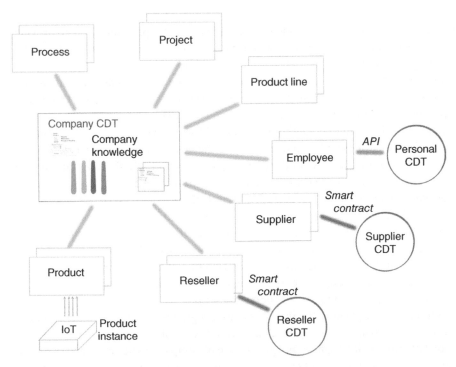

Figure 20.4 A framework representation of the relations among a company cognitive digital twin and the cognitive digital twins of its employees, suppliers, and resellers. Notice also that the company CDT is connected (owns the meaning) of the company's products, processes, and its various physical and logical components, like product lines and projects.

In this context, one has also to consider the creation of cognitive digital twins, as shown in the graphics, of their suppliers and resellers. In the coming years, one might expect that the suppliers/resellers themselves will be providing (an interaction with) their cognitive digital twin as a way for their client/provider to know what they can provide/need.

In the graphics, I also mention the possibility to develop a cognitive digital twin for each/some employees. In a way companies already have a description of each of their employees knowledge. This is essential in managing human resources (to allocate a person to a specific task or to train a person to prepare for a new job). What a cognitive digital twin can do is to provide more sophisticated representation of the owned knowledge and of the way this knowledge has been acquired (and potentially how it can be expanded).

How can an employee digital twin be created? The starting point, as for any other company's digital twins should be the set of existing data: HR has several data sets for each employee and a few of them (cv, training courses taken, career profile, roles played, projects involvement, etc.) can be part of that employee cognitive digital twin. Differently from an employee record, the CDT can provide an actionable model of that employee. It can be used to simulate the matching in a team, the contribution that can be provided by that person. It can contribute to the dynamic tapestry of that company knowledge and to identify gaps toward future evolution.

However, it should be clear that the profound difference between an employee record and that employee CDT is that a company record contains a description of that employee competences (experience and knowledge) whilst the CDT contains "the knowledge" of that employee. In this sense, the CDT could replace the employee in a given set of tasks, whilst, obviously, the employee record cannot do that.

This aspect of the CDT, the potential to replace employee knowledge, is a tricky one:

- who owns the CDT? At first glance, it should be owned by the employee that in turns can determine in what ways the CDT can be used/exploited.
- who is responsible for actions taken by using the CDT? The CDT is, most likely, not a perfect copy and anyhow decisions taken based on that knowledge might differ from the one that would be taken by the employee.
- once a CDT is used it can become more "experienced," i.e. knowledgeable than its physical counterpart. How can the person become aware of the increased knowledge of her CDT? Notice that the expansion of knowledge has a direct impact on responsibility since decisions will be take on that expanded set.
- would a company have rights on employees' CDTs since part of the embedded knowledge derives from direct experience gained on company's activity? Today most companies enforce non-disclosure agreements on their employees. What is going to happen once their CDTs acquires company's related knowledge? How could a CDT sign a non-disclosure agreement?
- would a company retain rights on their employees CDTs once they resign (or retire)?

This list is not – by any means – exhaustive; it is provided to highlight that the whole area of CDTs is fraught with ethical, social, and business issues. Nevertheless, it is self-evident the advantages that can be derived by creating and operating CDTs in an enterprise framework and how much the business can benefit from them.

Therefore, it is strongly suggested to consider the whole set of data related to company knowledge as a potential starting point to create company's CDTs. For some of them, like the knowledge based on "people," there is the need to address and regulate the ethical and business aspects and for sure there is a need for complete transparency.

It is also important to highlight that a company creating (or providing tools to create) CDTs of its employees can provide them as a benefit. The employee can "own" her CDT and use it in other contexts as well as in the future when she will not be any more connected to the company. Actually, we might envisage that by the end of this decade the present "cv" that is required when submitting a job application might morph into CDT. In the next decade, it is most likely, at least in some areas, that a CDT becomes a pre-requisite for being considered for a job and schools may actually develop CDTs for their students that will no longer receive a graduation but a certified CDT. This might well be a revolution in education!

Notice that a cognitive digital twin can play a significant role when delivering augmented reality since there is a need to contextualize the information, hence to be aware of the usage context (that comprises both the environment where the usage takes place as well as the way a product/service is being used and the familiarity of use.

20.4 Realistic Examples

In this section, I provide just a few examples of DR/DT applications in industry, out of the thousands that already exist. You may get plenty more examples on my blog.[3]

Mevea,[4] a Finnish company with one of the best business track record on the application of Digital Reality supported by Digital Twins, organizes a yearly seminar that is worth following to get the latest on industry applications. You can browse the material presented at these events from their website. These seminars are attended by over 200 companies from around the world, showing the high level of interest for this area.

The big message that comes across is that Digital Twins are for real, well past the hypercycle curve. They are a mature technology, which does not mean the technology is not evolving, actually, there are some application areas, like the ones on CDTs, that are still in research phase.

One application of Digital Twins that is mentioned over and over, through very specific examples, is their use as a training tool.

20.4.1 Siemens

An example is provided by Siemens: the use of a simulator for crane operators. In Figure 20.5, an operator training with the simulator.

A lot of emphases is put on the simulation of the environment through Digital Reality, including the sky and various position of the sun with different possible reflections as sun rays hit the container and the ship loading area.

Siemens has created a simulation platform that uses directly the software operating the crane. In this way, the operator trains on the real software he will be using in the field. If a change is made to this software, the change is mirrored in the simulator since it uses the same software.

Siemens is using digital twins throughout the whole lifecycle; it actually starts creating the digital twin "mock-up" in the specification phase. The same CAD that is used to create the model of the machine in the specification phase is also creating that machine digital twin mock-up. This is

---

3 Search for Digital Reality and Digital Twins to get up-to-date info on their application in industrial environment: https://cmte.ieee.org/futuredirections/category/blog
4 Watch this clip presenting Mevea use of DR and DT: https://www.youtube.com/watch?time_continue=1&v=fEI 5oz33Ia8&feature=emb_logo

Figure 20.5 A crane operator uses a Siemens simulator making use of Mevea Digital Twins to train on the use of a crane to load containers on ships. *Source:* Siemens.

further refined as specs are refined, and it keeps growing through the manufacturing phase. The manufactured product is associated with the generic digital twin of that type of machine that is instantiated to the specific product instance. Instances may differ even though they represent the same product since each instance has a specific digital thread that has recorded the various phases of manufacturing (including info on the supply chain, the specific components, what robots/human worker has participated in the manufacturing, etc.) and of course any specific customization required by the customer. This digital thread will continue to expand (and diverge from other instances of that product) as shadowing will bring in data from the actual use.

Siemens emphases the advantage of using digital twins for training. Cranes are complex machines operating in complex environment and training operators through simulators is both cheaper and more effective. As an example, through a simulator it is possible to inject errors that might occur and train the operator both to avoid them and to recover from them. Doing that in real life will lead to safety hazards and increase cost.

Another aspects emphasized by Siemens and other companies like Sandvik, is the high level of stress these machines are subject to, a stress that may reach peaks in a random way. Hence the usefulness of analyzing "shadowing" data and activate analyses that may lead to proactive maintenance. This is saving cost and most importantly decreases downtime.

20.4.2 Sandvik

An interesting application comes from Sandvik, a company manufacturing heavy machines applying DR and DTs to mining environment. The operating conditions are particularly demanding, often requiring the machine to operate in very narrow spaces.

Something one might not realize is the optical illusions that may occur when operating these machines underground with reflecting rocks that can alter the perception of available space. Sandvik along with Mevea developed extremely sophisticated assisted automation and full automated systems.[5] Indeed, Sandvik emphasis the goal of safety and how this can be better met using

---

5 An impressive demonstration by Sandvik: a machine moving, autonomously, in a glass labyrinth, with sensors subject to all kinds of reflections. It shows how an autonomous system can actually be better than a human controlled system. https://www.youtube.com/watch?v=TdYpxF9-8Ec

digital reality and digital twins. Simulation takes place in the cyberspace with the machine digital twin interacting within a virtual environment mirroring the one the machine will be operating in.

During the actual operation the machine's digital twin shadows the operation, and the data are being used by applications looking for possible anomaly, in which case corrective actions can be initiated.

Interestingly, a major use of operation data is for simulation of machine enhancement. These data are also used to simulate different tasks and operation in different environment. In other words, the simulation environment is constantly enriched with data derived from actual operation.

20.4.3 Infinity Foundry

Infinity Foundry is a Portuguese startup, founded in 2017, that is now in the scale-up phase having acquired over 30 large customers that use their platform. It has recently joined the scale-up program of EIT Digital.

What is very interesting in their platform is the support to a digital twin that mirrors a factory, a city, a healthcare institution, etc. The actual, physical entity is modeled into a digital twin, and then the DT can be used to train workers on the various operations and procedures. In this sense, it can also be used for training before the actual physical "thing" is available and this of course, if done in the design phase can be used to incorporate the feedback from the prospective users into the design.

Moreover, the DT can be used as a bridge to operate the physical entity from remote. The platform supports the use of virtual reality goggles so a technician can don the goggles and feel like being present at the plant and can operate the various components by acting on their virtual counterpart.

I find this approach fascinating. On the one hand, it will be a powerful enabler in change the way work is done you can operate equipment in noisy and potentially dangerous environment from the convenience of your living room (or comfortable office space). This is also decreasing the cost of plants since they would no longer need to be designed to accommodate human workers on site (on-site maintenance can be performed when the machines are switched off). On the other hand, this approach can extend to several other areas, including education. A student, or a professional engaged in continuous education, could get the same kind of involvement created by being on-site and yet take the course from faraway. I guess that the stumbling block, today, is the lack of really good VR devices that could provide the seamless immersion required to make the whole really felt "real." But I am sure it is just a matter of time. The rumored Apple glasses may provide a first step toward this seamless merging of the cyberspace with the physical space, that is the shift toward Digital Reality.

20.4.4 Digital Reality Applications

Digital Reality is becoming "Reality" also for processes. This is of particular interest, and significance, for the manufacturing industry and its value chain. Digitalization starts at the design phase and keeps evolving through the whole value chain. Co-design is performed on the digital model, simulation of single components, and of their resulting assemblage, is performed on the digital model. Most importantly, the tool design and the simulation of their deployment/operation are done at the digital level. Impact of failure in any component on the assembly line is simulated and evaluated at the digital level.

Furthermore, along the value chain some components (software ones) become part of the digital space and are seen both as digital representation and as real components.

The whole process results in the development of the final product's digital twin and is recorded in the digital twin's thread. At the delivery point (where the product is packaged and shipped) the digital twin is instantiated and becomes the mirror of that specific product instance.

The connection between the digital space and the workspace is supported by virtual reality and augmented reality technologies. Virtual reality comes into play in the design phase, where only a digital model exist and in the subsequent phases where no connection to the physical reality exist (e.g. looking at ways to organize the work of several robots over an assembly line, showing an end customer the intended product and the possible result of its customization, engaging the user in the operation of a product in a virtual space, etc.). Augmented reality comes into play when additional, digital information can enhance the understanding of a physical reality overlaying the former on the latter.

Industry 4.0, by making digital reality a fundamental part of its paradigm, will both benefit and stimulate the evolution of Virtual and Augmented Reality technologies, as well as all technologies that can leverage data, like data analytics, data rendering, artificial intelligence in all its forms, and, of course all data transmission technologies.

In a way, it will lead the digital transformation of the whole biz and society. Healthcare will follow suit taking human bodies into the "digital processing": predictive medicine and social medicine will be based on Digital Reality, on the creation and leverage of Digital Twins mirroring people.

The Industrial adoption of Digital Reality will include workers' skills and knowledge (Cognitive Digital Twins). These will be seen as resources mirrored in the digital space, to be managed and leveraged by the industrial processes.

The mirroring of people behavior in the digital space will become important in the modeling, planning, and operation of smart cities.

20.5 Summary and Conclusions

In closing this chapter a few consideration on the economic value of Digital Reality (the outcome of the Digital transformation) and of Digital Twins.

Part of the Digital Transformation result is the creation of a "mirror" world in the cyberspace that can be used to conduct business at a lower cost and without the constraints imposed by the physical world. As pointed out, this mirroring calls into play the concept of Digital Twins, software entities mixing data, processing and interactions with their siblings as well as with their corresponding physical entity and with other entities in the cyberspace. The Digital Twins derive from the adoption of CAD tools that create a digital model AND by the presence of sensors in products that can provide the operation data (the digital shadow). IoTs make sense as long as they generate data that are used and processed, and this basically calls for a Digital Twin, hence explains the rapid growth of Digital Twins. Platform support like the one of Azure, MS, – see Figure 20.6 – and Mindsphere, Siemens, foster the adoption of Digital Twins and their leveraging for the creation of value. According to ResearchAndMarket by 2025 89% of IoT platforms will include some form of Digital Twin management since they will be an integral part of the Industry 4.0 evolution.

As Digital Twins grow, more supporting tools become available, and it becomes easier to use and leverage Digital Twins. Also, these tools foster the adoption in Digital Twins beyond their core market (manufacturing). They are now being considered in real estate management, in ERP, in smart cities, in healthcare, and in education.

There is a whole ecosystem being developed on Digital Twins resulting in a market growing with a 28% CAGR over the next five years to reach B$20 in 2025 (with the North American market

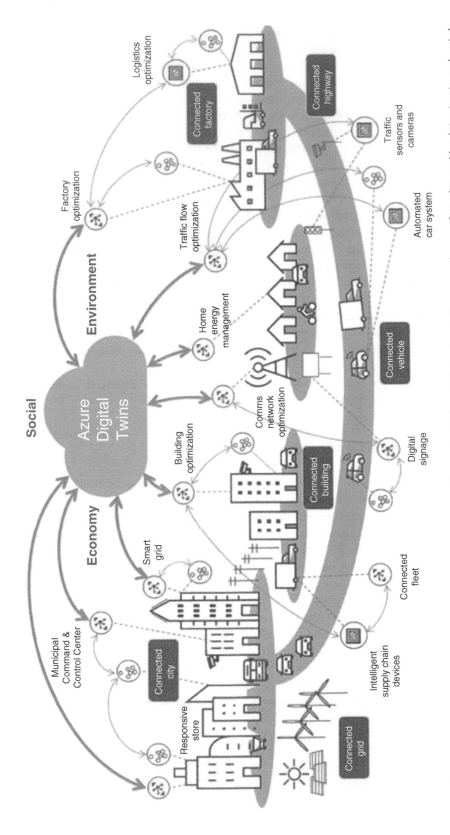

Figure 20.6 With Azure Digital Twins, environments of all types (offices, schools, hospitals, banks, stadiums, warehouses, factories, parking lots, streets, parks, etc.) can all become smarter, along with the electrical grids that connect them and the cities around them. *Source*: MS.

expected to grow at a 34% CAGR). Most recent estimates taking into account the acceleration of the Digital Transformation resulting from Covid-19 are pointing to a 45.4% CAGR in the next five years to reach a market value of 35. B$8 in 2025.

The market size of the tools supporting creation, operation, and exploitation of Digital Twins is just a part of the story, basically covering the cost companies have to sustain to deal with Digital Twins (cloud, data communications, data integration across different design, production, sales, operation/maintenance systems). Notice that these costs are an integral part of the new enterprise shifted in the cyberspace, so they should not be considered "on top" (in other words, if a company chooses not to adopt Digital Twins but still to move to the cyberspace it will still sustain –most – of those cost).

The other, most important part from an economic standpoint is related to leveraging Digital Twins to create and offer services.

This step is somewhat seamless for those companies that have been in the business of providing post-sales services like maintenance services. Using Digital Twins they can enhance those services through monitoring, analytics, and deliver proactive maintenance. However, for most companies this is not the case. The sale of a product is part of a business model that does not include interaction with the end user (sales is through a delivery chain that de-couples the end user from the manufacturer). Additionally, most companies are not in the business of services, they focus on products sale. In these cases a new business model, and supporting processes are required, and this is a challenge to many companies.

A further step would be to consider the Digital Twin as a separate entity that can be leveraged on its own, independently from its physical twin (the product). Take the case of Telecom Operators. Even though they operate on bits, and bits are the raw material from which they generate revenues, their business models are stuck to the world of atoms, i.e. copper and fiber plus SIM cards.

Telecom Operators have not been able, so far, to leverage on the huge amount of data they have and transport. The concern is on keeping their neutrality (although sometimes they claim for non-network neutrality to be able to deliver better quality – read: charge more some type of traffic), not getting involved in what is being transported – no responsibility – focus their business on the tools rather than on the product (they have mostly abdicated to create services leaving this economic space to OTTs).

Yet embracing the Digital Twin bandwagon, something that they would be ideally positioned to do, would propel them into this growing economic space. They could offer identity services, tailored communication paradigm fitting Digital Twins specific needs, hosting and mirroring, monitoring and authorization services, and so on. This transition is a difficult one since it means changing their mindset that is still tied up to twisted pairs (the fiber is a seamless evolution of a twisted pair, and the SIM is the equivalent for wireless communication – that is why most Operators hate the idea of a digital/virtual SIM!).

Digital Twins are evolving rapidly. It started in the last decade with them being used as digital "mock-ups," created by CAD systems. These mock-ups were refined till they could be used as digital specs for interaction with various groups in the company and with suppliers. The use of VR makes easier to visualize the Digital Twin of a future product.

Although that was not the case, one can find an economic value in these Digital Twins as "blueprints" that can be sold. This will be the case in this and the next decade as result of the market created by Digital Transformation with companies buying and using these digital specs.

The second step was the use of the Digital Twins to create their physical counterpart, through CAM systems (Computer Aided Manufacturing). In this way, a direct correspondence was established between a Digital Twin and its Physical Twin. Therefore one could use the Digital Twin as documentation of the Physical Object during operation and maintenance. The use of AR makes this correlation quite effective, since at the same time, a technician will be able to look at the physical engine and get information from its Digital Twin that can be overlaid on the physical one through real-time rendering.

This role in assisted operation and maintenance will become much more common as AR technology progresses. So far it is used in industrial environment but in the coming years it will become a common way of interacting with products and that will generate an additional economic benefit.

A further step was taken, and it is still being taken, enabling interaction between the digital twin and its physical twin. In this case, we are dealing with digital twin instance and physical instance of a product/object. The interaction (shadowing) ensure the alignment between the Digital and Physical twins. At this point, either the Digital Twin itself (in reality an extended digital twin) or applications interacting with the Digital Twin can deliver services, value, to the physical twin. There will be a progressive augmentation of a Digital Twin capabilities as it will not just interact with its physical counterpart but also with applications and services in the cyberspace. In this sense, the Digital Twin becomes a gateway between the physical counterpart and the cyberspace, a tool to deliver value. This is clearly a significant, transformational, step for companies making it possible to deliver services on a product in a seamless way. We are already seeing some examples of this happening. Remember when downloading a new version of an Operating System required the reinstallation of applications and some personal data? Now, this is no longer the case since our devices have a mirror image that is used when a new OS is installed. This mirror image is a sort of Digital Twin. A number of manufacturers, like GE, Mevea, Siemens, are now offering along with their products remote monitoring and proactive maintenance services based on the Digital Twin of their products. This changes the rules of the game since now the manufacturer has direct interaction with the end user (the flattening of the value chain that is one of the characteristics of Industry 4.0).

The next step will be the use of the Digital Twin to deliver additional functionality. In this case, the operation of a product requires its digital twin since some of its features are made possible through its Digital Twin. In practice, there is a function splitting among the physical and the digital twin. This, per sé, is nothing new: already today we have some functions split between the device and the cyberspace (voice recognition is often the case), so that a service may be partially be delivered through real-time interaction with a cloud.

This opens up the door to the "independent" exploitation of a digital twin by having it enabling functionalities to third parties (no longer, solely, to its physical twin). In other words, a Digital Twin may deliver services that in part are derived from its knowledge of its physical twin. An example would be a digital twin that has accumulated experience on the use of a car and can "sell" this experience to third parties interested in getting real feedback from the market. Another example might be the use of a Digital Twin as an avatar of a person (personal and cognitive digital twins as discussed previously).

Further Reading

General up to Date Innovation with DR and DT

Roberto Saracco, FDC blog: http://cmte.ieee.org/futuredirections/category/blog.
Digital Reality Initiative: https://digitalreality.ieee.org.

eBooks on Digital Twins downloadable from:

https://digitalreality.ieee.org/publications

Digital Twins evolution in manufacturing – Roberto Saracco.
Personal Digital Twins – Roberto Saracco.
The Future of Digital Twins – Derrich de Kerchkove, Roberto Saracco.
Digital Twins: Ethical and Societal Impacts – Derrich de Kerchkove, Roberto Saracco.

DT Supporting Platforms

Siemens Mindsphere: https://siemens.mindsphere.io/en/start?gclid=CjwKCAjwj8eJBhA5EiwAg3z0m
    wizhtmvpBCXkR0FYIhY4prPW1Fn8ZdZ7Cf0wG-Ogj0nGauIrC4ADhoCrREQAvD_BwE.
FiWare: https://www.fiware.org/news/digital-twin-consortium-and-fiware-foundation-join-forces-to-
    accelerate-digital-twin-technology-adoption.

Application to Industry

Arup: https://www.arup.com/expertise/services/digital/digital-twin.
Aveva: https://www.aveva.com/en.
Beamo: https://www.beamo.ai/?hsLang=en.
GE: https://www.ge.com/digital/applications/digital-twin?utm_medium=Paid-Search&utm_source=
    Google&utm_campaign=HORZ-DigitalTwin-MoF-EU-Search&utm_content=%2Bdigital%20
    %2Btwin.
Mevea: https://mevea.com/solutions/digital-twin.
Reply: https://www.reply.com/en/topics/internet-of-things/the-digital-companion-for-manufacturing.
Siemens: https://new.siemens.com/global/en/company/stories/research-technologies/digitaltwin/
    digital-twin.html.
15 application cases –2021: https://research.aimultiple.com/digital-twin-applications.

Section 9

Security

21

Bitcoin, Blockchain, Smart Contracts, and Real Use Cases

Ignacio Varese

Catholic University of Uruguay, Montevideo, Uruguay

21.1 Introduction to Bitcoin and Cryptocurrencies

Bitcoin was born in November 2008 from a paper (technical article) presented by Satoshi Nakamoto, a pseudonym used by a developer, or a group of developers. Bitcoin is defined as a digital cash and decentralized peer-to-peer payment system and started in production in January 2009. The biggest benefit is that it does not require third parties to operate and solves the problem of double spending on a distributed peer-to-peer network. For the first time in history, the concept of digital scarcity is incorporated, which can be exemplified as follows: If Pedro has a Bitcoin (digital asset) in his property, he can transfer that Bitcoin (digital asset) to Maria. As a result, the Bitcoin remains the property of Maria and it no longer belongs to Pedro. This transaction is recorded in the Bitcoin network permanently and immutable, where no one can modify or delete it. Before the existence of Bitcoin, infinite copies of digital assets could be made.

It is called a "cryptocurrency" because it is a digital currency that uses cryptography as a security basis. The Bitcoin network is public, open, and transparent, anyone can use it without permission, and in turn can consult all the data recorded on it. It is safe because it uses cryptography, that is, mathematics, to ensure the immutability of each data that is stored. Therefore, it is becoming popular for its security, transparency, and trust characteristics. Another of the key characteristics of the technology is that transactions can be carried out person to person without intermediaries.

After the creation of Bitcoin, some developers interested in the technology created new cryptocurrency options, taking the source code of Bitcoin, and modifying it, thus generating examples such as Litecoin, ZCash, Bitcoin Cash, and many more. New cryptocurrencies with their own technology, different from that of Bitcoin, also began to be created. Today there are more than 8000 cryptocurrencies on the market.

IEEE Technology and Engineering Management Society Body of Knowledge (TEMSBOK), First Edition.
Edited by Gustavo Giannattasio, Elif Kongar, Marina Dabić, Celia Desmond, Michael Condry, Sudeendra Koushik, and Roberto Saracco.

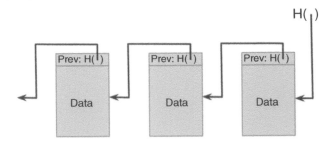

Figure 21.1 In blockchain, the previous block points to the next one.
Source: http://learningspot.altervista.org/hash-pointers-and-data-structures/.

21.2 Introduction to Blockchain

The concept of blockchain was born from Bitcoin, and it was called "blockchain technology" in the years 2013–2014 when it was beginning to be discovered that the technology can have other uses.

Blockchain is a set of technologies that makes Bitcoin viable. Cryptocurrencies are a use case of blockchain, within the multiple use cases that exists today. Blockchain can be used in different business solutions.

Blockchain is more than Bitcoin and cryptocurrencies, with blockchain you can build decentralized applications and many innovated business solutions.

Defining blockchain is not easy, there are many definitions and very confusing. As indicated by Dr. Alejandro Gómez de la Cruz, a Spanish lawyer specialized in blockchain, the definition depends on who we talk to, this will determine what meaning to use.

If we speak with someone technical, we define blockchain as "a distributed database to carry out value transactions"; if it is a businessperson, "it is an exchange network to move value between peers without intermediaries" and if it is a lawyer "it is a mechanism for validating transactions and interactions that does not require assistance from the intermediary."

Why is it called "blockchain"? Because the architecture of technology is literally a chain of blocks. In the figure behind, we can see that one block is linked to the other forming a chain of blocks (Figure 21.1).

Technically it works as follows: the network begins with a genesis block and every certain amount of time defined, in the case of Bitcoin every 10 minutes, a new block is created, the new block created points to or refers to the previous block, and so on, the chain of blocks or blockchain is formed. In each block the information of the transactions that are carried out in that period is stored, in turn, each transaction and each block is cryptographically sealed with a Hash or a cryptographic algorithm. The reference to the previous block that is stored in each block is the result of the previous block's Hash.

21.3 Blockchain Features

One of the characteristics of blockchain is that it can be operated without a trusted third party, it is operated peer to peer, without the need of intermediaries. At the same time, for the first time, I can represent value and send value over the Internet. When we talk about value, is in the form of money or in copyright, music, assets in the supply chain. We are now entering on the era of the Internet of value.

The main feature is security, since the information stored on blockchain cannot be deleted nor modified, the data only can be added to it.

Blockchain technology is secure and reliable due to three reasons: First, it is a distributed and decentralized network, it does not depend on a central node and each node always has an exact copy of the blockchain. This means that if a network node goes down or stops working, the blockchain network is resilient and continues to function (Figure 21.2).

Second, it uses cryptography, each block is cryptographically sealed, so if a piece of data is modified, the cryptographic result changes and therefore it will not be accepted by the rest of the network (Figure 21.3).

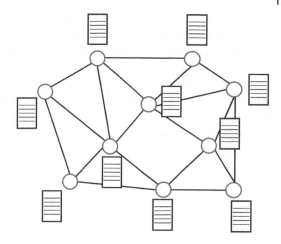

Figure 21.2 Distributed blockchain network.

Third, it uses consensus algorithms, which are rules that each node of the network knows and knows that it has to follow to accept the information; this makes each of the data stored in the blockchain be validated according to the rules defined by all nodes of the network.

21.4 Types of Blockchain

There are different types of blockchains, they can be public and open, private, or public permissioned.

The open public blockchains are in which anyone without permissions can access to the network, either to write or store information (transactions) on the network, or to be able to read it.

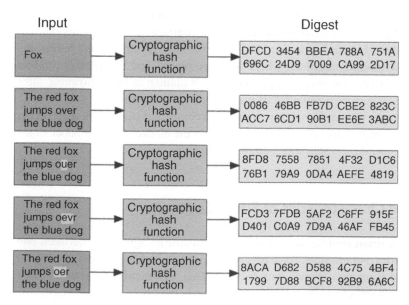

Figure 21.3 Operation of the Hash function. *Source:* https://en.wikipedia.org/wiki/Cryptographic_hash_function.

Any untrusted agent is allowed to participate with only an Internet connection. The transaction processing nodes, distributed around the world in a decentralized and distributed manner, invest financially to prevent fraud and spam. Examples of public blockchains are Bitcoin, Ethereum, Cardano, Solana among others.

Private blockchains require prior permissions to access, either for actions such as write or store information or to read data. This case is like a company intranet network, where only employees with the relevant credentials have access to the network. The permissioned private blockchains are used when it is required to set up a consortium of companies or organizations that do not trust each other but that want to exchange information and optimize their processes. Either for legal or business reasons they want to maintain a private network controlled only between those participants, but that none have central control. For example, they can be governments (among their entities) or the financial system, among others. Some of the most popular platforms to build permissioned private blockchains are Hyperledger Fabric, Hyperledger Besu, and Corda, among others.

A variant of the two previous ones is the permissioned public blockchains. In this case, users require prior permission from the consortium to write information on the blockchain, but its reading is public. In other words, anyone can see the data stored in it. This example can be used in a network of universities to register university degrees. Only universities with permissions can store information, but the reading can be done by anyone without permits, publicly. For example, in Spain, there is the Alastria network, which is a blockchain generated between several public and private organizations, where only they can write to the blockchain, but anyone can see the information. The platforms to build permissioned public blockchains are the same as the private permissioned ones: Hyperledger Fabric, Hyperledger Besu, and Corda, among others.

Depending on the business case that is necessary, one or the other solution can be used.

21.5 Smart Contracts

An important concept is Smart Contracts or Smart Contracts. This concept was originally created by Nick Szabo in 1994 when he described in an article that Smart Contracts are agreements that are executed automatically without human interaction; they only ran through software and hardware.

An example could be like vending machines, where coins are placed, and the selected product automatically comes out. At that time, this concept seemed a bit utopian to build, however, later with blockchain technology it became popular for its immutability characteristics.

Smart contracts are computer programs that run on the blockchain, that is, they are agreements between peers that are programmed, saved, and self-executed on the blockchain. Thanks of smart contracts, we can build decentralized applications on blockchain.

21.6 Examples and Characteristics of Smart Contracts

We can view an example of smart contract when we want to execute an agreement between two people without intermediaries. For example, Alice bets Bob that if Uruguay wins the Soccer World Cup will pay him $500, and otherwise Bob will have to pay it. This agreement is programmed in a Smart Contract, is stored, and self-executes on the blockchain.

The smart contract read the final champion of the World Cup through FIFA database, and depending on the winner, the money is given to Alice or Bob. Another example can be the rental of a property applied in the case of Airbnb: With a Smart Lock connected to the Internet, you can program a Smart Contract to accesses the Smart Key, and if the rent is paid and on date, the key is active, that is, the door can be opened and closed, and if the door is locked, it cannot be opened. Another example can be organizing a trip with friends, a Smart Contract is made to collect a certain amount of money to travel, but if cannot collect the total amount, all the money is returned. In this case, the Smart Contract locks the money until the conditions are met, if during certain days are not met, the money is returned, and if the agreed amount is reached, the tickets are purchased, all automatically.

Some other use cases are in different industries such as game of forecasts, inheritance, rent, loans, insurance, donations, and much more.

Smart Contracts are part of human evolution, humans improve and want to generate more certainties. At first the contracts were verbal, then they evolved into contracts signed on paper, and now with Smart Contracts they do not depend on a human, only on software and hardware that run contracts that are self-executing.

Smart Contracts are:

- **Descriptive:** It is what describes something (that is, that provides information so that people can represent it in their mind).
- **Prescriptive:** Prescriptive language, in this sense, is one that is used to tell an interlocutor what to do.
- **Deterministic:** It is completely determined by its initial conditions.
- Smart contracts give confidence, and as they are on the blockchain, they are executed without intermediaries.

21.7 Blockchain Benefits

Blockchain is a disruptive technology because the trust model changes. Currently, our society trust on central organizations, which have centralized information, and we all depend on these organizations to operate. But, what happens if we change the trust model? Instead of relying on organizations, we rely on technology that is based on cryptography. Trust model is changing from a centralized model to a distributed model. The importance of blockchain is amazing.

Blockchain is a **distributed** and **decentralized network**, where it can be operated without intermediaries, it does not depend on a central entity, therefore, using blockchain in business solutions allows us to:

1) Save costs.
2) Improve transparency.
3) Increase profits.
4) Optimize processes.
5) In addition to improving productivity among other benefits.

21.8 Blockchain for Business

Blockchain impact all industries in a transversal way, from the financial industry, insurance, logistics, pharmaceutical, and real estate, among others. More and more organizations are evaluating blockchain-based solutions, and there are already several projects implemented and in development around the world.

With blockchain independent parties can work collaboratively, because, although they do not trust each other, they trust technology, in mathematics. Business models change. It is a revolution in how companies change assets, how models change and how they share information.

In blockchain you can create your own cryptocurrency (token), make it smart and programmable. You can program your money with IoT devices, make voting systems, identity management, among others.

21.9 Decide to Build a Blockchain Solution

Before starting to develop a blockchain solution, we must ask ourselves if it really needs blockchain technology to that project. In this sense, we are going to use a flow chart based on the Uruguay Electronic Government Agency (AGESIC) to decide what type of blockchain do we need.

The first question is to know if "Is it a decentralized business model?" In this case, it is required to know if it is important that there is a central organization controlling the process or if it should be decentralized. If there needs to be a central entity, **DO NOT** use blockchain.

If it is a decentralized model, the second question comes, "Are there multiple writers?" At this point, you must know if there are different actors in the process who have conflicting interests, and who also write on the blockchain. If there is only one writer, **DO NOT** use blockchain, as it can be solved with a centralized model.

If there are multiple writers, we go to the third question, which is "can I always access a trusted third party?" In case, IF it can be accessed, then it is **NOT necessary** to use blockchain because the trust is already in a centralized entity.

If a trusted third party cannot be accessed, the fourth question comes, which is "are all the writers known?" In case they have NOT known, you **MUST USE** blockchain, and specifically **public and open blockchain**.

In case there are known writers, the fifth question is, "Are all writers trusted?" if answer is yes and the writers are trusted, it is **NOT necessary** to use blockchain because we have a centralized model that we trust.

Now if the writers are not trusted, then you **NEED** to use blockchain to ensure trust. Then you must decide what type of blockchain to use.

The sixth question is asked to know the type of blockchain, "is public verification needed? If public verification is NOT needed, use a **private blockchain**. If public verification is needed, use the **public blockchain**.

21.10 Considerations in a Project Based on Blockchain

When you decide to build a blockchain-based solution, you must consider certain project steps:

The first step is to analyze the organization's business processes, and study if it is for blockchain and what are the benefits of implementing the solution with blockchain.

The second step is to know if you really want to innovate, that is, to start a project with a new technology of these characteristics that can take time to learn and implement.

In the third step, you have to analyze the blockchain platform that best suits the solution. According to the flow diagram reviewed above, we can identify the type of blockchain (e.g. public or private permissioned), and then we must analyze the possible platforms, for example, if we have to use Bitcoin, Ethereum, Hyperledger, Corda, or others.

Regarding stakeholder analysis or project stakeholders, be prepared to expect all kinds of stakeholders between different organizations. Depending on the role, position, and participation within the project, it is necessary to analyze the strategies for each role. At the same time, it is important to involve them all and train them in the technologies that will be implemented.

Since with blockchain intermediaries and third parties are eliminated, there may be resistance in some roles, and therefore we expect many arguments against the value that blockchain gives. Resistance to change to a disruptive technology is to be expected, therefore it is important to mitigate this risk by educating and involving all necessary roles.

About defining the Scope, it is recommended to do the project in stages, go from less to more, do a Proof of Concept (PoC) since it is essential to start simple and then make it more complex.

Regarding implementation times, surely there is no old experience in the organization, therefore, it is recommended to save times for unforeseen events. Regarding the implementation team, in the case of not doing it with an outsourced company with experience, it should be known that there are very few people with previous knowledge in blockchain, therefore, in this case it is recommended to think of a training and training plan.

In addition, it is important to implement a good quality plan. A serious mistake with new technology is losing confidence due to errors that may arise especially when developing smart contracts. Regarding communications, it is advisable to have a good communication plan, since it is key internally in the team, for stakeholders and for external communication of the company. Good communication is very important for the process of a project with innovative technologies to be successful.

Carrying out a risk plan for the project is essential. There are many legal issues when implementing this technology.

21.11 Real Use Cases of Blockchain Solutions

Blockchain solutions can be implemented in all industries, providing many benefits.

If we are going to buy a luxury Rolex watch, how do we know that this product is not counterfeit? In general, we trust the brand, who made it, the business that sells it, or the person who delivers it. This is the current trust; we must trust in central organizations or people, as it is how we are used to. Now we have blockchain, the technology that ensures authenticity and transparency!

The cases of notarizations or digital certifications in the blockchain are the most used. For example, to authenticate and validate that a digital asset is not falsifiable. If I want to authenticate the information of a university degree, or a government digital certificate, it can be notarized and the information saved on the blockchain, and thanks to its immutability characteristics, that is, it cannot be modified or deleted, we can ensure that it is safe and reliable.

Another use case is in the traceability of the supply chain, the traceability of products with blockchain is carried out to know the origin, such as food, grains, medicines, and thus validate each stage to ensure that nothing has been altered, avoid fraud, and improve the efficiency of processes.

In the supply chain, it is applied to know the registry of origin, cut red tape in commerce, and use Internet of Things (IoT) devices. This can be done with traceability systems where each stage of the process can be stored in the blockchain and all the actors who do not trust each other, in this case do trust the blockchain, where the information is recorded.

An example of a traceability use case is the IBM project that started with Maersk, which is called "TradeLens," a blockchain solution for the transportation and shipping industry in international trade. It allows collaboration between multiple parties in international trade, currently more than

90 companies participate in the project. The objective is to have complete traceability of the supply chain where each of the actors participate: the exporter, the importer, maritime transport, port, terminal operators, land transport, customs operators, and government, among others. Each one at each stage is saving information on the Blockchain; In addition, the IoT sensor devices for temperature and weight control also store information on the blockchain. All records will be valuable information to obtain complete traceability, which will allow to improve product delivery times, reducing many costs.

Another application that can be built with blockchain is tokenization. Tokens are assets created on the blockchain, and the tokens can be Fungible or Non-Fungible. Fungible tokens are assets that are created on the blockchain, and all types of tokens are the same and can be interchangeable, for example a token you create that represents one dollar is equal to another token of the same type that represent a dollar, both are equal to one dollar. Another type of tokens are NFTs or Non-Fungible Tokens, which represent unique assets on the blockchain, such as representation of art or unique assets.

There are many business cases for fungible tokens, for example in art industry to distribute property rights through fractional ownership, we could "tokenize" an artwork. These are new models of ownership, where several people or investors can acquire the rights through the purchase of "tokens" that are represented on the blockchain. This would be like the issuance of shares of a company, each share would be a token. In this way, a larger audience base is reached, a more liquid distribution, since the market is much larger.

NFTs, or Non-Fungible Tokens, are used in different industries to identify something digitally unique. You could own an NFT to identify ownership of something, you could also trade or exchange it. For example, the most popular cases are NFT in digital art, collectibles, music, lands in real estate, in video game to represent unique tools, in sports, tickets, and many more. NFTs are being used for several new use cases at any given time.

Financial use cases could be the use of cryptocurrency to make payments and money transfers. In these cases, we have applications such as Bitpay, which is a payment gateway with cryptocurrencies, Pundi X, which is a POS that businesses can install to accept cryptocurrencies, or companies with an international money transfer service.

Some other examples that are being carried out are, in financial services, in the cases of Know Your Customer (KYC) or Anti Money Laundering (AML). In this case, a consortium of entities belong to a blockchain to share KYC/AML information, and this allows them to save costs together. There are also cases of loans and insurance notarizing transactions on the blockchain and ensuring reliability, and cases of transfers, using cryptocurrencies for international money transfer between countries, and thus saving implicit costs in the transaction.

Examples in the healthcare industry are supplier certification, patient registration, and the pharmaceutical supply chain. In these cases, the notarization of certificates and traceability are used, storing each step of the process.

21.12   Risk and Opportunities

Blockchain changes the way we do business; eliminate intermediaries and impact all industries. Therefore, this technology is going to bring big challenges on how to build blockchain projects. We have risk challenges that require considering the risks of a blockchain base solution project.

- **Regulations:** How the example topic will be regulated by the payment system.
- **Legal:** How to handle contracts and disputes to do it with smart contracts.

- **Data privacy:** This is an issue to be handled, how the data will be handled, in Bitcoin for example you can see all the transactions.
- **Security:** Blockchain is very secure, but you have to see how it is implemented and how it interacts with other applications.
- Governance, who is going to handle the updates and that is where the centralization begins, and it remains decentralized.
- **Reliability and Scalability:** depending on the solution and the implemented platform, it can change.
- **Integration:** Integration with other blockchains, and with other systems that adapt and integrate.

We have several challenges that generate risks for us and that we must analyze and mitigate them.

On the other hand, as we have seen, blockchain offers us new **opportunities** that will transform business models as we know them today.

Taking the opportunities and the benefits we can make a difference from today. We have been able to show that thanks to blockchain we can: Reduce costs, improve transparency, and Optimize processes, in addition to that we can have faster, safer, and more transparent solutions and, therefore, it is time to start thinking about how I can use it to my business.

21.13 Summary and Conclusions

As described above there are many business use cases using blockchain. This technology impacts all industries, some cases were presented in this article.

It is still in the early stages, many new projects are starting, and organizations are incorporating it. Leading companies use blockchain technology and many solutions are build, such as IBM, Microsoft, SAP, and Oracle, among others.

Blockchain changes the way we do business, and it will bring us many challenges, such as regulations in each of the areas, data privacy, governance (who will handle updates), reliability, scalability, and integration with other blockchains.

This technology is here to stay, it is a revolutionary change, it changes the way we do business, and we all must be prepared. Many opportunities are emerging where to build blockchain, so for its benefits, it is a great differential to apply it in business.

If you are in an organization that is not applying it, you can propose it, or you can join the very good opportunities that you can find in any organization that wants to implement it. When starting a blockchain project, it is important to start with small, simple projects that last weeks, carrying out some pilots to learn about the technology, and then adding complexity that makes it grow toward larger projects.

We are in a unique moment in history, blockchain undoubtedly changes how we perceive the world, it is a social and economic change, and we must take advantage of it.

I invite you to start to envision how your industries will perform in near future.

Join the blockchain revolution!

Further Reading

1 Nakamoto, S. (2008). Bitcoin: a peer-to-peer electronic cash system.

2 Andreas, M. (2017). *Antonopoulos: Mastering Bitcoin*, 2e. O'Reilly.

3 Andreas, M. (2017). *Antonopoulos: Internet of Money*, vol. 1. CreateSpace Independent Publishing Platform.

4 Arun, J.S., Cuomo, J., and Gaur, N. (2019). *Blockchain for Business*, 1e. Addison-Wesley.

5 Tapscott, D. and Tapscott, A. (2018). *Blockchain Revolution: How the Technology Behind Bitcoin is Changing Money, Business, and the World*. Portfolio Pinguin.

6 Gaur, N., Desrosiers, L., Novotny, P. et al. (2018). *Hands-On Blockchain with Hyperledger: Building Decentralized Applications with Hyperledger Fabric and Composer*. Packt Publishing.

7 Andreas, M. (2019). *Antonopoulos: Mastering Ethereum: Building Smart Contracts and Dapps*. O'Reilly.

8 Levy, S. (2010). *Hackers: Heroes of the Computer Revolution*. O'Reilly.

9 Preukschat, A. (2000). *Blockchain: The Internet Industrial Revolution*. Gestión.

10 Champagne, P. (2014). *The Book of Satoshi: The Collected Writings of Bitcoin Creator Satoshi Nakamoto*. e53 Publishing LLC.

22

Cybersecurity

Jon Clay

Trend Micro, Inc., Tokyo, Japan

22.1 Introduction

Malicious actors will continue to target businesses and individuals around the world with their attacks and cybersecurity will be a critical component of anyone's online experience. The shifts and changes in the threat landscape continue to require innovations and rigorous adherence to a cybersecurity implementation within your environment. This chapter will help anyone understand what cybersecurity is, how it works, and what best practices can help minimize the risk of compromise.

22.2 History of Cybersecurity

Crime is a crime and has been around since time began as we humans have had to deal with other humans trying to do harm against each other. So, whether we are talking about physical crime, like shoplifting or breaking into a bank and stealing money, or we are talking cybercrime where a malicious actor steals money from a bank account after stealing the victim's online account credentials, there will always be someone who wants to perpetrate a crime.

Technology has given us a tremendous advantage in making our lives better and allowed us to communicate with others around the world instantaneously. Technology has improved business processes and made organizations more nimble and more efficient in how they go to market. But technology has also allowed malicious actors to take advantage of its use too. Ever since Bob Thomas developed a computer program on APRANET in 1972 that could move across a network and leave a trail, and Ray Tomlinson modified the code to self-replicate, malicious actors realized they could take this concept and use it to their advantage. Ray also developed a program to detect his code which was the first anti-virus program.

Cybercriminals need a big pool of victims for it to make sense to target cyber as a mode to commit a crime. In the early years where not a lot of people and organizations had yet to adopt computers and applications there was not a lot of cybercrime and malware. There was also minimal ability to gain access to a lot of victims from a network perspective. When malware was shared using

IEEE Technology and Engineering Management Society Body of Knowledge (TEMSBOK), First Edition.
Edited by Gustavo Giannattasio, Elif Kongar, Marina Dabić, Celia Desmond, Michael Condry, Sudeendra Koushik, and Roberto Saracco.

floppy drives in the 1980s and early 1990s, there was not a lot of opportunity to target many victims, so cybercrime was not a big issue worldwide. But, as technology improved and the Internet started to see massive gains in people and businesses accessing it, criminals started shifting their activities toward cybercrime.

One of the biggest drivers of more widespread adoption of cybercrime was when email became the standard for business communications and individuals got themselves an email account. Malicious actors started using this to send spam in an effort to get recipients to go buy their offers. Email was also used to send malicious files (malware) as attachments that could infect computers when opened by the recipient. During this time, we saw a phenomenon where malicious actors tried to infect as many systems as possible and get bragging rights within their underground communities as infecting the most computers in an attack.

It was during this time that cyber security started to become a need for both Individuals and businesses and organizations in order to combat the rise in cybercrime and malicious attacks. In this section, we will take a look at many of the different areas within cyber security in an effort to give you some insights into this area of the technology landscape.

22.3 Who Are the Attackers

Before we discuss the different types of threats, we need to better understand who is perpetrating these attacks like those seen in Figure 22.1. As it turns out, there are actually many types of attackers, whose skills and motivations are quite different from each other.

In the early days of hacking, back in the early 1990s, hackers were computer enthusiasts who broke into systems for fun – for the ego and the challenge. Many of us have images of teenagers sitting in their parent's basement, finding their way into sensitive and highly secured locations. Many of these amateurs – or script kiddies – still exist today, wreaking havoc more for the challenge than for financial gain. The challenge we have is today's cybercrime can be very lucrative and

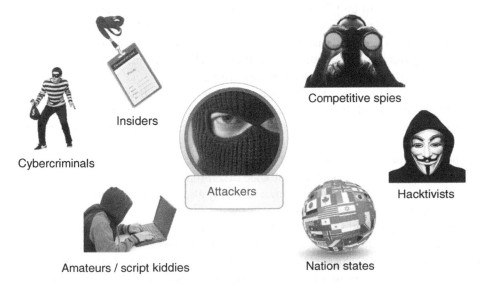

Figure 22.1 Cyber attackers. *Source:* Ardasavasciogullari/Adobe Stock Photos, Madia Krisnadi/Adobe Stock Photos, Tiero/Adobe Stock Photos, Dan Race/Adobe Stock, Szasz-Fabian Jozsef/Adobe Stock, Ctacik/Adobe stock, and Pasko Maksim/Adobe Stock.

access to technology and the Internet is quite easy around the world. As such, in many locations around the world where obtaining a well-paying job may be difficult, it is relatively easy to go into cybercrime and the profits from even one successful hack could be a year's salary for these hobbyist attackers.

Of course, there are criminals who are now finding it highly profitable and much easier to gain financially through cybercrime than through traditional break and entry and other types of physical crime. Many of the traditional crime syndicates have a cyber arm now and their professionalism is top-notch. Cybercrime has estimated to cost society $6T in 2021 and expected to be upwards $10.5T by 2025 [1].

There are employees and other insiders who seek financial gain, or perhaps some form of revenge against their employers. These malicious insiders are very difficult to detect as most have trusted access to systems and/or data that can be stolen by them. We also see many insiders recruited by outside cybercriminals to help with an attack and get a cut of the payout.

Competitive spies who are looking for an easy way to gain advantage through the theft of intellectual property (IP). This could be someone who wants to sell IP to competitors within an industry for profit, or it could be a nation-state who is looking to bypass the research & development (R&D) cycle of an industry they need improving in.

And at the other end of the spectrum, we have hacktivists, who are causing havoc to make a stand of some sort – free access for the people, political protest, etc. This could be individuals or a group like Anonymous who may recruit people for one of their attacks. In some cases, their acts may just be to display their message on systems or web pages, but in recent years we are seeing more disruptive or destructive attacks by these activists. More recently we are seeing hacktivists steal sensitive information of people or organizations and publish them for all to see.

And the one that many people fear most, nation-state or state-affiliated actors who are attacking other nation-states for the purpose of national security. These actors are typically well-funded and have excellent computer skills that allow them to be more successful than all the previous actors. They also have patience and are willing to try as often as they need to obtain access to the systems they are targeting. One aspect we must be aware of is that malicious actors can be harbored in areas of the world where law enforcement has difficulty arresting them. In many cases, if the actor does not attack citizens or organizations within their own territory, they can operate with impunity.

All of these "attackers" are actively seeking new ways to sneak into companies, government agencies, or consumer homes.

22.4 Malicious Actor Motivations

There are several motivations that the above actors tend to follow:

- **Financial gain:** This is by far the majority of malicious attacks today and most actors are looking to profit from their attacks. Whether it is selling stolen data within undergrounds, or extorting their victims, they want money. Some will sell items like malware or exploits as well as services within the cybercriminal undergrounds around the world. These actors supply other malicious actors with the means to launch and execute successful attacks. There are many services sold that allow anyone (i.e. script kiddies) to perform attacks on people or organizations. In recent years Ransomware as a Service (RaaS) has become popular in the underground markets.
- **Disruption or destruction:** The motivation to disrupt or destroy infrastructure tends to be the focus of a nation-state or hacktivist, but more and more we are seeing ransomware actors using

disruption of critical business systems that may motivate organizations into paying the ransom fee. The challenge with destructive attacks, if by a nation-state, is it could lead to cyberwar or even traditional war.

- **Espionage:** There are several groups that employ espionage, but note this is not always the traditional method of nation-states spying on each other. Today, espionage may include competitive businesses within an industry who want to spy on their competitors to learn how they do things, or what new products they will be developing.
- **Resources:** In some cases, malicious actors are simply looking to use the systems they occupy for the resources the systems offer. This could be to mine cryptocurrencies or launch a denial of service (DOS) attack, or send phishing and spam messages. Many businesses need to understand this as many believe they may not have data of interest to actors, but their systems are a lucrative target themselves.

Now that we have learned about the malicious actors and their motivations, let us look into the threats that have been used throughout time.

22.5 History of the Threat Landscape

Prior to the 1990s, most cyber threats were file-based and targeted the Windows operating system. The Morris Worm was released in 1988 and spread to 1/10 of all computers on the Internet. But as we mentioned above, as the Internet expanded and new technology like websites and email were adopted by more and more people, the malicious actors recognized they could target more victims. In the early days, the motivation was notoriety and trying to infect as many computers as possible in each attack. In the late 1990s, we saw the ILoveYou virus which used email to propagate. This utilized social engineering to get the victim to open an attachment. Many people fell for this and this technique is still used very successfully today. Human curiosity is a powerful weapon used by cybercriminals and keeps getting improved over the years. You will notice I did not mention notoriety in the motivation section above, as this did not last long as most malicious actors, even script kiddies, realized too much attention would get them in the sights of law enforcement. But as we moved into the 2000s we saw a shift in the motivation by malicious actors toward profiting from their attacks. The below graphic, Figure 22.2, shows a history of crimeware and the different threats developed by cyber criminals over the years to infect their victims.

From above you can see as technology evolved, so did the threats used. In the early 2000s, most were focused on web and email which were most utilized by people on the Internet. But in 2010 we started to see a shift from the one to many attacks to more targeted attacks. As mobile devices started to make their way into people's hands, attacks targeting this platform started to rise up. Ransomware started to really grow in 2015 and as we see today, it is still one of, if not the top threat targeting organizations. Since email is still the dominant communication method used within businesses, cybercriminals developed Business Email Compromise which was a bit unique in that the threat relied solely on a socially engineered email message to employees asking them to wire transfer funds to a fraudulent bank account. The uniqueness came from the email message not having any malicious payload, neither a weaponized attachment nor a malicious embedded link. It simply made it appear the message came from a superior within the organization and requested an urgent wire transfer.

With the rise of cryptocurrencies in 2018 and today, crypto-mining malware has been used on hijacked computers to mine bitcoin and other cryptocurrencies by malicious actors.

Threat landscape evolution

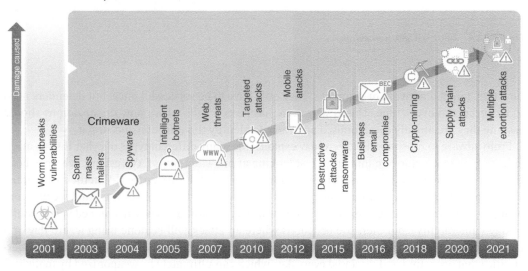

Figure 22.2 Cyber-threat landscape overview. *Source:* Trend Micro, Inc.

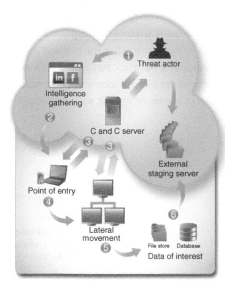

1. Intelligence gathering
Identify & research target individuals using public sources (LinkedIn, Facebook, etc) and prepare a customized attack.

2. Point of entry
The initial compromise it typically malware delivered via social engineering (email/IM or drive by download). A backdoor is created and the network can now be infiltrated.

3. Command & control (C and C) communication
Allows the attacker to instruct and control the compromised machines and malware used for all subsequent phases.

4. Lateral movement
Once inside the network, attacker compromises additional machines to harvest credentials, escalate privilege levels, and maintain persistent control.

5. Asset/data discovery
Several techniques and tools are used to identify the noteworthy servers and the services that house the data of interest.

6. Data exfiltration
Once sensitive information is gathered, the data is funneled to an internal staging server where it is chunked, compressed, and often encrypted for transmission to external locations.

Figure 22.3 Attack chain. *Source:* Trend Micro, Inc.

In Figure 22.3, we look at the lifecycle of an attack as today it is similar for most attacks. Originally Lockheed Martin came up with their Cyber Kill Chain® [2] which laid out how most advance persistent threats (APTs) followed.

Recently MITRE introduced a more complete view of the tactics and techniques used by adversaries using their ATT&CK [3] framework as seen in Figure 22.4.

As you can see, there are many ways malicious actors can target a victim and the biggest challenge defenders have today is that these actor groups are patient and persistent, and can change

Initial access	Execution	Persistence	Privilege escalation	Defense evasion	Credential access	Discovery	Lateral movement	Collection	Command and control	Exfiltration	Impact
8 techniques	8 techniques	16 techniques	11 techniques	22 techniques	14 techniques	21 techniques	7 techniques	14 techniques	16 techniques	8 techniques	13 techniques

Figure 22.4 Mitre att&ck framework.

how they target a network or computer. They only have to be successful once, but defenders need to be successful every time. This is why there has been a shift where now organizations should assume they are already compromised and work to identify the infection and stop it.

From adware to ransomware to zero-day exploits, the many different types of cyber threats available to malicious actors are immense. Just looking at malware, http://AV-Test.org, who has been tracking malware totals for years, currently shows over 1234 billion samples [4] in their repository. One area that has been used for many years is vulnerabilities found in software and weaponized using exploits. In most cases, malicious actors utilize an exploit of a known, patched vulnerability (*n*-day vulnerability), but there are occasions where an attack will be done using a zero-day or a previously unknown vulnerability against victims. There is a never-ending supply of new vulnerabilities that are found and disclosed every day, which can ultimately be weaponized by malicious actors. Exploit kits are available in the underground which utilize multiple exploits at the same time in hopes one of them is successful.

Another area we have recently seen used in attacks is the software supply chain process. Software vendors regularly need to supply their customers with updates to their code. Malicious actors recognized that these vendors have access to many customers/victims if they can compromise this update process. By implanting malware into the update process or compromising the process itself, has allowed attackers to laterally move from the software developer's network to the victim's network. Another name for this is Island Hopping whereby the attackers obtain access via an organization who has access to the ultimate victim either through direct access or via the software supply chain attack as seen in Figure 22.5.

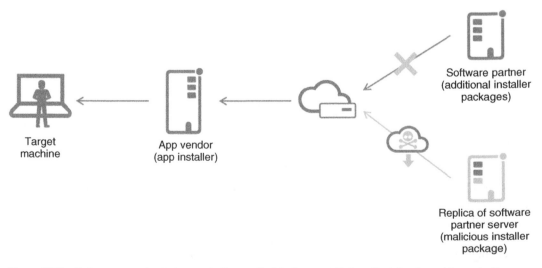

Figure 22.5 Software supply chain attack. *Source:* Redpirahna.net / https://redpiranha.net/news/software-supply-chain-emerging-attack-vector / last accessed 17 January 2023.

22.6   Attack Surface

The attack surface has changed over time allowing the attackers to shift the way they are able to compromise organizations and people's devices. Initially, we saw attackers utilize removable media like floppy disks to propagate malware, then email was successfully used, which is still in use today. Web downloads became a common attack surface as more people browsed websites around the world, including actors compromising legitimate websites with their malware and malicious scripts. As software started to be used everywhere, vulnerabilities became targets of exploitation which allowed actors to target many of the applications and operating systems used. Mobile devices started being targeted as the iPhone and Android devices became more and more popular. When AWS and Microsoft Azure clouds started to be utilized, actors looked for ways to target those infrastructures. The next horizon we are starting to see targeted are Internet of Things (IoT) and Industrial Internet of Things (IIoT) as billions of these devices are showing up within homes, offices, factories, automobiles, and cities. One commonality we have seen is regardless of the attack surface, the malicious actors use malware to infect devices, systems, and infrastructures.

22.7   Technology Shifts Cause Cyber Threat Shifts

Technology has changed over the years which has also allowed the attack surface to change. In the early years when not many people or organizations were connected, the threats were few and far between. Most viruses were transmitted via external media, like floppy disks, but when Novell and other networking technology started connecting systems together, the attack surface increased for threats to propagate much faster and offered more systems to be infected. Malicious coders developed worm-based malware that had technology to propagate itself across the network and find new systems to infect. This started the era where infecting the most systems was the goal and bragging rights were awarded to the actor who could do this. Email started becoming the main mode of communication around the world as the Internet expanded and connectivity between countries and continents gave access to more and more people and systems. During this time, web browsing and websites started to grow exponentially and gave people access to these sites for information. These two web mediums allowed malicious actors access to more and more victims. Phishing started to be used, where malicious actors would pop up a fake web account login page to get the victim to enter their credentials and allow them to access these accounts. During this time, we started to see the growth of cybercrime as well. Spam became the go-to threat to get people to buy goods and allowed spammers to generate profit from their activities. Email was also used to spread malware more quickly and to a far broader set of victims. Weaponized attachments were used within email messages and in order to get victims to open the attachment the use of socially engineered messages started to become utilized. As previously mentioned, one of the biggest attacks at the time was named, I Love You, where the email seen in Figure 22.6 enticed the victim's to open an attachment that was a love letter from the sender. The subject line said, "I Love You" and we saw millions of people infect themselves as they wanted to see what the person sent them.

During this time, Microsoft Office applications were being adopted by most organizations and people began using them every day. Part of this was technology called "Macros," which allowed multiple commands and instructions to be grouped together and automated, which improved efficiency. Malicious coders were able to weaponize this technology and macro viruses were created and used in many of these attacks. Most of the malware within email messages were macro virus attachments that would infect the victim's computer. As mentioned, websites started to become

Figure 22.6 I love you example.
Source: Zdnet.

utilized by the masses and malicious actors took note. They started infecting legitimate websites with scripts and malware to infect visitors to these sites. In many cases, the malicious scripts (Cross-Site Scripts – XSS) redirected the users system to other web pages where malicious code was stored. This redirection was done in the background so the victim never saw this occur within their browsers. This again increased the attack surface and the pool of potential victims grew exponentially again. Another phenomenon that came into being was the invention of botnets. Botnet herders were a group of malicious actors who looked to infect and recruit many computers into their network of compromised systems. This network of "bots" allowed them to do many things, from sending out massive spam runs, to launching denial of service attacks. Hacktivists were a big user of these as their dedicated denial of service (DDOS) attacks allowed them to disrupt their victim's networks or websites. They also used website hacking to post their activist messages on many websites of their victims. Today we are seeing very large botnets of IoT devices, with one of the biggest IoT botnets being Mirai.

Mobile phone technology has also improved over the years from simple phones to now being computers with access to all the activities that can be done on PCs. This represented a new attack surface for malicious actors and so we started to see malware and other threats targeting these devices. The same threats we saw on PCs were similar to what was developed and targeted on mobile devices. As mobile applications (apps) were available on mobile devices attackers started installing malicious apps on app stores like Google Play, the Apple store, and many third-party app stores. Phishing mobile devices started cropping up whether through SMS texting (Smishing) or via voice (Vishing) technology. Spyware was also targeting mobile devices as it allowed attackers to record voice conversations, identify places the person would be via their calendars and even record video from their phones. Ransomware also showed up as encrypting the data on phones (pictures, videos, etc.) allowed financially motivated actors to charge a ransom and make money from their victims. The IoT is a different type of mobile device. Smart devices like wearables are fast becoming the most used device around the world. With numbers expected to be in the billions, more devices than people on earth, these will certainly become a favored attack surface for actors in the future.

Business are also adopting this technology, IIoT through robotics and other smart devices. Smart factories, smart vehicles, and smart homes are all being equipped with technology that will connect them to people, places, and things that is the perfect storm to allow malicious actors to wreak havoc around the world. The monetization of these devices through attacks is also very attractive to these cyber criminals.

Virtual computing, whether on clients or servers, led the way toward using cloud-based infrastructure. Physical data centers are expensive and in many cases slow to change, and as such, the adoption of cloud services like Amazon Web Services (AWS) or Microsoft Azure started to grow. This too became a new attack surface for malicious actors to target in their attacks. One interesting area of revenue that has grown over recent years is crypto-currency mining. The growth in the values of cryptocurrencies has not gone unnoticed by financially driven actors and so mining for coins has grown as well. But many malicious actors do not want to invest in their own systems to do the mining for them, and as such, we have seen an increase in using crypto-mining malware to infect victim's systems that have the technology resources (CPU, Memory, GPU) to manage the mining process for them. Cloud-based systems are ripe for this type of activity so at the time of this writing we are seeing this threat being very active against business's cloud infrastructures.

The challenge over the years is that technology moves extremely fast and new advances seem to come daily. New features come out all the time, and between this and the new technology makes it very difficult for people to stay ahead and on top of how all this stuff works. As such, misconfigurations can be taken advantage of by malicious actors in finding ways to compromise networks and devices. This will continue to be a challenge for the world as I do not see any slowdown of new technology in the future. Quantum computing is promising, but if that does become reality, the entire encryption technology that is used today to protect data, could be rendered useless. That would mean new technology needs to be developed to protect data then. Since the beginning, there has been a cat and mouse relationship between security vendors and malicious actors. These actors will take advantage of new technology as well, and security vendors will need to develop new ways to protect against these advances. As you saw in Figure 22.2, over the years new threats that continue to be implemented. Each of these requires new security solutions to be developed. This process will continue for the time being, but we know that change is the constant factor in the world and progress is inevitable. Let us look at how cybersecurity has evolved like technology over the years.

22.8 Evolution of Cybersecurity

In the early days of computing, file-based malware was most prevalent and the ability to detect a virus (malware) was to scan the file with a pattern file, or signature/hash, to determine if the file was known bad. The malicious file had to be known and a hash of the file was developed and used to identify if that particular file was present. Scanning the system could be done in real-time, via a scheduled scan or on-demand. The challenge with this was the pattern/signature file needed to be regularly updated on the system in order to detect the newest malware. This process in the early days included physically mailing a floppy disk, or other forms of media, and copying the file to the system. As modems became utilized, the file could be transferred to the customer using this method. But as more and more malware started being produced, the ability to stay ahead of this became difficult. As technology improved and the Internet allowed more constant connections between security vendors and their customers, the pattern/signature file was moved to the cloud. Customers' systems would query the signature file in real-time to see if a file was bad. This allowed vendors to improve the speed with which they could keep their customers' protected. Signatures are still utilized but they are not as effective due to the enormous amount of malware being released every day. More than 10 million per month per http://AV-Test.org.

Early firewalls also started to be used as a way to filter traffic inbound and outbound within an organization. This allowed organizations to keep a lot of unauthorized access out of their networks using filters that also needed to be updated regularly however these were like a single guard at the door and did not check the contents of communications, and they only did basic traffic control.

As email became a threat vector, security vendor's developed technology to scan messages for malware. Initially, the attachments were scanned using pattern/signature files to determine if it was malicious. When embedded URLs were included within the message, these were also scanned to determine if the link was malicious. Anti-spam technology that analyzed the message headers, subject, and body of the message was developed to stop this threat.

Speaking of URLs, when webpages became used in attacks, the technology needed to be developed to scan a webpage for malicious content. This could include scanning files included on these pages for users to download, which initially used pattern/signature files on these files. When scripts were introduced these had to be scanned for malicious actions or behaviors. Malicious advertisements (Malvertisements) started being developed and included in pages where ads were present, and this required technology to detect these. Figure 22.7 shows a form of malicious domain creation, typosquatting [5], which is a technique used by criminals to register domain names that are similar to legitimate business's domains, required technology to detect. Even soundsquatting [6] (see Figure 22.8) is utilized where the registered domain sounded like the legitimate domain needed detecting as well.

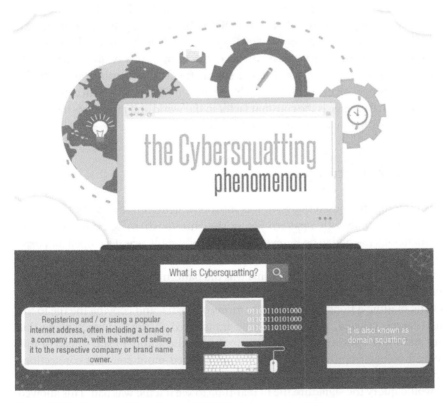

Figure 22.7 Typosquatting. *Source:* IFSEC Global.

Figure 22.8   Soundsquatting. *Source:* Trend Micro, Inc.

Most of these threats being discussed here required a scanning technology, whether files, email messages, or webpages. As the malware became more sophisticated and prevalent, new scanning capabilities were required. Behavior-based scanning technology was created to analyze a file during the execution phase to identify if the behavior was malicious or even suspicious. This allowed us to add additional protections against malware that was within executable files. Other file types are used to distribute malware like Adobe files, even embedding malware within images (steganography) is used today. Scanner technology has had to evolve as malicious actors targeted all kinds of filetypes in their malware developments.

One technology evolution was the use of artificial intelligence (AI) and machine learning (ML) [7] to detect different types of threats. This technology had to be utilized due to the amount of data that had become overwhelming and made human analysis nearly impossible. Big Data analysis became a requirement and this is where AI/ML technology thrives. One of the first uses of AI/ML was in anti-spam which improved the ability to detect spam. AI/ML was needed to detect threats without the need of a pattern/signature file or the need to regularly update the engines or models used. This was a major change for the cybersecurity industry since the amount of new threats became too much for humans to develop protections for these threats. AI/ML allowed data scientists to develop protections that could detect new, unknown spam, phishing, malware, URLs, and many other threats. Because there are many different ways to implement AI/ML like Deep Learning, Boosted Tree, Natural Language Programming, SVM, and more gave researchers the ability to use the appropriate type of ML for the threat. This greatly improved the ability to rapidly respond to new threats. The future will use AI/ML in most cases and across a wide range of technology and solutions for detecting threats.

22.9   Risk-Base Approach to Patch Management

We have discussed exploits targeting vulnerabilities which requires users and organizations to regularly patch software bugs as quickly as possible. Because patching is not easy and regularly takes a very long time to process, Intrusion Detection & Protection Systems (IDS/IPS) were developed to support virtual patching of software vulnerabilities. This technology allows organizations to implement these virtual patches that will detect or protect against known exploits of vulnerabilities. IDS/IPS can recognize an attack trying to exploit a specific vulnerability for any unpatched systems and block it. This is like a Band-Aid until the system can be patched.

When it comes to vulnerability patching, organizations can struggle due to the number of new patches they have to deal with every day. As such, developing a risk-based approach adopting the following steps can help:

1) Continuously conduct an exposure assessment to determine what Common Vulnerabilities and Exposures (CVE) **past** and present are always in your environment.
2) Assess the **criticality** of those systems that contain those CVEs.
3) Conduct a continuous but simple **risk assessment** and build a matrix like seen in Figure 22.9:
   - Assess the Probability-likelihood of those identified CVEs are or will be exploited in the wild or have a proof-of-concept (POC) against the severity-impact of those CVEs used in an attack.

Application Control is a newer capability of firewall that can delve deeper into the traffic and limit what can be done. For example, prevent sharing attachments via Facebook. Application Safe-Listing is a technology that was developed to help organizations deal with threats on certain types of systems where they could control which applications are authorized to run on the system. If a

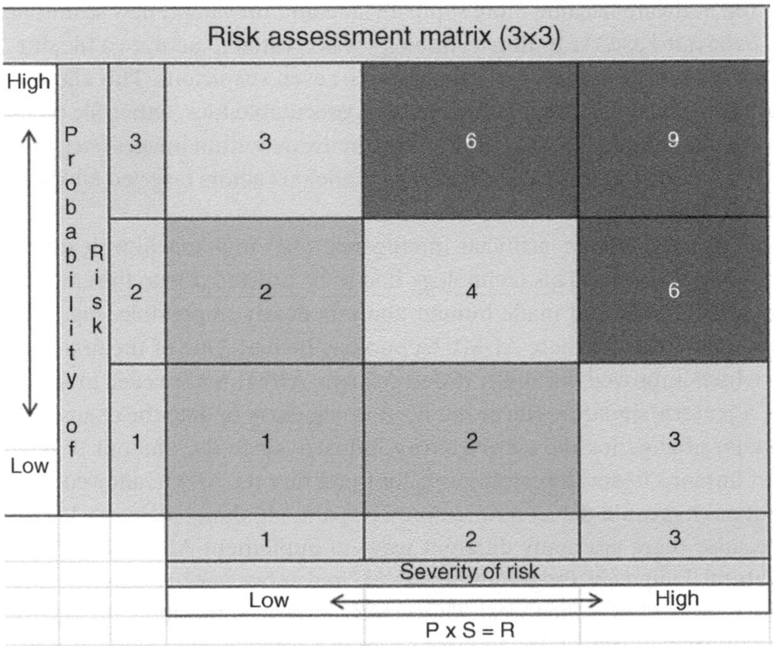

Figure 22.9   Probability vs. severity risk matrix. *Source:* Trend Micro, Inc.

file or application not authorized to run attempted to install itself it would be blocked. This is typically used in environments like financial (ATM, banking systems), manufacturing, and retail.

One technology that has helped over the last decade is sandboxing, where the sandbox would be built using the specific operating systems and applications running within the organization. Any suspicious files or URLs would be run within these sandbox environments to identify if any suspicious or malicious behavior is detected. The sandbox appears to be a standard Windows or Linux endpoint, and potentially hostile files and such are opened or interacted with so they would "detonate" or be hostile only to this simulated environment without hurting any live workstations or data.

One item that needs to be discussed is the practice by malicious actors to reverse engineer and use anti-forensics. Many of the security technologies discussed before have been analyzed by cyber criminals to identify ways to bypass them. This process has been done many times and requires security vendors to regularly update their technology to defend against this practice. We were even seeing this against AI/ML with malicious actors looking to poison ML models and utilize this technology in their attacks [8].

22.10 Cybersecurity Strategy and Models

Now that we have discussed some of the history, let us focus on how organizations can implement cybersecurity models and practices to minimize their risk of being compromised. Most will implement a defense in-depth model, which came from medieval times when castles were defended by a series of protections (see Figure 22.10).

As you can see in the image above, cybersecurity solutions can be implemented in many areas of the organization's infrastructure to ensure adequate coverage against threats that may be found there. An example would be to implement device security for PCs, Servers, mobile, IoT, IIoT, and any endpoint device. Also, solutions for the messaging layer, network layer, web layer, data centers, and in the Cloud. Even at each of these layers, there can be a defense in-depth model that

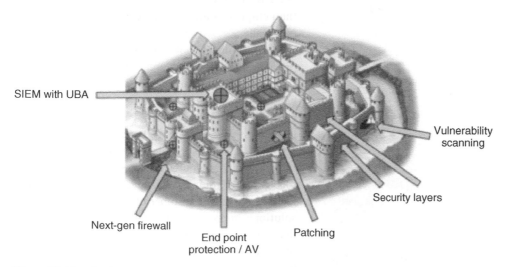

Figure 22.10 Castle defense in depth. *Source:* Atmosera https://www.atmosera.com/blog/defense-in-depth-a-castle-analogy/.

implements multiple technologies for that layer to improve detecting a threat. Network segmentation, especially micro-segmentation of the network, can be implemented to minimize the risk of spreading and lateral movement of an attack. One main reason this defense in-depth model is still relevant is that most attacks follow the APT model discussed earlier. These attackers will not stop and end at the endpoint, but the attack lifecycle will move across the entire network and as such, cybersecurity needs to be implemented where it can defend and identify all components of these attacks.

The US Secret Service interviewed many cyber criminals [9] they arrested over the years and found most looked for the following three areas to exploit when attacking an organization:

- Human error
- IT security complacency
- Technical deficiencies

22.11 Risk-Based Approach to Cybersecurity

As organizations build their cybersecurity strategy, it must include three areas as seen in Figure 22.11, which from above are areas of attack:

- People
- Process
- Technology

All three of these are important when building a cohesive strategy to defend against attacks.

When looking at people, organizations are currently challenged with lack of trained personnel. It is estimated that there are 3.5 million unfilled cybersecurity jobs around the world and 500 thousand in the USA alone. Ongoing training is critical to keep employees, developers, and administrators regularly updated on the latest technologies and security practices for their area of business. Burnout is a concern, so organizations need to address this with best practices for ensuring their employees are taken care of, both mentally and physically.

Cybercriminals look for weaknesses in security processes, so this is an area business need to constantly address and review. Patching, enabling new defense technologies, account privileges, account credentials are some of the areas where the process of creating and managing these areas can improve an organizations overall defense capability.

Technology is an area where many organizations focus most of their efforts, obtaining new technologies regularly. The number of security vendors and solutions an organization acquires can become burdensome and overwhelming for the employees having to manage and maintain them. Security sprawl is an area many CISO's are concerned about and need to identify any way to limit this.

People - process - technology framework

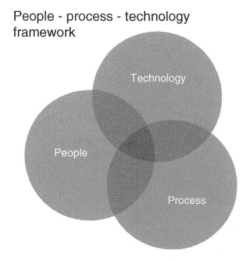

Figure 22.11   PPT framework. *Source:* https://www.smartsheet.com/content/people-process-technology.

All three of these areas need to be in balance to effectively manage the security strategy within an organization. When out of balance it can negatively affect the overall ability to defend against attacks.

22.12   Cybersecurity Frameworks

When implementing an effective cybersecurity strategy, many organizations look to the existing frameworks that are available to help them. As you see below from the National Institute of Standards and Technology (NIST [10]), a basic framework follows a process of identify, protect, detect, respond, and recover.

This framework is described in full in the Chapter 23.

A cybersecurity framework gives an organization a way to develop their strategy and be consistent in implementing it. A critical area that every business needs to develop is their incident response process. In the event of a data breach, speed of response is critical as there may be regulatory requirements, like HIPPA, GDPR, and others, that have to be responded to as well as contacting any external victims. Many States and Nations have data breach laws that have to be followed. Developing, reviewing, and testing the incident response regularly is something each organization should be doing.

The NIST framework is one of many that are available to organizations.

22.13   New Defense Ideas

As we mentioned before, many of the attacks today go across the entire network which has caused challenges for organizations whose security solutions do not co-exist and collaborate. Extended Detection and Response (XDR) is a newer way of integrating endpoint, messaging, network, and cloud layers to improve the visibility of the entire attack lifecycle.

The newest evolution in security architecture is called "Zero Trust." Since we are not winning the war against attackers by scaling up all the technologies we have discussed, making our digital businesses less vulnerable and harder to attack is the goal of this approach. Zero Trust means a highly skeptical design, that even authorized credentials are looked at closely to see how they have been used, and big security data is looked at before most transactions are allowed to proceed. It focuses on three core principles:

1) Verifying every user.
2) Validating every device.
3) Limiting access intelligently of 1 and 2.

Also, as part of the latest strategies and solutions for minimizing the risk of attack is Attack Surface Management (ASM) which is gaining interest as we mentioned the attack surface has been growing every year. ASM is intended to give organizations better visibility into their entire attack surface through identifying all their systems and devices on their network as well as all internet-facing systems.

So building an effective cybersecurity strategy requires thoughtful direction and requires the right balance of people, process, and technology that works seamlessly across the entire organization.

22.14   Summary

There is one constant that pertains to cybersecurity and that is change will occur. Whether it is shifts in technology or the changing attack landscape and how malicious actors commit their campaigns, people and organizations will have to adapt to these changing times. This means regularly educating yourself about the changing threat landscape and the technologies available. Regularly assessing your organizations defenses and incident response and working with your security vendors to ensure you have the latest and greatest capabilities to defend against the latest threats. This overview of cybersecurity was intended to give you a general view of this industry, from its history, to technologies, to frameworks that can help someone better understand it. There are many aspects that we were not able to go into but can be found on the Internet to build your knowledge of this exciting and challenging area of our world.

References

1 Morgan, S. Global cybercrime damages predicted to reach $6 trillion annually by 2021. Cybersecurity Ventures at https://cybersecurityventures.com/annual-cybercrime-report-2020 (accessed 12 August 2021).

2 Lockheed Martin, Inc. Cyber Kill Chain®. Lockheed Martin at https://www.lockheedmartin.com/ en-us/capabilities/cyber/cyber-kill-chain.html (accessed 12 August 2021).

3 Jeremy Singer. Mitre att&ck framework. MITRE at https://attack.mitre.org (accessed 12 August 2021).

4 AV-Test.org. AV-test malware totals. http://AV-Test.org at https://www.av-test.org/en/statistics/ malware (accessed 13 August 2021).

5 IFSEC Global. Typosquatting infographic. IFSEC Global at https://www.ifsecglobal.com/cyber-security/the-cybersquatting-infographic (accessed 13 August 2021).

6 Trend Micro, Inc. Soundsquatting infographic. Trend Micro, Inc. at https://www.trendmicro.com/ vinfo/us/security/news/cybercrime-and-digital-threats/soundsquatting-101-how-homophones-can-lead-to-risks (accessed 13 August 2021).

7 Polyakov, A. Machine learning for cybersecurity 101. Towards Data Science at https:// towardsdatascience.com/machine-learning-for-cybersecurity-101-7822b802790b (accessed 12 August 2021).

8 Trend Micro Research, United Nations Interregional Crime and Justice Research Institute (UNICRI), Europol's European Cybercrime Centre (EC3). Malicious uses and abuses of artificial intelligence. Trend Micro Inc. at https://documents.trendmicro.com/assets/white_papers/ wp-malicious-uses-and-abuses-of-artificial-intelligence.pdf (accessed 14 August 2021).

9 Verizon. Appendix A: transnational hacker debriefs. United States Secret Service at https:// www.verizon.com/business/resources/reports/dbir (accessed 14 August 2021).

10 National Institute of Standards and Technology (NIST). Cybersecurity framework. NIST at https:// www.nist.gov/cyberframework (accessed 14 August 2021).

23

Cybersecurity Standards and Frameworks

Santiago Paz

Interamerican Development Bank, Washington, DC, USA

23.1 Introduction, Cybersecurity Fundamentals

Data, information, and the systems that process them are currently very valuable assets for all organizations. However, it is often not identified as such by top management or other management positions.

When information existed only on paper, it was not possible to store and process it on a large scale – Big Data. However, it was also not possible for a hacker to attack these paper repositories. When repositories and information systems became digital (what is known as the exposure surface), there are more possibilities for a "hacker" to attack such systems, therefore protecting them digitally becomes a fundamental issue.

The technological development and digital transformation that has been experienced in recent decades have delivered multiple benefits to all people, like internet access, online shopping, digital government services, telemedicine, autonomous cars, and smart cities, among others. However, this has also significantly increased the attack surface, and cyber attackers have not waited.

We have seen attacks that have affected financial services and large corporations, but also attacks that affected basic services such as healthcare, energy, drinking water, and even the entertainment sector. Today, digital transformation needs cybersecurity to develop properly.

In this section, we will analyze the fundamentals of cybersecurity and its main bodies of knowledge.

Let us start by defining cybersecurity as such, for that there are multiple definitions around the world, and in many cases very thin borders separate cybersecurity, information security, and computer security. These boundaries are so thin that we could consider them almost unnecessary (see Figure 23.1).

Computer security focuses on all information stored in computer systems, leaving out information in physical format or those cybernetic systems that are not considered as computational.

Information security focuses on information as such, regardless of whether it is in digital or physical format, however unstructured information such as telemetry or control signals was often considered to be left out.

IEEE Technology and Engineering Management Society Body of Knowledge (TEMSBOK), First Edition.
Edited by Gustavo Giannattasio, Elif Kongar, Marina Dabić, Celia Desmond, Michael Condry, Sudeendra Koushik, and Roberto Saracco.

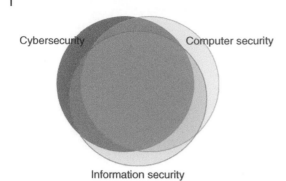

Figure 23.1   Terminology definition.

Cybersecurity, on the other hand, focuses on protecting cybernetic systems and cyberspace without going into further details.

As an example, the International Telecommunication Union (ITU) defines cybersecurity as:

> Cybersecurity is the set of tools, policies, security concepts, security measures, guidelines, risk management methods, measures, training, best practices, control and technologies that can be used to protect cyberspace and the organization, as well as user resources.
>
> *Source: www.itu.int.*

For the objective of this chapter, we will consider cybersecurity, as the preservation of confidentiality, integrity, availability, and authenticity of information and the systems that handle it. Additionally, the concepts of auditing and non-repudiation are also included, although they are not always used. This definition is wide enough to fit all possible definitions

Confidentiality: Only authorized users, systems, or entities can access de information
Integrity: Only authorized users can modify the information, or in other cases, will be possible to detect it.
Availability: The information is available in a good manner when needed. This means that the system is having good performance, and the information delay and jitter are ok.
Authenticity: The user or the receiver of information can verify the information is authentic, was created for the corresponding author, was sent by the corresponding sender, and was not modified.
No-repudiation: Is the characteristic that made that a user cannot repudiate an action on the information or a system, is very close with the authenticity characteristic, but it requires use unicity.
Auditing: All the actions and states of the system and information are tracked.

In case an incident jeopardizes any of these attributes, cybersecurity will be compromised and the organization will have some kind of impact. Since organizations have their businesses digitized, there is a risk then it is necessary to apply measures to prevent it, or in case it happens, to cause the least possible damage.

23.2   Business Impact and Risk Management

Information is a very valuable asset for organizations. Imagine the impact it can have for a laboratory that has the results of a research that took years to complete, or an oil company whose soil research results are altered and therefore drills in the wrong places, or an e-commerce company whose website is put out of service by an attacker, preventing organizations from selling their products. These three examples show cases where confidentiality, integrity, and availability were compromised respectively.

These types of attacks occur more often than we may think, as an example, in May 2021, a major oil pipeline in the United States suffered a cyberattack,[1] which caused the lack of fuel for several days on the East Coast increasing is a spectrum of USD 3 a gallon for the first time since 2014. Another case was that of the Dusseldorf University Hospital in September 2020,[2] which due to a cyber attack was forced to close its emergency room and had to transfer patients, resulting in the death of one of them.

To perform the risk analysis, the first thing the organization must establish is the context in which it is immersed, both external and internal context. This context will make it possible to capture the organization's objectives, stakeholders, risk tolerance, and top management's judgment.

The ISO IEC 31000 standard establishes a series of principles for managing risks effectively in an organization, in particular, establishes a framework for integrating this process into the organization's core processes. In this chapter we will not go into this standard in depth, see bibliography for ISO/IEC 3100 at the end of this chapter, we recommend that you read it.

The risk management process is an iterative, continuous process and must involve the entire organization, communication, and consultation is essential to understand the context and perform a correct analysis.

We introduce then the first concept of the risk approach, the Business Impact Analysis (BIA).

BIA is a very important concept because it measures the impact of a potential cybersecurity incident on the business, not the damage caused to the information system itself. As an example of the business impact in the U.S. pipeline example, it was not the cost of restoring computer systems, but the economic loss suffered due to not being able to sell fuel.

The challenge is the mapping between systems and processes, and the dependencies on systems. There are some tools to map it, and to model the business architecture, but is beyond the scope of this section.

Although the impact can be measured quantitatively, defining exactly the amount that may be affected, for simplicity's sake is usually measured quantitatively by defining scales of, for example, High – Medium – Low – Null.

It should be noted that it is very important to correctly define the meaning of the values in the organization, so that it is easy to be interpreted by senior management and business owners.

As an example, the following definition is given:

High: The business may have its operational continuity affected and may have to be rebuilt productively, organizationally, or financially. Economic losses greater than 1,000,000 dollars.
Medium: The business will be impacted from a reputational point of view, may represent loss of customers, breach of contracts, and economic impacts greater than 500,000 dollars.
Low: The business may see its reputation slightly affected, it may involve minor contractual breaches and the economic impact is not significant.
Null: Although the incident occurs, it does not have a significant impact on the business.

Once the scales have been defined, the analysis must be performed. This analysis MUST be done, not only by the technology and cybersecurity specialists but also with the business specialists and the C-Group.

---

1 https://cnnespanol.cnn.com/2021/05/11/ciberataque-oleoducto-estados-unidos-trax.
2 https://www.antena3.com/noticias/mundo/ciberataque-hospital-aleman-provoca-primera-muerte-mundo-ransom ware_202009185f64882041cb49000152a944.html.

Table 23.1 Risk matrix example.

BIA					Risk	
System	**Business processes**	**Confidentiality**	**Integrity**	**Availability**	**EF**	**Risk**
Customer database	Customer relationship	Med	Med	Low	Low	Mid
	Customer credit	Low	High	High		
Ecommerce website	Sales	Low	Low	Mid	Mid	High
	e-Payment	High	High	Low		

In order to do it, work meetings must be held, system by system, where scenarios are presented in which a cybersecurity incident affects the system (in confidentiality, integrity, availability, and authenticity). In these meetings, the cybersecurity specialist guides the discussion so that the technology team and the business can determine the impact levels of each system until the complete classification is reached.

The other key component of the risk analysis is the Exposure Factor (EF), that is: "How likely is it that such an incident will occur?". This has two components, when the threat could materialize and when the system could be vulnerable. While this analysis can be done in detail, it is generally not worth the effort and a high-level analysis should be performed. We will look at the standards recommendations for this later.

This exposure identification process is usually performed mostly with technology teams, and as with the BIA process, the different systems and the protection technologies they have been reviewed to identify how likely it is that an incident could occur. As with the BIA, a qualitative analysis is usually performed, with scales of, for example, High – Medium – Low.

Finally, the risk matrix is produced by crossing this information: Risk = EF × BIA and thus identifying the levels of risk that each of the systems represents for the organization. These risks are also usually measured qualitatively in, for example, High – Medium – Low.

A sample risk matrix is shown in Table 23.1.

Once the risks have been identified, it is necessary to decide how to deal with each one. A risk can be **eliminated, mitigated, transferred, or accepted.**

Eliminate a risk: To eliminate a risk, the way to do it is to eliminate the source of the risk. For example, if introducing a sales channel through a Mobile App generates a risk and you want to eliminate it, what you should do is to eliminate the App. This is clearly not always possible.

Mitigate a risk: The most common option for risk treatment is to mitigate it to an acceptable level. This usually involves incorporating controls and/or cybersecurity measures to reduce exposure, either by addressing the threat or the vulnerability.

Transfer a risk: Another risk treatment option may be to transfer the risk, usually through contracts with third parties or the purchase of insurance.

Accept a risk: Finally, another option is to accept the risk as it arises.

The ISO/IEC 27005 standard offers a methodology for dealing with information security risk management in an organization.

23.2.1 ISO/IEC 27005 Information Security Risk Management

The ISO/IEC 27005 (see Figure 23.2) standard belongs to the "27000" family of standards for information security. It focuses particularly on risk management. It is not a particular risk analysis methodology, but describes the risk management process, from the definition of the context to the review of the remaining risks.

As indicated in the previous section, the first step is to know the context in which the organization operates; it will be fundamental to understand the organization's objectives, risk tolerance levels, and main stakeholders. This process requires involving the entire organization through communication and consultation.

Then there is the risk assessment stage, which requires identify, analyze, and evaluate the risk. This is the phase that makes use of the BIA, EF, meetings with stakeholders and the IT team.

Finally, a decision must be taken: how to treat the risk? whether to accept, eliminate, mitigate, or transfer it.

This process is an iterative and continuous process, so it must be reviewed periodically, and risk treatment plans must be revised.

A remarkable point of ISO/IEC 27005 is that it offers two approaches to risk management, one high-level and one detailed.

There are several reasons, such as budget, time, organizational maturity, etc. that may indicate in advance that it is not possible to simultaneously implement all the necessary controls and that only the most critical risks can be considered in the risk treatment process. It is therefore inefficient to conduct detailed system-by-system studies, analyzing existing technological architectures and cybersecurity measures in detail.

Figure 23.2 Risk management process. *Source:* ISO/IEC 27005, www.iso.org.

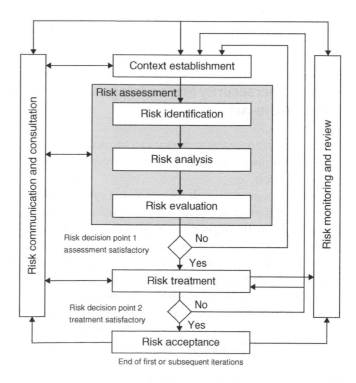

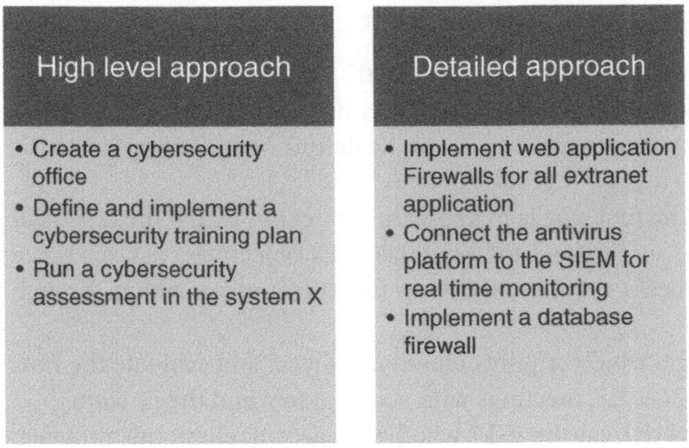

High level approach	Detailed approach
• Create a cybersecurity office • Define and implement a cybersecurity training plan • Run a cybersecurity assessment in the system X	• Implement web application Firewalls for all extranet application • Connect the antivirus platform to the SIEM for real time monitoring • Implement a database firewall

Figure 23.3 Risk treatment plan.

High-level risk approach characteristics:

- Does not analyze technological details, take systems as an overall whole, and focuses more on organizational and management aspects.
- The list of threats used is more limited and the threats are more general.
- Takes less time and effort to be implemented and requires less technological knowledge.
- Facilitates the development of a strategic cybersecurity vision.
- Optimizes the use of resources, prioritizing the most critical aspects.

This high-level approach is very appropriate at early cybersecurity maturity levels, where the organization still has little cybersecurity capability. The detailed analysis will offer greater protection for the systems, despite being more costly to perform.

In Figure 23.3, there are some illustrative examples of risk treatment plans depending on the approach.

23.3 Cybersecurity Methodologies and Standards: State-of-the-Art

There are many different tools that can be used by different organizations when implementing a cybersecurity program. These tools can be grouped into at least two groups: frameworks and cybersecurity controls.

Frameworks are tools to develop different cybersecurity activities in a structured, systematized, and controlled manner. Frameworks have different approaches, but in general, they provide mechanisms to define the organization's cybersecurity objectives and the implementation and maturity processes required to achieve them.

The risk approach is always present in these methodologies, as it guides the organization and justifies cybersecurity actions. In other words, the framework will tell us that we must manage risk but it will not tell us how. The same will happen with the cybersecurity measures, it will tell us that we must implement an identity management system, but it does not tell us how, there are other more detailed standards that indicate how to implement such controls.

Controls are technological or administrative cybersecurity measures that are intended to achieve certain objectives. For example, NIST SP 800-53 defines Identity Management activities as part of the measures focused on access control.

Finally, there are the guidelines, which are set in a given context and provide instruments for implementing cybersecurity controls for a particular issue. For example, the NIST SP 800-48 guide indicates how to protect IEEE 802.11 legacy wireless networks.

The following is a review of the most commonly used tools for each category.

23.3.1 Frameworks

23.3.1.1 NIST Cybersecurity Framework "Framework for Improving Critical Infrastructure Cybersecurity"

The NIST Cybersecurity Framework was first published in 2014 with the purpose of providing tools for critical infrastructure protection. That framework defines five cybersecurity functions: **Identify, Protect, Detect, Respond, and Recover**. For each of these functions, it groups a series of cybersecurity controls which it further groups by categories and subcategories (Figure 23.4).

- **Identify:** Develop an organizational understanding to manage cybersecurity risk to systems, people, assets, data, and capabilities. The activities in this function are very important to the effective use of the Framework. Understanding the business context, the resources that support critical functions, and the related cybersecurity risks enables an organization to focus and prioritize its efforts in alignment with its risk management strategy and business needs.
- **Protect:** Develop and implement appropriate safeguards to ensure the delivery of critical services. The protect function supports the ability to limit or contain the impact of a potential cybersecurity event.
- **Detect:** Develop and implement appropriate activities to identify the occurrence of a cybersecurity event. The detect function enables early discovery of cybersecurity events.

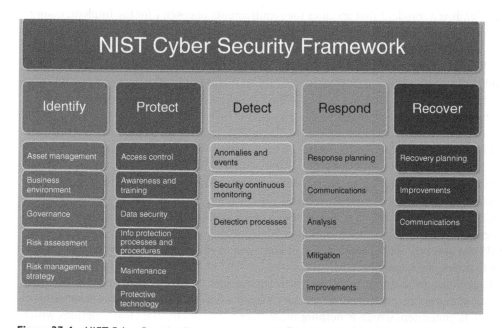

Figure 23.4 NIST Cyber Security Framework diagram. *Source:* www.nist.org.

- **Respond:** Develop and implement appropriate activities to take action on a detected cybersecurity incident. The response function supports the ability to contain the impact of a potential cybersecurity incident.
- **Recover:** Develop and implement appropriate activities to maintain resiliency plans and restore any capabilities or services impacted by a cybersecurity incident. The recovery function supports the timely recovery of normal operations to reduce the impact of a cybersecurity incident.

For example, in the Identify function we will find a category called "Asset Management (ID. AM)" with a sub-category called "Physical devices and systems within the organization are inventoried (ID.AM-1)" in that sub-category we will find all the cybersecurity controls recommended in that sub-category.

This is why this framework is very versatile, just the recommended controls are marked from multiple standards, both from NIST itself, as well as ISO/IEC, ISA, CIS, and other reference standards. For this reason, it adapts very well to multiple environments. The Figure 23.5 illustrates the template that accompanies the NIST framework.

On the other hand, it defines levels of implementation of the framework that progress from Level 1 (Partial), Level 2 (Informed Risk), Level 3 (Repeatable), and Level 4 (Adaptive), to reflect the management of cybersecurity risks based on the organization's risk management.

Finally, it defines profiles that identify the current state of the organization (Current Profile) and the target profile, the place where the organization wants to go, and which is aligned with the accepted risks.

23.3.1.2 ISO/IEC 27001 "Information Security Management Systems"

The ISO/IEC 27001 standard belongs to the ISO/IEC 27000 family of standards, created by the International Organization for Standardization as a tool for the development of information security.

In particular, it is a standard that specifies the requirements necessary for the implementation, maintenance, and improvement of an Information Security Management System (ISMS). The latest version existing during the preparation of this document is that of 2013, with corrections in 2015, which details 130 cybersecurity requirements found in Annex A of the same.

It should be noted that this is a certifiable standard so it allows organizations the possibility of being certified with an external agent that validates how their information security management system is implemented and executed.

The entire methodology is guided by risk management, i.e. risk analysis is carried out periodically and objectives are defined, controls to be implemented and risk treatment plans are designed.

This standard uses the Plan Do Check Act (PDCA) model to implement the Information Security Management System process (see Figure 23.6).

The first step in the planning process is the design of the ISMS including scope definition, initial risk assessment, and the first risk treatment plan.

The ISMS can cover the whole organization or just a part of it, e.g. ecommerce systems. It is very important to define the scope correctly.

Once the policies, procedures, required controls, and risk treatment plan are in place, they are implemented in the organization. At this stage, what is known as the Statement of Applicability is generated, which is nothing more than a document that clearly defines which

Function	Category	Subcategory	Informative references
	Asset management (ID.AM): The data, personnel, devices, systems, and facilities that enable the organization to achieve business purposes are identified and managed consistent with their relative importance to organizational objectives and the organization's risk strategy.		- CIS CSC 1 - COBIT 5 BAI09.01, BAI09.02 - ISA 62443-2-1:2009 4.2.3.4 - ISA 62443-3-3:2013 SR 7.8 - ISO/IEC 27001:2013 A.8.1.1, A.8.1.2 - NIST SP 800-53 Rev. 4 CM-8, PM-5
		ID.AM-2: Software platforms and applications within the organization are inventoried	- CIS CSC 2 - COBIT 5 BAI09.01, BAI09.02, BAI09.05 - ISA 62443-2-1:2009 4.2.3.4 - ISA 62443-3-3:2013 SR 7.8 - ISO/IEC 27001:2013 A.8.1.1, A.8.1.2, A.12.5.1 - NIST SP 800-53 Rev. 4 CM-8, PM-5
		ID.AM-3: Organizational communication and data flows are mapped	- CIS CSC 12 - COBIT 5 DSS05.02 - ISA 62443-2-1:2009 4.2.3.4 - ISO/IEC 27001:2013 A.13.2.1, A.13.2.2 - NIST SP 800-53 Rev. 4 AC-4, CA-3, CA-9, PL-8
		ID.AM-4: External information systems are catalogued	- CIS CSC 12 - COBIT 5 APO02.02, APO10.04, DSS01.02 - ISO/IEC 27001:2013 A.11.2.6 - NIST SP 800-53 Rev. 4 AC-20, SA-9
		ID.AM-5: Resources (e.g. hardware, devices, data, time, personnel, and software) are prioritized based on their classification, criticality, and business value	- CIS CSC 13, 14 - COBIT 5 APO03.03, APO03.04, APO12.01, BAI04.02, BAI09.02 - ISA 62443-2-1:2009 4.2.3.6 - ISO/IEC 27001:2013 A.8.2.1 - NIST SP 800-53 Rev. 4 CP-2, RA-2, SA-14, SC-6
		ID.AM-6: Cybersecurity roles and responsibilities for the entire workforce and third-party stakeholders (e.g. suppliers, customers, partners) are established	- CIS CSC 17, 19 - COBIT 5 APO01.02, APO07.06, APO13.01, DSS06.03 - ISA 62443-2-1:2009 4.3.2.3.3 - ISO/IEC 27001:2013 A.6.1.1 - NIST SP 800-53 Rev. 4 CP-2, PS-7, PM-11

Figure 23.5 NIST cybersecurity framework core (Excel template). *Source:* https://www.nist.gov/cyberframework/framework.

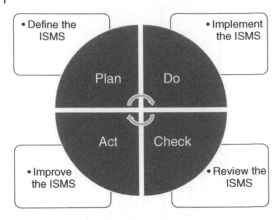

Figure 23.6 ISMS PDCA model.

controls must be implemented (from Annex A of the standard), but also how they will be implemented.

Depending on the maturity level of the organization, the Check stage may involve a general review and an internal or external audit. At this point, the ISMS performance is evaluated and improvement measures are identified.

Finally, in the Act stage, these improvements are implemented, the design is reviewed and risks are again identified and analyzed.

It places great emphasis on management commitment and business objectives. In that sense, one of the fundamental requirements is the definition of an information security policy approved by management.

It requires a high level of documentation, in particular a set of documents required for its correct implementation (Table 23.2).

23.3.1.3 COBIT "Control Objectives for Information and Related Technologies"

COBIT is published and maintained by the Information Systems Audit and Control Association (ISACA), a non-profit organization comprised of more than 450,000 professionals from more than 188 countries in various IT roles.

Table 23.2 ISO/IEC reference.

Document required	ISO/IEC ref.
Scope	4.3
Information security policy and objectives	5.2–6.2
Risk management methodology and assessment	6.1.2
Statement of applicability	6.1.3d
Risk treatment plan	6.1.3.e-6.2
Risk treatment report	8.2–8.3
Information security roles and responsibilities	A.7.1.2–A.13.2.4
Asset inventory	A.8.1.1
Acceptable use of assets	A.8.1.3
Access control policy	A.9.1.1
Operational procedures	A.12.1.1
Secure system engineering principles	A.14.2.5
Suppliers security requirement	A.15.1.1
Incident management procedures	A.16.1.5
Business continuity plan	A.17.1.2
Legal and regulatory compliance	A.18.1.1

Several versions of COBIT are currently active: COBIT 5, from 2012, and COBIT 2019. COBIT is a comprehensive framework that helps organizations achieve their goals and deliver value through effective governance and management of the organization's IT.

COBIT 5 is characterized by a strong emphasis on governance and fundamental principles meeting the needs of stakeholders

1) Covering the whole enterprise from end to end.
2) Application of a single integrated framework.
3) Ensuring a holistic approach to business decision making.
4) Separating the governance from the management.

Source: www.isaca.org.

Although this framework is not specific to information security, in its 2019 version it is aligned with different information security frameworks, controls, and guidelines. In particular, it is aligned with the ISO/IEC 27000 family, NIST Cybersecurity Framework v1.1

23.3.2 Cybersecurity Controls

In this section, we analyze the most commonly used cybersecurity control standards.

Cybersecurity controls are specific cybersecurity measures such as the application of a firewall, or the implementation of a two-factor authentication system. The standards that describe cybersecurity controls generally only list the range of possible cybersecurity controls in place but do not include any reference to which controls to apply in each organization or methods for identifying business risks.

A common mistake is to think that if you apply all possible cybersecurity controls you will have 100% security, that does not happen, 100% cybersecurity does not exist, and there will always be a remaining risk. Applying all possible controls, without any type of methodology, will only result in an unnecessary waste of resources. You will find next the description of the most commonly used cybersecurity controls standards.

23.3.2.1 ISO/IEC 27002 "Code of Practice for Information Security Controls"

This standard was first published in 2005. It was updated several times up to 2013 and corrections were made in 2015. It contains cybersecurity best practices grouped into 35 control objectives and 114 specific cybersecurity controls. All controls are arranged by subject area as follows:

5-Information Security Policy	12-Security of operations
6-Organization of Information Security	13-Communications
7-Security in human resources	14-Acquisition, development, and maintenance of systems
8-Assets management	15-Relations with suppliers
9-Access control	16-Information security incident management
10-Cryptography	17-Aspects of information security in business continuity management
11-Physical and environmental security	18-Compliance

The way of ordering the controls coincides exactly with Annex A of standard 27001, with the difference that 27001 only indicates the title, while 27002 explains the control in detail, even giving guidance for its implementation.

Table 23.3 NIST control families.

Id	Family	Id	Family
AC	Access control	PE	Physical and environmental protection
AT	Awareness and training	PL	Planning
AU	Audit and accountability	PM	Program management
CA	Assessment, authorization, and monitoring	PS	Personnel security
CM	Configuration management	PT	PII processing and transparency
CP	Contingency planning	RA	Risk assessment
IA	Identification and authentication	SA	System and services acquisition
IR	Incident response	SC	System and communications protection
MA	Maintenance	SI	System and information integrity
MP	Media protection	SR	Supply chain risk management

Source: www.nist.gov.

ISO/IEC 27002 is one of the most widely used and comprehensive standards as a "code of practice" for cybersecurity controls. It integrates very well with both ISO/IEC 27001 and NIST CSF.

23.3.2.1.1 NIST SP 800-53 Currently in version 5, its objective is to detail the different cybersecurity controls that can be applied in organizations to protect their systems.[3]

Like the ISO/IEC 27002 standard, it groups cybersecurity controls by subject area into 20 families. It should be noted that it manages over 1000 very specific controls. Table 23.3 shows the control families.

Each of the control families is identified with a two-letter identifier; for example, for access control it is AC. Within each family, specific controls are listed; for example, AC-2 is account management. For each control, a list of activities or tasks to be accomplished is defined, which are those listed in the ordered list (a, b, c, etc.).

It is a very complete and specific standard and integrates very well with the NIST CSF.

23.3.2.1.2 SANS CIS Critical Security Controls The Center for Internet Security (CIS) is an independent, non-profit organization with the mission of create confidence in the use of Internet.[4] It is made up of IT experts from different business verticals.

Version 8 of CIS Critical Security Controls defines a set of controls (20) with best practices and defense actions to mitigate the most frequent attacks on systems and networks.

One of the new features of CIS Critical Security Controls is the introduction of Implementation Groups. It defines three Implementation Groups according to the size of the organization, the criticality of the information handled, and the degree of maturity. Figure 23.7 shows the implementation groups.

---

3 National Institute of Technical Standards (NIST), 2020 (b).
4 Center for internet Security (CIS), 2019.

Implementation Groups

IG1

An IG1 enterprise is small to medium-sized with limited IT and cybersecurity expertise to dedicate towards protecting IT assets and personnel. The principal concern of these enterprises is to keep the business operational, as they have a limited tolerance for downtime. The sensitivity of the data that they are trying to protect is low and principally surrounds employee and financial information.

Safeguards selected for IG1 should be implementable with limited cybersecurity expertise and aimed to thwart general, non-targeted attacks. These Safeguards will also typically be designed to work in conjunction with small or home office commercial off-the-shelf (COTS) hardware and software.

IG2 (Includes IG1)

An IG2 enterprise employs individuals responsible for managing and protecting IT infrastructure. These enterprises support multiple departments with differing risk profiles based on job function and mission. Small enterprise units may have regulatory compliance burdens. IG2 enterprises often store and process sensitive client or enterprise information and can withstand short interruptions of service. A major concern is loss of public confidence if a breach occurs.

Safeguards selected for IG2 help security teams cope with increased operational complexity. Some Safeguards will depend on enterprise-grade technology and specialized expertise to properly install and configure.

IG3 (Includes IG1 and IG2)

An IG3 enterprise employs security experts that specialize in the different facets of cybersecurity (e.g., risk management, penetration testing, application security). IG3 assets and data contain sensitive information or functions that are subject to regulatory and compliance oversight. An IG3 enterprise must address availability of services and the confidentiality and integrity of sensitive data. Successful attacks can cause significant harm to the public welfare.

Safeguards selected for IG3 must abate targeted attacks from a sophisticated adversary and reduce the impact of zero-day attacks.

Figure 23.7 CIS Implementations Groups. *Source:* cisecurity.org.

Then depending on that, it indicates whether or not the control should be prioritized for implementation in that organization.

Each control is composed of sub-controls and mapped to the different cybersecurity functions so it integrates very well with the NIST CSF (see Figure 23.8).

The fact of having this mapping structure by Implementation Group of organization, type of asset, and cybersecurity function makes it very versatile.

CIS control 1: inventory and control of hardware assets

Sub-control	Asset type	Security function	Control title	Control descriptions	Implementation groups 1	2	3
1.1	Devices	Identify	Utilize an active discovery tool	Utilize an active discovery tool to identify devices connected to the organization's network and update the hardware asset inventory.		●	●
1.2	Devices	Identify	Use a passive asset discovery tool	Utilize a passive discovery tool to identify devices connected to the organization's network and automatically update the organization's hardware asset inventory.			●
1.3	Devices	Identify	Use DHCP logging to update asset inventory	Use dynamic host configuration protocol (DHCP) logging on all DHCP servers or IP address management tools to update the organization's hardware asset inventory.		●	●

Figure 23.8 CISC Control 1. *Source:* Adapted from https://www.cisecurity.org/.

Finally, for each case, a section with a diagram of system entity relationships, procedures, and tools is included to support the execution of the different activities.

23.3.2.1.3   OWASP: TOP 10, WSTGv4, ASVS y MASVS   The Open Web Application Security Project (OWASP) is a non-profit foundation that works to improve web application security.

Although in this case we are not dealing with general cybersecurity controls, web application development is fundamental nowadays and for this reason we decided to include it in this chapter.

OWASP generates multiple open-source projects including the OWASP Top 10 Vulnerabilities, OWASP Testing Guide, Application Security Verification Standard (ASVS), and Mobile Application Security Verification Standard (MASVS). These projects for web applications or mobile devices, respectively, provide a basis for testing technical security controls and also provide developers with a list of requirements for secure development.

OWASP TOP 10: It is a standard, open-construction guide, developed by cybersecurity professionals on the top 10 cybersecurity risks. It is a very good reference to verify on the application development and deployment.

OWASP Web Security Testing Guide: It is a guide, created by cybersecurity specialists, on what cybersecurity tests to perform on web applications.

Application Security Verification Standard: This is also a cybersecurity testing guide, but more oriented for the development process. There is a kind of overlapping but are very complementary.

23.4   Cybersecurity Building Blocks

Cybersecurity is not just a technology issue, nor is it an issue that affects only one sector; cybersecurity is a multidimensional discipline. Cybersecurity involves multiple dimensions:

- **People and organizational structures:** People, from the cybersecurity director to the end user, are the first protagonists of cybersecurity, they are the ones who at the end of the day will design and execute cybersecurity strategies and plans. Additionally, organizational structures are the fundamental instrument for mapping cybersecurity roles and responsibilities to organizations. In this sense, there are roles, such as the CISO, and organizational structures such as Security Operations Centers (SOCs), or Cybersecurity Incident Response Centers (CSIRTs) that have fundamental roles.
- **Policies and procedures:** Cybersecurity policies are those that reflect the will of the organization (as an abstract entity). It is the policies that indicate the value of cybersecurity for the top management, the objectives and risks they are willing to accept, as well as the fundamental strategic definitions (such as access control aspects). Procedures, on the other hand, outline how cybersecurity controls are to be implemented in a structured and repeatable manner. In summary, policies indicate the "what" and procedures the "how."
- **Technologies:** Last but not least is technology. In order to carry out effective protection, cybersecurity technologies will be required to implement the protection (Figure 23.9).

In this section, we will briefly discuss the most important components of each of the dimensions so that the reader understands what they are, how they work, and what they are for. The focus will always be from a management point of view.

23.4.1 Chief Information Security Officer

The chief information security officer (CISO) is the executive responsible for an organization's information and data security. As the name indicates, it should be part of the C-Group.

As mentioned in this chapter, cybersecurity is very strategic for the organization, so the CISO must be part of the strategic decision process.

However, in many organizations the position of the CISO is not actually in the C-Group, the exact position is not the most important, but the role of cybersecurity in strategic planning. So, the discussion should not be the position in the organigram but the role and reasonability.

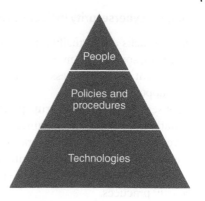

Figure 23.9 Cybersecurity dimensions.

23.4.1.1 The Role and Responsibilities of the CISO

The CISO is ultimately responsible and accountable for the cybersecurity of an organization. Although it is not the CISO who "signs" the cybersecurity policy, but rather the CEO or the board itself, it is the CISO who must ensure its implementation. For this reason, in addition to the responsibilities listed below, the CISO must have the necessary powers to carry them out.

The main responsibilities of the CISO are:

- **Cybersecurity strategic planning:** Responsible for incorporating cybersecurity into the organization's strategic planning processes.
- **Governance:** Making sure all cybersecurity initiatives are included in organizational process, the budget and needed resources are in place (human, technical and economic). Also, design and implement the ISMS
- **Risk management:** Must define risk management policies and define and implement evaluation methodologies. It is NOT who decides whether a risk is accepted or not, but it is responsible for evaluating it.
- **Cybersecurity engineering:** Is responsible for the cybersecurity architecture, and cybersecurity design of the systems.
- **Identity and access management:** Define the access control policy and supervise/implement the Identity and access control systems to ensure that only authorized people have access to information.
- **Security operations and incident response:** The CISO is responsible for security operations and incident response. The CISO has to implement the operation security policy, implement real-time analysis of immediate threats, and respond to an incident when it happens.
- **Compliance:** Ensure the required policies and regulations related to the topic are fulfilled.
- **Program management:** The CISO is the main responsible for the design and execution of the cybersecurity project portfolio. However, many projects will be executed by other teams such as IT, Human Resources, etc.

As can be seen, the CISO is the leader of a team of people who must design, implement, and oversee cybersecurity policies, projects, and operations. However, these teams do not always need to be very large; it depends on the organization's state of maturity and the organization's risk requirements.

23.4.2 Cybersecurity Incident Management

Cyber attacks are virtually impossible to avoid, given the openness of today's networks and the increasing sophistication of advanced threats. In the increasingly connected world, where there is insufficient user awareness of risks and threats, cyberattacks are no longer a question of "if they will happen," but when.

Consequently, cybersecurity practices must focus on ensuring that intrusion and business limitations do not result in damage or loss to organizations. We live in an era where cybersecurity prevention is not optional. Operational capabilities to detect and respond to attacks are key to mitigate their effect, or why not, to anticipate and prevent them. A cybersecurity incident is a violation or imminent threat of violation of computer security policies, acceptable use policies, or standard security practices.

Examples of incidents are:

- An attacker commands a botnet to send high volumes of connection requests to a web server, causing it to crash.
- Users are tricked into opening a "quarterly report" sent via email that is actually malware; running the tool has infected their computers and established connections with an external host.
- An attacker obtains sensitive data and threatens that the details will be released publicly if the organization does not pay a designated sum of money.
- A user provides or exposes sensitive information to others through peer-to-peer file-sharing services.

Improving threat detection and response requires an intelligence-driven approach to security and big data analytics that helps organizations use all available security-related information, from both internal and external sources, to detect hidden threats and even predict future threats.

23.4.2.1 The Cyber Security Operations Center

Cyber attacks have become increasingly sophisticated and difficult to prevent. So, we must focus on reducing their impact. To do this, we must understand how the attack occurs in order to stop it.

A cyber attack is composed of several phases, if we take as a reference for example the cyber killchain[5] we have seven phases shown in Figure 23.10.

As the attack evolves in phases, it generates more impact on the victim, which is why our objective will be to detect the attack in the earliest phases.

The problem is that in order to detect the attack in the earliest stages we must have "visibility" of the network and the different systems, and this is not always possible.

In other words, we must be able to identify the signs that something is happening in order to be able to respond in the shortest possible time.

Figure 23.10 Cyber killchain.

---

5 https://en.wikipedia.org/wiki/Kill_chain.

23.4.2.1.1 What Does a SOC Do? A SOC is a cybersecurity operations center that monitors all network systems to identify signs of a possible attack and respond to them.

In general, SOCs offer the following services:

- System monitoring and incident detection.
- Incident response.
- Information analysis (threat intelligence, Big Data).
- System vulnerability management.
- Operation of cybersecurity devices.

It is not necessary to offer all services at the same time, in fact, it is common to speak of different "generations" of SOCs according to the services they offer. However, the first two are considered the basic services.

Figure 23.11 illustrates an example of different SOC generations and their most typical services.

Regarding the service, it is possible to work 24×7 every day, at all hours, or 8×5 only during office hours. The effect of working 24×7 offers obvious advantages, since the objective is to detect the attack as early as possible.

How do you deal with this problem?

In order to provide this "visibility" of systems, SOCs implement cybersecurity event processing technologies known as Security Information and Event Management (SIEM).

These systems collect in real-time and in a centralized way the information from the different systems (logs) as well as from probes and sensors distributed in the network (such as firewalls, Intrusion Detection Systems /Intrusion Prevention Systems [IDS/IPS], Anti Malware, etc.). Once the information is collected, it is processed and correlated to identify if it is possible that we are facing an attack.

Additionally, threat intelligence information is collected from multiple sources and integrated with network information to identify known threats.

It is also common for vulnerability scanners from different systems to be integrated with this information to get a complete picture of the system.

Once a possible security problem is identified, it is analyzed by Tier 1 analysts, who validate the information and classify the incident, and if the problem cannot be solved, it is escalated to Tier 2 analysts and so on until the incident is closed.

In order to carry out this process, Case Management tools are used, and in the most mature SOCs tools for response automation known as SOAR (Security Orchestration Automatization and Response).

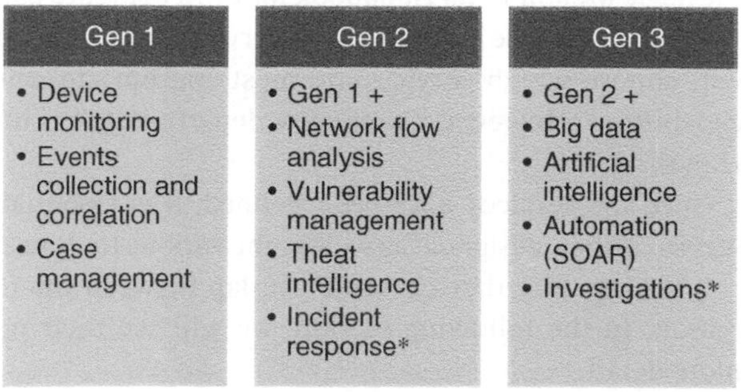

Figure 23.11 SOC generations.

Gen 1	Gen 2	Gen 3
• Device monitoring • Events collection and correlation • Case management	• Gen 1 + • Network flow analysis • Vulnerability management • Theat intelligence • Incident response*	• Gen 2 + • Big data • Artificial intelligence • Automation (SOAR) • Investigations*

23.4.2.2 Computer Security Incident Response Team

A Computer Security Incident Response Team (CSIRT) is an organization of people whose primary purpose is to provide computer security incident response to a given organization or community.

The concept was introduced in the late 1980s by Carnegie Mellon University with the creation of the Computer Emergency Response Team Coordination Center (CERT/CC). Since then, there have been multiple shades of these organizations, existing as a unit within a private organization, as an independent organization, in governments, universities, etc.

> The best way to describe them is just as Carnegie Mellon does: A CSIRT can most easily be described by analogy with a fire department. In the same way that a fire department has an emergency number that you can call if you have or suspect a fire, similarly a CSIRT has a number and an email address that you can contact for help if you have or suspect a computer security incident. A CSIRT service does not necessarily provide response by showing up on your doorstep (although some do offer that service); they usually conduct their interactions by telephone or via email.
>
> *Carnegie Mellon University*

CSIRT establishment

When creating a CSIRT, the first thing to define is its scope, i.e. the organization or community it will serve and the services it will provide.

The organization or community it will serve is known as constituency, this constituency must be well-defined in order to avoid confusion and leave systems unprotected.

Examples of constituency can be:

- All the systems of the university "X".
- All the systems of the company "Y," including the outsourced systems.
- All critical systems of a country, or all Financial Operators, etc.

As we can see, there are different levels of precision for these definitions, and that is why it is important to refine them as much as possible.

As an example, a CSIRT that protects the systems of company "Y," including its outsourced systems, must have the real capabilities to do so. That is, it must be able to detect and respond to an incident that actually occurs outside the company. For this reason, when defining the scope, all this casuistry must be analyzed and, if necessary, clauses must be incorporated in third-party contracts or, eventually, in government cases, regulatory aspects must be modified.

Once the constituency has been defined, the services that the CSIRT will provide to that constituency must be defined. A general definition of the different service options is also provided by Carnegie Mellon in its Handbook for CSIRT (Figure 23.12).

There are three categories of services, reactive, proactive, and value-added. Each CSIRT will choose which services are most appropriate, and in particular for which services it has capabilities. However, like the fire department, it is important that it can at least offer Incident Handling.

Once the services have been defined, it is essential that the organizational structure of the service is then designed, as well as the various tools needed to be able to provide services.

It is clear that there is some overlap between the functions of a SOC and a CSIRT. For this reason, in the following section, we will see their points of contact and main differences in more detail.

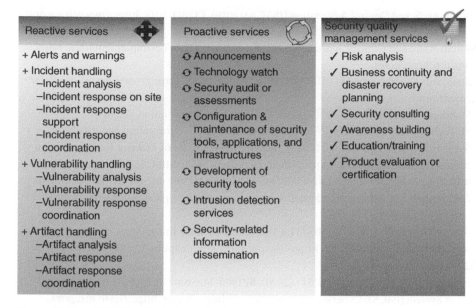

Figure 23.12 *CSIRT Handbook,* Carnegie Mellon.

23.4.2.3 CSIRT, SOC Comparison

Whereas a SOC may also serve as a CSIRT and vice-versa, SOC is commonly mainly occupied with monitoring the enterprise systems and often rely on third party for some of its services, especially for incident response.

This difference between CERTs and SOCs provides a complementary, integrative, and more holistic approach for handling cyber threats.

The incident management process can be divided into several parts. The first part, that is until an inci-

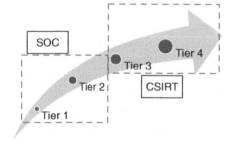

Figure 23.13 Risk incident tiers.

dent is reported or detected by monitoring systems, is commonly the constituents, namely falls under the responsibility of the SOC. The CSIRT part typically begins when it receives a report on a suspected incident, which is initially triaged followed by taking actions to remediate the incident.

This coordination is usually managed by Tiers of analysis, that Level 1 responds to more repetitive and low-risk incidents in the SOC and Tiers 3–4 are highly specialized responder of the CSIRT as illustrated in Figure 23.13.

Further Reading

Carnegie Mellon University (2003). Handbook for Computer Security Incident Response Teams (CSIRTs). https://resources.sei.cmu.edu/asset_files/Handbook/2003_002_001_14102.pdf.

Center for Internet Security (CIS) (2019). CIS Critical Security Controls. Versión 8. https://www.sans.org/critical-security-controls.

IBM (2020). Cost of a Data Breach Report 2020. https://www.ibm.com/security/digital-assets/cost-data-breach-report#/.

Information System Audit and Control Association (ISACA) (2019). COBIT 2019. https://www.isaca.org/resources/cobit.

Information System Audit and Control Association (ISACA) (2019). COBIT 5 implementation. https://www.isaca.org/resources/cobit.

International Standards Organization ISO/IEC. 3100 risk management. https://www.iso.org/iso-31000-risk-management.html.

International Standards Organization (ISO) (2013). ISO/IEC 27001. Information Security Management. Technical standard. https://www.iso.org/isoiec-27001-information-security.html.

International Standards Organization (ISO) (2013). ISO/IEC 27002. Code of Practice for Information Security Controls. Technical standard. https://iso.org/standard/54533.html.

National Institute of Technical Standards (NIST) (2006). FIPS-200. Minimum Security Requirements for Federal Information and Information Systems. https://csrc.nist.gov/publications/detail/fips/200/final.

National Institute of Technical Standards (NIST) (2018). Framework for Improving Critical Infrastructure Cybersecurity. Version 1.1. https://nvlpubs.nist.gov/nistpubs/CSWP/NIST.CSWP.04162018.pdf.

National Institute of Technical Standards (NIST) (2020). SP1800-24. Securing Picture Archiving and Communication System (PACS): Cybersecurity for the Healthcare Sector. https://csrc.nist.gov/publications/detail/sp/1800-24/final.

National Institute of Technical Standards (NIST) (2020). NIST Special Publications 800-53, revision 5. Security and Privacy Controls for Information Systems and Organizations. https://nvd.nist.gov/800-53.

National Institute of Technical Standards (NIST) (2020). NIST Special Publications SP 800-171, revision 2. Protecting Controlled Unclassified Information in Nonfederal Systems and Organizations. https://csrc.nist.gov/publications/detail/sp/800-171/rev-2/final.

Open Web Application Security Project (OWASP) (2016). Top 10 Mobile Application Security Risks. https://owasp.org/www-project-mobile-top-10.

Open Web Application Security Project (OWASP) (2017). Top 10 Web Application Security Risks. https://owasp.org/www-project-top-ten.

Open Web Application Security Project (OWASP) (2020). OWASP Application Security Verification Standard (ASVS). Version 4.0.2. https://owasp.org/www-project-application-security-verification-standard.

Open Web Application Security Project (OWASP) (2020). OWASP Mobile Application Security Verification Standard (MASVS). Version 1.3. https://github.com/OWASP/owasp-masvs.

Section 10

Data Science

24

Information-Enabled Decision-Making in Big Data Scenarios

Dario Petri[1], Luca Mari[2], Matteo Brunelli[1], and Paolo Carbone[3]

[1] Department of Industrial Engineering, University of Trento, Trento, Italy
[2] School of Industrial Engineering, Università Cattaneo – LIUC, Castellanza, Italy
[3] Department of Engineering, University of Perugia, Perugia, Italy

24.1 Introduction

The widespread diffusion of high-performance and low-cost devices with data acquisition, storage, processing, and transmission capabilities is increasingly driving the ubiquitous and massive collection and use of data in most human endeavors. The current interest in Data Science has been plausibly triggered by this "Big Data" phenomenon, thus explaining the emphasis on the development of mathematical and computational techniques that are efficient in dealing with large amounts of data, particularly in the contexts of Data Mining and Machine Learning. In the perspective of applications, *Data Science tools are enablers of effective Decision Making* (DM), that indeed requires information – in the following the distinction and the relations between data and information will be discussed – to characterize the current state of the system about which the decision has to be made and to establish a rational, though hypothetical, connection between the current state and the desired outcome: if the system under consideration is in the state x_i and the decision u_j is made, then it is predicted that the desired state x_k will be obtained; since through data acquisition, and possibly through measurement, the current state has been identified to be x_i, then it is inferred that u_j is indeed a good decision to be made.

Such *Information-Enabled Decision Making* (IEDM) is supposed to reduce the risk of wrong decisions, as compared to conclusions based only on intuition and/or untested beliefs (or other questionable criteria, such as reliance on random factors), assuming a sufficient quality (i) of the processes by means of which information is processed and decisions are made, and (ii) of the information that enables such processes to be performed. This sets IEDM as a core component of appropriate Problem Solving, where a problem is a situation that can be in some sense improved through an intervention that is supposed to be possible – otherwise, it is not a problem but just a fact – and problem-solving is the process of designing and performing such intervention. Indeed, solving a problem requires both finding a solution and implementing it (finding even an effective solution but not implementing it leaves the problem unsolved): this closes the loop and enables decision validation through the comparison of the expected and the actually obtained outcomes.

IEEE Technology and Engineering Management Society Body of Knowledge (TEMSBOK), First Edition.
Edited by Gustavo Giannattasio, Elif Kongar, Marina Dabić, Celia Desmond, Michael Condry, Sudeendra Koushik, and Roberto Saracco.

Such an application-oriented perspective suggests to critically reconsider the assumption that large amounts of data, fed into appropriate Data Science tools such as artificial neural networks, are always sufficient to perform effective DM, thus independently of any substantive theory or model of the system that generates the data, a position epitomized as "the end of theory" [1]. Since data as such inform us of correlation only, and correlation is not causation, the issue is whether this approach to DM – that may be called *Data-Driven* to contrast it to the traditional, Information-Enabled, one – may completely and definitely supersede the condition that some predictions, and then some decisions, require instead causal models because "data do not speak for themselves." In the perspective of Machine Learning, the endeavor of so-called *Explainable Artificial Intelligence* – what allows human users to understand how the results of the execution of a software system are produced and establish their reliability accordingly – is perhaps the most critical problem today in this context.

This is an open, fundamental challenge for Data Science today, as also some authoritative researchers in the field acknowledge (see, e.g. [2–4]). A plausible position is that statistical and Machine Learning methods that are designed to perform data fitting, not data interpretation, can lead to accurate predictions only in stationary contexts, when data are (at least approximately) independent and identically distributed (i.i.d.), but they are not sufficient to decide whether unexpected data (typically: distribution outliers) are physiological parts of the distribution tails or signals of out-of-distribution conditions. This is a delicate point, particularly given that to achieve a desired outcome DM often involves deliberate actions that may relevantly modify the context, and the prediction of the likely impact of these actions is essential to ensure optimal conclusions. According to this position, we discuss IEDM here in the evolutionary context of Data Science.

24.2 Knowledge Area Fundamentals

As a process of selection among alternative options with the aim of reaching a given target, DM plays an important role in many engineering management activities. When the target is achieved through a well-established, though possibly non-deterministic, process, as usually happens at operational level, *fully structured* DM can be performed. However, only *partially structured or unstructured* DM is sometimes possible, for example when not all significant alternatives or not all relevant constraints are identified, a usual situation in strategic decisions. Also in these cases, presenting IEDM as a structured, rational process provides some guidelines to design and perform effective IEDM processes.

According to an abstract model, as shown in Figure 24.1, IEDM is performed by following a *procedure* aimed at the selection of one out of two or more *alternatives* available in the given context through the optimization of an *objective function*, for example the minimization of a cost/loss

Figure 24.1 Abstract model of an IEDM process.

function. Usually the procedure is subjected to *constraints*, for example related to available resources or time limits, while the alternatives are associated to *feasible actions*. In addition, an at least approximate lower bound on the *acceptable confidence level* of the conclusion is specified, as related to the consequences of wrong conclusions. The actually *achieved confidence level* depends on various elements that will be discussed in the next section.

The structurally simplest kind of IEDM is such that [5]:

- there exists a single best choice (*single aimed process*);
- the decision process is guided toward the best choice by completely defined procedures (*fully structured process*);
- the available information suffices to achieve an unambiguous decision (*fully informed process*).

In this case, the selection of the best alternative can be completely automated by formally defining an objective function, and applying probabilistic methods also to evaluate the actually achieved confidence level. However, DM could be very different if:

- the available alternatives, the purpose of the decision, or the related objective function are not well-defined;
- data are not collected using fully-structured and validated processes, and information is extracted from data more as patterns discovered by applying Machine Learning methods or using ad hoc insights than through established models;
- the procedure to be followed and the constraints to be fulfilled are only partially specified.

This is a situation not uncommon in contexts such as health care, social, ecological, and earth sciences, in which a suitable process formalization is not easy to achieve and even a raw estimation of the risk of wrong conclusion can be hardly derived, and a combination of human heuristics and adaptive computer-based techniques could be the most effective approach. Indeed, DM remains a human activity, even if it can be computer-aided.

A structural analysis of the encompassing process of Problem Solving – like the one depicted in Figure 24.2 – may support the improvement of its core component of IEDM.

Five main stages can be distinguished in the process, as follows.

- **Understanding of the situation:** The problem at stake is analyzed, by acquiring information on the current situation and the constraints imposed by the relevant stakeholders.
- **Choice of the goal:** The possible goals are identified and one of them selected, by taking into account the preferences of the relevant stakeholders and the achievability of the identified goals, possibly through the definition of an objective function.
- **Choice of the strategy:** The possible ways to achieve the chosen goal are identified and one of them selected, by taking into account the conditions posed by the relevant stakeholders and the affordability of the identified strategies, possibly through the definition of a cost function.

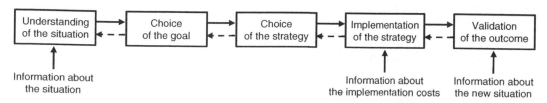

Figure 24.2 The main stages of a problem-solving process.

- **Implementation of the strategy:** The chosen strategy is operated, and information on the incurred costs is acquired.
- **Validation of the outcome:** Information on the situation produced by the implementation of the strategy is acquired, and exploited to compare the new situation with the chosen goal.

Problem-Solving generally includes feedback loops that help keeping a flexible management of the process. Furthermore, this structure emphasizes the importance of the availability of dependable information, such that IEDM could be considered in fact a condition for valid DM.

24.2.1 Data Processing and Causal Modeling

In the more and more common Big Data scenarios, a key issue is whether, and in the case under what conditions, data may be sufficient to accomplish the first stage of IEDM, thus actually leaving implicit and unexpressed a model of the situation. Indeed, there are cases of very effective model-free, Data-Driven DM, as the well-known example of automatic translation and more generally natural language processing. While this is still an open subject of research, a plausible thesis is that (large amounts of) data are sufficient for effective DM in three sorts of situations. First, if the context can be assumed to be stationary (and data obtained from independent and identically distributed (i.i.d.) random variables) so that Machine Learning algorithms can appropriately generalize from the training set. Second, if a trial-and-error approach is adopted, that could indeed be performed in a purely input–output/black box way, without an explicit understanding of the situation. Third, if the decision is not so critical, or however it is not socially required to explain/justify it. Instead, whenever out-of-distribution conditions apply – the rule more than the exception in real-life DM – and/or for some reasons trial-and-error is not affordable, and/or the decision must be transparently explainable, such purely inductive, bottom-up approaches appear to be unable to deal with the complexity of the object of the decision. In these situations, the first stage to understand the situation requires a substantive modeling activity, drawing from disciplinary knowledge and typically formalized as a causal model.

Indeed, the common saying that "correlation is not causation" assumes its most fundamental meaning in view of the distinction between data and entities that generate data, given that *correlation is a relation between mathematical variables*, and only indirectly between empirical properties modeled by such mathematical variables. Conversely, *causality is a relation between empirical properties*, and only indirectly between mathematical variables modeling such empirical properties. The fact that sometimes, for the sake of simplicity, variables and properties are not distinguished does not reduce their principled difference. In particular, sequences of values of empirical quantities, such as measurement results, may be correlated but they cannot be causes of effects, since it is the empirical property of an object that can be cause of effects, not its value. For example, a given deformation of a given elastic object could have been caused by a given applied force, not by 10 N, even if the applied force is precisely 10 N. As a consequence, the statement that correlation is not causality has first of all a categorical meaning (i.e. correlation is about data, causality is about entities that generate data); the conclusion that there are cases of correlations that are not causal is a consequence.

In an open-loop IEDM process (i.e. until the decision is not operated or however its effects are not considered yet), the difference between correlational information and causal information influences the structure of the process, which may be based on

- either a model of the behavior of the entity that is the object of the decision and that generates the data; such a model is called here *domain model*, given that it is grounded on knowledge that

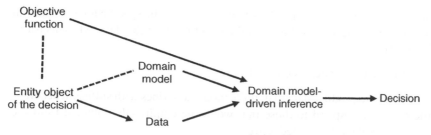

Figure 24.3 IEDM based on a domain, typically causal, model.

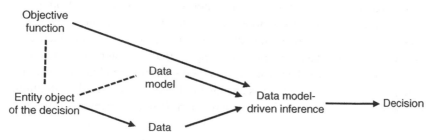

Figure 24.4 IEDM based on a data, typically correlational, model.

is related to the domain of the entity, such as physics, economics, etc.; it is often constituted of equations typically interpreted in causal terms (Figure 24.3).

- or a model of the data acquired in conditions deemed sufficiently similar to that involved in the considered decision; such a model, called here *data model*, relates to a context that currently would be referred to as "data science"; it is constituted of functions and algorithms of classification or regression, typically interpreted in correlational terms (Figure 24.4).

A sufficiently high-level description of the two strategies is the same: for example, having the objective of producing a given deformation in a given elastic object, the adopted model and the available data lead to the inference that the application of a force of 10 N would produce the required outcome. Where the two strategies diverge is about the model that enables the inference, since differently from the causal strategy, the correlational strategy is about "the description of data, not the process responsible for the data" [2, 6]. Hence:

- the *causal strategy* justifies the decision basing on a domain model and the available data on the entity; it interprets such a decision as a way to realize a cause that can produce the effect specified in the objective function; for example, in the conditions Z, a force X is cause of the deformation Y; being actually in the specific conditions z and willing to obtain the specific deformation y, apply the specific force x. Were an explanation of the decision requested, the domain model would provide it: we decided to apply a force of 10 N because according to a given physical law etc.;
- the *correlational strategy* assumes that the data model is adequately applicable to the available data on the entity (in statistical terms, that the training set and the test set are samples from the same population) and, on this, it grounds the hypothesis of validity of the decision; larger datasets lead to more corroborated validity, e.g. the regression of the data about the conditions Z and the forces X leads to reliable predictions on the deformations Y; being actually in the specific

conditions z and willing to obtain the specific deformation y, apply the specific force x. Were an explanation of the decision requested, nothing could be said but that the data led to decisions that so far proved to be valid.

The asymmetry between the two strategies is manifest:

- a domain model is sufficient to produce data that, if the model describes with sufficient accuracy the behavior of the entity, correspond to those that would be empirically obtained from the entity: in this sense, the causal strategy is deductive;
- a data model is instead not sufficient to produce causal knowledge on the entity: in this sense, the correlational strategy is inductive.

Of course, causal knowledge on empirical entities and properties does not generate from nothing, and it is rather the product of smart information processing, based traditionally on *randomized control trials*. Having observed a correlation between data on some empirical properties X and Y, a set of entities having those properties is split randomly, so that no systematic differences about X or Y are present between the parts. The property X is then somehow modified in only one of the parts, trying to guarantee that nothing else was changed in the meantime. In this situation, a systematic difference observed in the property Y between the parts (say, a deformation is present in the objects to which a force was applied, and only in them) makes credible the hypothesis that such a difference was produced by the modification of X, that is therefore causally, and not only correlationally, related to Y.

By embedding data processing in a causal context – by means of randomized control trials or other methods of causal inference from data [2, 6] – the outcomes of IEDM become explainable and their trustworthiness justifiable, clearly an important condition whenever the decisions are socially relevant.

24.3 Prominent Methods for IEDM Strategy Implementation

A number of methods to operationalize an IEDM problem has been developed and applied to a wide range of real-world problems. Whereas it is hard to find aspects common to all such methods, they all agree on considering a single goal but, at the same time, they acknowledge the multidimensional nature of alternatives by considering multiple decision criteria [7]. As a toy example, the choice of a 3D printer may be supposed to have the goal of maximizing the satisfaction of the user, but this depends on multiple criteria, related to the different characteristics of the candidate printer, e.g. deposition rate, accuracy, price.

Roughly speaking, IEDM methods consider the scores of alternatives with respect to a priori-defined criteria and then find a suitable aggregation rule such that the higher the score the better the alternative. Formally, the goal of IEDM methods is to identify a function v – often called "objective function," "value function," or "scoring function" – such that $v(x) \geq v(y)$ if and only if $x \geq y$, i.e. alternative x is better than, or indifferent to, alternative y [8].

The most widely acknowledged theory to objective function identification is the *Multi-Attribute Value Theory* (MAVT) [9]. As sketched in Figure 24.5 in the case of three criteria, MAVT provides the formal background to map alternatives into n-dimensional vectors and then to aggregate these normalized scores into a unique representative value.

One of the most useful results of MAVT is that, under mild conditions of independence among criteria, the objective function v is additive and therefore the level of fulfillment of the final goal

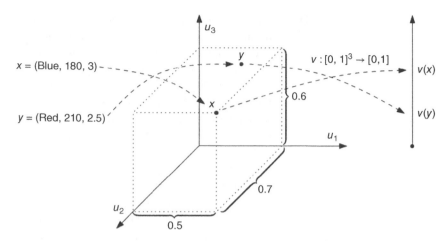

Figure 24.5 An example of MAVT: alternatives *x* and *y* are mapped into *n*-dimensional vectors *x* and *y* and eventually associated to overall scores *v*(*x*) and *v*(*y*).

can be formalized as a weighted sum of the scores of an alternative with respect to the criteria [8]. Besides its acknowledged normative role, MAVT has gained a prescriptive prominence: several methods have been proposed to operationalize it, so as to make the elicitation of the objective function *v* more friendly.

Unlike MAVT, the *Analytic Hierarchy Process* (AHP) proposed by Saaty [10] evaluates alternatives in light of a hierarchy of criteria. While grounded on a weaker axiomatization, it is perhaps the most used method in practical IEDM, possibly thanks to its simplicity and intuitive appeal. In the AHP, experts are asked to pairwise compare alternatives with respect to criteria by expressing their opinions using verbal scales. The results of the comparison are then translated into real numbers and used to obtain ratings of alternatives with respect to criteria, which are finally aggregated into a "priority vector" whose components represent the scores of different alternatives. The main benefit of using AHP lies in its divide and conquer strategy, according to which the decision maker only needs to compare two alternatives at a time under a specific point of view. Well-known techniques have been developed to reduce the theoretically large number of comparisons, so that the complexity of the method remains moderate [11]. Moreover, AHP makes it relatively simple to aggregate the opinions of multiple experts and thus to group DM problems. Finally, AHP is usually equipped with a module for the calculation of the consistency of the subjective information provided by the experts: when large inconsistencies are detected, the experts can be notified and asked to revise their judgments [12]. Real-world applications of the AHP have been object of literature reviews [13]. One representative application regards a software selection problem [14]. In this case, six criteria were considered in the evaluation of alternatives: Development interface, Graphics support, multi-media support, data file support, cost-effectiveness, and vendor support. Comparatively speaking, the AHP is at its best when the decision maker expresses judgments on the so-called "intangibles." In fact, going back to the example of the software selection, unlike physical characteristics, the fulfillment of the six criteria mentioned above cannot be measured objectively, and subjective assessments are necessary.

Among more heuristic methods that usually require less involvement of the decision maker, the *Technique for Order of Preference by Similarity to Ideal Solution* (TOPSIS) associates the set of alternatives with two, one ideal and one anti-ideal, fictitious alternatives [15]. After a normalization process, also aimed at weighing distances according to the relative importance of criteria, the

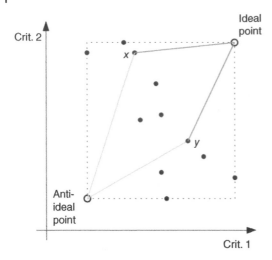

Figure 24.6 Geometry of TOPSIS in the case of two criteria to be maximized.

method suggests to choose the alternative that minimizes the distance from the ideal alternative and maximizes the distance from the anti-ideal one. The idea underlying TOPSIS is sketched in Figure 24.6 in the case of two criteria, where the chosen alternative should maximize the distance from the anti-ideal point and minimize the distance from the ideal point.

Recently, an approach called *preference learning* has been gaining popularity [16]. Preference learning techniques such as UTA [17] are based on MAVT, but do not ask direct questions about the objective function. On the contrary, they require the decision maker to indicate a reference set of alternatives on which he/she gives his/her holistic preferences, i.e. usually a ranking. These preferences are then used to estimate the objective function that best explains the decision maker's choice. The process of learning is usually done by means of simple linear programming problems [17]. Eventually, the estimated function can be applied to aid decisions on the initial set of alternatives, which may as well include new alternatives too. The procedure behind preference learning is sketched in Figure 24.7. While similar to Machine Learning, preference learning has the advantage that the resulting function is interpretable, so that the preferences given to criteria by the decision maker can be derived and analyzed.

Preference learning has often been used to support the definition of the objective function of problems modeled using MAVT, especially when to assess the function directly by asking questions to the decision maker is too complicated. One such case concerns the evaluation of risk related to the exposure to nanomaterials [18]. Among the considered criteria, one can find, for example particle size, toxicity, airborne capacity, detection limit, and exposure limit. In this case, experts were asked to rank 30 reference scenarios, from the most to the least risky. A best-fit function was then estimated and applied to rate and classify real-world scenarios.

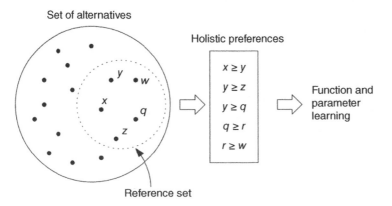

Figure 24.7 Preference learning: preferences of a reference set are used to learn preferences, which are then used on the entire set of alternatives. External and/or fictitious alternatives, like *r*, can also be considered.

So far, the set of alternatives, albeit possibly large, has been assumed known. However, in many cases, that set is not explicit, and only a region of feasible solutions of the optimization problem is provided by specifying a set of constraints.

Alternative feasibility can be guaranteed by means of a priori established lower and/or upper bounds on different criteria or by rules of thumb allowing to screen out undesirable alternatives [19]. In contrast to these "subtractive" methods of alternative generation, "constructive" techniques can be used: one of the most profitable proposals highlights the importance of focusing on the objectives of the decision [20].

While in many cases the structure of the optimization problem is simple enough (e.g. linear or convex), so that basic optimization algorithms can be

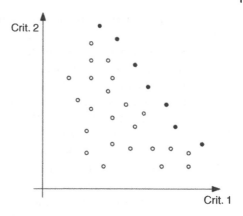

Figure 24.8 Pareto front (black circles) of dominating alternatives vs. dominated alternatives (empty circles) in the case of two criteria to be maximized.

used to solve it, real-world situations can be more complex. Not only the objective functions and the constraints may yield a non-convex optimization problem, but sometimes, as decisions may account for a number of conflicting objectives, it can be impossible to identify a unique objective function. While many objective functions can be combined into a single function by means of (a priori established) linear combinations, a posteriori interactive approaches are more and more preferred. They are aimed at finding the so-called "Pareto front," i.e. the set of non-dominated alternatives, that can be presented (also graphically, as in Figure 24.8 in the case of two criteria) to the decision maker, for a final decision.

Out of the many methods that can be used to estimate the Pareto front, NSGA-II is widely adopted [21]: like many other similar algorithms, it is based on genetic algorithms, in which a population evolves through generations trying to maximize some fitting functions [22]. As said, this type of multi-objective optimization assumes that the set of alternatives is defined by the constraints of the problem, and it becomes useful when the set of alternatives becomes huge, as for instance in combinatorial optimization problems. For this reason, a typical application may be the optimization of parallel machine scheduling, a typical combinatorial problem, where the criteria are total cost due to tardiness, deterioration cost, and makespan [23].

Other widely used methods consistent with the IEDM scheme are PROMETHEE [24] and ELECTRE [25]. State-of-the-art surveys of the different approaches were offered by Greco et al. [26] and Ishizaka and Nemery [27]. Interestingly, the choice of the most suitable method to model a decision problem is itself a decision problem. In spite of this impasse, there are significant contributions that can guide the choice of the method given the decision context (e.g. Roy and Slowinski [28] and Cinelli et al. [29]).

24.4 Quality of Information

Independently of the available amount of data, according to the so-called *garbage-in garbage-out* principle, information of non-sufficiently good quality may lead to wrong decisions, with potentially severe undesirable consequences. Indeed, the huge sizes of the datasets in Big Data scenarios make hard to assess and guarantee the quality of the used information, and sometimes the hypothesis is also (perhaps implicitly) assumed that the large quantity may compensate the poor quality. However, while the amount of data reduces the effects of random errors, any bias in them – due, for example, to a wrongly calibrated acquisition device – is unaffected by the sample size: the quality of information remains then an important topic also in Big Data scenarios.

According to the ISO 9000 series of standards, the quality of an object, such as a process, a product, or a service, is the "degree to which a set of inherent characteristics of an object fulfills requirements" [30]: without requirements there cannot be quality, nor non-quality in fact. A primary classification is then about *internal quality*, related to conformity to specifications, and *external quality*, about fitness for use and thus implying that the intended users of the entity are identified and their needs specified. In order to understand how quality of information can be assessed, a short introduction to the very concept of information is useful. While different definitions can be found in the literature, we propose here a characterization based on semiotics, that we apply in particular to that process of information acquisition that is measurement, and call *Quality of Measurement Information* (QoMI) [31]. Indeed, unlike other information processes, measurement is expected to produce information not only about properties of the empirical world but also about the trustworthiness of that information, so that it is formally possible to distinguish between "good" and "bad" measurement results.

24.4.1 Syntactic, Semantic, and Pragmatic Information

Even though different definitions can be found in the literature, information is characterized here according to a *semiotic approach*, with a hierarchical structure of three layers, traditionally called "syntactic," "semantic," and "pragmatic," related, respectively, to the form, the meaning, and the usefulness of the signs that vehicle information [32].

The *syntactic* layer only requires the definition of a set of elements, possibly together with some rules for comparison and combination among the elements. The selection of an element from the set results in the simplest and most fundamental kind of information. This standpoint has to do with a common use of the term "data," so that "data" and "syntactic information" are regarded as synonyms here. Thus, data are dealt with as purely formal entities, whose treatment (storage, retrieval, transfer, processing) does not depend on the meaning that might be associated with them.

However, the interest is usually for data as carriers of meaning, that is assigned by referring each element of the set to something else outside the set itself. In this way, through their meaning data stand for something else, typically entities of the context in which the data are generated. This embeds data into a second layer, in which the emphasis is on "the relations of signs to the objects to which the signs are applicable" [32]: data equipped with meanings provide *semantic* information.

The third semiotic layer is about *pragmatic* information, and takes into account the context in which data-with-meaning are used by some agents for some purposes. It is then about the usefulness that the agents attribute to signs [32].

While the quantity of syntactic information is well defined in Shannon's theory, only few suggestions have been advanced about the quantification of semantic information [33–36], and we are not aware of any general solution for the quantification of pragmatic information.

Each semiotic layer has its own quality criteria:

- syntactic information quality is about the *consistency* with the formal rules characterizing the set of elements;

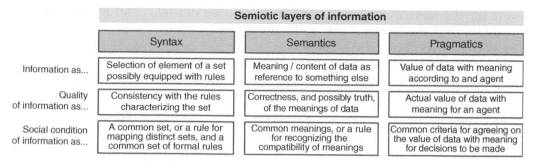

Figure 24.9 Summary of the semiotic layers of information. *Source:* Reproduced from Petri et al. [31]/IEEE/Public Domain CC BY 4.0.

- semantic information quality is about the *correctness* of the meanings associated with data, and thus, whenever this applies, to their *truth*, intended as the correspondence between the data and the state of the system that generates the data [37];
- pragmatic information quality is about the *relevance* of semantized data for some given purposes.

This layered structure of the quality of information suggests also the preconditions to ensure effective *information sharing* among IEDM actors. Indeed, in each layer, there could be a divergence (a gap) between the information delivered by the producer and that received by the user. If this occurs, the effectiveness of their communication may be compromised. Figure 24.9 summarizes the above aspects of the semiotic definition of information.

24.4.2 Measurement Information

Measurement is a process aimed at producing information and, as such, can be interpreted in a semiotic perspective [38]. The preliminary assumption is that the empirical world consists of objects (such as physical bodies, phenomena, events, and processes) that have properties, some of them with a quantitative structure [39, 40]. Measurement produces information about properties of objects: measurement results are in fact values attributed to the property intended to be measured, i.e. the measurand.

At the syntactic layer, *measurement information* (MI) corresponds to the *instrument indication*, i.e. the quantity produced in the measuring instrument as the outcome of its interaction with the object under measurement, thus for example the angular position of the needle on the scale of an analog instrument.

At the semantic layer, the relationship between instrument indications and measurand values is established, via instrument calibration that takes into account also the properties of the measurement context that influence the instrument behavior, so that MI becomes a *measurement result*. The non-ideality of the measurement may be acknowledged by choosing as measurement result, for example an interval of values, or a pair (measured value, standard measurement uncertainty), or even a probability distribution over the set of values [39].

The core of the pragmatic component of measurement is often encoded in *target uncertainty*, "specified as an upper limit and decided on the basis of the intended use of measurement results" [39, 41]: accordingly, a measurement result is considered to be useful to support DM if measurement uncertainty is less than target uncertainty. However, together with target uncertainty several other criteria concur to the pragmatic layer of MI and are essential to enable an effective IEDM process, as shown in Table 24.1. This layered structure of MI is summarized in Figure 24.10, which specifies Figure 24.9 to the case of measurement.

The semiotic layers discussed above provide an effective tool for understanding QoMI and supporting users in defining and assessing QoMI as the result of a top-down analysis, in which the compresence of multiple criteria, i.e. different aspects related to the fitness for use of MI, is

Table 24.1 QoMI framework: syntactic, semantic, and pragmatic general-purpose criteria.

Semiotic layers and related criteria	Explanation: degree to which measurement information (MI)...
Syntactic quality	... conforms to integrity rules related to instrument indications (meaning and value of MI are still not considered)
Domain integrity	... is reported with values belonging to the scales of instrument indications and with no missing values
Indication scale resolution	... allows the detection of small changes in the instrument indications
Semantic quality	... provides a trustworthy representation of the measurand; it pertains to the meaning of instrument indications as interpreted by the MI producer (value of MI is still not considered)
Object identification	... is univocally referred to the objects under measurement
Measurand identification	... is about the measurand, and therefore is independent of influence properties
Intersubjectivity	... has a meaning that is unambiguously interpretable by all intended users
Measurement scale resolution	... allows the detection of small changes of the measurands
Context-awareness	... provides an effective description of the empirical context in which measurements are performed
Domain consistency	... conforms to property-related domain conditions
Time consistency	... is coherent between repeated measurements of the same property
Conciseness	... is free of useless or misleading content (e.g. non-significant digits or a uselessly high reporting rate)
Pragmatic quality	... fulfills the requirements of the IEDM problem; it pertains to the value of measurement results as interpreted by the MI user; it depends on user's knowledge and IEDM context
Specificity	... is sufficiently specific to support IEDM because measurement uncertainty is less than target uncertainty
Relevance	... is relevant to support effectively IEDM
Sufficiency	... includes all components required to support IEDM
Confidentiality	... is accessible only by authorized users
Security	... is protected against unauthorized access, use, corruption damage, or modification
Timeliness	... is available to users within the time intervals required by IEDM
Accessibility	... is easily and quickly retrievable by authorized users
Immediate usability	... can be used directly to support IEDM, without a need of organizing or pre-processing it

Source: Reproduced from Petri et al. [31]/IEEE/Public Domain CC BY 4.0.

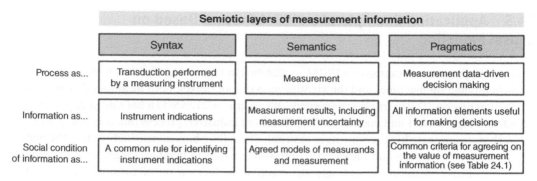

Figure 24.10 Summary of the semiotic layers of information in the case of measurement. *Source:* Reproduced from Petri et al. [31]/IEEE/Public Domain CC BY 4.0.

acknowledged [Ch. 14, 35], [42, 43]. The syntactic, semantic, and pragmatic criteria listed in Table 24.1 contribute to the overall, meta-criterion of validity of MI for a specific IEDM problem.

On this basis, methods and tools for QoMI assessment can be devised and then operationalized [31], and finally, the actually achieved confidence level of the IEDM conclusion (see Figure 24.1) assessed.

Observe also that the assessment of QoMI usually involves multiple actors, including the measurement process designers who choose the method, the procedure, and the measuring system [39], the agents who perform the measurement, the third-party auditors who validate the process, the decision-makers who use measurement results to support their conclusions, and the end users of measurement results, who can be also the society at large, as occurs for example in case of measurements for environmental protection. Each of these actors has its own, stated, or implicit, needs: QoMI evaluation and management is then aimed at meeting, and possibly anticipating, such needs [44–46]. Furthermore, the involvement of multiple actors adds to QoMI a critical requirement of internal coherence, i.e. the condition that information is not self-contradictory. To this purpose some redundancy in the acquired information can be helpful, by allowing more effective checks of internal coherence, as for example developed for the AHP method [10].

24.4.3 Factors Affecting Decision Confidence

Each IEDM stage described in Figure 24.2 may be affected by various errors that have an impact on the decision confidence, and that can be identified by means of a framework like the one introduced above. For example, the problem could be wrongly understood due to several causes, including wrong stakeholders identification, problem boundaries or scope set too narrowly, wrongly defined or conflicting constraints, causation inferred when only correlation exists in the relationships among the relevant variables. A poorly or wrongly understood situation is a major problem in IEDM, resulting in a wrongly or vaguely defined goal. Moreover, even when the problem has been rightly understood, the action identified to solve it could be wrong, for example if the identified action actually solves the problem but produces side effects that create other problems even possibly bigger than the original one. The wrong identification of the resources to be exploited for the implementation of the strategy is another possible error.

Of course, poor information quality or poor quality of the model with which the information is processed affects the decision confidence.

24.5 Application Example of an IEDM Process Based on Causal Modeling

This section proposes an example of IEDM, as sketched in Figure 24.2. It refers to the detection of faults in production machines and the identification of root causes, both meaningful goals in any production process. The example is presented with specific reference to a hot mill strip process, as depicted in Figure 24.11 and taken from [47]. This is an industrial process that uses a sequence of rolling bars to laminate high-temperature material. The rolling bars increasingly reduce the material thickness by applying and measuring controlled rolling and bending forces, so as to reach a target finishing dimension. Accordingly, the machine includes a set of finishing mills, whose actions are monitored to detect and isolate malfunctioning operations. The problems to be solved are to decide whether the final product meets the specified tolerances and quality characteristics and, if not, which finishing mills are responsible for the failed operations.

The main stages of this problem-solving process can be described as shown in the following with reference to Figure 24.2, where for the sake of simplicity, feedback loops among activities are not evidenced.

The first stage of *understanding of the situation* requires the analysis of the process to list the critical variables, their interrelation, and their effect on the product quality characteristic identified as critical for the producer. Faults and failure modes are classified, and their observability, severity and probability of occurrence are analyzed, possibly using FMEA and/or FMECA-based approaches, thus performing a risk analysis, in which in particular critical-to-measure parameters are listed in the case of hard-to-detect faults. Faults are prioritized with respect to predefined criteria, such as those based on the calculation of the risk priority number, and faults with impact on business and operations are taken as candidates for monitoring and maintenance activities.

In *deciding the goal*, the main issue is to understand the lamination process to guarantee the production of conforming products having an adequate shape quality. When it is decided that the process is malfunctioning, cascaded issues require the identification of the faulty process components and of the related root causes, and the choice of the methods to remove the faults and solve the causes, with the aim to achieve process recovery. In this example, correlation functions among features measured at each mill, such as forces and thicknesses of material, may become objective functions to be monitored. Alternatively, statistics associated with the usage of statistical process control procedures, such as control charts, may play a similar role. To understand the situation and reduce the risk of wrong decisions, this choice may require modeling the process behavior to identify causal relationships between controlled and uncontrolled variables and observed outcomes. Modeling also includes the definition of the quantities to be measured, and the instance of the QoMI framework to be applied to each measurement, so as to ensure validity of MI. An example of the application of this approach leads to the realization of the causal graph shown in Figure 24.12, reproduced from [47]. Accordingly, hot-temperature steel is moved by transfer bars in the milling

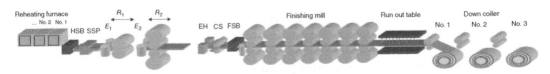

Figure 24.11 The hot milling lamination process illustrating the role of the finishing mills and additional processing components. *Source:* Reproduced from Dong et al. [47] (© [2017] IEEE).

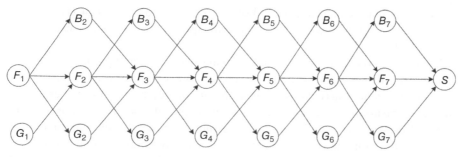

Figure 24.12 The causal graph showing the interrelation and causality relations among the roll gaps G_i, the rolling forces F_i, and the bending forces B_i, associated with each of the 7 mills. The outcome is represented by the product shape quality level S. *Source:* Reproduced from Dong et al. [47]/IEEE/Public Domain CC BY 4.0.

area, where seven stands provide the final thickness. Each roll is characterized by process control variables (roll gap G_i, rolling force F_i, bending force B_i) to obtain a final level of shape quality S. All process-controlled variables are measured. The model also relates variables values to possible milling malfunctions resulting in a poor product shape quality.

A *choice of the strategy* is made based on the knowledge and understanding of the process behavior and on the choice of the goal, and methods are defined to make decisions about possible malfunctions and associated causes. In the example, statistics are selected, and related thresholds defined that unambiguously allow DM about faulty or unfaulty process behavior. If a fault is detected, methods are adopted to find the causes. As an example, based on the MI, the causal graph in Figure 24.11 helps in identifying the root causes, by relating measured values to applied forces and by identifying the milling rolls that most probably resulted in the out-of-tolerance product status.

Once models and strategy are set, the *strategy is implemented*. The process is operated and the process variables (roll gap, rolling forces, bending forces) are measured. If the evidence is obtained of an out-of-tolerance situation, the causal graph is used to visualize cause-and-effect relationships among controlled variables and measured outcomes. A decision is made based on set statistics, in turn, calibrated using historical data and experience. Out-of-control rolling mills are detected and their in-control status reinstated.

Validation of the outcome is based on continuous monitoring of the produced decisions, to give feedback knowledge to improve operations and reduce risks of wrong decisions. Based on the outcomes, validation of the DM process is attained, as trust is progressively associated with its outcomes. Statistics may possibly be collected to provide a quantitative assessment of the process validity in terms of percentage of correct decisions.

24.6 Conclusions

Human beings are decision-makers, and engineering gives them methods and tools to become better decision-makers by effectively acquiring, processing, and presenting relevant information. Such IEDM reduces the risk of wrong decisions, as compared to conclusions based on intuition or untested beliefs, assuming that both the decision process and the exploited information are of sufficient quality.

Real-life problems are sometimes poorly understood and ill-structured, and this adds a layer of complication to their (high or low) complexity. The methods briefly presented here support the individual and, even more important, collective endeavor to understand (Section 24.2) and structure (Section 24.3) the decision processes and to structure and qualify the information exploited in such processes (Section 24.4).

Admittedly, when complex (i.e. non-single-aimed, partially-structured, and partially-informed) decisions are of concern, the best solution often remains the one ensured by a combination of human experience and mathematical (today typically computer-based) techniques. In these cases, the lack of a complete formalization is in itself a cause of complexity that leaves the doors open to systematic patterns of deviation from rational reasoning and makes a sufficient objectivity and intersubjectivity of the decisions harder to achieve. While entirely eradicating such cognitive biases might be impossible, the methods and tools of IEDM support the continuous improvement of a critical thinking mindset.

References

1 Anderson, C. (2008). The end of theory: the data deluge makes the scientific method obsolete. *Wired* 16: 1–3. https://www.wired.com/2008/06/pb-theory.

2 Pearl, J., Glymour, M., and Jewel, N.P. (2019). *Causal Inference in Statistics: A Primer*. Wiley.

3 Breiman, L. (2001). Statistical modeling: the two cultures (with comments and a rejoinder by the author). *Statistical Science* 16 (3): 199–231.

4 Schölkopf, B., Locatello, F., Bauer, S. et al. (2021). Toward causal representation learning. *Proceedings of the IEEE* 109 (5): 612–634.

5 Mari, L. and Petri, D. (2017). The metrological culture in the context of big data: managing data-driven decision confidence. *IEEE Instrumentation and Measurement Magazine* 20: 4–9.

6 Pearl, J. and Mackenzie, D. (2018). *The Book of why: The New Science and Effect*. Penguin.

7 Roy, B. (2013). *Multicriteria Methodology for Decision Making*. Springer Science & Business Media.

8 Keeney, R.L. and Raiffa, H. (1993). *Decisions with Multiple Objectives: Preferences and Value Trade-Offs*. Cambridge University Press.

9 Eisenführ, F., Weber, M., and Langer, T. (2010). *Rational Decision Making*. Springer.

10 Saaty, T.L. (2000). *Fundamentals of Decision Making and Priority Theory with the Analytic Hierarchy Process*. RWS Publications.

11 Ureña, R., Chiclana, F., Morente-Molinera, J.A., and Herrera-Viedma, E. (2015). Managing incomplete preference relations in decision making: a review and future trends. *Information Sciences* 302: 14–32.

12 Brunelli, M. (2015). *Introduction to the Analytic Hierarchy Process*. SpringerBriefs in Operations Research, Springer.

13 Vaidya, O.S. and Kumar, S. (2006). Analytic hierarchy process: an overview of applications. *European Journal of Operational Research* 169 (1): 1–29.

14 Lai, V.S., Wong, B.K., and Cheung, W. (2002). Group decision making in a multiple criteria environment: a case using the AHP in software selection. *European Journal of Operational Research* 137 (1): 134–144.

15 Yoon, K.P. and Hwang, C.L. (1995). *Multiple Attribute Decision Making: An Introduction*. Sage Publications.

16 Matsatsinis, N.F., Grigoroudis, E., and Siskos, E. (2018). Disaggregation approach to value elicitation. In: *Elicitation*, vol. 261 (ed. L. Dias, A. Morton, and J. Quigley), 313–348. Springer.

17 Siskos, Y., Grigoroudis, E., and Matsatsinis, N.F. (2016). UTA methods. In: *Multiple Criteria Decision Analysis*, vol. 233 (ed. S. Greco, M. Ehrgott, and J.R. Figueira), 315–362. Springer.

18 Kadziński, M., Martyn, K., Cinelli, M. et al. (2020). Preference disaggregation for multiple criteria sorting with partial monotonicity constraints: application to exposure management of nanomaterials. *International Journal of Approximate Reasoning* 117: 60–80.

19 Chen, Y., Kilgour, D.M., and Hipel, K.W. (2008). A case-based distance method for screening in multiple-criteria decision aid. *Omega* 36 (3): 373–383.

20 Keeney, R.L. (1996). *Value-Focused Thinking*. Harvard University Press.

21 Deb, K., Pratap, A., Agarwal, S., and Meyarivan, T.A.M.T. (2002). A fast and elitist multiobjective genetic algorithm: NSGA-II. *IEEE Transactions on Evolutionary Computation* 6 (2): 182–197.

22 Goldberg, D. (1988). *Genetic Algorithms in Search, Optimization and Machine Learning*. Addison-Wesley Professional.

23 Bandyopadhyay, S. and Bhattacharya, R. (2013). Solving multi-objective parallel machine scheduling problem by a modified NSGA-II. *Applied Mathematical Modeling* 37 (10–11): 6718–6729.

24 Brans, J.P., Vincke, P., and Mareschal, B. (1986). How to select and how to rank projects: the PROMETHEE method. *European Journal of Operational Research* 24 (2): 228–238.

25 Figueira, J.R., Greco, S., Roy, B., and Słowiński, R. (2013). An overview of ELECTRE methods and their recent extensions. *Journal of Multi-Criteria Decision Analysis* 20 (1–2): 61–85.

26 Greco, S., Ehrgott, M., and Figueira, J.R. (2016). *Multiple Criteria Decision Analysis: State of the Art Surveys*, International Series in Operations Research & Management Science. Springer.

27 Ishizaka, A. and Nemery, P. (2013). *Multi-Criteria Decision Analysis: Methods and Software*. Wiley.

28 Roy, B. and Słowiński, R. (2013). Questions guiding the choice of a multicriteria decision aiding method. *EURO Journal on Decision Processes* 1 (1): 69–97.

29 Cinelli, M., Kadziński, M., Gonzalez, M., and Słowiński, R. (2020). How to support the application of multiple criteria decision analysis? Let us start with a comprehensive taxonomy. *Omega* 96: 102261.

30 Standard ISO 9000: 2015 (2015). *Quality Management Systems - Fundamentals and Vocabulary*. Standard ISO 9000: 2015.

31 Petri, D., Mari, L., and Carbone, P. (2021). Quality of measurement information in decision-making. *IEEE Transactions on Instrumentation and Measurement* 70.

32 Morris, C. (1946). *Signs, Language, and Behavior*. New York: Prentice-Hall.

33 Bar-Hillel, Y. and Carnap, R. (1953). Semantic information. *The British Journal for the Philosophy of Science* 4 (14): 147–157.

34 Hintikka, I. (1970). On semantic information. In: *Physics, Logic, and History* (ed. W. Yourgrau and A.D. Breck), 147–172. Boston, MA: Springer.

35 Floridi, F. and Illary, P. (2014). *The Philosophy of Information Quality*. Cham: Springer.

36 Zhong, Y. (2017). A theory of semantic information. *Proceeding of Summit Digitalisation Sustainability* 1 (3): 1–12.

37 Tarski, A. (1944). The semantic conception of truth and the foundations of semantics. *Philosophy and Phenomenological Research* 4: 341–395.

38 Mari, L., Wilson, M., and Maul, A. (2021). *Measurement across the Sciences – Developing a Shared Concept System for Measurement*. Springer Nature.

39 Standard JGCM 200:2012 (2012). International vocabulary of metrology-basic and general concepts and associated terms (VIM), joint committee for guides in metrology. https://www.bipm.org/en/publications/guides/vim.html (accessed 21 March 2023).

40 Mari, L., Carbone, P., and Petri, D. (2012). Measurement fundamentals: a pragmatic view. *IEEE Transactions on Instrumentation and Measurement* 61 (8): 2107–2115.

41 Petri, D., Mari, L., and Carbone, P. (2015). A structured methodology for measurement development. *IEEE Transactions on Instrumentation and Measurement* 64 (9): 2367–2379.

42 Standard ISO 8000-8: 2015 (2015). *Data Quality – Part 8 – Information and Data Quality: Concepts and Measuring, Geneva*. Standard ISO 8000-8: 2015.

43 Standard ISO/IEC 25012:2008 (2008). *Software Engineering – Software Product Quality Requirements and Evaluation (Square) – Data Quality Model*. Standard ISO/IEC 25012: 2008.

44 Raghunathan, S. (1999). Impact of information quality and decision-maker quality on decision quality: a theoretical model and simulation analysis. *Decision Support Systems* 26 (4): 275–286.

45 Lillrank, P. (2003). The quality of information. *International Journal of Quality & Reliability Management* 20 (6): 691–703.

46 Gustavsson, M. and Wänström, C. (2009). Assessing information quality in manufacturing planning and control processes. *International Journal of Quality & Reliability Management* 26 (4): 325–340.

47 Dong, J., Wang, M., Zhang, X. et al. (2017). Joint data-driven fault diagnosis integrating causality graph with statistical process monitoring for complex industrial processes. *IEEE Access* 5: 25217–25225.

25

Data Management: An Enabler of Data-Driven Decisions

Cecilia Poittevin, Javier Barreiro, Gustavo Mesa, and Laura Rodriguez Mendaro

DAMA Uruguay, Montevideo, Uruguay

25.1 Introduction

The abundance of data and that they are "the oil of the twenty-first century" is no longer news to anyone. But while the abundance of data is an enabler of new initiatives in organizations, successful organizations will be those that manage and exploit it reliably. Grab information on the performance of your processes, get new insights for the strategy review or simulations on possible future scenarios are different benefits that reliable and well-managed data can provide.

But reliable data is not produced by accident. In today's complex world, well-managed data depends on planning and design, governance of business and technical processes, and your organization's commitment to high-quality results.

This chapter will introduce the main concepts for any organization to improve its data management capabilities. Among them, as stated in the book "Navigating the Labyrinth: an executive guide to Data Management" [1] of Data Management Association (DAMA) International [2], can be considered:

- **Data's ubiquity**
  Almost every business process uses and produces data. Most of this data is in electronic form, which means that data can be stored in large quantities, manipulated, integrated, and aggregated for different uses, including business intelligence and predictive analytics. It also provides evidence of an organization's compliance (or lack of compliance) with laws and regulations. The rapid growth of technology to produce, capture, and mine data for meaning has intensified the need to manage data effectively.
- **Data's value as an asset**
  The primary driver for data management is to enable organizations to get value from their data, just as effective management of financial and physical assets enables organizations to get value from those assets. Data is an asset, which means deriving value from data requires organizational commitment and leadership, as well as management.

IEEE Technology and Engineering Management Society Body of Knowledge (TEMSBOK), First Edition.
Edited by Gustavo Giannattasio, Elif Kongar, Marina Dabić, Celia Desmond, Michael Condry, Sudeendra Koushik, and Roberto Saracco.

- **Data management is not the same as technology management**
  IT focuses on technology, technological processes, the people who build applications, and the tools they use to do so. Habitually, IT does not focus on the data that is created by or stored in the applications it builds, despite the fact that many data management functions are part of IT. Though data management is highly dependent on technology and intersects with technology management, it involves separate activities that are independent from specific technical tools and processes.
- **The range of activities and functions involved with managing data**
  Data management involves planning and coordinating resources and activities in order to meet organizational objectives. The activities themselves range from the highly technical, like ensuring that large databases are accessible, performant, and secure, to the highly strategic, like determining how to expand market share through innovative uses of data. These management activities must strive to make high-quality, reliable data available to the organization while ensuring this data is accessible to authorized users and protected from misuse

The adoption of these concepts, as well as the implementation of best practices in data management, will be some of the most relevant differentials in those leading organizations that identify and generate evidence-based actions. Specially, evidence that their own reliable and well-managed data provides them.

25.2 Knowledge Area Fundamentals

Why should we manage the data of an organization? Because data is an asset, and as such, if well managed, it will be an opportunity for the business and the organization. But at the same time, they can become a problem because the large amount of data we collect can easily overwhelm us. It is as if we were on a shore and a tsunami was coming upon us . . . of data. More than ever, today obtaining data is a relatively simple task, but to obtain value from it, to really take advantage of it and turn it into a critical asset for the organization, we must work on it through data management practices.

Data management will allow us to have reliable and timely information to make better business decisions, for example, knowing the behavior of our clients or identifying trends. On the other hand, the large accumulation of data, in particular personal data, has made regulations evolve and become increasingly demanding. Data management is also necessary to comply with these regulations.

As we said, in that problem also lies an opportunity, and that opportunity is to have data to obtain value, but to obtain that value we must manage it. Data management will allow us to have reliable and timely information, which leads to make better decisions, knowing the behaviors and trends of our clients, as well as complying with the regulations.

Returning to the example of the tsunami, if we can manage this large amount of water, channel it, organize it, contain it, and control it, we will be able to obtain energy from water. In other words, through water management, we will be able to obtain value in the form of energy.

The same can be said about data. Today we have technology that allows us to produce an endless amount of data, to be able to store it, and process it quickly, but generating, storing, and processing data is not enough, to obtain value we must manage it correctly.

So, what does it mean to manage the data? A reference definition can be found in DaMa International's Data Management Book of Knowledge (DMBoK) [3]. The DMBoK defines:

> Data management consists of the development, execution and supervision of plans, policies, programs and practices that allow to deliver, control, protect and increase the value of data during its life cycle.

In other words, we could say that data management is a discipline that performs tasks (planning, development, execution and supervision) on different elements (plans, policies, programs, and practices) in order to deliver, control, protect, and increase the value of the data. In this way, we can begin to identify value in the data of our organization, always insofar as these tasks and elements are executed under a controlled process.

Continuing with the definitions, we can define "data" as the minimum unit of information with which we work; the data alone does not contribute anything, we cannot make decisions solely with it, regardless of the number of data we have. The data simply helps us to have an understandable representation of certain tangible or abstract characteristics. For example, a piece of data with the representation of a sequence of numbers such as "29042000" and another "10,200" are simply a sequence of numbers that does not tell us anything without a context. But, if we do know that the first sequence represents a date, and that this is the billing date, and the second number is the amount of sales for that day that data is transformed into information. Knowing the context is what transforms data into information.

In this way, the information is constructed by grouping, summarizing, visualizing, and characterizing the data in different ways. With the information we can answer questions such as *"What . . .? When . . .? Where . . .? How many . . .?"*

On the next level, we could understand knowledge as the application of our experience to information, to identify, for example, possible trends in customer or sales behavior, identifying and recognizing patterns. Our experience allows us to expand the information, comparing it with other similar situations and to be able to obtain value or corroborate the hypothesis that experience can give us. For example, if we compare our day of sale with the rest of the days and the amount that was sold, we can infer that 29 April 2000 is the day that is sold the least in the month, and if we compare with the rest of the months, we see that sales always drop at the end of the month. We have found a pattern and based on experience we can think that this happens because at the end of the month there are fewer purchases because the purchasing power of households decreases. We have found here the difference of basing decisions on data and not just on intuition. We can say then that Knowledge = Information + Experience. Experience tells us how or why.

Finally, wisdom is transforming knowledge into action, seeking not to repeat the same mistakes, improving our processes, increasing sales, in short: obtaining value from data. Although knowledge has a value, being able to exploit it in action is what gives the increase and the differential of its use.

Thus, wisdom allows us to modify business processes, improving them based on knowledge. Business processes are also the basis for knowledge, and we must know the current processes to understand the information in its context and be able to generate value. Last but not least, participating in this transformation process we have technology. Technology as a support for business processes, through which data is collected, data is transformed into information, and which allows information to be analyzed in order to carry out the knowledge process. All these components working in harmony give value to the data.

Data management is the discipline that makes these components (value chain + technology and processes) work in the best possible way, managing and identifying which data is necessary, why it is necessary, who executes each step, what is the meaning of the data and information, where the data is located and what transformations occur, what is the quality of the data, etc.

We should also note that obtaining value from data is closely associated with advanced data management functions. These include the functions that seek to analyze, interpret and obtain knowledge, which are the last stages of the process.

In turn, data management also consists of non-visible functions such as planning and design functions, among others. When we do not execute these functions correctly, we can still obtain value from the data, although it will surely take much more time, will have a greater cost and effort, will deliver less value, are less repeatable, and represent a greater risk.

Finally, as we obtain value and as the data begins to be used for decisions and by a greater number of stakeholders, the risks of exposure and vulnerability also increase, so we must protect them adequately; therefore, managing security and privacy is key.

In short, data management is the discipline that enables the functions that gives organizations the possibility to have reliable, quality, timely, and secure data to derive value from it.

25.3 Managing Approaches, Techniques, and Benefits to Using Them

When approaching a data management implementation, we must rely on reference frameworks. This can be understood as a standardized set of concepts, practices, and criteria to approach a particular type of problem that serves as a reference, to face and solve new problems of a similar nature.

The different existing frameworks in data management provide different perspectives on how to approach data management. These perspectives provide a vision that can be used to develop strategies, roadmaps, organize work teams and align their functions. A data reference framework provides organizations with tools to manage, organize, evaluate, improve data. It includes a set of tools, techniques, standards, processes, and good practices related to data management.

Its main objective is based on contributing to the systematization of data management in organizations. Within a data management framework, we can find:

- A conceptual framework that facilitates understanding, sorting ideas, and managing the different concepts of data management.
- A reference model that allows identifying those components involved in management.
- Processes that guide organizations regarding the relevant activities and roles to manage data.
- Some theoretical foundations and reference and support resources.

In order to put it into practice, it is necessary to know, master and adapt the framework to the needs of the organization that will adopt it.

For example, an organization might need a data observatory and for that to face data management from that perspective. In this context, it may not be necessary to generate a structure in the organization but rather some roles that can collect, and understand your data. The frameworks are generally comprehensive and tend to cover the widest possible scope. That is why it is very important to adapt them to our needs, to avoid incorporating processes, controls, or bureaucracy, without adding value.

There are several frames of reference for data management, the choice of which will depend on factors such as the industry where it is handled, the range of data it uses, the culture, the vision, the strategy, and the specific objectives that are used.

DAMA International, with the aim of supporting data management professionals, developed the Data Management Body of Knowledge (DMBOK) as a reference framework. This framework was

Figure 25.1 The DAMA-DMBOX2 guide knowledge area wheel.

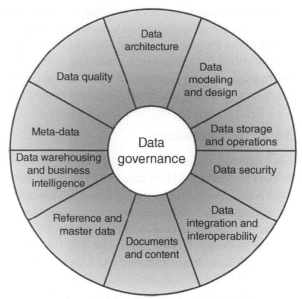

built based on the experience and knowledge of the community of professionals of DAMA to provide a reference for the implementation of Data Management business practices, including principles, practices, methods, functions, roles, deliverables, and metrics. In turn, it seeks to establish a common vocabulary for Data Management concepts and serve as a basis for best practices for Data Management professionals.

The frame of reference consists of three main elements:

- The DAMA Wheel that defines the Knowledge Areas of Data Management. Each Knowledge Area is a necessary part of a mature Data Management function.
- The Hexagon of Environmental Factors, which shows the relationship between people, processes, and technology. All activities are based on these three key pillars, which must be impacted with data management, and which are strongly interrelated.
- Finally, the Knowledge Area Context diagrams, which describe the detail of the Knowledge Areas, including details related to people, processes, and technology. They are based on the concept of a SIPOC diagram used for product management that defines Suppliers, Inputs, Processes, Products, and Consumers, adapted for data management.

These three artifacts are the framework on which to rely to generate data management initiatives. We will now describe in greater detail each of the areas of knowledge proposed by DAMA (Figure 25.1).

25.3.1 Data Governance

Data Governance is the area of knowledge that exercises authority and control over the management of data assets. All organizations make data decisions, those with a formal Data Governance function, do so "intentionally" or "consciously," and are better able to derive value from their data.

Data governance focuses on how decisions are made regarding data: how it is created, what people expect to do with the data, and the objectives that the processes associated with data management must meet. Like how accountants define the rules for working with financial assets, data governance professionals define the rules for working with data assets.

Data governance is key in establishing an effective data management program, and that is why it is at the center of the wheel because it is the function that coordinates and directs the efforts of the rest of the knowledge areas, to achieve common goals.

In the area of data governance, it must be defined what data governance means for the organization through the development of both the data strategy and the operational structure of data governance and the goals, principles, and policies that guide them.

25.3.2 Data Architecture

Data architecture describes how data should be structured, how data should be integrated into business processes, and how data should be controlled so that an organization can manage it effectively. A data architecture program serves as a blueprint for data management in the same way that a building plan provides a blueprint for the construction of a building.

A well-developed data architecture is essential for data to function as part of an organization's business processes. It is the responsibility of this area to document the design of the data architecture, which describes conceptual and logical entities and their relationships and business rules. This model is the fundamental roadmap for managing an organization's data.

Well-developed data architecture acts as a bridge between business strategy and technology execution. The data architecture is not independent but exists in relation to other organizational architectures such as business, applications, and technology. Each one depends on the others.

25.3.3 Data Modeling

Data modeling is a critical component of data management: it is the process of discovering, analyzing, and scoping data requirements. As a result of this process, data requirements are represented and communicated through a data model.

Data models are essential for effective management, as they:

- Provide a common vocabulary about the data used by several stakeholders in an organization.
- Capture and document explicit knowledge about an organization's data and systems.
- Serve as a primary communication tool during projects.
- Provide the starting point for configuring, integrating, or even replacing an application.

Data models enable an organization to understand its data assets and their movements through the various systems of the organization. Proper data modeling leads to lower support costs and increases an organization's ability to reuse data in future initiatives, which can reduce the costs of creating new applications.

25.3.4 Data Storage and Operations

Data storage and operations activities are crucial for organizations, business continuity is a critical driver of these activities. If a system becomes unavailable, operations can be affected or stopped altogether. A reliable data storage infrastructure for operations provides organizations with access and availability to data.

Data storage and operations include the design, implementation, and support of stored data to maximize its value throughout its lifecycle, from creation or acquisition to disposal.

This area of knowledge is a type of technology function, being one of the most mature areas since, nowadays, it is unthinkable to do without computerized data storage.

25.3.5 Data Security

Data Security ensures that data privacy and confidentiality are maintained that there are no data leaks, and that data is properly accessed.

Data security involves planning, developing, and executing security policies and procedures to provide adequate protection and access to data. It includes understanding and classifying data, controlling access to data, and auditing it. Data security practices work in accordance with privacy and confidentiality regulations, contractual agreements, and business requirements to protect an organization's assets.

Effective data security is essential to reduce the risk of organizational loss and maintain trust in the organization. The impact of security breaches and other incidents on organizations leads to a significant financial loss and a significant loss of customer trust.

While data management professionals have an important role to play in ensuring that data remains secure, effective data security requires the involvement of everyone who interacts with the data.

25.3.6 Data Integration and Interoperability

Data integration and interoperability is the area of knowledge that ensures that data is located where it is needed, available when it is needed, and in the way, it is needed. Data integration is the process of transforming data into consistent forms and migrating transformed data from source systems to destination systems. Data interoperability ensures that data is consistent so that it can be shared, migrated, and consumed by other systems. Data integration and interoperability describe the processes related to the movement and consolidation of data within and between data warehouses, applications, and organizations.

A primary driver for data integration and interoperability is the need to manage data exchange efficiently. As the amount and different types of data collected by organizations increase, so does the number of disparate data warehouses. If data is not properly integrated, the process of sharing data across multiple systems can overwhelm IT resources, drain budgets, and drain support services. Transforming and integrating data between multiple systems is an essential task of any organization to obtain value.

25.3.7 Document and Content Management

Document and content management focuses on controlling the capture, storage, access, and use of data and information stored outside of relational databases. Document and content management maintains integrity and enables access to unstructured or semi-structured documents and other information. Documents and unstructured content are expected to be secure and of high quality, requiring governance, trusted architecture, and well-managed metadata.

Both business and technical concerns drive the need for effective document and content management. The main drivers include:

- Laws and regulations that require organizations to keep records of certain types of activities. Organizational policies, standards, and best practices for record-keeping help ensure that records are properly maintained.
- Electronic discovery and litigation, often requiring documents to be provided.
- Business continuity, which requires planning, practice, and good records management to re-establish business operations.

Technological advancements in document management can help organizations streamline processes, manage workflow, eliminate repetitive manual tasks, and enable collaboration. These technologies make it possible for people to quickly locate, access and share documents, and they also help prevent documents from being lost.

25.3.8 Master and Reference Data

Master and Reference Data is the knowledge area that is responsible for the reconciliation and continuous maintenance of critical data in an organization to allow consistent use across systems and have a "single source of truth": an only accurate, timely, and relevant version of the data.

Reference data is data that is used to characterize and define other data (a simple example of reference data is a list that identifies which three-digit code a country represents). The goal of reference data management is to ensure that an organization has access to a complete set of accurate and current values for each concept represented in the master data. Using standard reference data, such as ISO codes, when possible, also simplifies interoperability with other organizations.

Master data is data about business entities (real-world objects, such as customers, products, employees, or suppliers) that provide the context for your organization's transactions and analysis. They are the critical entities that are shared between different areas and systems of an organization.

The goal of master data management (MDM) is to ensure that the values and identifiers of this data are consistent across systems and represent the most accurate and timely data about each entity.

Without reference data management and MDM, the same data can be represented in different ways. Not having a consistent understanding and agreement for both baseline data and master data will lead to inefficiency, leading to inconsistencies. Inconsistency leads to ambiguity and ambiguity presents a risk to an organization.

For example, identifying a customer by document number in one system and by customer number in another system makes it impossible to uniquely identify the same person since they can be represented in two ways. Also, if the country in one system is the ISO code and in another the telephone code implies an additional integration effort when unifying the data. That is why reference and MDM are key to facilitating data integration, interoperability, and use.

25.3.9 Data Warehousing and Business Intelligence

The data warehousing (DWH) and Business Intelligence area involves planning, implementation, and control processes to manage data to support decision-making and obtain value from the data through analysis and reports.

DWHs integrate and store data from a variety of sources in a common data model. Business intelligence is based on performing different types of analysis on the DWH in order to improve decision-making in an organization.

Organizations create DWH because they need to make reliable, integrated data available to a variety of stakeholders.

Supporting business intelligence is the primary rationale for a data warehouse, and business intelligence relies on extracting information from data warehouses through analytical processing.

The more a data warehouse can be aligned with business intelligence needs, the more data can be used to improve operations. Business stakeholders must be included in every stage of a data warehouse project, because they are the ones who know what data needs to be included for effective business intelligence.

Today we also talk about Data Lake and Data Science. These concepts are similar and complementary to DWH and Business Intelligence for data with higher volume, speed, and variety requirements, as well as other analytical techniques. This knowledge area also applies to the concepts of Data Lake and Data Science.

25.3.10 Metadata

The most common definition of metadata is "data and content about data." While accurate, the definition is deceptively simple.

Metadata can include information about technical and business processes, rules and constraints, and logical and physical data structures. The data source where information is obtained, its security, privacy, and preservation requirements, how is the process of creating that data, or how is the calculation method of an indicator. These are all examples of metadata.

Metadata is generally classified into three types: descriptive (describes the content of the data), structural or technical (information about the technical details and the systems that store the data), and administrative or operational (details about the processing and access of the data).

Like other data, metadata requires management. As organizations' ability to collect and store data increases, the importance of the role of metadata in data management increases.

The Metadata management area includes planning, implementation, and control activities to allow access to high quality integrated Metadata, including definitions, models, data flows and other information critical to understanding the data and the systems through which it is used. Create, maintain, and access them.

Mctadata helps an organization understand its data, its systems, and its workflows.

In an organization, different individuals will have different levels of knowledge, but no one individual will know everything about all the data. The information must be documented, or the organization runs the risk of losing valuable knowledge about itself.

25.3.11 Data Quality

For the data to be useful, it must be reliable, and of high quality, this means that it complies with what is expected according to the purpose of its use. The quality of the data depends on the context; the same data that is of high quality in one situation (that is, it meets a certain need) may be of low quality in a different situation (if it does not meet the need). For example, suppose a data set that contains information about customers and their contact details which we do not know how old they are, and which have completeness problems. This data is of quality if the purpose is to count the number of customers, but it is of poor quality if we want to contact them.

The quality of the data is essential to ensure that the result of any analysis based on that data is valid. If the data does not meet quality standards, the result of an analysis using that data would be unreliable or even invalid.

The Data Quality area includes the planning and implementation of quality management techniques to measure, evaluate and improve the suitability of data for use within an organization.

Data quality management is a recurring program over time that must identify critical data, its purpose, and rules that must be met, validate compliance, analyze problems to determine causes and possible improvements, perform data cleaning and enrichment, and ensure continuous monitoring to carry out a proactive management of them.

25.4 Realistic Examples

Examples will be presented in this section to convey how the practices mentioned above can add value to different projects.

Some time ago, with the objective of reaching reports with timely and accurate information, a well-known clothing brand embarked on a project that allowed it to cut in half the effort to generate its financial reports. Among its main needs to achieve its implementation, the need for a central repository of product and warehouse data with easy access by data administrators without significantly depending on the technology area of the organization was identified. These administrators are the people in the organization who know the data best, so it was necessary to give them the capabilities to add or modify data and attributes when appropriate, thus allowing the reports to be generated more efficiently.

This simple change allowed them to identify and eliminate erroneous and inconsistent data accumulated over the years in different types of applications, managing to focus on reliable data to drive analytics. In turn, with a continuous data management strategy to improve sales and financial analysis, it was possible to create a standardized data set in a central repository, available for multiple uses in all departments.

Also supported by a data management strategy, an American food chain managed after several years to build a reliable, clean, and accurate database in multiple systems. For years, his main concern was making decisions with incorrect data, since his operation includes the recording and harmonization of data from multiple legacy systems, including POS, touch devices, and various third-party tools. Among other limitations, these challenges meant they were unable to create recurring customer records to manage the loyalty program and drive better-targeted marketing campaigns.

After an orderly process supported by decision-makers, the American chain managed to establish a "data culture" in the organization, allowing significant savings, elimination of waste, timely decision-making, among others. As one of the pillars to fulfill these achievements, the creation of a master database (MDM) was identified, a basis for each of the different regions in which the chain operates to achieve, for example, make modifications to its gastronomic offer without generate inconsistencies when obtaining aggregate information at the corporate level.

As mentioned above, the MDM ensures that the values and identifiers of this data are consistent across all systems and that they represent the most accurate and timely data on each of the business entities. This type of data is used in multiple processes of the organization, providing a complete and consistent view of this data to make the appropriate decisions.

The management of reference and master data allows the organization to ensure that the most important data for the business is created with the necessary consistency and quality. In turn, this standardization was key to guarantee the integration and interoperability of the different systems.

The banking sector has also been one of the large sectors in which data management practices have been strongly incorporated, mainly driven by the regulatory need associated with the effective identification of risks to which the organization is exposed. For this, having quality and always accessible data is one of the basic requirements for operations. This includes not only internal data of the organization (customers, transactions, accounts) but also integration with various external sources, which can have high levels of volatility (quotes, reports from regulatory entities, etc.), so that the need for fresh data is crucial for decision making.

This need entails facing various challenges such as the lack of tools to ensure defined data quality policies, the need to have traceability mechanisms so that decisions can be audited, avoid the existence of incomplete data that affect calculations and estimates, etc.

For this, it is necessary to implement various initiatives, such as defining and disseminating clearly and visibly throughout the organization the policies and procedures to be applied to ensure acceptable data quality levels. Among other rules, these may include the use of international standards for reference data, facilitating reports related to data quality ratings, thus enhancing the trust in them by their users, defining and implementing mechanisms for reporting inconsistent data, as well as the processes necessary for its attention and correction, among others. In all cases, these processes, procedures, and rules must remain centralized in the organization, being easy to modify in the event of changes in the environment, whether associated with operational, strategic, or regulatory adaptation needs.

This type of action must be accompanied by a shared vision of data management activities, which must be accepted and adopted by all areas of the organization, thus allowing a common language to be achieved. In this way, a better understanding of the available business data can be achieved as well as more consistent governance throughout the organization.

There are also success stories related to data management in the government area. During the last decade, Uruguay experienced a period of significant, tangible, and visible digital development in various fields. Guaranteeing the population's access to digital technologies and bridging the digital divide has been a priority for the country, addressed through different public policies in order to guarantee the rights of all social sectors to the opportunities offered by the society of information and knowledge. Uruguay has developed a platform for digital government that defines and implements standards and solutions for the development of the digital transformation of government agencies.

The digital ecosystem is supported by a legal framework that regulates the exchange of information and security that all public entities must adopt. It also regulates the protection of personal data, mechanisms for accessing public information and establishes the principles for the correct management of data.

One of the main objectives pursued in the Uruguayan digital government strategy is the development of an intelligent government through the intensive use of data, predictive analytics for proactive services, and evidence-based decision-making. This requires extensive use of citizen data, personal data, and consent is a key issue in these data exchange model designs. It always depends on the purpose for which the personal data was provided, as well as the provisions and protection of the law.

The development of a National Strategy and Policy during 2018 [4] gave guarantees to Uruguayan public organizations in the use of data for the formulation of public policies. Not only in the adoption of a data-driven government, but also to establish fundamental principles, such as: Data as a government asset, Data quality management, Data interoperability, Privacy by design, Open data by default and Preservation of data.

Currently, one of the main concerns of the Uruguayan government regarding data is its ethical use. The objective of data ethics in the context of digital government is to promote its responsible and sustainable use for the benefit of society, thus guaranteeing that the knowledge obtained through the data is not used against the legitimate interests of an individual or group. This is done through the identification and promotion of standards, values, and responsibilities that allow guiding whether decisions or actions are appropriate, "correct" or "good."

Several governments have developed different initiatives in relation to data ethics [4]. For example, the UK developed a framework that guides public sector organizations on how to use data appropriately and responsibly when planning, implementing, and evaluating a new policy or service [5]. In the case of Denmark, the Danish Group of Experts on Data Ethics recommends that companies provide a statement about their work on data ethics in their business activities.

Another recommendation from the Danish Group of Data Ethics Experts is that company managers and employees who work with data take an oath on data ethics [6]. The Government of Canada has created the Algorithmic Impact Assessment (AIA) is a questionnaire designed to help assess and mitigate the impacts associated with the implementation of an automated decision system [7]. Finally, Germany has adopted a risk-tailored regulatory approach. The German Data Ethics Commission recommends taking a risk-tailored regulatory approach for algorithmic systems. The Data Ethics Commission recommends that the potential of algorithmic systems to harm people and/or society be uniformly determined based on a universally applicable model [8].

In this context, Uruguay is evaluating the conformation of an ethics committee that can recommend on the use of data in public management, preparing user guides for the different profiles of public officials who work with citizen data, and in relation to with building capacities in a digital citizenry that is capable of understanding and applying the best practices in that discipline.

25.5 Recommended Best Use and Focused Decisions to Identify the Most Adequate Approaches and Techniques for Particular Cases

It is widely known that, when working with reference frameworks or methodologies, the "one size fits all" is practically a utopia. Therefore, when selecting the best approach on how to use these practices and add value, different adaptations arise to the different frames of reference.

For data management and the framework proposed by the DMBoK, this is not the exception, and there are various adaptations recognized as good practices for use in particular cases. These adaptations present a set of advantages that they provide when prioritizing and ordering work in the different areas of knowledge.

A possible adaptation to the DAMA framework is the Aiken Pyramid which organizes the DMBoK knowledge areas to describe a phased implementation path that enables data analytics on well-managed and reliable data.

Considering that most organizations begin to manage data before having the possibility of defining a strategy, the pyramid shows a trajectory allowing an organization to obtain benefits from Data Governance and Management from the first steps.

Another alternative is based on proposing a set of relationships between knowledge areas. This frame of reference begins with the guiding purpose of data management: "to provide value from data." This requires data lifecycle management, so related functions are at the center of the diagram. It includes planning and designing activities for reliable and high-quality data and establishing processes and functions by which data can be enabled for use and maintenance, and finally used in analysis for the generation of value. In turn, the direct functions of the life cycle must be supported in foundational activities, such as risk management, metadata, and data quality management.

To ensure that activities are carried out with discipline, data governance is established. This allows an organization to be data-driven, establishing a stewardship strategy, principles, policies, and practices that ensure that the organization recognizes the value it derives from its data.

Another possible adaptation is to understand the functions on the target that act and how they are related to each other, dividing between data and technology. Technology functions are those related to storage, integration, interoperability, security, BI/DWH platforms, and related to data architecture, modeling, reference, metadata, and quality. This way of looking at them also makes it possible for us to define responsibilities within the organization.

Finally, another adaptation alternative is to see the knowledge areas and how they are related to each other, dividing the activities according to the data and technology roles that carry them out. The technology roles oversee the knowledge areas related to storage, integration, interoperability, security, and data warehouse platforms. The data roles are the areas of architecture, modeling, reference, metadata, and quality. All this framed by data governance to align the roles of data and technology. This way of looking at them also makes it possible for us to define responsibilities within the organization.

25.6   Summary and Conclusions

As highlighted throughout the chapter, the primary goal of data management is to enable an organization to get more value from its data. In a world that is increasingly reliant on data, having reliable data management practices are critical to realizing value. An organization can begin to understand the value of its data by recognizing both the costs of low-quality data and the benefits of having high-quality data.

Data should be viewed as an asset, but it also represents a risk, as it has unique characteristics that make it challenging to manage. The best approach to address these challenges is to manage data throughout its lifecycle and take an organizational perspective.

Data management encompasses a heterogeneous group of activities that require different skills, from strategic to highly technical, but all necessary for effective data management. These include governance activities, foundational activities, and data lifecycle management. These practices are also constantly evolving since business needs, and technological capacity are also constantly evolving.

Data governance is an ongoing program that oversees all other data management functions, articulating strategy, setting frameworks, setting policies, and enabling data to be used throughout the organization. We must note that data governance is not an end in itself, but a means to achieve the organization's objectives. Data governance requires leadership commitment, and that commitment will also allow other data management functions to be more successful.

Other fundamental aspects for effective data management include data architecture and modeling, data security and privacy, metadata, and quality management.

Data architecture is understood as fundamental for an organization to understand its systems, its data, and the relationship between business and technical processes. Data architecture focuses on enabling an organization to understand and capture explicit knowledge about its own data.

The metadata created and managed through data architecture processes is critical to using and managing data over time.

Data modeling is critical to data management as data models define the entities that are important to the organization, concisely capture data requirements, and clarify the rules and relationships necessary to manage data and quality.

Managing data security is critical to the success of data management. Adequate data protection is necessary to meet stakeholder expectations, and it is also the right thing to do for an organization. Data that is managed according to data management best practices is also easier to protect, as it can be classified and tagged with a high degree of reliability. These practices include taking an organizational approach to security planning, establishing a trusted security architecture, and managing security-related metadata. Strong and demonstrable data security practices can be a differentiator by building trust.

Metadata management is key to data management. You cannot manage the data without the metadata. Metadata is a means by which an organization captures explicit knowledge about its data to minimize risk and increase value. Most organizations do not manage their metadata well and pay the price in hidden costs. They increase the long-term cost of managing the data by creating unnecessary readjustments with each new project, as well as the operational costs of trying to locate and use the data. We must consider metadata as data and as such have a life cycle and must be managed based on that cycle. Different types of metadata will have different specific life cycle requirements.

Data quality management is another fundamental and necessary aspect since having poor-quality data is expensive, and in contrast having high-quality data has many advantages. The quality of data can be managed and improved, as can the quality of physical products, and the cost of obtaining correct data from creation is less than obtaining incorrect data and then trying to fix it. Managing data quality requires a broad skill set and organizational commitment that has leadership committed to improving data quality by raising awareness of the importance of data and prioritizing changes needed for improvement.

Although data management is complex, it can be done effectively and efficiently. As a first step, the current state must be known through an evaluation that allows an understanding of where the organization is, what its objectives are and from that point on, plan improvement activities.

Changes in the way of managing data will change the way people work, which is why it is necessary to carry out a formal change management to achieve the cultural changes that will allow success.

References

1 Sebastian-Coleman, L. (2018). *Navigating the Labyrinth: An Executive Guide to Data Management.* DAMA International.

2 DAMA Internacional The global data management community. http://www.dama.org/cpages/home (accessed 1 December 2021).

3 DAMA Internacional (2020). *DAMA-DMBOK: Data Management Body of Knowledge.* DAMA Internacional.

4 Agesic (2018). Política y estrategia de datos para la Transformación Digital. https://www.gub.uy/agencia-gobierno-electronico-sociedad-informacion-conocimiento/comunicacion/publicaciones/politica-estrategia-datos-para-transformacion-digital (accessed 5 September 2021).

5 Rodriguez Mendaro, M.L. (2021). Los datos en una administración pública digital – Perspectiva Uruguay. https://www.damauruguay.org/novedades/los-datos-en-una-administracion-publica-digital-perspectiva-uruguay (accessed 5 September 2021).

6 U. Government (2020). Guidance for data ethics framework. https://www.gov.uk/government/publications/data-ethics-framework/data-ethics-framework-2020 (accessed 1 December 2021).

7 Ministry of Industry, Business and Financial Affairs – Denmark (2018). Data for the benefit of the people. https://eng.em.dk/media/12190/dataethics-v2.pdf (accessed 1 December 2021).

8 Treasury Board of Canada Secretariat (2021). Algorithmic impact assessment tool. https://www.canada.ca/en/government/system/digital-government/digital-government-innovations/responsible-use-ai/algorithmic-impact-assessment.html. (accessed 7 September 2021).

Section 11

Legal and Ethics

26

Innovating with Values. Ethics and Integrity for Tech Startups

Germán Stalker

Center for Intellectual Property of the University of San Andrés (C-PINN- UdeSA), Victoria, Argentina

26.1   Introduction

Innovation is a requirement for economic growth. There are studies that link innovation and technology to better economic performance [1]. The global knowledge and innovation markets are growing exponentially, packed with challenges and opportunities. Knowledge has become a critical resource and valued asset, fueling innovation and has a positive relationship with development [2]. Management and transfer of technology strategies protected by the intellectual property are also key for successful progress. Advanced innovation lowers transaction costs and fosters technology transfer (TT) and collaboration.

However, it is not enough for economic development. Commitment with the rule of law, strengthening of institutions, and working with common values are factors of sustainable economic development, as well. Latin-American innovation initiatives co-exist with high levels of corruption and a lack of ethical behaviors in the ways of doing business and in the relationship between companies and governments.

Innovation is a key factor in the diversification of production and companies. It can also contribute to fostering economic growth. Sometimes, innovation comes from inside the company. In other cases, it is the result of TT strategies from scientific centers or universities to enterprises. It can also be the result of *spin-off startups* launched in accelerators and incubators from universities and centers of research. All along the Americas universities and scientific institutions work together with industries, entrepreneurs, and investors to drive innovation, employability, and economic growth. Scientists and scholars working together with entrepreneurs generate *startups* with high levels of innovation that impact in the digital era. From the south of Patagonia to Canada, we can find several clusters of innovation.

Nevertheless, they face challenges regarding the legal framework, ethical behaviors, lobbying, and conflicts of interest. This is the reason why *integrity* and *compliance* programs are needed. When tech companies are negotiating with investors, other companies or governments face ethical risks. In particular, *startups* launched by public universities or public scientific bodies deal with issues regarding conflicts of interest, property rights, and collusive practices.

IEEE Technology and Engineering Management Society Body of Knowledge (TEMSBOK), First Edition.
Edited by Gustavo Giannattasio, Elif Kongar, Marina Dabić, Celia Desmond, Michael Condry, Sudeendra Koushik, and Roberto Saracco.

The goal of this paper is to demonstrate that by incorporating ethics by design, integrity in the business model, and compliance practices in the transfer of technology process, tech companies will be more accountable and generate trust.

This essay is the result of an investigation held in 2019 for a master's degree thesis titled: "Integrity in technology-based companies for economic growth." In that work we have done a qualitative survey in innovative ecosystems. We concluded that there is a field of knowledge to be addressed: ethics in innovation environments.

In this paper, we will see that ethical behavior of *startups* improves their reputation. Together with innovation generate trust in the value chain. Trust, recognized as the glue of strategic partnerships between universities, private investors, and industries generate more accountable ecosystems. We conclude that innovation plus integrity will lead to a more sustainable economic model of development.

In order to do so, we first analyze the literature that links innovation with growth and corruption with growth. Second, we describe innovation ecosystems. Third, we analyze ethical emerging risks. Finally, we propose the main themes a code of ethics should contain and a compliance and integrity program for *startups*.

26.1.1 Innovation and Ethics for Economic Development

After the worst pandemic history can remember, humanity faces new risks and challenges. Unexpectedly we are getting used to managing different kinds of risks and protocols in many routines of our lives. New risks and emerging technologies are diffusing faster than ever and provoke profound and systematic changes. The planet we live in is warming at unprecedented levels, the intimacy and personal data protection are threatened by big data, income inequality and gender gap are still broad. These are some examples of the risks that require innovative responses with an ethical way of doing business. Responses to these risks and the extent society embraces innovation conditions the ability of mankind to transform crisis into opportunities.

There is evidence that innovation fosters economic growth [1]. That discovery of new ideas, products, and processes throughout innovation make industries more competitive [3], p. 473. Besides, experience says that governments play a key role in promoting technological progress [4]. Innovation in the economies is so important that it has been measured for many years. The Global Innovation Index[1] is carried out by the World Intellectual Property Organization around the world as a tool to assess innovation and improve national innovation strategies in countries [2].

These approaches consider economic development as a result of innovation and the creation of new products, news businesses based on the intensive use of technology, or based on new needs. Disruptive innovation breaks the *status quo*, improving productivity, creating new markets, or delivering new services worldwide. It is proved that innovation is the only way economies can grow stable, even more so when that innovation comes from *startups* [4], p. 21.

Venture capitals are the financial resources provided by investors to fund *startups* and small businesses that show potential for long-term growth. In the last few years, they have become an important source of capital for entrepreneurs. Venture capital investments have a high level of risk as only a small portion of the companies they invest in have become successful and high-return businesses (Statista, 2021).

---

1 *Global Innovation Index (GII).*

Latin America represents around 6% of the global venture capital market. Nevertheless, the world of venture capital and high-risk investments has set its eyes on this region. In fact, venture capital investments in Latin America have skyrocketed in the last decade, especially in 2019 (Statista, 2021). In that year, investment in startups was estimated at 4.6 billion US dollars, 300% more than in 2017, when it totaled at just 1.1 billion dollars.

One indicator of the growth of innovative markets is the presence of *unicorn* companies. A *unicorn* company is defined as a privately held company worth 1 billion US dollars or more. By 2021, Latin America is home to 16 unicorns. These unicorns are headquartered in Bermuda, Brazil, Colombia, Argentina, Mexico, and Uruguay, and most belong to the finance and insurance fields.

Two of the leading sectors in innovation and technology markets are finance and logistics. Fintech and logistics *startups* are the sectors with the largest amount of venture capital investment deals in the region. In 2019, they accounted for more than half of the transactions closed in Latin America. Following this trend, the largest funding amount by venture capital firms in Latin America was awarded to the Colombian *startup* behind the delivery app *Rappi*,[2] at a whopping 1.5 billion US dollars. The Brazilian-challenger bank *Nubank*[3] and the Argentinian financial management app *Ualá* were also among the startups that received the highest value investments.

Most of those unicorns were once *startups*. Some are the fruit of entrepreneurship accelerated by accelerators. Others were incubated in public research centers or universities. *Startups* may emerge in different ways.

Thus, innovation flows in many diverse directions. It can come from inside the companies that have I + D labs to improve industrial processes or equipment. It can create new products or devices, find more efficient ways to develop new technologies, or improve customer's experience.

Innovation can also come from outside the company. From research centers or universities – both public or private – to companies. In these cases, the transfer of technology is crucial. The ability to identify where developments or technologies can be found is the first step for innovation applied to companies. The technological *mashup* is the holy grail of innovation [4], p. 208.

Innovation generates value, creates high-quality work, reduces poverty, and fosters economic development. It may come from technology-based companies. These kinds of companies are the vehicle for the transfer of technology in universities and centers or research to industry. The innovation ecosystem is where *startups* develop as companies that apply intensive technology or knowledge to improve existing products or services.

As we know innovation is a necessary condition but is not sufficient to explain economic development. Economic development is a multi-causal concept that needs other causes. Academics argue that malfunctioning of institutions constitute an obstacle to investment, entrepreneurship, and innovation.

There is evidence that corruption is linked to low indexes of GDP, equal economy, and human development [5–7]. Rose-Ackerman argues that corruption may have its roots in culture and history. But it is most of all an economic and political problem. Economic, because it produces inefficient and unequal distribution of costs and benefits. Political, because it is a symptom that the system is not working for the public interest [8].

---

2 *Rappi – Wikipedia.*
3 *Neobank* headquartered in Brazil, was the most valued Latin American unicorn in May 2021, with a market capitalization value of around 25 billion US dollars, followed by the Mexican automotive e-commerce platform, *Kavak*, with a market value of 4 billion US dollars (Statista, 2021).

To sum up, corruption affects economic growth because:

- **Harm-free competitiveness** between companies. If we start from the idea that competitive markets are those where industries compete in equal conditions, behavioral misconduct interferes in these conditions and impacts free competition.
- **Increases costs** of big companies and state big enterprises.
- **Inhibits innovation**. Why should a company innovate or improve a product or a service if it has the demand warranted by illegal means?
- **Hinders foreign investment.** Due to the obstacles generated with bureaucratic regulations to income capitals, reduces the market efficiency.
- **Deepens inequality,** increasing poverty.

Overall, corruption generates negative incentives for the economy, affects the innovation ecosystem and development of economies. Corruption and instability may be intrinsically linked in the sense that they are a consequence of the same coordination problems of the ruling elite [6], p. 705. If institutional inefficiency persists over time, bad institutions in the past may have played a role in low economic growth [6], p. 706.

There are some complex practices such as collusion and colonization of public agencies that have persisted for many years. Castellani has shown that these kinds of habits enhance privilege accumulated spaces (PAS) as a permanent source of rent-seeking for companies [9], p. 2. Finally, we cannot avoid the analysis of corruption between companies [10], p. 3. Collusive practices between enterprises, misuse of company assets, detriment of investors, dubious accounting practices, conflict of interests, traffic of privilege information, *lobby,* and finance of political campaigns. These are some of corporate misconduct that justify an ethical approach for doing business [11].

26.2 Startups in Innovative Ecosystems

Transfer of technology may come from inside the company or flow from universities and research centers to enterprises. Sometimes this flow happens in special environments: the innovative ecosystems.

Innovative ecosystems are a set of institutions, people, and conditions that promote the development of businesses, *startups,* and technology-based companies [12]. They are designed with the idea that generation of value emerges from innovation based on knowledge collaboration [13].

In the ecosystem, small companies can catch up with better innovation processes, can validate technologies, and test products. The environment provides everything that can help *startups* to bring ideas from the lab to the market. The usefulness of the ecosystem depends on the features of each environment. Once it is settled, the ecosystem itself takes its own identity and evolves.

Companies not only optimize resources and reduce costs, but they also capture research and technological developments made in universities and scientific centers through *transfer of technology offices* (TTO) and different tools that link groups of academics with entrepreneurs, investors, and technological managers. They build capacities to face the challenges of industries and society. Ecosystems are the meeting point between supply and demand of technology. That point of contact is where know-how innovation creates value. But this innovation spark does not emerge magically nor spontaneously. It is the result of a process that needs patience, transparency, ethics, and constancy.

Besides, the ecosystem can generate financial stimulus, tax incentives, or common public services for new technological developments. The ecosystem itself is a generator of conditions for innovation in the whole value chain. It generates a business environment, capable of incorporating and sharing innovation. Through the creation of *startups*, public–private partnerships, or alliances between universities and companies, the innovation ecosystem grows.

There are many successful stories of innovative ecosystems around the world. Israel constitutes one of these cases. Israel's economy is *startup* centered and the technological realm is half of the country's exports. "It can be that tech companies are small but they are the catalyzers of a dynamic system" [4], p. 264, [12], p. 8. To analyze these experiences, one must consider the cultural behavior and patterns of each country or each region. Not always, best practices can be replicated in other countries.

Depending on their degree of development and from an evolutionary perspective there are two kinds of ecosystems: mature and in development. Mature models have companies, individuals, government agencies, and research centers in permanent interaction. Institutions together make possible joint public–private alliances for midterm and long-term projects. Some ecosystems can coordinate efforts and initiatives to plan priorities and align incentives in a specific region or market. Often, ecosystems reinforce energies that create a virtuous circle where *startups* constantly emerge. One feature of the Israeli ecosystem is the leadership of the government during various decades, where the platform of technological services is based on public-partnership alliances with public and private components [12], p. 8. Examples like Israel, replicated in Ireland, south of Canada, Massachusetts, Silicon Valley, Minneapolis, Chicago, and North Caroline and some of these ecosystems took more than 25 years to consolidate.

In Latin America, there are many innovative ecosystems at an early stage. Although they do not produce the same quantity of *startups*, or exports of knowledge, they have the potential to do so in upcoming years.

In the innovation ecosystems institutions and people from different realms interact. Investors, scientists, technological managers, entrepreneurs, customers, and users relate in different ways. To understand these relationships, it is necessary to know the main features of each stakeholder (as shown in Table 26.1).

Table 26.1 Roles and main features of stakeholders in innovative ecosystems.

Role	Main characteristic	What to expect?	Function in the ecosystem
Entrepreneur	Attitude	Start the business	Implement the idea
	Tireless	To compete	Knows the market
	Risky	Take risks	
Researcher	Curious	Research	Invent new developments
	Conservative	Build know-how	
	Pertinence instinct		
Technological manager	Translate	Solve problems	Listen
	Patient	Link lab with market	Link
	Knows different knowledge areas	Understand science and market	Pitch
Investor	Goal: earn money	Funding	Finance
	Transcendence	Long term vision	Vision
	Aim to leave legacy		

Source: Adapted from Stalker [14].

A *startup* is a company designed to commercialize goods or services with a huge technological component. The main goal of a *startup* is to transform scientific ideas into great products, services and to deliver new products or services that are a result of innovation, scientific research, and technology.

There are cases in which *startups* are the best vehicle for the transfer of technology generated in universities and scientific centers. Some scientific developments require a superior wager than the one a university can provide. In other cases, the intellectual property work plays a key role in the commercialization area, or sometimes the value proposal is already verified and is ready to be scalable. In these cases, the institutional design of a *startup* fits better than any other to commercialize the development.

The creation of *startups* is an alternative tool that strengthens the commercialization of technologies at a global level. They have a permanent innovation spirit in their business. *Startups* are agile to commercialize the technologies and to grow as the dynamic of the business changes. *Startups* generate a spill effect in their environment. They create quality employees, pay high salaries, and demand technology and services. Some of these *startups* are later acquired by bigger ones or receive injections from investors.

Startups are designed as a device for the commercialization of technology. *Startups* focus on costs and value proposals, besides technical features of technology. *Startups* protect intellectual property of the technology, patents the invention or the utility model, register copyright, protect commercial secrets, and develop a commercial strategy.

Another feature of *startups* is that they do not have a history. No commercial or financial background; they only have a real present and a promising future. But they do not have antecedents as a company so it is not easy for them to access the financial market. And not getting funds on the stage they need is one of the major causes of their death. Traditional financial institutions are faraway from innovative ecosystems.

The *startup* generation has at least two stages: the genesis of the idea in the lab and the emergence of entrepreneurship. All along the process, they face risks and threats.

Inaccurate legal framework, access to funds, internal changes in the team, unexpected high costs, threats of competitors, the business model, bad marketing strategy, or not protecting the intellectual property: these are some of the reasons why there is a high rate of failures of *startups*. International experiences show that most of the failures happen in *Death Valley*, a metaphor that links the period between the genesis of the *startup* to its failure with the geographic region in the desert of South California where pioneers died looking for gold.

In the process of scaling up *startups*, incubators, and accelerators play a key role. *Incubators* and *accelerators* walk through labs and universities to find out new developments and ideas that can change an industry or a market. Some *accelerators* meet scientist groups and analyze in depth their promising projects, from which they select the best ones and in parallel, search for business entrepreneurs. Then, they team up scientists with entrepreneurs and investors. Some connect *startups* with venture capital and a network of international accelerators and investors. They can also execute follow-up investments in their seed and *series A* rounds. Acceleration of technology-based enterprises is the boost they need for a global impact.

Incubators can be public or private, or can be the result of a strategic partnership between public agencies and private funds. The public sector plays a key role in the acceleration of *startups*.[4] With policies that promote innovation, technological linkage, financing, and generation of incentives for projects with social impact. International multi-state agencies are also key actors from public

---

4 Senor and Singer [4].

sectors. There is a lot of cooperation between regional and multilateral organizations to foster innovative *startups* and environments.[5]

This role is also played by international and regional financial organizations, as well as by large corporations. The formation of innovation consortia and accelerators generates a circle of trust between corporations and entrepreneurs. *Wayra*,[6] the accelerator of the *Telefónica* Company and *Natura Startups*,[7] the accelerator of the Natura cosmetics company, are some of those examples.

Shifting ideas from the lab to the market faces specific hazards and threats. There are some systemic risks and examples that show why ethics are necessary and can be materialized to bring value. What happens in the innovative ecosystem, in the *black box*, is nonlinear. There are layers of complexity driven by the problems society faces.

26.3 Emerging Risks and Threats

Systemic threats are risks that sustain in time and are common to all ecosystems and *startups*. As innovative ecosystems evolve, risks also change. Commercial market ecosystems are a complex network of persons and businesses with different interests. Some of the threats we found in our research are:

1) **Lack of knowledge about ethics and compliance in startups:** One of our research findings is that *startups* and tech companies, in general, do not know much of compliance and integrity policies. They are also unfamiliar with the ethical standards, directives, and good practice guides of regulation of companies. Some of the *startups* even see potential obstacles to the flow of innovation, incubation, and acceleration of tech companies with greater regulations.
   Besides, there are no recommendations or guidelines on how to deal with ethical dilemmas in innovative ecosystems.
2) **Researchers have low exposure to the entrepreneurial field and to the productive sectors:** The scientific environment values research more than entrepreneurship. Evaluation systems in the scientific realm do not encourage entrepreneurship or the launch of *startups*. There is a need to promote a greater link through incentives between researchers and scholarship holders with the entrepreneurial ecosystem.
3) **Mutual understanding is complex:** Entrepreneurs, scientists, technology managers, and investors talk in different languages. Words have a different connotation according to knowledge and contexts. They understand each other differently, according to each of their abilities and expectations. This makes networking extremely complex and risky. Misunderstandings might be the cause of conflicts of interest and appear along the TT process. Articulation and collaboration are also hard challenges.
4) ***Startups* generally do not have financial history:** They have only the present and future based on innovation and the work team. These are their two main assets. Having no business background as a company, it is difficult for them to access the capital to scale the product. That is one of the reasons for failure. Capital must arrive at the right time, be sufficient for escalation, and facilitate market access. As long as the money is needed it can appear conflicts of interests, lobbying, and misconduct.

---

5 *OAS: Department of Economic Development.*
6 *Wayra.*
7 *Startups|Natura Argentina.*

5) **Legal framework is usually outdated:**

Entrepreneurs often break the *status quo*. Because most of the time innovation requires thinking of necessities from a different point of view or outside the box. Legal frameworks appear as obstacles to innovation in some cases. Every new idea or development goes through and beyond regulatory frameworks that become obsolete. In these cases, the risk of *lobbying* with regulators increases. Gifts and courtesies to influence them in the decision-making process may occur.

Sometimes, to accelerate the process of scaling up the company, *startup*s face institutional arrangements obstacles. For example, the use of artificial intelligence in health, transport, finance, or entertainment markets may imply the issuance of new rules. Smart contracts – programs to do specific tasks in any *blockchain* database and run functions when triggered by transactions- are sometimes presented as a way to bypass regulation and enforcement.

The storage, use, and reutilization of databases had an exponential growth during pandemic years. With the expansion of the digital economy, risks also grow. Bias, fallacies, personal data, privacy, and security are threats that *software startups* and companies must assess as potential risks. The risks generated to intimacy, privacy rights, and personal data of people are high, and regulations across the world are changing fast.

We can find *specific risks* in each market. The finance market has *high-tech* infrastructure challenges and regulatory issues, or the biotech has regulatory issues, among other examples.

Real cases can explain these risks better.

The first one is the *Theranos case*.[8] In 2013, Elizabeth Holmes launched a *startup* called "*Theranos.*" While she was a student at Stanford University, Holmes developed a technology that promised to rocket the clinical trial analysis market. In a very short period, her company received more than 700 million US dollars, and in its peak, the company was valued at more than 10,000 million US dollars. However, in 2015 thanks to a journalistic investigation, the veracity and efficacy of the technology were questioned. None of the investors had seen if the device was really effective because of a lack of transparency. Three years later, the company was declared as a massive fraud by the Security Exchange Commission[9] and ceased its operations. In December 2021, she was convicted of defrauding investors after a trial in California[10] and now she faces the possibility of decades in prison.

The second example was described by Rachel Botsman in her book [15]. Jack Ma, the founder of *Alibaba*, discovered in 2011 a set of financial bribes inside the company in exchange for allowing vendors to skip stages in the creation and verifying process of their accounts throughout two years. This resulted in hairny more than 2000 of low-quality vendors or even fake products being verified as *gold vendors*.

Ma knew that the controls had failed and that he had to send a clear message to maintain the trust of *Alibaba* users. Employees in charge of setting up the accounts and those who knew about this situation and did nothing, were fired. Finally, even though they were not involved in the fraudulent maneuvers, the CEO and COO of the company resigned. In an interview published in The Guardian, Ma stated: "We must send a strong message that it is unacceptable to compromise our values and culture."[11]

---

8 *US v. Elizabeth Holmes, et al.*
9 *Elizabeth Holmes and Theranos, Inc.*
10 *Elizabeth Holmes: Theranos founder convicted of fraud – BBC News.*
11 Alibaba is one of the greatest B2B marketplace in the world. *Alibaba.com chief executive resigns|Business.*

These cases show the importance of having integrity and values in a tech company. In some cases, exemplary measures and strong leadership are necessary. The cost of losing the trust of the *startup's* customers is critical.

Ethics in innovative ecosystems is about behaviors linked to rules. Entrepreneurs, scientists, and investors, interact with their interests to develop *startups*. How do we know when they are *lobbying* instead of *advocating* for good goals? How do we know when an agreement between companies can be a collusive practice? How do we know if a present or a courtesy can be a bribe? These are the kind of risks we have to anticipate.

To fight corruption and misconduct it is necessary to prevent it. Everyone would agree that a Code of Ethics and an integrity program are essential tools that improve behaviors in the company. So, how to do it?

26.4 Ethics by Design

Ethics is a discipline-oriented to the causes and explanation of human behavior. When we talk about professional ethics, we talk about applied ethics. In philosophy and psychology, there is a debate to determine if ethics is situational – the ethics of a person depend on the environment in which he/she is – or if personal features and honesty are intrinsic and affects all the actions of a person [16]. From this point of view, personal ethics, the way of doing business, and corporate conduct are closely related [17].

In any case, ethics is the discipline oriented to explain human behaviors [18]. In applied ethics, the ethical criteria are conceived not to build their own theory, nor to validate philosophical theories, but rather they are thought and designed to indicate the principles and norms to be applied in human behavior. There is a tension between ethics of responsibility – which points to the development of behaviors governed by its consequences called "*consequentialism*" and the ethics of conviction – motivated by compliance with moral principles called "*deontologist*."

How do we apply ethics in the innovation ecosystem? Which are the main variables that must be considered to design a Code of Ethics and how an integrity program for *startup* should be implemented?

One of the first steps is to know where to start from. From the very beginning, at the first team meeting, it must be clear that the *startup* will incorporate ethical behavior.

26.4.1 The Beginning Set Trend

From the very beginning, we must consider the ethical risks. When we assess the breakthrough market, it is necessary to have the values incorporated to generate an ethical culture. These will be the minimum values the *startup* will seek.

Mark the baseline for each and every one of the *startup's* activities. Identify the real limits that we will face in each activity: during the incubation process, searching for financing or when pitching for them, in the first drafts of business plans.

From there, we need to establish the hierarchy of values. Prioritize which ones are going to guide the business model. First, in abstract, think of each principle as an aspiration of the tech company. Which are those values that the company going to prioritize? Later, these values will need to be concrete in behavioral guidelines when doing the risks map.

For example, the launch of a new product needs a previous market assessment. A product must not only comply with local regulations and guarantee safety, but must also meet the expectations of an increasingly demanding consumer. What are the ethical risks that could impact the way my *startup* meets its objectives? What are the unethical actions that it could face? Is there any tool that detects consumer's preferences for the technology we want to sell?

26.4.1.1 The Process Is Incremental

We are innovating not only by developing a new product. We are also innovating on the *how:* the way the product is designed, manufactured, marketed, and sold.

Therefore, the ethical way of doing business must be assimilated and shared by the entire team. Also, by the people closest to the *startup*: colleagues, family, suppliers, investors, scientists. Everyone should be aware that this innovative idea is implemented in accordance with certain ethical values.

26.4.2 Managing Risks

The risk-based approach to an integrity program starts with a premise: the higher the risks, the more rigorous the measures to prevent them must be.

First, we must know what the risks are. But what is an ethical risk?

It is the possibility of an event occurring that affects values and the achievement of our goals. It is a contingency that can go against the values that guide our actions and can cause eventual damage. The ethical risk may or may not happen.

Zero risk does not exist.

Hence the question of how to manage risks becomes important. How does the *startup* design, produce and market its products and services? What are the behavioral risks they face?

Drawing the risk map helps us identify threats to be faced by the company, depending on the nature of the markets the products belong to. Assists us to identify in which link of the value chain we most likely to find behaviors that put our values at risk. Recognize those stakeholders with whom the *startup* interacts. Detect the interactions with suppliers, investors, clients, employees, competitors, public officials, or intermediaries that can cause ethical dilemmas.

Once the risks are identified, they are determined for other decisions that will come later. If you anticipate these, you can prevent and manage them.

The risks are situations of uncertainty that the *startup* faces in order to meet their objectives. They are events that can have consequences. For example, potential conflicts of interest, gifts and courtesies, influence peddling, and lobbying [19].

For example, in some Asian countries, it is customary to give presents as a way to cultivate business relationships, to show respect and appreciation for the other party. It does not matter if the person we meet with is a supplier, customer or public official. In China, this tradition in the business environment is usually nuanced according to the hierarchy of the interlocutor.

In some Latin American countries, giving presents is also a sign of respect. However, when it comes to public officials, it is prohibited by corruption prevention regulations. Complimentary gifts are only allowed in certain cases.

The general rule of thumb is that public officials cannot accept gifts of any kind and, should they be faced with such a situation, they should kindly return them.

In some countries, officials who receive courtesy gifts, traveling invitations, or benefits must register them in public records when they exceed a certain value.

Features of the risk map:

- It has to be coherent with the heart of the company business since no *startup* faces the same risks.
- It has to be coherent with the economic dimension and capacity of the *startup*.

We must measure the size of the *startup*. We will include both the target market and the size of the company. In both cases, current and projected. Then the structure of the company, the main roles, and the processes for making decisions, the sector in which it operates, the nature, scale, and complexity of the startup.

With a tailored identification scheme, the risk map is prepared according to these principles. Then, we have to classify the risks, depending on the mission and objectives of the venture. Finally, measure them in terms of probability and impact.

Objective of the risk map: Identify potential ethical conflicts and misconduct of the *startup*. Discern when there is a legal duty and when there is an ethical failure.

Who will be responsible: We have to recognize the person or persons with decision-making functions in the startup that may be affected by the ethical dilemma. Even if they are partners, founders, or employees.

When the risks are identified, we will know what to do, who and when to act to prevent them. It is important to know the context of the startup's business. Know the characteristics that can help to achieve our purpose and those that can affect the achievement of the objectives. Knowing the actors, we interact with helps us being more effective to prevent and map ethical risks.

Once the dilemmatic issues have been identified, we will be able to assess the consequences of behaviors and act timely. Also, we will know how to recognize the necessary resources to implement an Integrity Program.

26.5 Writing the Code of Ethics for My Startup

A Code of Ethics is a declaration of the integrity policy of a company. It settles the principles, values, and virtues of the *startup*. It also approves institutional commitments and personal behavioral rules that must be followed by all employees [17].

Ethical practices are defined once risks have been identified.

In other words, the Code of Ethics makes explicit the values shared by members of the *startup*. Then, it is broken down into the *startup's* commitments at the corporate level and the individual level: it guides how each member's reaction should be when interacting with people. It must guide the *startup's* reaction in case of ethical dilemmas according to the risk map that we established. The production process can be extended depending on each case. Finally, the code must be validated by all the members of the *startup*.

Its content should be simple, with clear and accessible language. The drafting process should include the survey and analysis of the regulation of the sector, the risks that they have already reviewed and consider best international practices.

The development process must guarantee the greatest possible participation and the contribution of the various members of the *startup*. To do so, consultation and fine-tuning meetings should be held. The more participative the process, the better practices they are going to have and the better examples of dilemmas arising from the reality of the industry in which the startup develops they will find.

26.5.1 The Compliance and Integrity Program

Compliance is a term that was originated with the idea of adjusting behaviors to standards. It is the activity of obedience to the norm. Whether it is agreed or imposed.

It is the integrity policy of the *startup* focused on the *startup's* legal obligations with application mechanisms and a subsequent evaluation strategy.

Compliance policies can cover different areas, not just the prevention of corruption. Both in the company's internal processes, as well as in the relationship with third parties. The protection of personal data, the environment, or competition are some examples. For compliance programs to be effective they must be seen as more than just a series of routine exercises and questions being asked.

In recent years, in the heat of corruption scandals of multinational companies, the compliance effectiveness policies have boomed as a mechanism to mitigate corruption within organizations. It involves not only regulatory compliance, but it also involves integrity culture and the practice of preventing ethical risks associated with the business. In addition to compliance with current regulations, the implementation of integrity policies and processes should focus on promoting values and ethical behavior by the members of the organization.

Some Latin American countries[12] passed laws on corporate criminal responsibility. These regulations establish that legal entities are responsible of crimes of corruption and require them to implement *Integrity* and *Compliance Programs* that include internal mechanisms to promote ethics and integrity and prevention models (see Table 26.2).

Ultimately, an ethics and compliance program are a management model. Attached to the heart of the *startup*, it must guide behaviors according to business standards. It should include internal mechanisms for promoting ethics and integrity. It is desirable that its impact can be assessed and

Table 26.2 Integrity program.

Variable	Goal	Main features
Risk map	To identify potential threats to the Code of Ethics, misconducts, and ethical dilemmas	Suited to the size of the startup and the market
Complain Channel	Whistleblower protection and fraud prevention	Anonymous, safe
Code of ethics	Align behaviors to values, mission, and vision of the startup	Contains values that orient the conduct of the startup employees. How to act in specific cases
Compliance officer	In charge of implementing the Integrity Program	Depending on the size of the startup, it can have other functions or can be external.
Training strategy	Train in integrity, values, and ethics	The training program in company conflict of interests, lobbying, and ethical dilemmas.
Assessment	To know the impact of the program and if it is useful to prevent misconducts and adjusts details	Reasonable variables according to the core business of the startup
Due diligence	To know antecedents of clients and suppliers	Perform prior to negotiations with suppliers and customers

Source: Adapted from Stalker [14].

---

12 Argentina, Law No. 27401, Chile, Law No. 20393.

that it contains supervision and control mechanisms aimed at preventing acts of corruption. It should include:

1) A Code of Ethics or conduct applicable to all directors, administrators, and employees.
2) A risk approaches. A risk matrix and map.
3) Specific rules and procedures to prevent illegal activities. In the field of contests and bidding processes, in the execution of administrative contracts, or in any interaction with the public sector.
4) Periodic training for directors, administrators, and employees.

It must have the leadership and commitment of the top management of the *startup*. The person or group of people who direct and control the organization must be focused on transparency and ethics.

Then, the best way to convey the values of the startup is through example. In the implementation of the Code of Ethics, leadership is key. The *startup* founders do not just have the job of making it work. They have to be *ethical influencers*, teaching by example inspiring followers in an ethical manner. This commitment of the leaders, middle and senior management is called "*tone from the top*" in anti-corruption literature.

Startups leaders must be aware that every decision they make must be oriented to the values of the *startup*. And that the beginning of the *startup* sets the trend of what it will be in the future. Aligning thought with action is easier after everyone in the company orients their actions to that purpose. A compliance officer is a position that can be covered when the *startup* grows.

26.6 Summary and Conclusions

Many factors explain economic development. Innovation, intellectual property protection, and commercialization of technologies are key drivers. But they are not enough. Rule of law, ethical behavior, and institutional strengthening are necessary as well for economic development.

Innovative ecosystems are the arenas in which *startups* can develop. As vehicles for change, *startups* have a key role in spreading compliance practices. Innovation fostered by *startups* must be guided by integrity practices when doing business. Innovative environments change in time. When they evolve, they face emerging threats. Systemic threats sustain in time: mutual understanding, lack of knowledge in compliance practices, legal framework, lack of financial history, and researchers with low exposure to the entrepreneurial field are some of them.

There are some specific ethical risks in the ecosystem as well. *Startups* face ethical dilemmas that need to be assessed in order to scale up. Conflict of interests, lobbying, or financing high-tech infrastructure. The use of big data, artificial intelligence, programs, algorithms, and the integration of technologies present specific challenges for tech companies [11].

Being responsible when doing business also implies respect for the environment, migration, poverty, and inequality. Respect for human rights is an emerging value of the negative consequences of global capitalism that is increasingly concerned by academia [20]. Besides, from a consumer and corporate social responsibility (CSR) perspective, it is becoming increasingly important for companies to focus on ethics and long-term reputation [21].

To deal with ethical dilemmas in the innovation ecosystems, we have to know the risks. Risk-based approach to an integrity program starts with a premise: the higher the risks, the more rigorous the measures to prevent them must be. An ethical risk is the possibility of an event occurring that affects values and the achievement of our goals.

Having an integrity strategy at the very beginning, during incubation and escalation becomes an intangible asset for *startups*. Hand in hand with innovation, it increases trust and improves their reputation [22]. An ethical program aligned with the business strategy generates transparent companies. *Startups* that are ethically linked in their value chain distribute trust horizontally, generating a sustainable business environment.

If the relationship between universities, industries, policymakers, and investors is based on ethical behaviors, it will generate trust. Good practices inspired by values conduct business in a sustainable way. It sparks trust in the value chain and confidence in innovative ecosystems.

Finally, ethics, rule of law, and integrity are key drivers to make economic development more accountable and sustainable in the long run.

References

1 Schumpeter, J. (1934). *The Theory of Economic Development*. Cambridge, MA: Harvard University Press.

2 (a) WIPO (2017). World Intellectual Property Report 2017 Intangible Capital in Global Value Chains. (b) WIPO (2021). Global innovation index. https://www.wipo.int/global_innovation_index/en/2021.

3 Stiglitz, J. (1993). *Economics*. United States of America: W.W. Norton & Company Inc.

4 Senor, D. and Singer, S. (2009). *"Start-Up Nation" La historia del milagro económico de Israel*. Publiexprés S.A.

5 Lucic, D., Radišić, M., and Dobromirov, D. (2016). Causality between corruption and the level of GDP. *Economic Research-Ekonomska Istraživanja* 29 (1): 360–379. https://doi.org/10.1080/1331677X.2016.1169701.

6 Mauro, P. (1995). Corruption and growth. *The Quarterly of Journal Economics* 110 (3): 681–712.

7 Transparency International (2020). Corruption perception index 2020. https://www.transparency.org/en/cpi/2020/index/nzl (accessed 23 October 2019).

8 Rose-Ackerman, S. (1999). *Corruption and Government. Causes, Consequences and Reform*. Cambridge University Press.

9 Castellani, A. (2012). Privileged accumulation spaces and restrictions on development state-business relations in Argentina (1966–1989). *The American Journal of Economics and Sociology* https://doi.org/10.1111/j.1536-7150.2012.00864.

10 Svenson, J. (2005). Eight questions about corruption. *Journal of Economic Perspectives* 19 (3).

11 Lin, T.C.W. (2016). Compliance, technology, and modern finance. *Brooklyn Journal of Corporate, Financial & Commercial Law* 11: 159–182.

12 Kantis, H. (2018). Mature and developing ecosystem: a comparative analysis from an evolutionary perspective. Working Paper PRODEM.

13 Fagerberg, J. and Srcholec, M. (2008). National innovation system, capabilities and economic development. *Research Policy* 37 (9): 1417–1435. https://doi.org/10.1016/j.respol.2008.06.003.

14 (a) Stalker, G. (2020). *Integridad en EBT (Empresas de Base Tecnológica) para el crecimiento económico*. Victoria: Departamento de Derecho, Universidad de San Andrés http://hdl.handle.net/10908/18579. (b) Stalker, G. (2018). Transparency Policies and Access to Information Law. Alternatives for assessing its impact. GIGAPP Working Papers, ISSN 2174–9515. Núm.110, págs. 651–670. Publication: 15 September 2018. (c) Stalker, G. Evaluating the impact of transparency policies", in the Journal of Public Policy Studies No. 5/17, published by the Center for Public

Policies of the Catholic University of Asunción, Paraguay, 2017. https://issuu.com/jorgegarciariart/docs/armado (accessed 3 August 2019).

15 Botsman, R. (2018). *Who Can You Trust? How Technology Brought us Together and Why it Could Drive us Apart*. UK: Penguin Random House.

16 Griffin, J.M., Kruger, S.A., and Maturana, G. (2017). Do personal ethics influence corporate ethics? https://ssrn.com/abstract=2745062 (accessed 19 July 2019).

17 Adelstein, J. and Clegg, S. (2016). Code of ethics: a stratified vehicle for compliance. *Journal of Business Ethics* 138 (1): 53–66.

18 Bautista, O.D. (2007). *Ética y Politica: Valores para un buen Gobierno*. Universidad Autónoma del Estado de México.

19 Krimsky, S. (2003). Small gifts, conflicts of interests and the zero-tolerance threshold in medicine. *The American Journal of Bioethics* 3: https://doi.org/10.1162/15265160360706589.

20 Wettstein, F., Giuliani, E., Santangelo, G.D., and Stahl, G.K. (2018). International business and human rights: a research agenda. *Journal of World Business*. https://research-api.cbs.dk/ws/files/58219777/grazia_santangelo_et_al_international_business_and_human_rights_acceptedversion.pdf.

21 Stanaland, A.J.S., Lwin, M.O., and Murphy, P.E. (2011). Consumer perceptions on the antecedents and consequences of corporate social responsibility. *Journal of Business Ethics* 102 (1): 47–55. Stable URL: https://www.jstor.org/stable/41476000.

22 Atakan, A. and Ekmecki, M. (2012). Reputation in long run relationships. *The Review of Economic Studies* 79 (2): 451–480. https://doi.org/10.1093/restud/rdr037.

Index

IEEE Technology and Engineering Management Society Body of Knowledge (TEMSBOK), First Edition.
Edited by Gustavo Giannattasio, Elif Kongar, Marina Dabić, Celia Desmond, Michael Condry, Sudeendra Koushik, and Roberto Saracco.
© 2024 The Institute of Electrical and Electronics Engineers, Inc. Published 2024 by John Wiley & Sons, Inc.

Printed and bound by CPI Group (UK) Ltd, Croydon, CR0 4YY

27/10/2024

14580678-0004